INTRODUCTION TO
THE PRINCIPLES OF
HETEROGENEOUS CATALYSIS

INTRODUCTION TO
THE PRINCIPLES OF
HETEROGENEOUS CATALYSIS

J. M. THOMAS
DEPARTMENT OF CHEMISTRY, UNIVERSITY COLLEGE OF
NORTH WALES, BANGOR, WALES

AND

W. J. THOMAS
DEPARTMENT OF CHEMICAL ENGINEERING, UNIVERSITY
COLLEGE OF SWANSEA, SWANSEA, WALES

1967
ACADEMIC PRESS LONDON · NEW YORK

ACADEMIC PRESS INC. (LONDON) LTD
24-28 Oval Road
London NW1 7DX

U.S. Edition published by
ACADEMIC PRESS INC.
111 Fifth Avenue
New York, New York 10003

Library of Congress Catalog Card Number: 66–30323
SBN 12–688650–4

Printed in Great Britain by
Lowe & Brydone (Printers) Ltd, Thetford, Norfolk

Preface

Brevity, the soul of wit, may become the very body of untruth. In defending this statement, Aldous Huxley argued that the dangers of attempting a short appraisal of a difficult and complex subject are very considerable. Omission and simplification help us to understand—but help us, in many instances, to understand the wrong thing; for our comprehension may be only of the abbreviator's preconceived notions, not of the "vast, ramifying reality from which these notions have been arbitrarily abstracted". These sentiments convey, we hope, the reason for our writing a lengthy book. They do not, however, explain why, in the first place, we felt that another book on heterogeneous catalysis was necessary. Our reasons, which established protocol requires us to divulge, are as follows.

As teachers of, and researches in, surface chemistry and chemical engineering, we have felt the need for a text which concentrates throughout on the *principles* of heterogeneous catalysis. A text in which as much emphasis is given to the experimental principles as to the theoretical ones; where nonmetals and oxides receive as much prominence as metals; and where the full range of heterogeneous catalysis in practice—from the laboratory experiment to the design of catalytic reactors—is considered. It is our conviction that undergraduates and young research workers, to whom this book is primarily aimed, ought to know as much about how the results are obtained as they do about the various theories that unify the results. This is the main reason for including one comprehensive chapter on the experimental aspects of the study of catalysis and adsorption. All other chapters, and their accompanying lists of references, are deliberately selective and, we hope, synoptic.

Although we have strenuously endeavoured, during the course of rendition, to present as uniform a style as possible, we have been compelled by our respective ineptitudes and predilections to resort to the principle of the division of labour. One of us (J.M.T.) is largely responsible for Chapters 1, 2, 3, 5 and 6; the other (W.J.T.) for chapters 4, 7, 8 and 9.

We have been influenced, for the better, by authors of previous textbooks on catalysis, particularly by Rideal and Taylor, Schwab, Emmett, de Boer, Bond and Ashmore. And we are greatly indebted to a large number of scientists who tolerantly entered into numerous stimulating discussions and correspondences with us. We are profoundly aware of the kind influence of Professor K. W. Sykes, who first aroused in us an interest in surface chemistry, and of the constant support of our respective Heads of Department, Professors S. Peat and J. F. Richardson. We are grateful for the advice and support of Academic Press and, not least, we acknowledge with gratitude the patience and encouragement of our wives.

Bangor and Swansea   J. M. THOMAS
January 1967   W. J. THOMAS

Contents

CHAPTER 4

The Significance of Pore Structure and Surface Area in Heterogeneous Catalysis

CHAPTER 5

The Role of Lattice Imperfections in Heterogeneous Catalysis

CHAPTER 9

Design of Catalytic Reactors

List of Most Frequently Used Symbols†

A	Pre-exponential factor	moles $L^{-2}T^{-1}$ or moles $L^{-3}T^{-1}$
A_c	Cross-sectional area	L^2
a	Constant in Temkin isotherm	—
a	Radius parameter	L
B^*	Two-dimensional virial coefficient	moles L^{-2}
b	Constant in Langmuir isotherm	$M^{-1}LT^2$
c	Constant in BET equation	—
c_i	Concentration of species i	moles L^{-3}
c_0	Concentration at pore mouth	moles L^{-3}
c_g	Concentration of gas molecules	(No. of molecules) L^{-3}
c_i	Surface concentration of species i	moles L^{-2} or (No. of molecules) L^{-2}
c_{ads}	Concentration of gas molecules at surface	(No. of molecules) L^{-2}
c_s	Total concentration of active surface sites	(No. of molecules) L^{-2}
c_v	Surface concentration of active vacant sites	moles L^{-2} or (No. of species) L^{-2}
$c_{\ddagger}$	Surface concentration of transition state complex	(No. of species) L^{-2}
D	Molecular diffusivity	L^2T^{-1}
D_B	Bulk diffusion coefficient	L^2T^{-1}
D_e	Effective diffusivity	L^2T^{-1}
D_K	Knudsen diffusion coefficient	L^2T^{-1}
D_P	Diffusion coefficient for forced flow	L^2T^{-1}
d	Diameter	L
d_P	Particle diameter or size	L
E	Activation energy	cal mole^{-1}
E'	Electron affinity	MT^{-2} or volts
F	Flowrate	MT^{-1} or moles T^{-1}
F	Ratio of activity of poisoned pore to activity of unpoisoned pore	—

† Where appropriate, these are defined in terms of the fundamental dimensions of mass M, length L and time T. In this list the symbol θ denotes temperature.

F	Helmholtz free energy per mole (constant volume and temperature)	cal mole^{-1}
F_i	Partition function of gas per unit volume ($=f_i/V$)	L^{-3}
f	Fraction of surface available for reaction (equivalent to effectiveness factor η for isothermal conditions)	—
f_i	Total partition function of one molecule of species i in the total volume V of the system	—
$f^{\ddagger}$	Partition function of transition state complex	—
$f^{\ddagger}_m$	Partition function of mobile transition state complex	—
$f_{\ddagger}$	Factorized value of $f^{\ddagger}$ (defined by $f^{\ddagger} = \dfrac{kT}{h\nu}f_{\ddagger}$)	—
G	Mass flowrate per unit area	$ML^{-2}T^{-1}$ or moles $L^{-2}T^{-1}$
G	Gibbs free energy per mole (constant pressure and temperature)	cal
H	Enthalpy per mole	cal
$H_{C,R,T}$	Height of catalytic, reactor or transfer unit	L
H_0	Magnetic field strength	gauss
ΔH	Heat of reaction at constant pressure	cal mole^{-1}
h	Planck's constant	ML^2T^{-1}
$h_{n,m}$	Thiele modulus. First subscript denotes reaction order, second subscript denotes number of reaction	—
h'_1	Thiele modulus for two concurrent first-order reactions	—
h_T	Thiele modulus for reaction occurring under non-isothermal conditions	—
h_θ	Thiele modulus for adsorption	—
h_p	Thiele modulus for pellet	—
I	Ionization potential	volts
j_D	Mass-transfer factor	—
K	Equilibrium constant for homogeneous reactions	units depend on reaction
$K_{A,B}$	Adsorption-desorption equilibrium constant ($=k_{A,B}/k'_{A,B}$) for species A,B etc.	$M^{-1}T^2L$

K_s	Equilibrium constant for surface reaction $(=k_s/k'_s)$	units depend on reaction
k	Boltzmann constant	$ML^2T^{-2}\theta^{-1}$
k_0	Zero order rate constant per unit surface area. A second subscript refers to number of reaction	moles $L^{-2}T^{-1}$
k_1	First-order rate constant per unit surface area. A second subscript refers to number of reaction	LT^{-1}
k_2	Second-order rate constant per unit surface area. A second subscript refers to number of reaction	moles$^{-1}L^4T^{-1}$
$k_{A,B}$	Rate constant for adsorption of A,B, etc.	$M^{-1}LT$
$k'_{A,B}$	Rate constant for desorption of A,B, etc.	T^{-1}
k_D	Mass transfer coefficient	$L^{-1}T$
k_s, k'_s	Rate constant for forward and reverse surface reactions	units depend on reaction. If first order T^{-1}
k_g, k_w	Heat transfer coefficients for solid–gas interface and wall respectively	cal $L^{-3}T^{-1}\theta^{-1}$
k_v	First-order rate constant per unit volume	T^{-1}
L	Length	L
$L(r)$	Length distribution function	—
M	Molecular weight	—
m	Mass	M
N	Avogadro's number	—
N	Rate of mass transport per unit area	$ML^{-2}T^{-1}$
$N_{C,R,T}$	Number of catalytic, reactor or transfer units	—
N_g	Number of molecules in gas phase	—
N_s	Total number of surface sites	—
N_v	Number of vacant surface sites	—
$N_{\ddagger}$	Number of single transition state complexes at surface	—
n	Number of moles	—
n	Number of Frenkel defects	—
n_0	Number of carriers in semiconductor	—
n_s	Number of Schottky defects	—
n_p	Number of pores per unit surface area	L^{-2}

P	Total pressure	$ML^{-1}T^{-2}$
(Pe)	Peclet number	—
p	Partial pressure	$ML^{-1}T^{-2}$
p_0	Saturated vapour pressure	$ML^{-1}T^2$
$p_{a,d}$	Pressure on adsorption or desorption loop of isotherm	$ML^{-1}T^{-2}$
Q	Heat of adsorption	cal mole^{-1}
Q_g	Rate of heat generation per maximum quantity of heat available	—
Q_c	Rate of heat loss per maximum quantity of heat available	—
R	Gas constant	cal mole^{-1} or $ML^2T^{-2}\theta^{-1}$
R	Radius	L
(Re)	Reynolds number	—
R'	Drag force per unit area	$ML^{-1}T^{-2}$
$R_{1/2}$	Rate of reaction in single half pore	moles T^{-1}
R_0	Rate of reaction in pore in absence of diffusion	moles T^{-1}
R^0	Rate of reaction at pore mouth	moles T^{-1}
R_p	Rate of reaction per pellet	moles T^{-1}
$r_{a,d}$	Rates of adsorption and desorption	moles $L^{-2}T^{-1}$
r_b, r_n	Radii of pore body and neck	L
r_k	Kelvin radius	L
r_m	Rate of reaction per unit mass	moles $M^{-1}T^{-1}$
r_0	Overall reaction rate	moles $L^{-3}T^{-1}$
r_s	Rate of surface reaction	moles $L^{-2}T^{-1}$
$\bar{r}_{L,V}$	Mean pore radius based on length or volume parameter	L
S	Entropy per mole	calθ^{-1}
S	Surface area	L^2
S_g	Surface area per unit mass	$M^{-1}L^2$
S_v	Surface area per unit volume	L^{-1}
S_x	External surface area	L^2
(Sc)	Schmidt number	—
T	Absolute temperature	θ
T_a	Adiabatic temperature rise	θ
t	Time	T
t	Thickness of adsorbed layer	L
U	Internal energy	cal mole^{-1}
$\bar{u}$	Mean Maxwellian velocity	LT^{-1}
u	Volumetric velocity	L^3T^{-1}
u_l	Superficial velocity or linear velocity	LT^{-1}

V	Reaction volume	L^3
V	Potential	volts
$\bar{V}$	Molar volume	$mole^{-1}L^3$
V', V''	Rate of slowest forward reaction step (single prime) and the reverse step (double prime) associated with it	moles $L^{-3}T^{-1}$
V_b	Height of potential energy barrier	MT^{-2} or volts
V_g	Volume of gas adsorbed per unit mass of adsorbent	$M^{-1}L^3$
V_p	Volume of catalyst pellet	L^3
v	Volume of gas adsorbed	L^3
v', v''	Rate of forward and reverse process	moles $L^{-3}T^{-1}$
v_m	Volume of gas adsorbed in a mono-layer per unit mass of adsorbent	$M^{-1}L^3$
v_p	Pore volume	L^3
v_s	Volume of pore occupied by adsorbate at its s.v.p.	L^3
W_F	Work required to form single Frenkel defect	cal
W_S	Work required to form Schottky defect	cal
X	Electronegativity	eV, cal mole^{-1} or $ML^2T^{-2}\theta^{-1}$
x	Relative pressure ($=p/p_0$)	—
x	Fractional conversion to undesired product	—
y	Fractional conversion to desired product	—
z	Number of nearest neighbour-occupied sites	—
z	Reactor length	L
z_0	Distance from surface where energy is zero	L
z^*	Distance from surface of two-dimensional gas	L
α	Elovich constant	mole^{-1}
α	Contact or wetting angle	radians or degrees
α	Fraction of surface covered with poison	—
α	Rate of addition to end carbon atom	moles $L^{-2}T^{-1}$
α_n	Probability that chain will grow beyond nth carbon atom	—

β	Dimensionless parameter $(=c_0 \Delta H D_e/\kappa T)$	—
β	Rate of addition to penultimate carbon atom	moles $L^{-2}T^{-1}$
β_n	Ratio of probability that chain will terminate at nth carbon atom	—
β_L	Factor by which $2V_p/S$ differs from $\bar{r}_L$	—
β_L	Factor by which $2V_p/S$ differs from $\bar{r}_V$	—
γ	Dimensionless parameter $(=E/RT)$	—
γ	Rate of desorption of chains from surface	moles $L^{-2}T^{-1}$
Δ	Electronegativity correction	eV, cal mole^{-1} or $ML^2T^{-2}\theta^{-1}$
δ	Difference in number of moles between product and reactant per mole of reactant $(=\nu-1)$	—
δ	Thickness of stagnant gas film surrounding particle	L
ϵ	Potential energy	cal molecule^{-1}
ϵ	Eddy diffusivity	L^2T^{-1}
ζ	Fraction of total surface area occupied by one component of composite catalyst particle	—
η	Viscosity	$ML^{-1}T^{-2}$
η	Effectiveness factor	—
θ	Fraction of surface covered by adsorbate	—
κ	Thermal conductivity	cal $L^{-1}T^{-1}\theta^{-1}$
λ	Wavelength	L
λ	Mean free path	L
μ	Dipole moment	(e.s.u.). L or Debye unit
μ	Magnetic moment of nucleus	(e.m.u.). L or ergs gauss^{-1}
μ	Chemical potential	cal mole^{-1}
μ^0	Standard chemical potential at unit pressure	cal mole^{-1}
Π	Surface pressure	MT^{-2}
ξ	Dimensionless reactor length	—
ξ	Fraction of reactor volume occupied by one component of a composite catalyst	—
ν	Frequency	T^{-1}

ν	Number of moles of product in stoichiometric equation	—
$\underline{\nu}$	Velocity of reaction	Units depend on reaction
ρ	Density	ML^{-3}
ρ_b	Bulk density of catalyst	ML^{-3}
ρ_p	Pellet density	ML^{-3}
Σ	Area occupied by adsorbed molecule	L^2
σ	Condensation coefficient	—
σ	Surface tension or energy	MT^{-2}
τ	Residence time $(=V/u)$	T
τ	Tortuosity or roughness factor	—
τ	Average time elapsing before re-evaporation of molecule	T
τ'	Halting-time	T
Φ	Dimensionless parameter $(=\phi_s^2\eta)$	—
ϕ	Dimensionless temperature	—
ϕ_s	Thiele modulus for spherical particle	—
ϕ	Work function	MT^{-2} or volts
χ	Dimensionless adiabatic temperature rise	—
χ	Paramagnetic susceptibility	—
ψ	An eigenfunction	—
ψ	Fraction of particle volume which is void space $(=$ porosity$)$	—
ψ_b	Fraction of packed bed which is void space	—
ω	Solid angle	radians
ω	Angular velocity	radians T^{-1}

CHAPTER 1

Introductory Remarks

1.1. Definition

A catalyst is a substance that increases the rate at which a chemical reaction reaches equilibrium; catalysis is the word used to describe the action of the catalyst. These terms were invented by Berzelius in 1835 in his renowned appraisal of the researches of Edmund Davy (1820), J. W. Döbereiner (1822) and others (see Ref. [1]).

Berzelius endowed catalysts with some mysterious quality—he talked of a recondite catalytic force—and only in comparatively recent times [2] has the aura of the occult been finally exorcized from discussions on catalysis. The phenomenon has been extensively studied for nearly a century and a half, and unconsciously used for a much longer period: it may not, like Melchizedek, have existed from eternity, but it was certainly harnessed by the ancients in the pursuit of some of their primitive arts [1]. Today, catalysts make possible about nine-tenths of the chemical manufacturing processes in use throughout the world, and research into the mechanisms of catalysed reactions continues to increase at a feverish pace [3, 4].

Just as we classify certain reaction mechanisms into two main types— homogeneous and heterogeneous—so also do we distinguish between homogeneous and heterogeneous catalysis. Heterogeneous catalysis describes the enhancement in the rate of a chemical reaction brought about by the presence of an interface between two phases, and it may, as has recently been suggested, be thought of as a kind of homogeneous catalysis in two dimensions.

In our discussion of the principles and study of heterogeneous catalysis, we shall give overwhelming preference to those reactions occurring at interfaces between solids and gases, although we shall occasionally deal with reactions at other interfaces. This preference is dictated partly by the enormous industrial significance of catalysis by solid surfaces of reactions between gases, and partly by the wealth of largely uncoordinated information relating to such catalysis.

1.2. Historical Outline of the Study of Heterogeneous Catalysis

In order that we may, in subsequent chapters, portray the complexities of heterogeneous catalysis in better perspective, it is not inapposite that we outline, at this stage, some of the landmarks reached in previous studies of the subject.

During the first decade of this century, chiefly as a result of the kinetic studies of Bodenstein and Ostwald, and the thermodynamic principles enunciated by van't Hoff, it became apparent that the following statements were valid. (i) A catalyst can increase the rate of only those processes that are thermodynamically favourable; it cannot initiate reactions that are not thermodynamically feasible. (ii) Any increase that a catalyst brings about in the velocity constant of the forward reaction is accompanied by a corresponding increase in the velocity constant of the reverse reaction. That is why we say a catalyst facilitates the approach to equilibrium of a given chemical change. (iii) For a given reactant or group of reactants there may be several reaction paths, and by the appropriate choice of catalyst any one of these paths may be "selected". For example, formic acid may suffer two alternative decompositions:

$$HCOOH \rightleftharpoons H_2O + CO$$

and

$$HCOOH \rightleftharpoons H_2 + CO_2$$

The first of these changes is catalysed by oxides such as alumina, the second by various metals (see Chapter 6). (iv) To talk about theories of catalytic reactions *in general* is profitless, if not illusory, because the methods of operation of catalysts are as varied as the modes of chemical change.

By 1920 several significant contributions, some theoretical, others of a more practical nature, had been made to the study of heterogeneous catalysis. Sabatier [5] had already suggested that a metal such as nickel, which was known to catalyse hydrogenation reactions, possessed its activity because it could readily form an intermediate hydride which, in turn, decomposed to regenerate the free metal. Haber had synthesized ammonia, catalytically, and worked out the thermodynamics of the system. Moreover, Langmuir had almost completed his demonstration of the inadequacy of the Nernst theory—which had satisfactorily explained the kinetic features of the dissolution of solids in liquids—to account for gas reactions at surfaces. In due course, after Langmuir, Rideal, Hinshelwood and their associates had concentrated upon the study of the kinetics of heterogeneously catalysed reactions, it became possible to formulate some generalized principles to account for the various rate-pressure relationships that had been observed experimentally. Thus, the Langmuir-Hinshelwood mechanism for catalysed processes postulated that the rate of a heterogeneous reaction is controlled by the reaction of the adsorbed molecules, and that all adsorption and desorption processes are in equilibrium. The Rideal-Eley mechanism, on the other hand, envisaged that a heterogeneous reaction could take place between strongly adsorbed atoms (that is, those chemisorbed) and molecules held to the surface only by weak, van der Waals forces (that is, those physically adsorbed).

In the next decade several notable concepts were propounded. H. S. Taylor [6] advanced cogent reasons for believing that preferential adsorption on a catalyst surface would take place at those atoms situated at peaks, fissures and other crystalline discontinuities. Moreover, it was inferred that such atoms would also have the highest catalytic activity. This constituted the notion of "active sites" or "active centres", terms which, along with "catalyst poisoning" to which they are related, are still widely used (but with somewhat greater precision, see Chapter 5) in the current literature. Taylor also suggested [7] that the process of chemisorption frequently involves an activation energy, an idea which proved fruitful in connection with the interpretation of adsorptive and catalytic phenomena. In this period also, A. L. Marshall [8] discovered one of the earliest examples of chain reactions initiated at surfaces, a topic of considerable interest today; also Balandin [9] proposed his mutiplet hypothesis in which he interpreted catalytic activity in terms of the geometric arrangement of atoms at the catalyst surfaces. By about 1940 a flood of ideas ran through the literature of heterogeneous catalysis. The theory of absolute reaction rates; the suggestion that for metallic and semiconductor catalysts, forces other than chemical interaction between surfaces and adsorbed species were responsible for their action (for example, their ferromagnetism or electrical properties); the availability of reasonably reliable methods of measuring the surface areas of catalysts—all these made their impact. The work of Beeck [10] on orientated metal films, with which he demonstrated the possible correlation of lattice spacing or percentage d-character and catalytic activity, together with that of Schwab [11] on metal alloy systems, where the electron concentration seemed to be linked to catalytic activity, had also reached fruition.

During the past twenty years an enormous effort has been made to concentrate both upon the direct study of the interaction of gas molecules and atoms at the catalyst surface and upon the structure of the catalyst surface itself. Extremely powerful techniques have been employed, and it has emerged that, although most of the simple correlations of the past have proved either inadequate or erroneous, there now exists the rich promise that heterogeneous catalysis can be brought closer to homogeneous catalysis, which has, in the past, proved much more tractable. It has become apparent that, rather than relying too much on such notions as the overall or general electron concentration of the solid, what is required is an intelligent use of some of the more important aspects of band theory, ligand field theory and the general principles of organic and inorganic chemistry in assessing the properties of the transition complex formed at the surface.

1.3. Sequence and Organization of Ensuing Chapters

The ultimate objective in this book is to acquire a degree of understanding sufficient to enable us on the one hand to become fully acquainted with the mechanisms of the more important types of heterogeneously catalysed reactions and, on the other, to appreciate the main factors involved in the design of catalytic reactors. Irrespective of whether our prime interest is in reaction mechanisms (a topic more likely to appeal to the academic and the researcher) or in catalytic reactors (which are usually of greater concern to the pragmatist and the chemical engineer), it is necessary to consider carefully a number of important concepts and principles.

The first of these is adsorption, which is discussed in Chapter 2. Various types of adsorption isotherm will be considered in relation to the dependence of the magnitude of the heat of adsorption upon surface coverage; also, the energetics of adsorption will be discussed in terms of potential energy diagrams. From the equations of the theory of absolute reaction rates, the velocities of heterogeneous and homogeneous reactions will be compared. Such comparisons enable us quantitatively to appreciate how the activation energy of a reaction is lowered in the presence of a heterogeneous catalyst.

Since the role of adsorbed species in catalysis is still incompletely understood and much debated [3], it is important that we become familiar with the wide range of techniques that can be used in adsorptive and catalytic studies. In Chapter 3, accordingly, much attention is devoted to such topics as the preparation of surfaces, measurement of adsorbed amounts and the estimation of surface areas. Also considered is the way in which information derivable from heats and entropies of adsorption throws light upon the structure and mobility of intermediates on catalyst surfaces. In addition, an exhaustive survey is conducted of the various approaches—for example, isotope exchange, spectroscopic methods, work function measurements, electron diffraction and electron microscopy—that enable a direct study to be made of the structure of catalysts or adsorbed intermediates.

Because the extent of surface that a catalyst pellet may present to any gaseous reactant is bound up with the pore structure of the solid, it is important to be able experimentally to deduce the total amount of accessible surface which is contained within a given pellet. Chapter 4 consequently deals with such topics, as well as the geometry of pore structures and the surface area of the pores, since both the latter are of vital importance in determining the performance of a catalyst. It also contains a discussion of the geometry of pore structures, models of which are used in interpreting various types of experimental isotherm.

The next two chapters dwell on the numerous theories that have over the past half-century sought to relate the catalytic activity of solids to some more completely understood properties of such solids. In Chapter 5 the role of lattice imperfections is investigated. After outlining the salient features of the various types of defect known to exist in solids, together with the energetics of interaction at the surfaces of semiconductors, specific examples are given illustrating the part played by dislocations and point defects in catalysed reactions. This chapter also deals with the exciting new fields of stereo-regular polymerization (Ziegler-Natta catalysis) and radiation catalysis, since their interpretation rests heavily upon the concept of imperfection in the solid state. It will be shown that it is profitable to unite such fields as inorganic coordination chemistry, modern organic chemistry and solid-state physics if a deeper understanding is to be gained, of, for example, Ziegler-Natta catalysis.

Chapter 6, on the other hand, is concerned with the more traditional attempts to interpret heterogeneous catalysis. Reference will be made to the present status of the comparatively well-known electronic and geometric factors in catalysis, and some prominence will be given to the re-emergence of the intermediate compound theory for certain reactions.

The dynamics of selective and polyfunctional catalysis are considered in Chapter 7. It has been known for some time that a catalyst is seldom entirely specific, in the sense that one reaction only proceeds in the presence of the catalyst. As a rule, the catalyst gives rise either to alternative, but simultaneous, reaction paths that yield two or more products, or to successive kinetic steps that lead ultimately to an end-product but in which one (or more) intermediate may be isolated. In order to be able to exercise control over a catalyst so that a maximal yield of some product may be obtained, it is necessary to consider those factors (such as the characteristics of pore structure and catalyst composition) which operate in such a way as to be selective in their function. Such factors are discussed in Chapter 7.

The concepts and principles discussed in the previous chapters are exemplified, using a number of typical examples, in Chapter 8. Catalytic oxidation, hydrogenation, cracking, synthesis and decomposition are considered, as also are some aspects of electrocatalysis and a somewhat unusual kind of heterogeneous catalysis, the influence of solid impurities on the oxidation of elemental carbon. Our treatment of catalytic reactors in Chapter 9 embraces the rudiments of mass transfer, heat transfer and gas diffusion, some design calculations for adiabatic and isothermal conditions, and the optimization of temperature and concentration profiles in various types of reactor bed.

References

1. E. K. Rideal and H. S. Taylor, "Catalysis in Theory and Practice", Ch. 1. Macmillan, London (1926).
2. S. C. Lind, "The Chemical Effects of Alpha Particles and Electrons", 2nd Ed., p. 69. The Chemical Catalog Company Inc. (1928).
3. Discussions of the Faraday Society, "The Role of the Adsorbed State in Heterogeneous Catalysis", Liverpool (1966).
4. G. C. Bond, *Rep. Chem. Soc.* **61**, 99 (1964).
5. P. Sabatier, "La Catalyse in Chemie Organique". Librairie Polytechnique, Paris (1913).
6. H. S. Taylor, *Proc. R. Soc.* **A108**, 105 (1925).
7. H. S. Taylor, *J. Am. chem. Soc.* **53**, 578 (1931).
8. A. L. Marshall, cited by H. S. Taylor in *Rev. Phys. Chem.* **13**, 10 (1962).
9. A. A. Balandin, *Z. phys. Chem.* **132**, 289 (1929).
10. O. Beeck, A. E. Smith and A. Wheeler, *Proc. R. Soc.* **A177**, 62 (1941).
11. G. M. Schwab and E. Shwab-Agallidis, *Naturwiss enschaften* **31**, 322 (1943).

CHAPTER 2

Adsorption: Energetics, Isotherms and Rates

2.1. The Nature of Adsorption

From the definition of heterogeneous catalysis, it follows that at least one of the reactants must become attached in some way, and for a significant period, to the surface of the solid catalyst. As the process of attachment is of central importance in catalysis, it is imperative that we first concern ourselves with the precise details of the nature of this process. Indeed, there is little hope that we can ever fully appreciate the mechanism of heterogeneous catalysis unless we first comprehend the characteristics of adsorption.

The forces acting at the surface of a solid, be it a catalyst or not, are unsaturated. Hence, whenever a fresh surface is exposed to a gas a higher concentration of gas molecules will result on the surface compared with that in the gas phase. This preferential concentration of a molecule at the surface is termed adsorption. Although the strength of adsorption, that is, the tightness with which molecules of the adsorbate are attached to the adsorbent, together with the extent of adsorption, may vary widely from system to system, it is nevertheless possible to divide all adsorptions into two main types: physical adsorption and chemisorption.

2.1.1. Physical adsorption and chemisorption

In principle, the distinction between these two types of adsorption is clear cut. Physical adsorption is caused by the forces of molecular interaction which embrace [1] permanent dipole, induced dipole and quadrupole attraction. For this reason, it is frequently designated van der Waals adsorption. Chemisorption, on the other hand, involves the rearrangement of the electrons of the interacting gas and solid, with consequential formation of chemical bonds. In other words, physical adsorption is akin to the condensation of a vapour to form a liquid, or to the liquefaction of gases, and chemisorption can be regarded as a chemical reaction which is restricted to the surface layer of the adsorbent. It is evident from these definitions that as our concept of what constitutes a chemical bond alters, so also will our concept of chemisorption.

Experimentally, it is frequently possible to distinguish between the two types of adsorption, but in some instances the distinction is so ill defined that several criteria have to be employed before a decision can be reached. The magnitude of the *heat of adsorption* forms the basis of one, and probably the best (but see Section 3.2.8), of these criteria. During physical adsorption the heat liberated per mole of gas adsorbed is generally in the region of 2–6 kcal, but values as large as 20 kcal mole^{-1} have been reported [2]. For example, the heats of physical adsorption of argon [3] and krypton [4] on graphitized carbon black are, respectively, 2·70 and 3·90 kcal mole^{-1}, the heats of liquefaction being 1·55 and 2·31 kcal mole^{-1} at the corresponding temperature. Seldom does the heat of physical adsorption exceed the heat of liquefaction of the gas in question by more than a factor of two or three (but see Table 1 of this chapter). During chemisorption larger values for the heat of adsorption are usually encountered; the heat of chemisorption of oxygen on some metals may, for instance, be a few hundred kcal mole^{-1}. Generally, heats of chemisorption are rarely less than 20 kcal mole^{-1}, but values as low as those associated with physical adsorption are known, and evidence has recently come to light to demonstrate the occurrence of endothermic adsorption (see Section 2.2.3).

A second criterion used to distinguish experimentally between chemisorption and physical adsorption is the *rate* at which the process occurs. It is argued that, since physical adsorption simulates liquefaction, the same dispersion forces being at work, it should, like liquefaction, require no activation and therefore occur very rapidly. Chemisorption, on the other hand, should, like most chemical processes, require activation. This criterion of rate proves to be a useful guide but, just as for the heat of adsorption, it can be misleading if it alone is the sole criterion used. Thus, chemisorption of hydrogen and oxygen on many clean metal surfaces is essentially a non-activated process, as evidenced by the fact that, at liquid nitrogen temperature, the process takes place extremely rapidly. Moreover, there is another difficulty which may render this criterion untenable. Physical adsorption on a porous solid, such as hydrocarbons on a silica-supported alumina catalyst, may take place very slowly, if diffusion of the adsorbate along the pores is the rate-determining step. This could easily be mistaken for activated chemisorption.

As well as the rate of adsorption, the *rate of desorption* is frequently a useful guide in deciding which type of adsorption obtains. It will emerge later (Section 2.2), when the energetics of adsorption are discussed, that the activation energy of desorption from the physically adsorbed state (assuming an adsorbent of low porosity) is rarely more than a few kilocalories per mole, whereas the activation energy for desorption from the chemisorbed state is generally greater than 20 kcal mole^{-1}, being almost invariably greater than or equal to the heat of chemisorption.† It is for this reason that the ease of desorption noted on warming a system from liquid air temperatures to $-78°C$ or to room temperature is often employed [5, 6] as a preliminary means of ascertaining the nature of the adsorption.

Another useful criterion is based on the *temperature range* over which the adsorption occurs. Physical adsorption should occur only at temperatures close to the boiling-point of the adsorbate at the operative pressure, whereas chemisorption, being associated with much stronger forces, should be capable of occurring at temperatures well above the boiling-point of the adsorbate at the operative pressure. Although unquestionably useful, this test can also prove misleading if the adsorbent is highly porous. On such a solid, physical adsorption may be quite extensive, even though the ratio p/p_0 of the pressure of the gas to the saturated vapour pressure (s.v.p.) at the operative temperature is as low as 10^{-5} or less; that is, even though the temperature is very much in excess of the boiling-point of the adsorbate. The boiling-point of argon is $87·4°K$ and p_0 at $160°K$ is $\sim10^5$ mm Hg. Therefore, on the basis of previous experience, we can expect physical adsorption to take place at an equilibrium pressure of 1 mm Hg when the temperature is $160°-87·4°=72°$ above the boiling-point of argon. To appreciate why this is possible, it is instructive first to consider, as did Zsigmondy [7], the classical Kelvin (Thomson [8]) equation (see Section 4.3.1), which relates the vapour pressure above a planar liquid surface to the vapour pressure above a curved surface, as of a meniscus of liquid inside a cylindrical capillary. The Kelvin equation is:

$$\ln\left(\frac{p}{p_0}\right) = \frac{-2\,\sigma\,\bar{V}\cos\alpha}{rRT} \tag{1}$$

† Substances which are desorbed in the ionic form may have heats of desorption less than the heats of adsorption. The respective values for caesium on tungsten are 54 and 63 kcal mole^{-1} [2].

where p is the vapour pressure above a surface the radius of curvature of which is r, p_0 is the vapour pressure above the planar surface, σ the surface tension of the liquid, $\bar{V}$ the molar volume of the liquid, α the angle of contact, R the gas constant and T the absolute temperature. Now, if we suppose that a highly porous adsorbent consists of numerous cylindrical pores of mean radius r, and that the adsorbate within the pores has properties characteristic of the bulk liquid phase, the system should conform to equation (1), p/p_0 now being the ratio of the equilibrium adsorption pressure and the s.v.p. of the adsorbate. Clearly, the smaller the value of r, the smaller the ratio p/p_0 and the greater the temperature, above the boiling-point, at which physical adsorption can take place.

A fourth criterion which helps to distinguish adsorption types is based on the extent of *specificity* in gas–solid interaction. Since chemisorptions are chemical reactions confined to the surfaces of solids, and since chemical reactions are specific, so also are chemisorptions in the sense that, if a gas is chemisorbed by a given solid under certain conditions, it does not follow that the same gas will be chemisorbed by another solid of equivalent surface cleanliness and under identical conditions. Obviously, the chemical potentials of the interacting substances and of the possible surface products govern the feasibility of the chemisorption. Physical adsorption, however, is effectively the process whereby condensation of gas occurs, the adsorbate being capable of building up to one or many layers on an inert solid. Like all the other criteria that have been discussed thus far, this one also has its limitations. There is little doubt [9] that the adsorbent is not always "chemically" inert, and that specificity, to some degree, may obtain even in physical adsorption.

Although no single foolproof test is available to assess the type of adsorption in any given system, the above-mentioned criteria, taken collectively, leave little doubt as to the type which prevails. There are also other, rather special, criteria (see Section 3.3.1) which, taken singly even, afford sufficient evidence to indicate the category of the adsorption. For example, a substantial change of magnetic susceptibility of an adsorbent during adsorption may constitute convincing evidence that the process is chemical [10]. Again, if the surface area of the solid is known, the extent of adsorption can sometimes be used as an indication, for it is doubtful whether chemisorption can ever exceed more than a monolayer, whereas physical adsorption can extend to multilayers [11].

2.2. The Energetics of Adsorption

Since it is well established (see Section 2.4.1.3) that heterogeneous catalysis is usually a consequence of the fact that the activation energy of a reaction is lowered as a result of chemical combination between the catalyst and at least one of the reactants, it is profitable to discuss the general nature and extent of the energy changes involved in adsorption. To appreciate the way in which the energetics of chemisorption are normally expressed, we shall first consider the simpler problem of the energetics of physical adsorption, which are also relevant to the question of catalysis.

2.2.1. Depiction of adsorption processes by potential energy diagrams: heat of adsorption

It was shown by Lennard-Jones [12, 13] over thirty years ago that, by calculating the van der Waals fields outside solid surfaces, it is possible to estimate the magnitude of the heats of physical adsorption of certain gases on the surfaces of the solids. In the intervening years numerous theoretical studies have been carried out (see Refs. 1, 2, 14) to compute the dependence of the potential energy of an adsorbate molecule upon its distance from the adsorbent surface. It has emerged [1] that different theoretical approaches are required for different systems if satisfactory agreement is ever to be reached between the calculated and observed heats of physical adsorption. For example, a certain equation may be applicable for the interaction between an inert gas and the surface of an alkali halide, but be quite inappropriate for the same gas and a metal or other good electrical conductor. It is not here necessary to enumerate the various types of physical interaction which are now known [2] to demand distinct theoretical treatment; all that is necessary is to appreciate that the interaction energy responsible for physical adsorption can, in principle, and for most systems, be represented by an equation similar to that which Lennard-Jones successfully employed for the potential energy of gas molecules as they approach each other:

$$E = - ar^{-m} + br^{-n} \tag{2}$$

In equation (2), a, b, m and n are constants, and r is the separation distance. The first of the terms on the right-hand side arises from attractive forces, and the second from repulsive forces, between the interacting species. Lennard-Jones chose [13, 15] values of 6 and 12 for m and n respectively. The first of these choices was theoretically justified in view of London's work [16] on dispersion forces, which were shown to fall off with the seventh power of distance; the second, $n=12$, had no theoretical significance and was chosen for computational convenience. The constant a presented relatively little difficulty, its value being calculable from such properties as the atomic susceptibilities and polarizabilities. However, much uncertainty, which still persists, attached to the calculation of the "repulsive" constant b. Largely because of this, it is not yet possible to calculate, a priori, the complete potential energy equation for gas molecules. The current procedure is to adjust the value of b so that a "correct" value is obtained for the separation of two gas molecules in the equilibrium condition when the attractive and repulsive forces between the molecules are equal [17]. Thus, at the equilibrium distance, r_0, we have that $dE/dr=0$, so that equation (2) yields:

$$mar_0^{-m-1} - bnr_0^{-n-1} = 0 \qquad (3)$$

or

$$b = \frac{m}{n} ar_0^{(n-m)} \qquad (4)$$

The total decrease E_0 in potential energy at the equilibrium distance is therefore given by:

$$E_0 = ar_0^{-m} \left(\frac{m}{n} - 1\right) \qquad (5)$$

Clearly, the theoretical value of E_0 can be made to coincide with the observed energy of interaction by adjusting b through equations (4) and (5).

If the theoretical difficulties in evaluating potential functions are quite considerable for gas phase molecules, the problems encountered when the interaction involves a condensed phase are not likely to be easier. When an adsorbate molecule approaches a surface, it is, effectively, interacting with a large number of atoms simultaneously, and the total potential energy of the interaction is an infinite series composed of terms similar to those contained in equation (2), with the appropriate value of r being used in each case. Although the form of the potential energy function, which is dependent upon the structure of the solid, cannot be written down in such a way as to be applicable to all known systems of physical adsorption, and although precise calculations are still impossible, we can at least assert that an equation such as equation (2) gives, in most instances, the correct shape of the potential energy diagram depicting physical adsorption, and frequently leads to values of heats of adsorption in reasonable agreement with those obtained experimentally [18]. To take a specific example, Kiselev [18] and his associates calculated the potential energies (and, hence, the heat of adsorption) of vapours of non-polar substances on graphitized carbon black, using the following potential-equation:

$$E = - C_{i1} \sum_j r_{ij}^{-6} - C_{i2} \sum_j r_{ij}^{-8} + B \sum_j \exp\left(-\frac{r_{ij}}{\rho}\right) \qquad (6)$$

where i is the force centre of the adsorbate molecule; the dispersions C_{i1} and C_{i2} are calculable from magnetic susceptibility and polarizability data using formulae previously derived by Kirkwood [19] and Müller [20]; ρ is an exponential repulsion constant; the constant B is adjustable in much the same way as constant b in equation (2). Table 2.1 shows some of the results quoted by Kiselev. Apart from showing that equation (6), (i) predicts the correct sequence of adsorption energy values, and (ii) leads to satisfactory agreement between observed and "computed" values, Table 2.1 also shows the type of energies we can anticipate in physical adsorption.

TABLE 2.1. Energies of adsorption (kcal mole^{-1}) on the surface of graphitized carbon blacks at half coverage (after Kiselev [18])

Adsorbate	Energy of adsorption Calculated (equation (6))	Observed
Hydrogen	0·90	0·91
Deuterium	0·95	0·95
Nitrogen	2·6	2·8
Propane	6·8	6·5
n-Hexane	12·4	12·5
Benzene	10·3	10·0
n-Octane	16·1	16·0

Figure 1 represents the general shape of the potential energy curve relating to the physical adsorption of a gas molecule M_2 on the surface of a solid S. It can be seen from this figure that the heat of adsorption is given by the difference in energy between the states X where the molecule is infinitely removed and Y where it is at its equilibrium distance from the surface. The potential energy increases rapidly with decreasing distance from the surface, since repulsive forces must dominate ultimately (state Z) to satisfy the physical requirement that two species cannot occupy the same space.

By analogy with what has already been said concerning physical adsorption, we would expect the potential energy diagram depicting the chemisorption of a reactive atom, such as atomic hydrogen, on a solid surface to be as

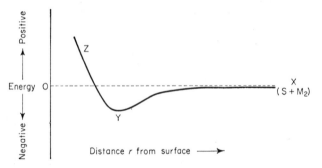

FIG. 1. Potential energy curve relating to the physical adsorption of a molecule M_2 on the surface of a solid S.

shown schematically in Fig. 2. Such a Lennard-Jones plot, as it is sometimes designated, is reminiscent of the Morse curve used [21] to represent the electronic energy of a molecule. At separation distances r not far removed from r_0, the equilibrium separation, the curve takes a parabolic form given by:

$$E = + Q \left\{- 1 + \sigma (r - r_0)^2 + \ldots\right\} \tag{7}$$

where σ is a constant, and Q is the heat or energy of adsorption (see Fig. 2). The heat of adsorption (per g atom) is again equal numerically to the difference between the corresponding levels X and Y, and it is to be noted that the heat of desorption is equal to the same difference in energy levels.

Having outlined the present status concerning the calculation of the

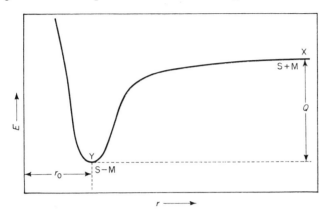

FIG. 2. Schematic representation of the chemisorption of a reactive atom M on the surface of a solid, S. Q is the heat or energy of adsorption, and r_0 the equilibrium separation distance.

potential energy of physical adsorption, the question naturally arises as to the feasibility of calculating the potential energy curve for chemisorption. It has to be stated, at the outset, that a complete theoretical computation, from first principles, is not possible. This is partly because of the inadequacy of our present knowledge of the forces involved in chemisorption, partly because of our ignorance of the precise structure, electronic and otherwise, of the surface layers of solids, and not least because of—to use the language of the quantum mechanist—our inability to solve the many-body problem [22]. However, the *general shape of the potential energy curve* may be constructed (with greater or lesser precision, depending upon the nature of the chemisorbed link with which we shall deal later) provided that we know the heat of adsorption and some other information, such as the bond dissociation energies of the adsorbate. It follows, therefore, that the task of constructing a realistic potential energy curve depicting a particular chemisorption devolves upon our obtaining a reliable value for the heat of that adsorption. It will emerge shortly that the potential energy curves for chemisorption are almost invariably constructed from the observed, rather than the calculated, heat of adsorption (that is, the curves are derived semi-empirically), the prime reason for this being, again, our inability to calculate heats of adsorption with the necessary precision.

It would be wrong, however, to assume that the calculation of heats of adsorption plays an insignificant role in our study of chemisorption and its relevance to catalysis. Apart from enabling us to draw acceptable Lennard-Jones plots, calculations of the energy or heat of adsorption play an important part in the assessment, and in the prediction,. of the type of bonding involved in a particular chemisorbed link. For these reasons, we shall now adumbrate the present status of the theoretical calculations of heats of adsorption before continuing with the main aim of deriving potential energy plots. Greater emphasis will be given in the following section to chemisorption on metals, as the relevant theoretical work on other adsorbents is extremely scant.

2.2.1.1. Theoretical Calculation of Heats of Chemisorption

The first step is to postulate a model on which the calculation is to be based. From our knowledge of chemical bonds generally, we can expect three types of chemisorption: (i) the pure ionic, in which the adsorbate acts as an electron donor or acceptor, (ii) the pure covalent, embracing the coordinate link, or (iii) the mixed type of bond.

Suppose the adsorbed bond is one hundred per cent ionic, a state of affairs which is certainly known to exist when a refractory metal such as tungsten interacts with alkali metal or alkaline-earth metal vapours [23]. The energy associated with this type of adsorption is readily calculable [24] from the known ionization energy eI of the alkali atom (e being the electronic charge); the electron affinity of the refractory metal as a whole (this is synonymous with the work function $e\phi$ of the adsorbent metal); an "image" energy $e^2/4R$ which can be regarded as the electrostatic energy of attraction of the ionized metal adsorbate atom by its image charge in the adsorbent metal, $2R$ being the distance between the charge of the ion and its image. The energy E of adsorption† is therefore given by:

$$E = e\,\phi - e\,I + \frac{e^2}{4R} \qquad (8)$$

The correspondence between energies of adsorption calculated from this equation and those observed for sodium, for rubidium, and for caesium on tungsten is satisfactory; but calculated values for the energy of adsorption of hydrogen as an H^+ ion at each adsorption site on the surface of the metal leads to values which are not only vastly different from the observed value, but so highly endothermic that great doubt prevails as to the likelihood of the existence of an ionic bond of the type S^-H^+. Equations similar to equation (8)

---

† Since ϕ and I in equation (8) are expressed in stat volts and e, the electronic charge, in e.s.u., the r.h.s. of the equation should be multiplied by $5\cdot02 \times 10^{23}/4\cdot2 \times 10^7$ to give E in ergs mole^{-1}.

also reveal that the same conclusion is valid for the adsorption, as positive ions, of oxygen atoms and nitrogen atoms on metal surfaces. On the other hand, ionic adsorption, as O_2^-, is very likely on caesium and on silver (see Section 2.2.3).

If the adsorbed bond is assumed to be covalent, it is not too difficult to estimate first the magnitude of the strength of the adsorbed bond, and thence the heat of adsorption. Again we shall restrict our discussion to adsorption on metals, and outline briefly the calculations of Eley [25, 26], Stevenson [27] and Eyring and his co-workers [28].

In the dissociative chemisorption of a hydrogen molecule on a metal:

$$2S + H_2 \longrightarrow 2S\text{---}H \tag{9}$$

the implication is that one hydrogen atom is held covalently by one metal atom only. Eley calculated the strength of the covalent bond, $D(S\text{---}H)$, by invoking Pauling's approximation [29] which leads to:

$$D(S\text{---}H) = \tfrac{1}{2}\left\{D(S\text{---}S) + D(H\text{---}H)\right\} + \Delta \tag{10}$$

$D(S\text{---}S)$ and $D(H\text{---}H)$ being the strengths of the metal–metal bond and of the hydrogen molecule, respectively; Δ, the electronegativity correction, effectively takes cognizance of the fact that the covalent bond may be partly ionic. $D(H\text{---}H)$ is well known; the magnitude of $D(S\text{---}S)$ is directly related to the sublimation energy of the metal, being one-sixth of the latter for face-centred cubic metals, since each atom is coordinated twelvefold to nearest neighbours in the lattice ($D(S\text{---}S)=2E_s/12$ where E_s is the sublimation energy). It is in the computation of Δ that the approaches of Eley and of Stevenson diverge. Instead of using the Pauling scale of electronegativities ($\Delta=23\cdot06\,(X_S-X_H)^2$ with X_S and X_H denoting the Pauling electronegativities of the respective elements), Stevenson utilized the theoretically more reliable Mulliken electronegativity values X^M given by:

$$X^M = \tfrac{1}{2}\,(eI + eE') \tag{11}$$

where eI and eE' are, respectively, the ionization energy and electron affinity. For a metal, X^M is exactly equal to the work function $e\phi$.

Bearing in mind that the heat of adsorption Q of hydrogen (per mole) is:

$$Q = 2\,D(S\text{---}H) - D(H\text{---}H) \tag{12}$$

it follows from equation (10) that:

$$Q = D(S\text{---}S) + 2\,\Delta \tag{13}$$

Using the Eley-Stevenson approach, the correspondence between observed and calculated bond energies $D(S\text{---}H)$ is moderate as evidenced by Table 2.2

TABLE 2.2. Comparison between the observed and calculated values of the metal–hydrogen bond energies (kcal mole^{-1})

Metal	W	Ta	Fe	Ni	Cr	Cu
Calculated value	73·4	67·6	60·1	60·2	59·5	64·4
Observed value	74·1	71·1	67·6	67·1	74·1	65·6

Although the deficiencies of this approach are by no means trivial—the value of $D(S—S)$, for example, derived from the sublimation energy is erroneous [30] because the excess surface energy of the solid is ignored—there is little doubt that it has gone a long way towards demonstrating that covalent bonding is a dominating feature in the chemisorption of certain gases such as hydrogen and nitrogen, on metal surfaces.

Eyring and his co-workers [28], in their calculation of heats of adsorption, adopted a quantum-mechanical approach, which can embrace all three adsorption types: pure ionic, pure covalent and mixed-type bonding. The eigenfunction ψ of the bond in question, S—M, where M represents an atom of the adsorbate, is written as:

$$\psi = c_i \psi_i + c_c \psi_c \qquad (14)$$

in which ψ_i and ψ_c are the eigenfunctions of the ideal ionic and covalent bonds, respectively, and c_i and c_c are constants. By solving the wave-equation:†

$$H\psi = E\psi \qquad (15)$$

where H and E are the total Hamiltonian operator and the bond energy, respectively; the fraction of ionic character (the ionicity) c_i^2 was found to be given by:

$$\frac{1}{c_i^2} = 1 + \frac{(E - H_{ii})}{(E - H_{cc})} \qquad (16)$$

the quantities H_{ii} and H_{cc}, the separated energies for the pure ionic and for the pure covalent bond, being given by:

$$H_{ii} = \int \psi_i H\psi_i \, dv \qquad (17)$$

and

$$H_{cc} = \int \psi_c H\psi_c \, dv \qquad (18)$$

dv being an element of volume.

Eyring et al. [28] calculated H_{cc} and H_{ii} semi-empirically from:

$$H_{cc} = \tfrac{1}{2} \left\{ D(S—S) + D(M—M) \right\} \qquad (19)$$

† For an introduction to the concepts used in this section see Coulson [31].

and from an equation for H_{ii} analogous to equation (8). They obtained c_i from the equation:

$$\mu = c_i^2 \, e \, r_{\text{S-M}} \tag{20}$$

in which μ and $r_{\text{S-M}}$ are the dipole moment and bond length, respectively, of S—M. The bond energy E is therefore calculable from equation (16) and so, in turn, is the heat of adsorption from equation (12). The agreement between observed heats of adsorption and those calculated in this way, even for systems in which c_i^2 is as high as 0·97 (as it is for barium on tungsten) or as low as 0·02, is satisfactory, but by no means perfect.

Serious consideration has recently been given [32, 33] to tackling the very difficult task of calculating, quantum-mechanically, the strength of the chemisorbed bond. The subject is still, however, laden with difficulties, and it is agreed [22] that much of the theoretical work required for a fuller theoretical understanding of chemisorption remains to be accomplished.

2.2.2. Potential energy diagrams describing chemisorption

Having discussed, at some length, the difficulties encountered in calculating heats of adsorption, we now return to the original theme of depicting chemisorption by means of Lennard-Jones plots, it being understood that, so long as heats of adsorption are known, it is always possible to evaluate the shape of these plots and, if certain additional information is available, this can be done with considerable precision.

We may begin by recalling the work of Eley and Rossington [34], which dealt with the dissociative adsorption of hydrogen on the surfaces of Group IB metals. They observed the heat of adsorption Q of hydrogen on copper to be 8·0 kcal mole^{-1}, and the activation energy of desorption of hydrogen molecules from the surface to be 13·0 kcal mole^{-1}. They could therefore draw the Lennard-Jones plot shown in Fig. 3. Curve A refers to the physical adsorption of molecular hydrogen, and curve B to the chemisorption of atomic hydrogen to form two covalent bonds:

$$H_2 + 2Cu \rightleftharpoons 2Cu\text{—}H \tag{21}$$

The heat of adsorption of hydrogen as two atoms, 8·0 kcal mole^{-1}, is clearly equal to $2Q_a - D(\text{H—H})$, where Q_a stands for the heat of adsorption of the atomic hydrogen, and $D(\text{H—H})$ the bond dissociation energy of the hydrogen molecule.

Figure 3 shows how, in energetic terms, hydrogen can be dissociatively adsorbed by a metal without the prior dissociation of the molecule in the gas phase: so long as the physically adsorbed molecule can acquire adequate activation, that is, get sufficiently near to the metal surface so as to reach the point X on curve A, it can transfer to curve B and become chemisorbed in the dissociated state. The height of X above the potential energy minimum is clearly the activation energy of desorption, 13·0 kcal mole^{-1}, and the height of X above the level of zero energy is the activation energy of adsorption.

Because the chemisorption of nitrogen on iron, as well as that of hydrogen on silver and gold, is an activated process, the Lennard-Jones plots in these

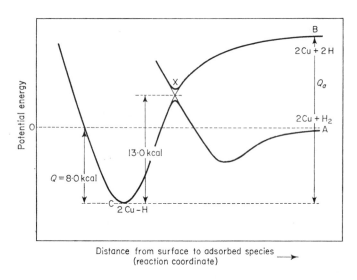

Fig. 3. Lennard-Jones curves for the copper–hydrogen system [34]. Curve A refers to the physical adsorption of molecular hydrogen, and curve B to the chemisorption of atomic hydrogen. The intersection of the two curves is slightly rounded, as expected from the non-crossing rule [31] between two states.

cases would also be drawn such that the height of the intersection point X above the energy zero would be equal to the corresponding activation energy of adsorption. For surfaces which have been stringently cleaned, chemisorption of gases at liquid air temperatures (ca −193°C) may occur at an extremely fast (frequently immeasurably fast) rate, thus revealing that there is no activation energy of adsorption. Trapnell [35] gives a comprehensive list of metals and oxides which display this behaviour towards hydrogen, carbon monoxide, carbon dioxide, oxygen, ethylene and other gases. One example is the chromia–carbon monoxide system and the nickel–hydrogen system is another. In the potential energy diagrams depicting these non-activated chemisorptions, the point X must fall below the level of zero energy.

Suppose that experimental evidence indicated that a particular adsorbent surface was heterogeneous in the sense that the strength of the adsorbed link, and hence the magnitude of the heat of adsorption, for certain sites on the surface took two or more values. For the state of affairs where just two

distinct types of chemisorption bonds form, the Lennard-Jones diagram
would be as shown in Fig. 4. From the way in which these curves have been
drawn, we may deduce that:

(i) when the molecule M_2 is dissociatively adsorbed to produce the surface
bond $(S—M)_y$, the heat of adsorption is greater than if it were dissociatively
adsorbed as $(S—M)_x$;

(ii) to convert physically adsorbed molecules of M_2 into the chemisorbed
$(S—M)_y$ state requires activation, but to convert physically adsorbed molecules via the $(S—M)_x$—chemisorbed state to the $(S—M)_y$ state does not require activation, since Z and Y fall below the level of zero energy; and

(iii) if the solid were exposed to the gas M_2 at temperatures near to the
boiling-point of the latter, the extent of adsorption would be considerable,
and would consist of both physically adsorbed and chemisorbed $[(S—M)_x$
and $(S—M)_y]$ material. Upon increasing the temperature, the physically retained species would be driven off from the surface. Further increase in temperature would provide sufficient activation for more gas to be chemisorbed
(preferentially as $(S—M)_y$), and an even greater increase in temperature

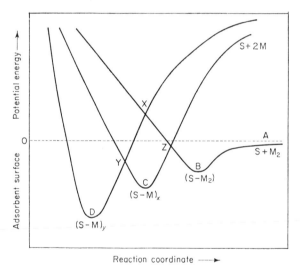

Fig. 4. Lennard-Jones curves for a system in which there are two distinct sites for
dissociative chemisorption.

would lead to desorption from the chemisorbed state. If the results were expressed as an adsorption isobar (p. 32), there would be a minimum and a
maximum as shown schematically in Fig. 5(a).

Had the Lennard-Jones curves in Fig. 4 been drawn in such a way that
both the intersection points Y and Z were above the level of energy zero, we

would conclude that the adsorption isobar would be akin to that shown in Fig. 5(b) (the route of desorption being DYCZBA, Fig. 4).

Gundry and Tompkins [36] interpreted their results on the kinetics of hydrogen adsorption on evaporated iron films in terms of diagrams similar to Fig. 4, there being evidence [37] that one chemisorbed state involved solely

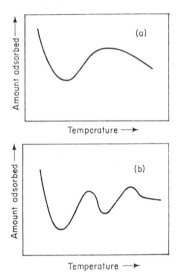

FIG. 5. (a) Schematic illustration of the adsorption isobar expected if adsorption of M_2, over a range of temperatures, followed the route ABXD of Fig. 4. (b) Isobar showing two maxima.

the d-orbitals of the surface metal atoms, and the other hybridized bonds of the dsp-type.

With regard to Fig. 5(a, b), it is instructive to recall that, in a crucial paper dealing with the emergence of potential energy diagrams, Lennard-Jones [13] explicitly states that one of his prime aims was to formulate a theoretical explanation for the occurrence of adsorption maxima and minima which had so perplexed investigators up to about 1932 [38–41]. This aim appears, in retrospect, to have been amply fulfilled.

2.2.2.1. *Activation Energy of Adsorption and the Shape of the Potential Energy Diagram*

Since it is obvious that a knowledge of the activation energy of chemisorption is related to the general question of the energetics of surface-catalysed reactions, and since we can make certain predictions concerning the dependence of the activation energy of adsorption upon the energetics of the participating species, we shall enumerate the factors that determine the magnitude of the activation energy of adsorption. The procedure adopted here follows that used successfully in gas-phase kinetics [42] in which the various factors influencing the shape of Eyring-Polanyi diagrams [43, 44] are considered.

Taking Fig. 3 as our basis of discussion, let us suppose that a gas molecule M_2 is being dissociatively chemisorbed on to the solid surface. The height of the intersection point X is the direct measure of the magnitude of the activation energy in question, and it is patent that four factors come into consideration [2] so far as elevating X (that is increasing the activation energy) is concerned: (i) separation distance between adsorbate and adsorbent atoms at the minimum value of C; (ii) difference between levels B and C; (iii) difference between levels A and B (the dissociation energy); and (iv) steepness of the curve BC. (The steepness is greater the smaller is the ionic character of the adsorbed link.)

Clearly, the activation energy of chemisorption will increase the smaller the separation distance and the smaller the difference between B and C. Furthermore, a large difference between A and B, and a steeper curve BC will also serve to increase the activation energy.

2.2.3. Endothermic adsorption

The view is frequently expressed that endothermic adsorption cannot occur. This is tantamount to stating that all adsorptions are exothermic processes. The problem of ascertaining the veracity of this statement is not simply a matter of academic interest. As will be shown (Section 5.4.3), endothermic adsorption may play a prominent role in certain heterogeneously catalysed reactions, and the importance of endothermic adsorption has been recently emphasized by de Boer [45] and by Schwab [46]. We shall deal with the reasons for initially accepting, and later retracting, the notion that endothermic adsorptions are impossible.

The justification generally produced for the statement that adsorptions are always exothermic is a thermodynamic one: since the Gibbs free energy G must decrease for any spontaneous process, and since chemisorption is one such process which is always accompanied by a decrease in entropy (the number of degrees of freedom in the adsorbed state being less than the number in the gaseous state), then, according to the second law of thermodynamics, which leads to the equation

$$\Delta G = \Delta H - T\Delta S, \tag{22}$$

ΔH, the enthalpy or heat of adsorption, must be negative; that is, adsorption is always exothermic. (An exothermic adsorption is one that, in the field of chemisorption and heterogeneous catalysis, is said to have a positive heat of adsorption. For example in Fig. 2, $Q = -\Delta H$, where $-\Delta H$ is another notation for the heat of adsorption.)

For the overwhelming majority of adsorptions, this argument would be valid. It may well be that the argument applies invariably to physical adsorptions, where it can be shown [47], using equations derived from statistical mechanics, that, even when the adsorbed species is extremely mobile, the entropy of a species in the adsorbed state is always less than the entropy in the gaseous state. That endothermic physical adsorption has never been reported is, therefore, not surprising.

For several chemisorptions, however, there is cogent evidence that endothermic processes have been observed [45, 46, 48–51]. In the chemisorption of molecular hydrogen on glass, the heat of adsorption is negative [50]. The heat of adsorption is also negative (that is ΔH is positive) when hydrogen is dissociatively adsorbed on iron surfaces contaminated with sulphide ions [45]. There are other instances of chemisorption where endothermicity is strongly suspected, but not yet incontrovertibly established. Molecular hydrogen at 0°C on Group 1B metals and on cadmium is one example [52], molecular oxygen on silver is another [45, 49].

Before considering the detailed energetics of endothermic adsorption, it is pertinent to consider, in general terms, how it is that we can account for the absence of endothermic physical adsorption while also accepting the existence of endothermic chemisorptions. In physical adsorption, the entropy change is always negative, since we have essentially a condensation process, analogous to liquefaction, on the adsorbent. Because the number of degrees of freedom of the adsorbed species is less than the number it possessed prior to adsorption, and because the entropy of the adsorbent is unaltered (since there is no chemical interaction, but see Ref. 52), the entropy change ΔS is negative, and so, from equation (22), ΔH must be negative. In chemisorption, on the other hand, a surface reaction takes place involving the rupture and creation of chemical bonds. In the same way that some chemical reactions are endothermic, so also it may be expected that some surface reactions can be endothermic.

In an adsorption process which is endothermic, the entropy change must be not only positive, but the $T\Delta S$ term must also exceed numerically the ΔG term (equation (22)). The implication of this conclusion meets with some scepticism in view of the popular conception that the entropy of adsorption is always negative. Consequently, it is useful, for heuristic purposes, to visualize the following hypothetical system. A gas molecule M_2 is dissociatively chemisorbed on the surface of a solid S. Suppose the strength of the S—M bond is equal to half the strength of the M—M bond. Now, if the adsorbed atoms M have complete two-dimensional mobility, it would then follow that a positive entropy change ΔS would result, corresponding to a net gain of one degree of freedom, and the free energy change associated with this thermally neutral process ($\Delta H=0$) would be equal solely to the $T\Delta S$ term.

However, positive values for the entropy of adsorption are also likely to originate from another source: the change in entropy of the adsorbent itself. Thus, during chemisorption, even if, as is usual, the entropy of the species adsorbed decreases on adsorption, due largely to reduction in the number of degrees of freedom, this decrease may be exceeded by a concomitant increase

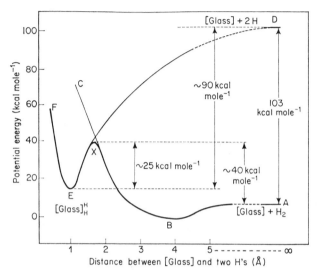

FIG. 6. The endothermic chemisorption of hydrogen on glass: the heat of adsorption is about -15 kcal mole^{-1} [2].

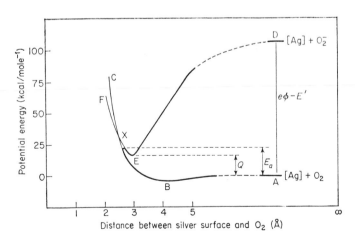

FIG. 7. The endothermic chemisorption of molecular oxygen, as O_2^-, on a silver surface [45].

in the entropy of the adsorbent. This, in turn, would lead to an endothermic process if the $T\Delta S$ term were greater numerically than the ΔG term.

Energetically, it is not difficult to envisage endothermic adsorptions. Figure 6, which is taken from the work of de Boer [2], shows two potential energy curves; one refers to the chemisorption of atomic hydrogen on glass, the other to the physical adsorption of molecular hydrogen on the same adsorbent. The energy level of the two hydrogen atoms adsorbed on glass is above that of *gaseous* molecular hydrogen (for exothermic adsorptions the situation is reversed, compare Fig. 3). The path of endothermic adsorption is ABXEF. Figure 7 is one way in which the suspected endothermic chemisorption of molecular oxygen on silver or copper can be drawn [45, 49]. The relatively large value of the work function ($e\phi$) for these metals and the slightly positive electron affinity (E') of molecular oxygen make it likely that the chemisorption of oxygen as O_2^- ions (just as in the anions of the lattice of KO_2 or RbO_2) on these metals is endothermic. The size of the adsorbed O_2^- ion is thought to be less than the distance of the physically adsorbed oxygen from the surface, and the potential energy minimum E is assumed to fall on the right-hand side of the curve BC. The path of endothermic adsorption is ABXEF.

One further aspect of endothermic adsorption merits some comment. The statement that the activation energy of desorption is equal numerically to at least the heat of adsorption, which is valid for most exothermic adsorptions (see Section 2.1.1) also holds good for endothermic adsorptions, as can be judged from Figs. 6 and 7.

2.3. Adsorption Isotherms

Experience has shown that the equilibrium distribution of adsorbate molecules between the surface of the adsorbent and the gas phase is dependent upon pressure, temperature, nature and area of the adsorbent, and the nature of the adsorbate. An *adsorption isotherm* shows how the amount adsorbed depends upon the equilibrium pressure of the gas at constant temperature. An adsorption *isobar* relates how the amount adsorbed varies with temperature at constant pressure; and an *isostere* relates the equilibrium pressure to the adsorption temperature for a stipulated amount of gas adsorbed.

The isotherm is by far the most convenient method of defining an experimentally determined adsorption equilibrium. Moreover, theoretical treatments of adsorption, for certain assumed models, usually arrive at adsorption isotherms rather than isobars or isosteres. Adsorption isobars are, however, useful in ascertaining whether two or more adsorption types are involved in a given system—see, for example, Fig. 5(a, b). Adsorption isosteres are often utilized when heats of adsorption are determined from adsorption measurements at two or more temperatures (see Section 3.2.8). It is necessary for us to become acquainted with the salient features of a number of well-known isotherms before we can begin to appreciate the kinetic interpretation of heterogeneously catalysed reactions. The isotherms, whether derived theoretically or empirically, can frequently be represented by simple equations which convey directly how the concentration of the adsorbed species varies with the gas pressure.

TABLE 2.3. A selection of adsorption isotherms†

Name	Isotherm equation	Applicability	Equation number
Langmuir	$\dfrac{v}{v_m} = \theta = \dfrac{bp}{1 + bp}$	Chemisorption and physical adsorption	(23)
Freundlich	$v = kp^{1/n}, (n > 1)$	Chemisorption and physical adsorption	(24)
Henry	$v = k'p$	Chemisorption and physical adsorption	(25)
Slygin-Frumkin‡ (Temkin)	$\dfrac{v}{v_m} = \theta = \dfrac{1}{a} \ln c_0 p$	Chemisorption	(26)
Brunauer-Emmett-Teller (BET)	$\dfrac{p}{v(p_0 - p)} = \dfrac{1}{v_m c} + \dfrac{(c - 1)}{v_m c} \dfrac{p}{p_0}$	Multilayer physical adsorption	(27)

† v is the volume adsorbed (usually expressed in cm^3 or ml at n.t.p.); v_m is the volume required to cover the adsorbent surface with a monomolecular layer of adsorbate; θ is the fraction of the monolayer covered at an equilibrium pressure p; p_0 is the s.v.p. of the adsorbate; all other symbols in equations (23) to (27) are constants.
‡ This equation was first derived theoretically by Slygin and Frumkin [54], but Temkin and Pyzhev [55] popularized it in their classical interpretation of the decomposition of ammonia on platinum and tungsten surfaces.

It is not necessary here to recapitulate the full details of how all the isotherm equations in Table 2.3 were first or subsequently derived. We will, however, devote some attention to the Langmuir and the BET equations. In heterogeneous catalysis, the Langmuir equation is ubiquitous because, as will be shown in Chapter 8, it figures eminently in numerous kinetic interpretations of surface reactions. It has also been used to estimate surface areas. The BET equation warrants comment in this chapter because it is the most successful equation that seeks to account for physical adsorption from the submonolayer to the multilayer stage. Its appearance in 1938 marked a turning-point in the history of catalysis, since it meant that there became available, for the first time, a reliable means of measuring the surface area of solid catalysts. Furthermore, its status in the theory of physical adsorption is as central as that of the van der Waals equation in the theory of liquids [56].

2.3.1. The Langmuir isotherm

The original derivation of Langmuir [57] was a kinetic one. Implicit in the derivation were the following assumptions [58]: (i) the adsorbed entities are attached on the surface of the adsorbent at definite, localized sites; (ii) each site can accommodate one and only one adsorbed entity; and (iii) the energy of the adsorbed entity is the same at all sites on the surface, and is independent of the presence or absence of other adsorbed entities at neighbouring sites.

Langmuir visualized the dynamic equilibrium between adsorbate molecules in the gas phase at a pressure p, and the adsorbed entities in the surface layer, the fraction of the sites covered being θ. The number of molecules impinging on the adsorbent surface per unit area in unit time is proportional to the pressure. The rate of adsorption is, therefore, proportional to $p(1-\theta)$, and the rate of desorption is proportional to θ only. At equilibrium, these two rates are equal, so that:

$$k_1 p (1 - \theta) = k_{-1}\theta \qquad (28)$$

where k_1 and k_{-1} are the rate constants for adsorption and desorption respectively. Hence,

$$\frac{\theta}{1 - \theta} = \frac{k_1}{k_{-1}} p = bp \qquad (29)$$

or

$$\theta = \frac{bp}{1 + bp} \qquad (30)$$

which coincides with equation (23).

This derivation tends to conceal the significance of the constant b. The full kinetic derivation [59], which is much more enlightening, is therefore considered next. Consider the general case of chemisorption at the surface of a solid. The velocity of chemisorption will clearly depend on (i) the pressure (since this determines the number of collisions with the surface), (ii) the activation energy of chemisorption E_a (since this determines the fraction of the colliding molecules which possess the necessary energy to be chemisorbed), (iii) the fractional coverage of the surface $f(\theta)$ ($f(\theta)$ is equal to $(1-\theta)$ only when the adsorbed molecule occupies one surface site), and (iv) a steric factor or condensation coefficient σ which may be defined as the fraction of the total number of colliding molecules *possessing the necessary activation energy E_a* that results in adsorption. (The condensation coefficient is not to be confused with the sticking probability s defined as the fraction of *all* colliding molecules that is adsorbed.) The velocity v_1 of chemisorption is therefore given by:

$$v_1 = \frac{\sigma p}{(2\pi mkT)^{1/2}} f(\theta) \exp\left(\frac{-E_a}{RT}\right) \qquad (31)$$

$p/(2\pi mkT)^{1/2}$, from the kinetic theory of gases, is the number of molecules of mass m striking unit area of surface in unit time. The velocity of desorption v_{-1} may, likewise, be written

$$v_{-1} = k_{-1}\,\phi(\theta)\,\exp\left(\frac{-E_d}{RT}\right) \tag{32}$$

where k_{-1} is the specific rate constant for desorption, $\phi(\theta)$ represents an appropriate function of the fractional number of sites available for desorption ($\phi(\theta)=\theta$ when the desorption occurs from sites which are occupied by a single adsorbed molecule), and E_d stands for the activation energy of desorption.

At equilibrium $v_1=v_{-1}$ so that,

$$p = (2\pi mkT)^{1/2}\,\frac{k_{-1}}{\sigma}\,\exp\left(\frac{-E_d + E_a}{RT}\right)\frac{\phi(\theta)}{f(\theta)} \tag{33}$$

Since the difference between the activation energies of desorption and adsorption equals the heat of adsorption Q (see Fig. 3), we have, for the special case when one molecule occupies a single site on the adsorbent surface ($\phi(\theta)=\theta$ and $f(\theta)=(1-\theta)$), that

$$p = (2\pi mkT)^{1/2}\,\frac{k_{-1}}{\sigma}\,\exp\left(\frac{-Q}{RT}\right)\frac{\theta}{1-\theta} \tag{34}$$

Reverting now to the basic assumptions of the Langmuir isotherm, if the third, and most important, of these is valid, viz. that the heat of adsorption Q is constant for all sites, then equation (34) may be rewritten as:

$$p = \frac{1}{b}\frac{\theta}{(1-\theta)} \tag{35}$$

where

$$\frac{1}{b} = \frac{k_{-1}}{\sigma}\,(2\pi mkT)^{1/2}\,\exp\left(\frac{-Q}{RT}\right) \tag{36}$$

Equation (35) is synonymous with the Langmuir equation (30) previously derived.

Using statistical thermodynamics, the value of the constant b in the Langmuir equation has been shown by Fowler [58], Sexl [60], Rushbrooke [61] and others [62] to be given by

$$\frac{1}{b} = \frac{(2\pi m)^{3/2}\,(kT)^{5/2}\,f_g(T)}{h^3 f_a(T)}\,\exp\left(\frac{-Q^1}{RT}\right) \tag{37}$$

where $f_g(T)$ and $f_a(T)$ are the internal partition functions for a molecule in the gaseous and adsorbed state, respectively, h is Planck's constant, and Q^1 is the energy required to transfer a molecule from the lowest adsorbed state to the lowest gaseous state. Q^1 is, in other words, the heat of adsorption at absolute zero. Since it is reasonable to assume that, for a specific type of adsorption, the heat of adsorption is virtually temperature independent† we may write $Q = Q^1$. The ratio k_{-1}/σ of the desorption and adsorption constants can then be calculated, using equations (36) and (37), provided that the values of $f_g(T)$ and $f_a(T)$ are, in turn, calculable. Evaluation of $f_g(T)$ presents few problems for simple molecules (see p. 49 and Glasstone *et al.* [63]), and $f_a(T)$ is also tractable, although its value, for a given molecule, varies according to whether the adsorbed molecule is mobile or immobile [47].

If, on adsorption, each molecule dissociates into two entities and each entity occupies one site—a phenomenon frequently encountered in chemisorption—then the Langmuir equation becomes:

$$\theta = \frac{(bp)^{1/2}}{1 + (bp)^{1/2}} \tag{38}$$

Under these circumstances, $f(\theta)$ and $\phi(\theta)$ become equal, respectively, to $(1-\theta)^2$ and θ^2. When the adsorbed molecule dissociates into n entities each of which occupies a surface site, we have [64]:

$$\theta = \frac{(bp)^{1/n}}{1 + (bp)^{1/n}} \tag{39}$$

The Langmuir isotherm for two gases adsorbed simultaneously, and without dissociation, on the same adsorbent is of considerable utility in the interpretation of catalysed reactions involving two substances. It follows readily that, if θ_A and θ_B refer to the fraction of sites covered by molecules of type A and by the type B, respectively, then

and

$$\left.\begin{aligned}\theta_A &= \frac{b_A p_A}{1 + b_A p_A + b_B p_B} \\[2em] \theta_B &= \frac{b_B p_B}{1 + b_A p_A + b_B p_B}\end{aligned}\right\} \tag{40}$$

where b_A and b_B are the Langmuir constants for substances A and B and p_A and p_B are their respective partial pressures. In the general case when i substances are being simultaneously adsorbed, we have

$$\left.\begin{aligned}\theta_A &= \frac{b_A p_A}{1 + \sum\limits_i b_i p_i} \\[2em] \theta_B &= \frac{b_B p_B}{1 + \sum\limits_i b_i p_i}\end{aligned}\right\} \tag{41}$$

† But see Section 3.28, p. 105.

2.3.2. Limiting forms of the Langmuir equation

Taking the most widely encountered form of the Langmuir equation (equation (30)) and bearing in mind that we may write $\theta = v/v_m$, we have:

$$v = \frac{v_m \, bp}{1 + bp} \tag{42}$$

an equation which, at very low pressures ($bp \ll 1$) predicts, as does Henry's isotherm (equation (25)), a linear relationship ($v = v_m bp$) between amount adsorbed and equilibrium pressure. As the pressure approaches infinity, equation (42) predicts that v approaches v_m asymptotically. Figure 8 shows, schematically, the type of adsorption isotherm to be expected if the Langmuir equation were obeyed. The slope of the straight part of this curve (at low pressure) is determined by the constant b, the magnitude of which, as seen from equations (36) and (37), is largely determined by the ratio Q/RT. For a fixed temperature, the adsorption isotherm for the case of strong adsorption (large Q) will be of the type shown as curve (1) in Fig. 8, whereas the isotherm for weak adsorption will be more like that illustrated by curve (2), Fig. 8. For a specific adsorption, as the temperature increases, b decreases, and a gradual transition from type 1 to type 2 occurs.

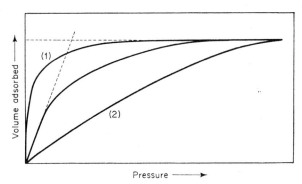

FIG. 8. General shapes of isotherms which conform to the Langmuir equation.

There are a few more points concerning the limiting forms of the Langmuir equation. At intermediate pressures where bp is neither very much greater nor very much less than unity, we may write

$$\theta = bp^m \tag{43}$$

where m steadily decreases from unity to zero as the pressure increases. Also, at high pressure, equation (35) can be modified to

$$(1 - \theta) = \frac{1}{bp} \tag{44}$$

since θ approaches unity. The amount of the surface that is uncovered is inversely proportional to the pressure.

2.3.3. Applicability of Langmuir isotherm

Equation (23) may be transformed into three forms suitable for testing:

$$\frac{p}{v} = \frac{1}{bv_m} + \frac{p}{v_m} \tag{45}$$

$$\frac{v}{v_m p} + \frac{bv}{v_m} = b \tag{46}$$

and

$$\frac{v_m}{v} = \frac{1}{bp} + 1 \tag{47}$$

Clearly, plots of p/v against p; v/p against $-v$; and $1/v$ against $1/p$ should all yield straight lines from which the values of v_m and b are calculable. The last-named method of plotting should be avoided, because too much weight is placed on the low-pressure part of the isotherm (and adsorption at low coverages is rarely representative of the entire surface). Either of the other two methods is satisfactory, though the second is probably the best, because values of b are given by both the slope and the intercept of the straight line.

The procedure for testing the applicability of the Langmuir equation is best illustrated by a specific example. In their study of the physical adsorption of krypton on evaporated carbon films, Sykes and Thomas [65] reported the isotherm shown in Fig. 9. Upon plotting p/v against p, the straight line shown in Fig. 10 was obtained. But it would be premature to conclude that the entire carbon surface was energetically uniform towards krypton simply because of apparent conformity to the Langmuir equation. In fact, at lower coverages ($\theta = 0.1$–0.5) the physical adsorption data did not obey the Langmuir equation, and there was good reason to believe (compare Ref. 66) that the carbon surface was energetically heterogeneous.

Numerous adsorption systems appear, at first, to conform to the Langmuir equation, but stringent testing generally reveals that, especially when we are considering chemisorption, adsorbent surfaces are far from uniform. What are these tests? In the first place, the Langmuir plot should be linear from $\theta = 0$ to $\theta \to 1$, or at least for a broad range of θ values. Secondly, and more important, the values of v_m and b should be acceptable. It is easy to decide from equations (36) and (37) whether the values of v_m and b derived from the Langmuir plot are acceptable. Moreover, upon applying the Langmuir equation to the adsorption data over a series of temperatures and the appropriate range of θ, it should emerge, if the surface sites are energetically uniform, that (i) v_m should be temperature independent and (ii) b should decrease exponentially with increasing temperature.

In Langmuir's own work [67] on the adsorption of carbon monoxide, carbon dioxide, oxygen, nitrogen and other gases on glass and mica adsorbents, reasonably linear p/v against p plots were obtained. Jacobs and Tompkins [68] have, however, made the interesting observation that Langmuir's values of v_m, deduced from the linear plots, ranged from 3 to 86% of

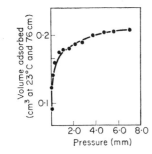

FIG. 9. Adsorption isotherm of krypton adsorbed on an evaporated film of carbon at $-183°C$ [65].

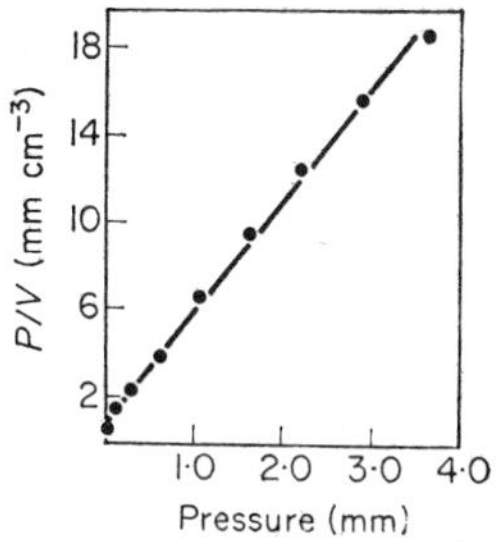

FIG. 10. The Langmuir plot obtained from the data of Fig. 9.

the measured geometric area which, in turn, was known to be appreciably less than the actual surface area. Langmuir's theory is, therefore, not applicable to his own results.

It is important to emphasize that very few chemisorptions are known to conform to the Langmuir equation over the entire range of surface coverage, but that several systems are known in which the Langmuir model is valid for a restricted range of θ. This is another way of stating that most surfaces are not uniform, and adsorptions noncooperative, in the sense implied in assumption (iii), page 34. The picture is much the same for physical adsorption also, although it has been predicted theoretically [69] and confirmed experimentally [66] that, even when the shape of the adsorption isotherm at low coverage indicates significant mutual interaction between the adsorbed molecules, the adsorption isotherm at higher coverages may obey a Langmuir equation. The reason for this apparently paradoxical state of affairs is that, at a certain coverage, a pattern of uniform, localized sites may be established on the adsorbent surface by the adsorbed molecules themselves. In this new adsorption régime the heat of adsorption differs from that characteristic of the original surface, a *constant interaction energy* being operative and an equation such as (48), rather than equation (36), being valid:

$$\frac{1}{b} = \frac{k_{-1}}{\sigma}(2\pi mkT)^{1/2}\exp\left\{-\frac{(Q + zE)}{RT}\right\} \tag{48}$$

E is the interaction energy between adsorbed molecules and z is the number of occupied sites adjacent to any given site on which adsorption can take place. The data in Table 2.4 show how, during the physical adsorption of argon on graphitized carbon black at low temperatures and relatively high coverage (θ=0·60 to θ=0·90), the Langmuir equation is satisfactory.†

TABLE 2.4. Results of adsorption of argon on graphitized carbon showing conformity to Langmuir equation†

Constant	Defining equation	At temperature	
		77·8°K	90·1°K
b torr^{-1}	(23)	5·95	0·593
v_m (cm^3 at s.t.p.)	(23)	3·23	3·12
Q (cal mole^{-1})	(48)	2650	
zE (cal mole^{-1})	(48)	380	

† Ross and Winkler [66].

2.3.4. Significance of the failure of the Langmuir model

We shall, in subsequent chapters, delve more deeply into the properties of surfaces and surface layers, but it is as weil, at this stage, to delineate how the feeble success of the Langmuir model itself illuminates the chemistry of the adsorbed state. Our aim is to arrive at the reasons why the Langmuir theory was found wanting.

Brunauer [70] has stated succinctly two objections to the unmodified Langmuir model. First, apparent saturation of a surface is often observed when only a small fraction of the area of the adsorbent is covered. Secondly, the adsorption maximum at the so-called saturation, viz. at the v_m value, frequently increases with decreasing temperature. Brunauer pointed out that such anomalies can be explained if it is accepted that the surface is hetero-geneous: they can also be explained [71] if there is interaction between mole-cules which are adsorbed close to one another on a surface.

That surfaces of catalysts, and solids in general, may be energetically heterogeneous was first emphasized by Taylor [72] and by Constable [73]. Taylor argued that atoms situated at promontories on the surface and along crystal edges or at crystal corners will be the most unsaturated, so that the adsorption energy will be greater at these sites than at atoms situated in smooth regions of the surface. His views on the existence of "active centres" were corroborated by many examples in which a reduction in the rate of a catalysed reaction through "poisoning" had been observed. Pease and Stewart [74], for instance, reported that mercury vapour diminished the amount of adsorption of ethylene on copper to about 80% of the normal value, and that of hydrogen to about 5% of the normal value. Furthermore, the rate of ethylene hydrogenation was diminished to about 0·5% of the

† See, however, Ross and Olivier, "On Physical Adsorption", Chapter 5. Interscience, New York (1964).

original value. These results could be understood if it were conceded that only a small fraction (that is, the "active sites") of the copper surface was involved in the catalysed reaction.

Before pursuing the notion of surface heterogeneity, it is useful to note that experimental evidence pointing to the operation of repulsive forces between adsorbed species came much to the fore shortly after the concept of "active centres" was formulated. For example, Roberts [75], in his meticulous work on stringently cleaned tungsten filaments, found it reasonable to ascribe the steadily decreasing heat of adsorption of hydrogen with increasing coverage to repulsion between adsorbed species. Over the past thirty years abundant evidence has come to light to confirm that both surface heterogeneity and surface interaction are factors of considerable importance in the surface chemistry of solids. Since in no facet of our discussion is the relative importance of these factors brought more sharply into focus than in the assessment of the reasons for the fall in the heat of chemisorption with increasing coverage, this topic will be summarized next.

2.3.5. The decrease in the heat of adsorption with increasing coverage [59, 71]

If the Langmuir theory were universally valid, we should expect the heat of adsorption to remain constant with increasing values of the coverage θ (see curve 1, Fig. 11, p. 43). Although it is not unknown [76] for the heat of adsorption to be independent of θ, direct determinations of the Q–θ dependence reveal that almost invariably there is a decrease: the results of Beeck [77], summarized in Fig. 12, p. 45, are particularly apposite. For the decrease there are four possible explanations,† all of which have been discussed in detail by de Boer [53] and by Gnudry and Tompkins [24].

The first explanation is based on the concept of the work function. If it is assumed that, in the formation of a chemisorbed layer, electrons are extracted from the solid, then, as coverage increases, more energy will be expended in forming the chemical bond since extra work will have to be done against the progressively increasing work function. Although this effect may sometimes be significant, it is not a very convincing explanation as correlation between calculated and observed falls are frequently poor [2, 53]. A second explanation attributes the decrease to changes in bond type which may occur with increasing coverage. We have already seen (Section 2.2.2) how the chemisorption of hydrogen on iron or nickel films was interpreted as taking place on two, energetically different, surface sites. In more recent years infrared absorption studies of the adsorbed state indicate [78] that both carbon monoxide and ethylene alter their type of bonding to metal surfaces as coverages increase. This explanation has not yet been fully tested.

† Recently Blyholder has formulated another explanation based on the notion that the extent of hybridization within a transition metal alters during the course of chemisorption.

Interaction of adsorbed molecules, mentioned above, has long been cited as responsible for the fall in heat of adsorption with coverage. The suggestion here is that the heat of adsorption decreases because of the mutual repulsion of oriented dipoles in the adsorbate, it being well established that the bond in chemisorption possesses a dipole moment [79]. From the measured magnitude of the dipole moment it is possible to calculate the total interaction energy of the long-range electrostatic and short-range repulsive forces. It emerges that, for hydrogen on tungsten, for example, the calculated fall in Q ($= -\Delta H$) is about 2–3 kcal mole^{-1} compared with an observed decrease of 25–30 kcal mole^{-1}. Consistently poor agreement for other systems shows that this is not the main reason for the fall. However, it would be wrong to infer that the contribution of repulsive forces is always trivial. Kummer and Emmett [80], by studying the adsorption of ^{14}C-labelled carbon monoxide on an iron catalyst used in the synthesis of ammonia, concluded that about half of the surface was uniform, and that the molecules adsorbed thereon gave rise to a substantial degree of mutual repulsion.

Of the four explanations, the one based on the concept of surface heterogeneity appears to be the most likely, especially when the fall of Q at low coverages is rapid. An ingenious test was devised by Roginskii and his co-workers [81, 82] to ascertain the relative importance of the once competing hypotheses of surface heterogeneity and surface repulsion. Keier and Roginskii [82] demonstrated the surface heterogeneity of nickel in the following manner. Hydrogen was adsorbed on freshly prepared nickel so as to cover a fraction of the surface only. Deuterium was then adsorbed on the remaining sites. Desorption of the adsorbate gases was then effected by raising the temperature (pumping at very low dynamic pressures would also have sufficed). Bearing in mind that the strength of the adsorbate–adsorbent bonding is not significantly different for hydrogen and deuterium, the tendency for deuterium to appear first in the gas phase signifies that the surface is energetically heterogeneous. Had the surface been energetically uniform, and the fall in Q the result of repulsive interactions, the ratio of the deuterium to the hydrogen in the desorbed gas would be in strict conformity with the composition of the adsorbed layer. Subsequent use of this technique, notably by Kummer and Emmett [80] and by Walker and Vastola [83], has shown that surface heterogeneity, as well as some degree of repulsive interaction, is the predominant cause for the decrease in heat of chemisorption with coverage.

In the last decade direct evidence concerning the actual structure of active sites at solid surfaces has emanated from such techniques as transmission and replica electron microscopy [84], field-ion microscopy [84, 85], and electron-probe microanalysis [84] (see Chapter 3). It is becoming increasingly apparent that it is now possible to obtain a more complete physical picture of the heterogeneous solid surface. Regions of the surface where the heat of chemisorption is likely to be different from that at smooth parts of the surface are (in addition to those suggested by Taylor and Constable): (i) the points of emergence of screw and edge dislocations, twin boundaries and bend planes; (ii) point defects, such as Frenkel and Schottky defects, situated in the outermost layers of the solid; (iii) metallic and nonmetallic impurity centres occurring singly or in agglomerates. Such imperfections are discussed in Chapter 5.

2.3.6. The Freundlich isotherm

We now have abundant evidence for the fall of the heat of adsorption with increasing coverage. Let us suppose that this fall is logarithmic as represented diagrammatically in Fig. 11 (curve (2)). What we are really supposing is that the adsorption sites are distributed exponentially with respect to the energy (heat) of adsorption, and this coincides with the basic assumption made by Zeldowitch [86] in his classical derivation of an adsorption isotherm for a heterogeneous surface. The isotherm he finally arrived at, after a series of relatively minor assumptions, was synonymous with the Freundlich equation (equation (24)) which, hitherto, had been regarded as an empirical equation. The essence of Zeldowitch's derivation is contained in the following simplified version due to Laidler [71].

The surface sites are subdivided into several types, each type possessing a characteristic heat of adsorption (that is, the Langmuir model is assumed to be

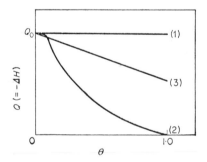

FIG. 11. Schematic diagram illustrating (1), constant heat of adsorption; (2), logarithmic fall of heat of adsorption; and (3), linear fall of heat of adsorption with increasing coverage.

valid for the sites within one type, there being no repulsion forces). Hence, from equation (35),

$$\frac{\theta_i}{1 - \theta_i} = b_i p \qquad (49)$$

If the only property that causes the parameter b to vary is the heat of adsorption Q, and none of the pre-exponential terms such as k_{-1} and σ (see equation (36)), then equation (49) may be written:

$$\frac{\theta_i}{1 - \theta_i} = b^1 \, p \, \exp\left(\frac{Q_i}{RT}\right) \tag{50}$$

where Q_i is the heat of adsorption for the ith type of site, and b^1 is the factorized part of b. Upon rewriting equation (50) we have:

$$\theta_i = \frac{b^1 p \, \exp\left(\dfrac{Q_i}{RT}\right)}{1 + b^1 p \, \exp\left(\dfrac{Q_i}{RT}\right)} \tag{51}$$

If there are N_i sites of the ith kind, the overall fraction of sites covered is:

$$\theta = \frac{\sum\limits_i \theta_i N_i}{\sum\limits_i N_i} \tag{52}$$

By combining equations (51) and (52) and replacing the summations by integrations we have:

$$\theta = \frac{\displaystyle\int_0^\infty \left[N(Q)b^1 \, p \, \exp \, (Q/RT)/\{1 + b^1 p \, \exp \, (Q/RT)\} \right] \, \mathrm{d}Q}{\displaystyle\int_0^\infty N(Q)\mathrm{d}Q} \tag{53}$$

where $N(Q)$ is some function of Q. This is where Zeldowitch introduced his crucial assumption:

$$N(Q) = a \, \exp\left(- \frac{Q}{Q_0}\right) \tag{54}$$

a and Q_0 being constants. An exact integration of the expression in equation (53) was still not possible, but, for fairly small values of θ, the solution is:

$$\ln \theta = \frac{RT}{Q_0} \ln p + \text{constant} \tag{55}$$

which can be recast into $\theta = kp^{1/n}$, the form in which the Freundlich equation is normally written.

The Freundlich equation is, therefore, no longer to be regarded merely as a convenient form of representing the Langmuir equation at intermediate values of θ. Moreover, the method of derivation disposes of the criticism that the Freundlich equation predicts a progressively increasing coverage with increasing pressure: the isotherm is expected to be valid only at low coverages.

In several systems involving the chemisorption of gases on metals and oxides, the Freundlich equation has been found to be strictly applicable (see, for example, Refs. 87 and 88).

2.3.7. The Slygin–Frumkin or Temkin isotherm

Slygin and Frumkin [54] supposed that the fall in heat of adsorption with increasing coverage was linear, and not logarithmic as in a system conforming to the Freundlich equation. This was a reasonable supposition, as many systems, especially at low or medium coverages, tend to exhibit a linear

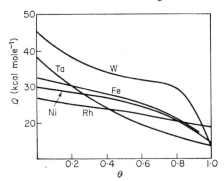

FIG. 12. Heat of adsorption Q plotted against fractional coverage θ for hydrogen on evaporated films of metals [77].

$\Delta H/\theta$ dependence. (The results of Beeck, see Fig. 12, vindicate this remark.) The linearity can arise, as explained in Section 2.3.5, from a variety of causes: from repulsion between adsorbate species on a uniform surface as much as from intrinsic surface heterogeneity.

Derivation of the isotherm equation in a form which can be transposed into equation (26) has been accomplished by Brunauer, Love and Keenan (see Section 8.4.1). Thus, whenever a plot of ln p against the amount adsorbed (or fractional coverage) is linear, it may fairly be deduced that the heat of chemisorption falls linearly with increasing coverage. Such a state of affairs obtains [90] when certain iron catalysts adsorb nitrogen at elevated temperatures.

2.3.8. The Brunauer–Emmett–Teller isotherm

Almost every review of physical adsorption published during the last two decades has referred to the five main types (see Fig. 13) of physical adsorption isotherms which were first clearly recognized by Brunauer [70] in 1945. The most conspicuous feature of these diagrams is that, apart from type I adsorption, all the isotherms, particularly types II and III, imply that the extent of adsorption does not reach a limit corresponding to the completion of a monolayer. The first attempt to account for the S-shaped adsorption isotherms typified by type II was made in the polarization theory of de Boer and Zwikker [91], who assumed, incorrectly, that a process of successive induction of dipoles occurred between juxtaposed layers in the multilayers.

This theory failed, as did the generalized Langmuir [57, 92] theory designed to embrace multilayer adsorption, to account for S-shaped isotherms. Success came with the Brunauer-Emmett-Teller (BET) theory, which, quite apart from its theoretical significance, is of interest here, as it has received prodigious attention from surface chemists faced with the problem of estimating surface

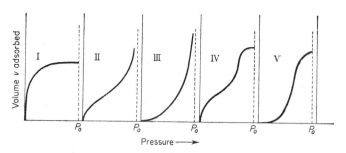

FIG. 13. The five types of physical adsorption isotherm [70].

areas and adsorption heats. It is for this reason that the theory has been responsible for considerable progress in the field of heterogeneous catalysis.

The BET theory [93, 94] leads generally to a "two-constant" equation (equation (27)) by starting from assumptions identical with those made by Langmuir (Section 2.3.1), but by assuming in addition that the localized monolayer treatment can be extended to multilayers in the following manner. Each adsorbed species in the first layer serves as a site for adsorption into the second layer, and each species in the second layer serves as a site for adsorption into the third layer, and so on. (It is pertinent to note that lateral interactions are ignored and that a homogeneous surface is assumed.) Although numerous elegant derivations [1, 95–98] of the BET equation have appeared since the classical paper of Brunauer, Emmett and Teller, the fundamental structure of the theory is best judged by citing some of the major steps reported in the original publication [93]. It was argued that at equilibrium the rate of condensation (adsorption) on the bare surface is equal to the rate of evaporation (desorption) from the first monolayer, so that two expressions similar to the r.h.s. of equations (31) and (32), referring respectively to the condensation and evaporation processes, can be equated. In a like manner, expressions for the rate of condensation on the first monolayer and the rate of evaporation from the second monolayer can also be equated, and so on. The next step in the theory is crucial. As a simplifying assumption, it was stated that the heat of adsorption (or condensation) in the second layer, and in each of the subsequent layers, are equal to one another, and equal also to the heat of liquefaction of the gas. The heat of adsorption in the first layer was assumed to be different from the heat of liquefaction. All this is tantamount to stipulating that the condensation–evaporation characteristics of the second and subsequent layers are the same as those of the surface of the bulk adsorbate (or liquid).

The above assumptions lead to the so-called "simple" or "∞-form" BET equation

$$v = \frac{v_m c p}{(p_0 - p)\left\{1 + \dfrac{(c-1)p}{p_0}\right\}} \tag{56}$$

which can be transposed to the more familiar equation

$$\frac{p}{(vp_0 - p)} = \frac{1}{v_m c} + \frac{(c-1)}{v_m c}\frac{p}{p_0} \tag{27}$$

and which reduces to the Langmuir equation when p/p_0 is very low and when c is large.

If the BET theory is applicable, then a plot of $p/[v(p_0-p)]$ against p or p/p_0 should yield a straight line. If we write $x = p/p_0$ then we have:

$$\frac{1}{v(1-x)} = \frac{1}{v_m} + \frac{1}{v_m c}\frac{(1-x)}{x} \tag{57}$$

The intercept of a plot of the left-hand side of this equation against $(1-x)/x$ is the reciprocal of v_m; this is probably the simplest method [99] of estimating the monolayer volume, and hence the surface area of an adsorbent (Chapter 3).

The value of the constant c in the BET equation turns out to be:

$$c = \frac{a_1\,b_2}{a_2\,b_1}\exp\left(\frac{H_1 - H_L}{RT}\right) \tag{58}$$

where H_1 and H_L are the heats of adsorption in the first layer and the heat of liquefaction, respectively, and the constants a_1, b_1, and a_2, b_2 embrace the pre-exponential terms in the expressions (see equations (31) and (32)) for condensation and evaporation to and from the first layer and the second layer respectively. It is often conveniently assumed that $a_1\,b_2/a_2\,b_1 \approx 1$, so that equation (58) becomes

$$c = \exp\left(\frac{H_1 - H_L}{RT}\right) \tag{59}$$

The difference between the heat of adsorption H_1 and the heat of liquefaction H_L is known as the net heat of adsorption [100, 101], and it is clear how this quantity can be evaluated, together with the heat of adsorption if H_L is known, from the magnitude of c. One of the reasons why heats of adsorption derived in this way are often wildly erroneous [102, 103] is that, as shown by Hill [96], the ratio $a_1\,b_2/a_2\,b_1$ may be far removed from unity. There are adsorbent–adsorbate systems, however, which, because of the similarity in internal degrees of freedom for the adsorbate molecule in the first adsorbed layer and in the bulk liquid, the ratio approaches unity, and heats of adsorption estimated in the manner described above are, consequently, acceptable.

If, instead of imagining an infinite number of monolayers stacked up one on top of the other, it is supposed that multilayer adsorption is restricted to n layers (n being finite)—this could well happen when adsorption occurs on a porous solid—the BET treatment [104] leads to the equation:

$$v = \frac{v_m c x \left\{1 - (n + 1)x^n + nx^{n+1}\right\}}{(1 - x)\left\{1 + (c - 1)x - cx^{n+1}\right\}} \tag{60}$$

This equation also has been of value in the estimation of surface areas.

2.3.8.1. Current Status of the BET Theory

The BET equations, when fed with the appropriate values of the constants c and v_m (and, for porous solids, n), are capable of [1, 104, 105] yielding the shapes of all five isotherm Types shown in Fig. 13. This fact alone commends the theory. But we have already noted some of the deficiencies, and in recent years there has been a growing awareness of the inadequacy of the theoretical basis of the BET approach—any theory which ignores lateral interaction and surface heterogeneity possesses built-in obsolescence. Consequently, numerous investigators have suggested a variety of modifications to the theory. Full accounts are given elsewhere [1, 56, 106]; suffice it to say here that the crucial assumption of the BET theory, viz. that the heats of adsorption in the second and subsequent layers all equal the heat of liquefaction, has turned out to be one of its greatest fallacies. However, it is not too drastic an assumption to treat the *third* and subsequent layers as liquid-like [106], and this approach is, indeed, the basis of the theories of multilayer adsorption developed by Hill [56], Halsey [107] and Steele [108].

Not surprisingly, the BET theory has been supplanted by many others, most of which have yielded useful isotherm equations if only under somewhat restricted conditions. Many of these equations do not concern us here, but a few are of interest, notably those associated with the names Hüttig and Harkins and Jura. Such equations, along with others which are of a more transcendent nature—Dubinin's application of the Polanyi potential theory—are discussed in Chapter 3.

2.4. Rates of Adsorption and Desorption

The overall rate of a heterogeneously catalysed reaction is liable to be determined by the rate of one (or more) of the following consecutive steps: (i) migration or diffusion of the reactant molecules to the catalyst surface;

(ii) adsorption of the reactant(s) on the surface; (iii) rearrangement of (or reaction in) the adsorbed layer; (iv) desorption of the product molecules; and (v) diffusion of the products away from the vicinity of the surface into the gas phase.

If the catalytic activity of the surface is high, it may well turn out that the process of diffusion of molecules to, or from, the catalyst surface becomes rate determining. This state of affairs often prevails in catalytic reactors (see Chapters 4 and 9). Quite often, however, it is the desorption (which, for convenience, can be taken to mean steps (iii) and (iv) combined) or the adsorption process which dictates the rate of a catalysed reaction. It is therefore prudent that we acquaint ourselves with the theoretical means of estimating the rates of these processes. Unquestionably, one of the most useful methods of estimating these rates is via the theory of absolute reaction rates (63), which extends the concepts of chemical equilibrium into the realm of reaction kinetics by statistical mechanical methods.

2.4.1. The statistical mechanics of chemical equilibrium [109, 110, 111]

According to the precepts of statistical mechanics, which we shall now adumbrate, the equilibrium constant K of the general reaction

$$aA + bB \rightleftharpoons cC + dD \tag{61}$$

where†
$$K = \frac{[C]^c[D]^d}{[A]^a[B]^b}$$

can be written as

$$K = \frac{\left(\frac{f_C}{V}\right)^c \left(\frac{f_D}{V}\right)^d}{\left(\frac{f_A}{V}\right)^a \left(\frac{f_B}{V}\right)^b} \exp\left(-\frac{\Delta E_0^0}{RT}\right) \tag{62}$$

where the f's are known as the total partition functions of one molecule of the respective species in the volume V of the system, and where ΔE_0^0 is the increase in internal energy, or the endothermicity, at the absolute zero when a moles of A react with b moles of B to form c moles of C and d moles of D, with all the four substances being in their standard states. The partition function of a given molecule, per unit volume, is, effectively, a measure of the probability of occurrence of that molecule in the specified volume. Mathematically, we define the partition function by

$$f = \sum_i g_i \exp\left(\frac{-\epsilon_i}{kT}\right) \tag{63}$$

where ϵ_i is the energy, with respect to the zero-point energy, for a given energy

† The square brackets [ ] denote concentrations.

level ϵ_i of the molecule, and g_i is the number of states corresponding to that level. The summation is taken over all states: electronic, translational, rotational and vibrational. To a first approximation, it may be assumed that these various types of energy are independent of one another, so that the total energy ϵ_i may be written

$$\epsilon_i = (\epsilon_e)_i + (\epsilon_t)_i + (\epsilon_r)_i + (\epsilon_{vi})_i \qquad (64)$$

$(\epsilon_e)_i$, $(\epsilon_t)_i$, etc., referring, respectively, to the electronic energy, translational energy, etc., of the ith state. Equation (63) may, therefore, be written in factorized form as

$$f = f_e f_t f_r f_{vi} \qquad (65)$$

where f_e, f_t, f_r and f_{vi} are the separate partition functions, each referring to one type of energy.

For most molecules, at ordinary temperatures, the electronic partition function reduces to unity, $f_e = 1$. The values of the other partition functions depend upon the mass m of the molecule, the absolute temperature T, the moment(s) of inertia I, the frequency of vibration(s) ν of the molecule and also upon the fundamental constants R and k.

The translational partition function of an ideal gas is

$$f_t = \left(\frac{2\pi m k T}{h^2} \right)^{3/2} V \qquad (66)$$

(If we were interested in the translational partition function of an ideal two-dimensional gas on a catalyst surface, the value would be $(2\pi m k T/h^2)\,A$, where A would be the total surface area of the catalyst. The value of the translational partition function in one dimension X would be $(2\pi m k T/h^2)^{1/2}\,X$.)

The rotational partition function for a linear molecular, which has but two degrees of rotational freedom, is given by

$$f_r = \frac{8\pi^2 I k T}{h^2} \qquad (67)$$

If we were dealing with a non-linear molecule, the r.h.s. of equation (67) would be

$$8\,\pi^2\,(8\,\pi^3\,I_a I_b I_c)^{1/2} \left(\frac{kT}{h^2} \right)^{3/2}$$

I_a, I_b and I_c being the moments of inertia about any three mutually perpendicular axes.† The partition function for vibrational motion is equal to

$$f_{vi} = \prod_i \left\{ 1 - \exp\left(-\frac{h\nu_i}{kT} \right) \right\}^{-1} \qquad (68)$$

† If the molecule possessed certain types of symmetry, this expression would require further modification.

where ν_i is the vibrational frequency, and Π stands for the product of the terms $\left\{1 - \exp\left(-\dfrac{h\nu}{kT}\right)\right\}^{-1}$ taken over all the modes of vibration, there being $(3N-6)$ modes if the molecule is nonlinear, N being the number of atoms in the molecule. For a simple diatomic molecule, the vibrational partition function would be simply

$$\left\{1 - \exp\left(-\frac{h\nu}{kT}\right)\right\}^{-1}$$

In general, the translation partition functions are very large, the rotational partition functions are much smaller, and the vibrational ones are close to unity, unless the vibration happens to be particularly weak or the temperature very high.

2.4.1.1. The Theory of Absolute Reaction Rates

So far, we have utilized statistical mechanical concepts in considering the equilibrium of a general reaction. Equation (62) yields a value of the equilibrium constant after feeding into the equation the various pre-exponential terms, which are, in turn, calculable from equations (63) to (68). The value of ΔE_0^0 is also calculable from the total partition functions of the participating molecules (see Refs. 109–111). To derive a velocity of adsorption, it is necessary to visualize, once more, the potential energy of a reacting system as a function of the reaction coordinate.

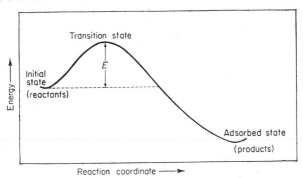

FIG. 14. Potential energy diagram illustrating the relative energies of the initial state (or reactants), the transition state (or transition complex), and the adsorbed state (or products). E is the activation energy.

Before adsorbate molecules are actually adsorbed, they must, in moving from the gas phase to a solid surface, pass over a potential energy barrier, the height of which, by definition, is the activation energy (see Fig. 14). Now, one of the tenets of the theory of absolute reaction rates when applied to gas–solid reactions is that there are molecules at the top of the barrier in a transition (i.e. in an activated) state, in statistical equilibrium with molecules in the gas phase and with vacant sites at the solid surface. It is important to emphasize that even when equilibrium is not established between the reactants and the products of reaction (Fig. 14), the transition state, or transition complex as it is sometimes designated, is in equilibrium with the reactants. Another cornerstone of the theory of absolute reaction rates is that when the transition complex vibrates perpendicular to the surface (with a frequency v say) no restoring force operates. The frequency will therefore be low and the corresponding partition function will be equal to $kT/h\nu$. This vibration will therefore result in desorption so that v is synonymous with the frequency of decomposition of the transition complex.

Now reconsider the general reaction described by equation (61), and let the transition complex for this reaction be denoted by $X^{\ddagger}$. If $X^{\ddagger}$ is in statistical equilibrium with the reactants A and B, it follows from equation (62) that the appropriate equilibrium constant, $K^{\ddagger}$, given by,

$$K^{\ddagger} = \frac{[X^{\ddagger}]}{[A]^a [B]^b} \tag{69}$$

can be written as

$$K^{\ddagger} = \frac{\left(\dfrac{f^{\ddagger}}{V}\right)}{\left(\dfrac{f_A}{V}\right)^a \left(\dfrac{f_B}{V}\right)^b} \exp\left(-\frac{\Delta E_0^0}{RT}\right) \tag{70}$$

ΔE_0^0 is now the difference between the zero-point energy per mole of the transition complex and that of the reactants; and as this energy corresponds to the amount required by the reactant molecules at $0°K$ before they can react, ΔE_0^0 is also the activation energy at this temperature. In equation (70) the term $f^{\ddagger}$ stands for the total partition function for the transition complex. But one of the vibrational modes (that perpendicular to the surface) which the complex undergoes is exceptional, and for this vibrational degree of freedom which experiences no restoring force the partition function is, as stated above, $kT/h\nu$. We can legitimately factorize $f^{\ddagger}$ such that

$$f^{\ddagger} = \frac{kT}{h\nu} f_{\ddagger} \tag{71}$$

where $f_{\ddagger}$ refers to the partition function of the transition complex taking cognizance of $[3(N_A+N_B)-7]$ degrees of vibrational freedom, assuming the complex to be nonlinear. Equations (69) and (70) therefore yield

$$\frac{[X^{\ddagger}]}{[A]^a [B]^b} = \frac{\left(\dfrac{f_{\ddagger}}{V}\right)}{\left(\dfrac{f_A}{V}\right)^a \left(\dfrac{f_B}{V}\right)^b} \left(\frac{kT}{h\nu}\right) \exp\left(-\frac{\Delta E_0^0}{RT}\right) \tag{72}$$

which rearranges to

$$\nu[X^{\ddagger}] = [A]^a\,[B]^b\,\frac{kT}{h}\,\frac{\left(\dfrac{f_{\ddagger}}{V}\right)}{\left(\dfrac{f_A}{V}\right)\left(\dfrac{f_B}{V}\right)}\,\exp\left(-\frac{\Delta E_0^0}{RT}\right) \tag{73}$$

Since ν is the frequency of decomposition, and $[X^{\ddagger}]$ the concentration, of the transition complex, the l.h.s. of equation (73) is therefore the rate of reaction. The importance of equation (73) is that it enables us to calculate, in principle at any rate, the velocity of any reaction from fundamental properties, such as the mass, moment of inertia and frequency of vibration, etc., of the reactant species and of the transition complex. Without going into details, it can be said that, even for simple molecules reacting solely in the gas phase, it has proved extremely difficult to calculate theoretically the rates of reaction from equations such as (73). One of the major obstacles is ignorance of the precise structure of the transition complex. How can the partition function be evaluated if the molecular structure is not known? Another difficulty is the recalcitrant one of calculating, by quantum-mechanical methods, the activation energy ΔE_0^0. Despite these discouragements we should not lose sight of the fact that the theory of absolute reaction rates yields an equation which, if it were soluble, would yield the theoretical value for the rate of any reaction.

Now we shall proceed from the general to the particular by considering next the absolute rates of surface reactions. Following the approach pioneered by Laidler [71, 112] used so extensively [35, 113, 114] in the field of surface chemistry, we shall consider the equilibrium:

$$\begin{array}{ccc}\text{Gas molecule} + \text{Vacant surface} & \rightleftharpoons & \text{Surface transition} \\ \text{sites} & & \text{complex}\end{array} \tag{74}$$

For convenience, let the symbols c_g, c_v and $c_{\ddagger}$ represent the concentration (per cm³) of gas molecules, the concentration (per cm²) of vacant surface sites potentially capable of adsorbing a gas molecule, and the concentration (per cm²) of transition complexes on the surface, respectively. The adsorption equilibrium constant is

$$K^{\ddagger} = \frac{c_{\ddagger}}{c_g c_v} \tag{75}$$

Equation (75) can be rewritten as

$$K^{\ddagger} = \frac{\left(\dfrac{N_{\ddagger}}{S}\right)}{\left(\dfrac{N_g}{V}\right)\left(\dfrac{N_v}{S}\right)} = \frac{N_{\ddagger}}{\left(\dfrac{N_g}{V}\right)N_v} \tag{76}$$

if N_g is the total number of molecules in the gas phase, the volume of which is V; N_v is the total number of vacant surface sites; $N_{\ddagger}$ is the total number of transition complexes on the surface; and if S is the area of the surface ($c_g = N_g/V$; $c_v = N_v S$; and $c_{\ddagger} = N_{\ddagger}/S$). Moreover, from the theory of absolute reaction rates outlined above, the rate of adsorption (r_a) is given by

$$(r_a) = c_g c_v \frac{kT}{h} \left(\frac{f_{\ddagger}}{F f_v}\right) \exp\left(-\frac{\Delta E_0^0}{RT}\right) \tag{77}$$

where f_v and $f_{\ddagger}$ represent the complete partition functions for the relevant species and sites (except that, in $f_{\ddagger}$, the function for the vibration perpendicular to the surface equal to $kT/h\nu$ has been factorized out), and F stands for the partition function of the gas per unit volume: $F=f/V$. Again ΔE_0^0 is the activation energy for the adsorption at $0°K$.

Laidler argued that since the adsorbed molecule constituting the transition complex is virtually immobile, its partition function could, with impunity, be taken as unity. The vacant surface sites have vibrational energy only, so that the partition function f_v may also be taken as unity. For F, of course, it is sufficient to consider only the translational and rotational factors, the vibrational contribution being negligible.

It has been pointed out by de Boer [115] that there are many adsorption processes likely to be of importance in heterogeneous catalysis where the transition complex has surface mobility. The complex should be capable of undergoing activated jumps to neighbouring surface sites. If, for example, it is assumed that the complex possesses two degrees of translational freedom within the area S/N_v occupied by a site, and if it has the same vibrational and rotational freedom as in the gas phase, then, instead of equating $f_{\ddagger}$ with unity, we would have:

$$f_{\ddagger} = \left(\frac{2\pi mkT}{h^2}\right) \frac{S}{N_v} f_{vi} f_r \tag{78}$$

f_{vi} and f_r refer to the vibrational and rotational contributions to the partition function. Their product, at room temperature, for small molecules, is approximately 10^4.

The rate of adsorption given by equation (77) has been derived on the assumption that the gas phase molecule is adsorbed as one entity on the surface sites. Since, frequently, adsorption occurs dissociatively, it is instructive to examine the rate equation for such a reaction. If the gas molecule G_2 is adsorbed on two sites, we can represent the process as:

$$
G_2 + \underset{\text{Reactants}}{-S-S-} \rightleftharpoons \underset{\text{Transition complex}}{\overset{G-G}{-S-S-}} \rightleftharpoons \overset{G-G}{-S-S-} \tag{79}
$$

By subjecting the equilibrium between the reactants and the transition complex to the theory of absolute reaction rates, that is, by regarding the rate of formation of the transition complex as the slow step, we arrive at a rate of adsorption given by:

$$(r_a) = c_g c_{s_2} \frac{kT}{h} \frac{f_\pm}{F f_{s_2}} \exp\left(-\frac{\Delta E_0^0}{RT}\right) \tag{80}$$

Where the subscript s_2 refers to bare adjacent surface sites. This equation is naturally analogous to equation (77).

2.4.1.2. Comparison of Theory with Data

In view of the extreme theoretical difficulties encountered in the quantum-mechanical calculations of the activation energies ΔE_0^0 of reactions—difficulties which are intensified by our ignorance of the detailed electronic structure of solid surfaces—it is immediately evident that, when we talk of comparing experimental data with theoretically predicted adsorption rates, certain, rather severe, limitations are implied. Since values of ΔE_0^0 are, in general, theoretically inaccessible, the procedure is to accept the experimental value of the energy of activation, to substitute this value into the "absolute" rate equations (such as equations (77) and (80) above) and then to compare the "calculated" result with the experimentally observed rate. The exercise is obviously more empirical than absolute. For the sake of completeness, however, it ought to be mentioned that the question of comparison of theory with experimental data could be approached in another way, for there are several empirical means available [116–119] for estimating the activation energies of heterogeneous reactions, and these estimates could, though with even greater error, be incorporated into the rate equations derived from the theory of absolute reaction rates.

When comparison can be made—data are sparse—between the observed rates of adsorption and those calculated from equations (77) and (80) in the manner explained, agreement is satisfactory, especially in view of the approximations involved in the theory and of the uncertainty attaching to the experimentally determined energy of activation. For example, Laidler [71] quotes that the calculated rate of adsorption of N_2 on a promoted iron catalyst at 544°K is 35.8×10^{16} molecules sec^{-1} cm^{-2} compared with the observed [90] rate of 3.4×10^{16} molecules sec^{-1} cm^{-2}.

2.4.1.3. Comparison of Heterogeneous and Homogeneous Reactions

One of the most striking ways in which we can illustrate why certain reactions proceed more rapidly on a solid surface than in the gas phase is by utilizing the theory of absolute reaction rates.

Suppose that two substances A and B can react homogeneously in the gas phase, the rate of reaction being ν_{hom}. Let us further suppose that, in the presence of a suitable catalyst, the molecules A and B can both be adsorbed at adjacent surface sites and that reaction proceeds, heterogeneously, according to the scheme

$$
A + B + {\overset{A}{\overset{|}{-}}} {\overset{\ }{\overset{|}{S}}} - {\overset{B}{\overset{|}{S}}} - \rightleftharpoons - {\overset{A}{\overset{|}{S}}} - {\overset{B}{\overset{|}{S}}} - \rightarrow \text{Products} \qquad (81)
$$

(Such a scheme is an example of the Langmuir-Hinshelwood mechanism referred to in Chapter 1.)

From the theory of absolute reaction rates, we have that:

$$
\nu_{\text{hom}} = c_A c_B \frac{kT}{h} \frac{F_{\ddagger}}{F_A F_B} \exp \left\{ - \frac{(\Delta E_0^0)_{\text{hom}}}{RT} \right\} \qquad (82)
$$

where c_A and c_B are the respective gas phase concentrations and the F's are the respective partition functions per unit volume. (As is customary, the function $kT/h\nu$ for the vibration along the reaction coordinate has been factorized out from the partition function of the transition complex.) If c_{s_2} refers to the concentration of bare but adjacent surface sites, then the rate of heterogeneous reaction is:

$$
\nu_{\text{het}} = c_A c_B c_{s_2} \frac{kT}{h} \frac{1}{F_A F_B} \exp \left\{ - \frac{(\Delta E_0^0)_{\text{het}}}{RT} \right\} \qquad (83)
$$

(Note that, as explained in Section 2.4.1.1, the partition functions is unity for an immobile transition complex and for vacant surface sites.)

From equations (82) and (83), the ratio of the rates of reaction is

$$
\frac{\nu_{\text{het}}}{\nu_{\text{hom}}} = \frac{c_{s_2}}{F_{\ddagger}} \exp \left\{ \frac{(\Delta E_0^0)^1}{RT} \right\} \qquad (84)
$$

the term $(\Delta E_0^0)^1$ referring to the amount by which the activation energy of the homogeneous reaction exceeds that of the heterogeneous one.

For an essentially smooth surface most solids have a value of c_{s_2} close to 10^{15} per cm². The value of $F_{\ddagger}$ ranges from 10^{24} to 10^{30}, so that we have

$$
\frac{\nu_{\text{het}}}{\nu_{\text{hom}}} \approx 10^{-12 \pm 3} \exp \left\{ \frac{(\Delta E_0^0)^1}{RT} \right\} \qquad (85)
$$

Bearing in mind that, in these estimates, the heterogeneous rate refers to 1 cm² of surface and to 1 cm³ of gas phase, and the homogeneous rate to 1 cm³ of the gaseous reactants, it transpires that the heterogeneous reaction would be very much slower if the activation energies for the two reactions were the same. Clearly the heterogeneous reaction will tend to become more significant if (i) $(\Delta E_0^0)_{\text{het}}$ is less than $(\Delta E_0^0)_{\text{hom}}$, and (ii) the surface area is increased. By far the more important is the activation energy difference—for a surface area of 100 m²g⁻¹, there would still remain a pre-exponential factor of $10^{-6 \pm 3}$. And it follows from equation (85) that, for the rates of the heterogeneous and homogeneous reactions to be equal at room temperatures, the activation energy of the heterogeneous reaction must be less by 16–17 kcal mole^{-1} than the homogeneous reaction. Obviously, as the reaction temperature increases, the disparity in activation energy must increase in order to maintain parity of rate. For example, at $500°K$ $(\varDelta E_0^0)^1$ must be 27·6 kcal mole^{-1}. There is ample experimental evidence [120–123] which bears out these conclusions; and it has consequently become one of the shibboleths of surface chemistry that a catalyst lowers the activation energy of a reaction.

2.4.1.4. *Rates of Desorption*

Since, in later chapters, particularly when mechanisms of catalysis are discussed, it will be necessary to utilize some theoretical equations for desorption as well as adsorption, some of the more important results will be outlined of the theory of absolute reaction rates applied to desorption.

In describing the desorption process, we again consider the statistical equilibrium between adsorbed species and the transition complexes. When the adsorbed species and the desorbed gas have the same molecularity, the absolute rate theory gives

$$\text{`} (r_d) = c_{\text{ads}} \frac{kT}{h} \frac{f_{\ddagger}}{f_{\text{ads}}} \exp\left\{-\frac{(\varDelta E_0^0)_{\text{des}}}{RT}\right\} \tag{86}$$

where c_{ads} is the concentration of the adsorbed species, f_{ads} and $f_{\ddagger}$ are the partition functions of the adsorbed species and the transition complex, and $(\varDelta E_0^0)_{\text{des}}$ the activation energy of desorption. Equation (86) takes different forms depending upon whether the chemisorption is dissociative, the transition complex is mobile or immobile, and upon the precise value of the molecularity. Laidler, Glasstone and Eyring [124] have shown that (i) when the adsorbed species and transition complexes are diatomic, both for mobile and immobile adsorption, equation (86) rearranges to:

$$(r_d)_{\text{diatomic}} = \tfrac{1}{2}c_s\theta \frac{kT}{h} \exp\left(-\frac{(\varDelta E_0^0)_{\text{des}}}{RT}\right) \tag{87}$$

and (ii) when the adsorbed species are immobile and dissociatively chemisorbed, and the complexes diatomic and immobile, equation (86) becomes

$$(r_d)_{\text{immob}} = 2c_s\theta^2 \frac{kT}{h} \exp\left(-\frac{(\varDelta E_0^0)_{\text{des}}}{RT}\right) \tag{88}$$

In both equations (87) and (88) c_s is the total concentration (cm^{-2}) of single surface sites when the surface is completely bare, and θ is the fractional coverage.

When the rate of desorption is determined not by the rate of formation of the transition complex, as has been tacitly assumed above, but by the rate of desorption of the complex, the rate differs from that expressed by equations

(86), (87) and (88). In other words, the second step in the following scheme (cf. the scheme represented by equation (79)) is rate-determining:

$$
\begin{array}{ccc}
\text{G} \quad \text{G} & \text{G....G} & \\
| \quad | & \vdots \quad \vdots & | \quad | \\
\text{—S—S—} \;\rightleftharpoons\; \text{—S—S—} & \rightleftharpoons & \text{—S—S—} + \text{G}_2 \\
& \text{Transition} & \\
& \text{complex} &
\end{array}
\tag{89}
$$

This is an example of the Temkin-Pyzhev [55] mechanism of heterogeneously catalysed reactions. If the desorption of the transition complex occurs as a diatomic molecule, the rate equation is

$$
(r_a)_{\text{diatomic}} = \frac{c_{\text{ads}}^2 \, c_{s_2}}{c_s^2} \frac{kT}{h} \frac{f_{\pm} f_s^2}{f_{\text{ads}}^2 f_{s_2}} \exp\left(-\frac{(\Delta E_0^0)_{\text{des}}}{RT}\right)
\tag{90}
$$

where c_s and c_{s_2} refer to the concentration of bare single and bare adjacent sites, respectively, and the f's have their customary significance.

2.4.1.5. *Rates of Adsorption on Non-uniform Surfaces*

In this subsection we shall be concerned with the kinetics of adsorption on energetically non-uniform surfaces. If a catalytic reaction takes place on such a surface, it is no longer possible to postulate a specific rate determining step, for now there is available a whole spectrum of sites with various energies, some of which will favour one reaction step and others a different reaction step. One of the main features displayed by a heterogeneous catalyst is an enhancement in reaction rate of a given process by virtue of a diminution in the activation energy. This is inextricably bound up with the adsorption energy of the activated complex: the stronger the adsorption, the lower the activation energy for adsorption, but now a higher activation energy is required for the desorption of products. On a certain part of the surface only, the respective activation energies will have their optimum values. Thus, the energy of adsorption of the activated complex on an optimum site will be sufficiently large to allow surface reaction to occur and yet not be too great to impede the desorption of products.

Following Halsey [125], let us consider an energetically non-uniform surface and suppose that the Gibbs free-energy of activation for the adsorption process is G_1 while that for the desorption process is G_2. In the vicinity of optimum conditions a slight decrease in G_1 will cause a corresponding increase in G_2 and we may therefore write,

$$
\frac{dG_1}{dG_2} = -n
\tag{91}
$$

and, provided that n is constant,

$$
G_1 + nG_2 = G_0
\tag{92}
$$

where G_0 is a constant.

On a given part of the surface, a fraction θ_1 of the surface will be covered by species involved in the adsorption rate process, while a fraction θ_2 of surface will be occupied by rearranged species prior to desorption as product. It is assumed here that other parts of the surface are unoccupied and that the coverage θ_1 is determined by a Langmuir type adsorption isotherm

$$\frac{\theta_1}{1 - \theta_1 - \theta_2} = k_1 p_1 \tag{93}$$

in which p_1 is the partial pressure of the component responsible for determining the rate of adsorption.

Now the absolute rate theory predicts that the rate of a surface reaction may be given by

$$\frac{kT}{h} \frac{f_{\ddagger}}{f_{\text{ads}}} \theta_{\text{ads}} \exp\left(-\frac{\Delta E_0^0}{RT}\right) \approx \frac{kT}{h} \theta_{\text{ads}} \exp\left(-\frac{\Delta G_0^0}{RT}\right) \tag{94}$$

where $f_{\ddagger}$ and f_{ads} are, respectively, the partition functions for the activated complex and the adsorbed gas, θ_{ads} the fraction of surface covered by adsorbate, ΔE_0^0 the enthalpy of activation (see Section 2.4.1.1) into which the logarithm of the partition function ratio $f_{\ddagger}/f_{\text{ads}}$ is absorbed for convenience to constitute ΔG_0^0 the free energy of activation. Therefore, in terms of the rate of occupation of type 2 sites, we have, in the steady state,

$$\frac{d\theta_2}{dt} = \frac{kT}{h} \left[\theta_1 \exp\left(-\frac{G_1}{RT}\right) - \theta_2 \exp\left(-\frac{G_2}{RT}\right) \right] = 0 \tag{95}$$

Hence, from equations (93) and (95),

$$\theta_2 = \frac{\left[\dfrac{k_1 p_1}{(1 + k_1 p_1)}\right] \exp\left(-\dfrac{G_1}{RT}\right)}{\left[\dfrac{k_1 p_1}{(1 + k_1 p_1)}\right] \exp\left(-\dfrac{G_1}{RT}\right) + \exp\left(-\dfrac{G_2}{RT}\right)} \tag{96}$$

Since the reaction velocity per site will be

$$\nu = \frac{kT}{h} \theta_2 \exp\left\{-\frac{G_2}{RT}\right\} \tag{97}$$

the overall rate may be calculated from the equation

$$r = \int F(G_2)\nu\, dG_2 \tag{98}$$

where $F(G_2)$ is a function describing the distribution of the energy G_2 amongst sites. Halsey assumes an exponential type of energy distribution,

$$F(G_2) = c \exp\left(-\frac{G_2}{G_M}\right) \tag{99}$$

in which c is a constant and G_M is a constant modulus, and proceeds to evaluate, from the above equations, the optimum value of the reaction velocity per site by setting dv/dG_2 equal to zero. If G_M/RT is large, and this implies that the distribution of energy in the vicinity of optimum conditions is sufficiently broad to assume that the number of sites of different kinds are equally numerous, then the result obtained is

$$v_{max} = \text{Const} \times \left(\frac{\theta_1}{1 - \theta_2}\right)^{n/2} \tag{100}$$

This equation implies that the optimum reaction rate is determined by a function of the ratio of the fraction of surface sites involving one of the rate-limiting processes (say 1) to the fraction of surface unused by species involved in the other rate determining process (say 2).

This result may be compared with that which would be obtained for an energetically uniform surface. Suppose the reaction under consideration is one in which a reactant is adsorbed, rearranges to give another molecular structure by a surface reaction and the product ensuing from this rearrangement is desorbed. If the surface reaction (step (iii), p. 49) is the rate determining step and, by comparison, desorption is fast, then the reaction rate is directly proportional to the fraction of surface covered by reactant. From the Langmuir isotherm (93), the rate is then

$$k\theta_1 = \frac{kk_1 p_1}{1 + k_1 p_1}(1 - \theta_2) \tag{101}$$

and, if θ_2 is small in comparison to the total amount of available surface, the rate is proportional to $k_1 p_1/(1 + k_1 p_1)$. On the other hand, if desorption is the rate-determining step, the reaction rate is directly proportional to θ_2. If it is supposed that the product of the surface reaction is more strongly adsorbed than the original reactant, then the reaction rate is independent of the partial pressure p_1. Thus, it emerges that, on an energetically non-uniform surface, the behaviour is somewhere between these two extremes for, if the isotherm (93) is substituted into equation (100), v_{max} is proportional to $\{k_1 p_1/(1 + k_1 p_1)\}^{n/2}$.

2.4.2. The Elovich equation

No treatment of rates of adsorption, however much abbreviated, is complete without reference to the Elovich equation. This equation is widely applicable, and is one of the most convenient ways of describing chemisorption kinetics, especially if the chemisorption is slow.

In its differential form, the equation [126, 127] is written as

$$\frac{dq}{dt} = a \exp(-\alpha q) \tag{102}$$

where q is the amount adsorbed at time t, and a and α are constants for each system at a fixed temperature. In its integrated form, the equation can be written as:

$$q = \frac{1}{\alpha} \frac{\ln (t + t_0)}{t_0} \tag{103}$$

where $t_0 = 1/a \, \alpha$. Clearly if a plot of log $(t+t_0)$ against q, or of log (dq/dt) against q is linear the Elovich equation is obeyed. It has been found that the slow chemisorption processes which so often follow an initial rapid uptake obey the Elovich equation. Sometimes the adsorption process is slow and obeys equation (102) from the very beginning. For example, Scholten et al. [128, 129] in their analysis of the rate-determining step in ammonia synthesis, found that, up to about 25% coverage, chemisorption of N_2 on the iron catalyst could be perfectly described by the Elovich equation.

Precisely what is implied by the validity of the Elovich equation is still obscure, as discussed by Low [127]. Taylor and Thon [130] were the first to point out that systems which obeyed equation (102), and which consequently call for a relation

$$-\frac{dN}{dq} = \alpha N \tag{104}$$

(where N is the number of surface sites) are inconsistent with the classical concept of an adsorbent surface containing a fixed number of adsorption sites. The treatment of Taylor and Thon [130] extended by Landsberg [131] is based on the notion that, when a gas and an adsorbent first come into contact, a certain number N_0 of sites is formed, the number gradually decaying as adsorption continues. Such views can be accommodated by the recent theories of Volkenstein [132] and the more established views of Semenov [133] as explained by Low.

2.4.3. Time of adsorption

de Boer [2, 134] has extended the early views of Frenkel [135] dealing with the concept of the average length of time τ, that elapses before an adsorbed species is re-evaporated during the course of dynamical adsorption equilibrium. Frenkel in 1924 gave an equation

$$\tau = \tau_0 \exp \left(\frac{Q}{RT} \right) \tag{105}$$

which relates τ to the time of oscillation τ_0 of the molecules in the adsorbed state referring especially to vibrations perpendicular to the surface. Q is the heat of adsorption of the adsorbed molecules. The identification of τ_0 with the time of oscillation led to the conclusion, via an equation derived by Lindemann, that τ_0 is approximately 10^{-13} sec for most systems, this time being the order of magnitude of the time of vibration of atoms of the adsorbents. de Boer has shown [2], however, that, although τ_0 is indeed frequently close to 10^{-13} sec, the reasons for this magnitude have nothing to do with the time of oscillation of molecules.

If, for the time being, we accept that τ_0 is $\sim 10^{-13}$ sec, it follows from equation (105) that the magnitude of the heat of adsorption influences τ profoundly. For example, taking τ_0 to be 10^{-13} sec and $T = 300°K$, we have that, as the heat of adsorption Q increases from 1·5 to 15 and then to 25 kcal mole^{-1} the corresponding values of τ increase from $1·3 \times 10^{-12}$ sec to $1·8 \times 10^2$ sec to 6×10^5 sec. The consequences of equation (105) are even more strikingly demonstrated when we realize that, as the heat of adsorption of a gas on a surface falls from say 30 kcal mole^{-1} (for a clean surface) to 10 kcal mole^{-1} (for a completed monolayer), the time of adsorption decreases from about one century to roughly a millionth of a second.

de Boer [2] and Kruyer [136] have evaluated the values of τ_0 by means of statistical mechanics, using the partition functions of the gaseous and adsorbed species. It emerges that, depending on the degrees of freedom possessed by the adsorbed species, the value of τ_0 can vary from about 10^{-12} to 10^{-16} sec. These workers show that, not unnaturally, the greater the mobility of the adsorbed species, the greater is the time of adsorption (for a given heat of adsorption).

Connected with the time of adsorption is the so-called [2] halting time τ', which is defined as the average time an adsorbed species resides at a particular site before "hopping" to another surface site. Clearly, during the time of adsorption, the adsorbed species will "hop" τ/τ' times and so migrate over the surface. Just as the time of adsorption was related as in equation (105) to the heat of adsorption, so the halting time can be related to the activation energy E_m of the hopping motion:

$$\tau' = \tau'_0 \exp \left(\frac{E_m}{RT} \right) \tag{106}$$

de Boer [2] has shown that the constant τ'_0 is of the same order of magnitude as τ_0, and also points out that E_m may be regarded as the difference between the heat of adsorption when the species is adsorbed on a preferential site of the normal surface and the heat of adsorption of the same species adsorbed on a spot just in between two such preferential sites. It is the relative magnitudes of E_m and RT which largely determine the freedom of surface migration of the adsorbed species. If $E_m \gg RT$, the migration will be minimal. As the temperature increases, migration becomes more pronounced and, when $RT \gg E_m$, the adsorbed species will migrate freely over the surface during the adsorption time t, the number of "hops" being calculable from equations (105) (106) and the ratio τ/τ'. In field-emission and field-ion microscopy (see Chapter 3), it is possible to follow directly the surface migration whose energetics and kinetics are discussed here.

References

1. D. M. Young and A. D. Crowell, "Physical Adsorption of Gases". Butterworths, London (1962).
2. J. H. de Boer, *Adv. Catalysis* **8**, 17 (1956).
3. R. A. Beebe and D. M. Young, *J. phys. Chem.* **58**, 93 (1954).
4. C. Amberg, W. B. Spencer and R. A. Beebe, *Can. J. Chem.* **33**, 305 (1958).
5. J. J. Kipling and D. B. Peakall, *in* "Chemisorption", ed. by W. E. Garner, p. 59. Butterworths, London (1957).
6. K. W. Sykes and J. M. Thomas, *in* "Proceedings of the Fourth Conference on Carbon", p. 29. Pergamon Press, Oxford (1960).
7. R. Zsigmondy, *Z. anorg. Chem.* **71**, 356 (1911).
8. W. Thomson, *Phil. Mag.* **42**, 448 (1871).
9. D. J. C. Yates, *Adv. Catalysis* **12**, 265 (1960).
10. P. W. Selwood, "Adsorption and Collective Paramagnetism". Academic Press, New York (1962).
11. B. M. W. Trapnell, "Chemisorption", Chapter 1. Butterworths, London (1955).
12. J. E. Lennard-Jones and B. M. Dent, *Trans. Faraday Soc.* **24**, 92 (1928).
13. J. E. Lennard-Jones, *Trans. Faraday Soc.* **28**, 333 (1932).
14. J. H. de Boer, *Adv. Colloid Sci.* **3**, 5 (1950).
15. J. E. Lennard-Jones, *Physica* **4**, 941 (1937).
16. F. London, *Z. Phys.* **63**, 245 (1930).
17. G. D. Halsey, "1961 Transactions of the Eighth Vacuum Symposium and Second International Congress", p. 119. Pergamon Press, New York (1962).
18. A. V. Kiselev, *Q. Rev.* **XV**, 99 (1961).
19. J. G. Kirkwood, *Phys. Z.* **33**, 57 (1932).
20. A. Muller, *Proc. R. Soc.* **A161**, 476 (1937).
21. P. M. Morse, *Phys. Rev.* **34**, 57 (1929).
22. T. B. Grimley, *Adv. Catalysis* **12**, 1 (1960).
23. J. H. de Boer, "Electron Emission and Adsorption Phenomena". Cambridge University Press (1935).
24. P. M. Gundry and F. C. Tompkins, *Q. Rev.* **XIV**, 257 (1960).
25. D. D. Eley, *Disc. Faraday Soc.* **8**, 34 (1950).
26. D. D. Eley, "Catalysis and the Chemical Bond". University of Notre Dame Press (1954).
27. D. P. Stevenson, *J. chem. Phys.* **23**, 203 (1955).
28. I. Higuchi, T. Ree and H. Eyring, *J. Am. chem. Soc.* **79**, 1330 (1957).
29. L. Pauling, "The Nature of the Chemical Bond". Cornell University Press, Ithaca (1939).
30. G. Ehrlich, G.E. Res. Lab. Rep. No. 59-RL- 2205 (1959).
31. C. A. Coulson, "Valence". Oxford University Press, London (1952).
32. J. Koutecky, *Trans. Faraday Soc.* **54**, 1038 (1958).
33. P. M. Gundry and F. C. Tompkins, *Trans. Faraday Soc.* **56**, 846 (1960).
34. D. D. Eley and D. R. Rossington, *in* "Chemisorption", ed. by W. E. Garner, p. 137. Butterworths, London (1957).
35. B. M. W. Trapnell, "Chemisorption". Butterworths, London (1955).
36. P. M. Gundry and F. C. Tompkins, *in* "Chemisorption", ed. by W. E. Garner, p. 152. Butterworths, London (1957).
37. D. A. Dowden, *in* "Chemisorption", ed. by W. E. Garner, p. 3. Butterworths, London (1957).

38. A. F. Benton and T. A. White, *J. Am. chem. Soc.* **52**, 2332 (1930).
39. H. S. Taylor and H. McKenney, *J. Am. chem. Soc.* **53**, 3604 (1931).
40. W. E. Garner and M. Kingmann, *Nature, Lond.* **126**, 352 (1930).
41. J. W. McBain "The Sorption of Gases and Vapours by Solids", p. 306. Routledges, London (1932).
42. C. N. Hinshelwood, "The Kinetics of Chemical Change", 4th Ed., p. 63. Oxford University Press, London (1940).
43. H. Eyring and M. Polanyi, *Z. phys. Chem.* **B12**, 279 (1931).
44. M. G. Evans and M. Polanyi, *Trans. Faraday Soc.* **34**, 11 (1938).
45. J. H. de Boer, *Adv. Catalysis* **9**, 472 (1957).
46. G. M. Schwab, *Adv. Catalysis* **9**, 496 (1957).
47. C. Kemball, *Adv. Catalysis* **2**, 233 (1950).
48. S. H. Bauer, *Adv. Catalysis* **9**, 496 (1957).
49. J. H. de Boer, *Adv. Catalysis* **8**, 17 (1956).
50. J. H. de Boer, *Bull Soc. chim. Pays-Bas Belg.* **67**, 284 (1958).
51. J. M. Thomas, *J. chem. Educ.* **38**, 138 (1961).
52. D. J. C. Yates, *Adv. Catalysis* **12**, 265 (1960).
53. J. H. de Boer, *in* "Chemisorption", ed. by W. E. Garner, p. 27. Butterworths, London (1957).
54. A. Slygin and A. Frumkin, *Acta phys.-chim. URSS* **3**, 791 (1935).
55. M. I. Temkin and V. Pyzhev, *Acta phys.-chim. URSS* **12**, 327 (1940).
56. T. L. Hill, *Adv. Catalysis* **4**, 211 (1952).
57. I. Langmuir, *J. Am. chem. Soc.* **40**, 1361 (1918).
58. R. H. Fowler, *Proc. Camb. phil. Soc.* **32**, 144 (1936).
59. B. M. W. Trapnell, "Chemisorption". Butterworths, London (1955).
60. T. Sexl, *Z. Phys.* **48**, 607 (1928).
61. G. S. Rushbrooke, *Trans. Faraday Soc.* **36**, 1055 (1940).
62. J. E. Lennard-Jones and A. F. Devonshire, *Proc. R. Soc.* **A156**, 6 (1936).
63. S. Glasstone, K. J. Laidler and H. Eyring, "The Theory of Rate Processes". McGraw-Hill, New York (1941).
64. A. Ganguli, *Kolloidzeitschrift* **60**, 180 (1932).
65. K. W. Sykes and J. M. Thomas, "Proceedings of Fourth Conference on Carbon", p. 29. Pergamon Press, London (1960).
66. S. Ross and W. Winkler, *J. Colloid Sci.* **10**, 319 (1955).
67. I. Langmuir, *J. Am. chem. Soc.* **40**, 136 (1918).
68. P. W. M. Jacobs and F. C. Tompkins, *in* "Chemistry of the Solid State", ed. by W. E. Garner, p. 91. Butterworths, London (1955).
69. R. Peierls, *Proc. Camb. phil. Soc.* **32**, 471 (1936).
70. S. Brunauer, "The Adsorption of Gases and Vapours". University Press, Princeton (1943).
71. K. J. Laidler, *in* "Catalysis", ed. by P. H. Emmett, Vol. 1, p. 75. Reinhold, New York (1954).
72. H. S. Taylor, *Proc. R. Soc.* **A108**, 105 (1925).
73. F. H. Constable, *Proc. R. Soc.* **A108**, 355 (1925).
74. R. N. Pease and R. Stewart, *J. Am. chem. Soc.* **47**, 1235 (1925).
75. J. K. Roberts, *Proc. R. Soc.* **A152**, 25 (1935).
76. A. F. Ward, *Proc. R. Soc.* **A133**, 506 (1931).
77. O. Beeck, *Disc. Faraday Soc.* **8**, 118 (1950).
78. R. P. Eischens and W. A. Pliskin, *Adv. Catalysis* **10**, 1 (1958).
79. F. C. Tompkins and R. V. Culver, *Adv. Catalysis* **11**, 68 (1959).

80. J. T. Kummer and P. H. Emmett, *J. Am. chem. Soc.* **73**, 2886 (1951).
81. S. Z. Roginskii and O. Todes, *Acta phys.-chim. URSS* **21**, 519 (1946).
82. N. P. Keier and S. Z. Roginskii, *Izv. Akad. Nauk SSSR Otd. khim. Nauk* 27, (1950).
83. P. L. Walker and F. J. Vastola, remarks made at Faraday Society Informal Discussion on Graphite, London, 1962.
84. J. B. Newkirk and J. H. Wernick (eds.), "Direct Observations of Imperfections in Crystals". Interscience, New York (1962).
85. J. H. de Boer (ed.), "Reactivity of Solids". Elsevier, Amsterdam (1961).
86. J. Zeldowitch, *Acta phys.-chim. URSS* **1**, 961 (1935).
87. W. G. Frankenburg, *J. Am. chem. Soc.* **66**, 1827 (1944).
88. W. J. Thomas, *Trans. Faraday Soc.* **53**, 1124 (1957).
89. S. Brunauer, K. S. Love and R. G. Keenan, *J. Am. chem. Soc.* **64**, 751 (1942).
90. P. H. Emmett and S. Brunauer, *J. Am. chem. Soc.* **56**, 35 (1934).
91. J. H. de Boer and C. Zwikker, *Z. phys. Chem.* **B3**, 407 (1929).
92. E. C. C. Baly, *Proc. R. Soc.* **A160**, 465 (1937).
93. S. Brunauer, P. H. Emmett, and E. Teller, *J. Am. chem. Soc.* **60**, 309 (1938). See also Ref. 7.
94. P. H. Emmett and T. W. de Witt, *Ind. Engng Chem. analyt. Edn* **13**, 28 (1941).
95. A. G. Foster, *J. chem. Soc.* 769 (1945).
96. T. L. Hill, *J. chem. Phys.* **14**, 263, 268 (1948).
97. J. Shinokawa, *Busseiron Kenkyu* **58**, 8 (1953).
98. T. Keii, *J. chem. Phys.* **22**, 1617 (1954).
99. T. Keii, T. Takagis and S. Kanetaka, *Analyt. Chem.* **33**, 1965 (1962).
100. A. S. Coolidge, *J. Am. chem. Soc.* **49**, 708 (1927).
101. F. G. Keyes and M. J. Marshal, *J. Am. chem. Soc.* **49**, 156 (1927).
102. R. T. Davis and T. W. de Witt, *J. Am. chem. Soc.* **70**, 1135 (1948).
103. G. L. Kington and J. G. Aston, *J. Am. chem. Soc.* **73**, 1934 (1951).
104. S. Brunauer, L. S. Deming, W. E. Deming and E. Teller, *J. Am. chem. Soc.* **62**, 1723 (1940).
105. D. C. Jones, *J. chem. Soc.* 1464 (1951).
106. D. H. Everett, *Proc. chem. Soc.* 38 (1957).
107. G. D. Halsey, *J. chem. Phys.* **16**, 931 (1948).
108. W. A. Steele, *J. chem. Phys.* **25**, 819 (1956).
109. G. S. Rushbrooke, "Introduction to Statistical Mechanics". Oxford University Press, London (1949).
110. R. H. Fowler and E. A. Guggenheim, "Statistical Thermodynamics". Cambridge University Press (1939).
111. S. Glasstone, "Theoretical Chemistry". van Nostrand, New York (1944).
112. K. J. Laidler, "Chemical Kinetics", p. 167. McGraw-Hill, New York (1950).
113. G. C. Bond, "Catalysis by Metals". Academic Press, London (1962).
114. P. G. Ashmore, "Catalysis and Inhibition of Chemical Reactions". Butterworths, London (1963).
115. J. H. de Boer et al., "The Mechanism of Heterogeneous Catalysis". Elsevier, Amsterdam (1960).
116. A. Couper and D. D. Eley, *Disc. Faraday Soc.* **8**, 172 (1950).
117. A. Couper and D. D. Eley, *Proc. R. Soc.* **A211**, 536, 544 (1952).
118. K. E. Shiller and K. J. Laidler, *J. chem. Phys.* **17**, 1212 (1949).
119. J. O. Hirschfelder, *J. chem. Phys.* **9**, 645 (1941).
120. C. N. Hinshelwood and C. R. Pritchard, *J. chem. Soc.* **127**, 1552 (1925).

121. C. N. Hinshelwood and R. E. Burk, *J. chem. Soc.* **127**, 2896 (1925).
122. C. N. Hinshelwood and C. R. Pritchard, *Proc. R. Soc.* **A108**, 211 (1925).
123. G. M. Schwab and E. Pietsch, *Z. phys. Chem.* **121**, 189 (1926).
124. S. Glasstone, K. J. Laidler, and H. Eyring, "The Theory of Rate Processes". McGraw-Hill, New York (1941).
125. G. D. Halsey, *J. phys. Chem.* **67**, 2038 (1963).
126. G. Tamman and W. Koster, *Z. anorg. allg. Chem.* **123**, 196 (1922).
127. M. J. D. Low, *Chem. Rev.* **60**, 267 (1960).
128. J. J. F. Scholten, P. Zwietering, J. A. Konvalinka and J. H. de Boer, *Trans. Faraday Soc.* **55**, 2166 (1959).
129. J. J. F. Scholten, J. A. Konvalinka and P. Zwietering, *Trans. Faraday Soc.* **56**, 262 (1960).
130. H. A. Taylor and N. Thon, *J. Am. chem. Soc.* **74**, 4169 (1952).
131. P. T. Landsberg, *J. chem. Phys.* **23**, 1079 (1955).
132. Th. Wolkenstein, *Adv. Catalysis* **12**, 189 (1960).
133. N. N. Semenov, *Usp. Khim* **20**, 673 (1951).
134. J. H. de Boer, "The Dynamical Character of Adsorption". Oxford University Press, London (1953).
135. J. Frenkel, *Z. Phys.* **26**, 117 (1924).
136. S. Kruyer, *Proc. K. Ned. Acad. Wet.* **B58**, 73 (1955).

CHAPTER 3

Experimental Aspects of Adsorption and Allied Phenomena on Catalyst Surfaces

3.1. Introduction

From what has been said in Chapter 2, the reader will conclude that adsorption is inextricably related to heterogeneous catalysis, and that an understanding of the mechanism of adsorption is a prerequisite to the understanding of mechanisms of catalysis. He may also realize that the process of physical adsorption can be put to good advantage in the study and assessment of catalysis, and that much valuable information about the nature of catalyst surfaces can be gleaned from studies of chemisorption and the physical and chemical changes of adsorbent or adsorbate accompanying chemisorption.

It is intended in this chapter to consider firstly the reasons for, and the experimental conditions necessary for, a direct study of adsorption on solid surfaces. Secondly, methods of estimating surface areas of solids from adsorption data are discussed. (Other methods of estimating surface areas together with means of evaluating the porosity and pore structure of catalysts are deferred until Chapter 4.) Then follows a comprehensive review of the experimental methods that are available for the study of the adsorbed phases on, and the surfaces of, solid catalysts.

3.2. The Study of Adsorption

3.2.1. The place of adsorption studies in the investigation of catalysis

Physical adsorption constitutes a very important means of investigating the properties of heterogeneous catalysts. Indeed, with the appearance and application of the BET theory (Section 2.3.8), the study of heterogeneous catalysis was given a significant impulse, because a reliable means of estimating the specific surface areas (surface area per unit mass) of a wide range of industrial catalysts became available. At present, the most widely used methods for estimating surface areas are still those based on the measurement of the physical adsorption of a gas close to its boiling-point, although certain refinements, to be discussed below, have been incorporated. Physical adsorption also forms the basis of most of the numerous methods of measuring the porosity of solids: this topic is considered in detail later (Section 4.3). Less widely appreciated is the fact that physical adsorption can also enable deductions to be made concerning the fractional amount of a solid surface which is energetically heterogeneous. For example, Graham [1], from physical adsorption isotherms which departed from linearity at low coverages, concluded that a certain fraction of the graphitized carbon black studied by him consisted of topographical irregularities which provided more than one point of interaction with an adsorbate molecule. This procedure, which will not be discussed further here, has been used successfully recently [2] to assay the variation in the surface heterogeneity of various graphites.

Chemisorption on a catalyst surface is in principle capable of yielding directly a quantitative measure of the adsorbability of the surface. Clearly, it goes one stage further than physical adsorption, in that it furnishes information about the "active", rather than the total, surface area: in some cases the active and total surface areas are synonymous. So far as the study of chemisorption and its role in catalysis goes, two extreme approaches are possible. One is to take the solid surface in the condition known to be effective catalytically and study the adsorption of the appropriate gases on such a surface. This approach is essentially empirical. It usually rejects or ignores all the precepts so often preached by surface chemists, namely that outgassed, clean, reproducible surfaces be used in studies of chemisorption. Quite obviously, adsorption studies carried out under arbitrarily imposed conditions on a multitude of imperfectly defined, often mysteriously produced, industrial catalysts are more likely to lead to confusion than to an elucidation of the mechanisms of catalysis. Yet much valuable information, largely empirical, has been assembled in this way. An alternative approach is to go to the other extreme. To take great pains to produce stringently cleaned surfaces; and to study the interaction of simple molecules with such surfaces under conditions chosen with a view to obtaining information of fundamental importance. In the words of Gwathmey and Cunningham [3], "this is equivalent to picking the 'battleground' on one's own terms, and there are decided advantages in doing so". Although this approach is the one more likely to yield fruitful results, it has its dangers, too. Not least amongst these dangers is the tendency to recede farther away from the problems of primary concern in heterogeneous catalysis, and to move towards ever more simple systems that are customarily regarded as the domain of the surface physicist. Be this as it may, the second approach has led to the study of several distinct types of surface, quite apart from those of powdered catalyst. This is the reason why it is logical for us, in this chapter, to consider adsorption on a variety of well-defined surfaces.

3.2.2. Preparation of surfaces for the study of adsorption

Upon realizing that attempts to disentangle the intricacies of catalysis are being made by conducting adsorption studies on powders, filaments, foils, evaporated films and single crystals of a wide variety of solids, it will be appreciated that few, if any, methods of producing chemically clean surfaces are universally applicable, and that several *ad hoc* procedures have to be evolved.

Before summarizing recent methods of generating reproducible or so-called "clean" surfaces, let us make a simple estimate of the time required to contaminate a surface under specified ambient conditions. The rate of collision of gas molecules at a pressure p with unit area of a surface is equal to $p/(2\pi mkT)^{1/2}$, where T is the absolute temperature and m the mass of the gas molecule. This formula enables us to conclude that, at room temperature, at a pressure of 1 torr, approximately 4×10^{20} molecules (assumed to be nitrogen) collide with each cm^2 of surface per sec. If it is supposed that the sticking coefficient, which is the probability that a molecule striking a surface will actually be adsorbed on it, is about 0·25 (which is a reasonable value [4]), it can be seen that the surface will be covered by a monolayer of contaminant in about a microsecond. In a vacuum of 10^{-6} torr, the contamination time is increased to approximately 1 sec, whereas, in an ultrahigh vacuum of 10^{-10} torr, the corresponding time is about 10^4 sec. It is therefore evident that, in order to minimize the rate of contamination of a solid surface, the lower the pressure the better. But it would be a mistake to assume that any experiment carried out on a surface which has been prepared under a modest ultimate vacuum (say 10^{-6} torr) is invalidated because of surface contamination: the actual number of residual gas molecules which are potentially capable of being adsorbed by the surface is vital [4, 5]. In a vacuum of 10^{-6} torr there are roughly 3×10^{10} molecules per cm^3 (at room temperature). Now if a fresh surface is generated, in a *closed* system, at a pressure of 10^{-6} torr, and the area of the surface is as little as 100 cm^2, it follows that, even when *all* the residual molecules are finally adsorbed from the gas phase, less than 0·03% of the surface will be contaminated. Clearly there would be no necessity to utilize ultrahigh vacuum techniques in this study.

Nevertheless, numerous adsorption studies are carried out where it has been imperative to produce the surfaces in ultrahigh vacuum systems. Pressures less than 10^{-9} torr can now be routinely obtained [4], and several investigators [6–9] have attained pressures as low as 10^{-13} torr using titanium evaporation pumps and liquid-helium pumps. The processes limiting the ultimate pressures attainable in vacuum systems [10], together with the various ionization gauges and mass spectrometers required for the measurement of very low pressures, have been thoroughly discussed [11–15].

3.2.2.1. Classification of Methods Used to Clean Surfaces

If it is our intention merely to investigate the adsorptive properties of an industrial catalyst under conditions which simulate actual working conditions, the question of stringently cleaning the surface beforehand does not usually arise, and the well-known [16] techniques of catalyst preparation are employed. If, however, we aim to work with clean surfaces as frameworks for our study of catalysis, then several possibilities exist. But what do we mean by a clean surface? Obviously, there must be some impurity left at the surface, or in the bulk, of any highly purified solid. (It is a salutary thought to recall that a substance containing a total impurity content of $10^{-5}\%$ has approximately 10^{15} impurity atoms per cm^3.) Any definition of surface cleanliness must needs be arbitrary. One recent, widely accepted [17], definition is that of Allen *et al.* [18]. These workers state that an atomically clean surface is "one free of all but a few per cent of a single monolayer of foreign atoms, either adsorbed on or substitutionally replacing surface atoms of the parent lattice". Straightaway we appreciate that some evaluation techniques must be available to assay the surface cleanliness of a solid. The principles of the structure-sensitive phenomena used will be discussed fully later in this chapter; all that need be done here is to enumerate the most important techniques that have been employed. One of the earliest and most widely used techniques of surface analysis is the chemisorptive capacity of the adsorbent. Trapnell [19] showed that the amount of hydrogen chemisorption at liquid-air temperatures is a satisfactory criterion for assessing the degree of cleanliness of a number of metal surfaces. Roberts and Sykes [20] used essentially the same criterion for evaluating the surface purity of extensively reduced nickel powder. Another technique capable of detecting small fractions of a monolayer of surface contamination is that of low-energy electron diffraction [21–24]. Other techniques include field emission of electrons or field ionization of a gas such as helium [25, 26], photoelectric emission [27], secondary electron yield [28], Auger electron emission [29] and, not surprisingly, catalytic reactivity. It will become increasingly apparent later that the catalytic behaviour of evaporated metals is dependent upon the cleanliness of the evaporated surface. Roberts has shown [30, 31] that the catalysed low-temperature decomposition of ethane on rhodium and iridium films is greatly inhibited if there is adsorbed oxygen or carbon monoxide present.

We may divide the available methods of cleaning surfaces into two categories [32, 33]. The first category (A) consists of those which form a surface that is initially clean; the second (B) those which remove contamination from surfaces which are initially contaminated. The seven methods which have been used separately or in combination are subdivided as follows.

Category A: Vacuum evaporation
Vacuum cleaving
Vacuum crushing

Category B: High-temperature outgassing
Ion bombardment (sputtering)
Chemical reaction
High-field evaporation

3.2.2.2. *Vacuum Evaporation*

For the production of clean metal, and some non-metal, surfaces with quite large surface areas that can be usefully and reproducibly studied using conventional high-vacuum systems, the evaporated film technique introduced by Beeck, Smith and Wheeler [34], and later extended by Trapnell [19], Tompkins [35], Suhrmann [36] and others [37–39] is of immense value. The film is prepared by thermally evaporating the thoroughly outgassed metal, which may be in the form of a filament, or a bead held in a refractory boat, on to the walls of the containing vessel—usually glass—the outside of which is maintained at a low temperature. If liquid air is used to cool the vessel, surface area-to-film weight ratios of about $1 m^2 g^{-1}$ can be obtained. But since the films so produced have considerable stored energy [40], it is necessary to stabilize them by repeated heating to higher temperatures before commencing adsorption studies. The same considerations apply to the production, by evaporation, of films of compound semiconductors which are evaporated from inert crucibles [41].

Usually, the films produced by evaporation are polycrystalline, although it is possible, by varying the conditions of preparation, to produce single crystal films on the one hand and amorphous films on the other. By appropriate choice of substratum (for example, mica or rock salt rather than borosilicate glass), it is possible, owing to the phenomenon of epitaxy or oriented over-growths [42], to produce films which exhibit electron diffraction patterns characteristic of definite faces of a single crystal of the solid. Pashley [43] demonstrated that it was possible to prepare an atomically smooth single crystal of evaporated silver on a mica substratum. However, the vacuum conditions were rather poor, and it is almost certain that the film was heavily contaminated.

Insufficient attention has been given to the bulk and surface structure of evaporated films; but the recent excellent electron microscopic studies of Anderson, Baker and Sanders [44] have already gone a long way towards rectifying this state of affairs (see Section 3.3.9). These workers produced films of nickel and tungsten at pressures in the region of 10^{-7} torr on to glass substrata. From transmission electron micrographs, crystal widths were obtained and the presence of gaps between the crystals was clearly estab-lished. Upon progressive sintering, the intercrystalline gaps were first seen to be removed, then, in a later stage, surface asperities were removed, and finally crystal growth, which could eventually lead to one single crystal [45], occurred.

One serious limitation associated with the evaporated film technique is that, unless different workers preparing ostensibly the same type of evaporated film ensure that preparative conditions are identical—an extremely difficult condition to fulfil—the concentration of defects may vary significantly and this is likely [32] to exert a profound influence on the surface properties of the film. In fact, the influence of defects and surface structure on the catalytic activity of evaporated films is a topic of great current interest [46] and will be discussed in Chapter 5.

3.2.2.3. Vacuum Cleaving

If it is intended to study a crystallographically well-defined surface of quite low area (less than a square centimetre), a convenient method of producing the surface is to take a highly purified and annealed single crystal of the material and to cleave in vacuo at low temperatures. The cleaving is customar-ily effected by forcing a wedge into a small orientated slot in the crystal. Van-derslice and Whetten [28], in their studies on cleaved alkali halide crystals, used a razor blade mounted on a rod connected to a glass tube through flexible bellows and a glass–metal seal, to cleave their crystals. Sometimes special wedge-anvil arrangements are required to effect cleavage [17].

As with vacuum evaporation, cleavage may yield a surface which has con-siderable strain and a largely unknown concentration of defects. Proof that strain and defects introduced during the act of cleavage may prove significant is illustrated by the results of Palmer, Morrison and Dauenbaugh [47]. These investigators found that surfaces of silicon prepared by vacuum cleaving exhibited properties which were quite different from those displayed by silicon which had been prepared by ion-bombardment and subsequent annealing.

Although this technique has found considerable use [48–51], it is pertinent to cite a few of its limitations. In the first place, one is limited to working with the permitted cleavage faces (cleavage is obviously favoured between successive layers when the interlayer bonding is weak), so that an exhaustive study of the variations in surface properties with crystalline orientation for a particular solid is not feasible. Secondly, the method is limited to materials which are brittle and which fracture along given cleavage planes. Thirdly, a clean surface can be regenerated only by another cleavage. Lastly, the very minute areas (often no more than a few square millimetres) may, under the prevailing vacuum conditions, be far too readily contaminated.

3.2.2.4. Vacuum Crushing

The technique of impounding or comminuting solids so as to produce fine powders and a consequent increase of surface to volume ratio has been known for a long time [52, 53], but it has not been widely used in adsorption studies until recently. As with the cleaving technique, the clean surface is produced at low temperatures so that diffusion of impurities to the surface, which may occur in high-temperature heating procedures, is greatly alleviated. Unlike the cleaving technique, however, this method does not produce any definite crystal surface. On the contrary, a number of different crystallographic planes and fracture surfaces are exposed. Here again there is also uncertainty concerning the role of strains and defects introduced into the solid during crushing. The method has been extensively used, chiefly by Green and his co-workers [54–57], in the study of the surface properties of germanium.

3.2.2.5. High-temperature Outgassing

This is the method that has been most widely used in attempts to produce clean surfaces [19]. In principle, it is the simplest of methods: in practice, it is fraught with difficulties. If the strength of the adsorbent–adsorbate bonds is smaller than the strength of adsorbent–adsorbent bonds, then it should follow that, upon gradually increasing the temperature, all adsorbate molecules should be desorbed. When the adsorbent has a high melting-point (for example, tungsten or molybdenum), heating in a high vacuum to temperatures just below the melting-point generally suffices to purge the surface of impurities. In fact, with the refractory metals, the "flash-filament" technique [19] can be utilized.

The production of clean surfaces by heating alone often leads to difficulties. In the first place, diffusion of impurities from within the bulk of the adsorbent to the surface is facilitated at elevated temperatures. Thus trace quantities (*ca* 0·01%) of carbon in metals such as molybdenum, tantalum and tungsten migrate to the surface when these metals are heat treated *in vacuo* [*58–60*]. Diffusion can also occur in the opposite direction, that is to say from the containing vessel on to, or into, the surface. Certain types of Pyrex systems are particularly prone to "desorb" oxygen [*61*] and boron [*18, 62, 63*] on heating and these impurities may end up on the adsorbent surface. Another problem that may be encountered with this technique is that thermal etching [*64*] or thermal faceting [*65, 66*] may occur at elevated temperatures. Effectively, this converts the surface into a collection of crystallographically distinct planes, all of which may be different from the crystal plane known to predominate on the solid surface prior to heat treatment. Such surface rearrangements cannot be countenanced when it is intended to study the influence of crystal orientation upon adsorbability or catalytic activity.

Yet another difficulty with the present technique is the fact that the chemical properties, particularly the surface characteristics, of some materials can be very significantly altered as a result of certain heat treatments. Both the adsorbability and the catalytic activity of nickel surfaces can be modified by varying the rate of cooling after high-temperature heating [*67*]. The catalytic activity of nickel and copper wires can be enhanced by several orders of magnitude, simply by flashing these metals at high temperature [*68*]. This is how catalytic superactivity (see Section 5.3.2.1) is induced.

3.2.2.6. *Ion-bombardment or Sputtering*

When a metal semiconductor, or dielectric material is bombarded with inert gas ions above a certain critical energy, atoms of the solid are removed, and the process is known as sputtering. This phenomenon has been known for several years; Langmuir utilized it in his studies of the removal of thorium from thoriated tungsten filaments [*69*]. Farnsworth and his co-workers have used this technique extensively in producing atomically clean metal and semiconductor surfaces. Moore [*70*] has also applied this technique with some success.

Despite the fact that surface contamination, which is difficult, if not impossible, to remove by other means, can be removed by sputtering, the bombardment by the ions breaks up the surface extensively, and generally some of the bombarding gas molecules become embedded in the solid lattice [*71*]. It is therefore prudent to heat the solid *in vacuo* subsequent to bombardment. This serves the dual function of annealing lattice damage and releasing trapped gas [*23, 32, 72*]. Many cycles of heating, bombardment, and annealing are usually advisable; it is imperative to ensure that the bombardment conditions are such as to minimize ion-bombardment etching [*73*] for reasons similar to those given in the previous section. See also Section 5.3.1.

3.2.2.7. Chemical Reaction

Reduction of partially oxidized metal surfaces with molecular hydrogen at high temperatures has traditionally been one of the most favoured methods of producing reasonably clean surfaces of metal powders. Roberts and Sykes [20] took this procedure to its logical limit. By subjecting nickel powder, previously purified by forming and subsequently decomposing the carbonyl, to prolonged reduction in stringently purified hydrogen, they were able to produce surfaces which in part simulated the behaviour of evaporated nickel films as judged by the avidity with which the surface chemisorbed hydrogen at −183°C. Significantly, a rather large proportion of the surface still remained contaminated. This work illustrates well the difficulty generally encountered with this method. Impurities are given ample opportunity to diffuse from the bulk of the adsorbent to the surface: this process occurs concomitantly with the reduction of the surface.

It has been shown that carbon can be removed from tungsten and nickel tips (used in field-emission work, see p. 138) by annealing in hydrogen or oxygen [25]. Furthermore, it is possible [74] to remove oxides from the surface of minute iron whiskers by heating the samples in a hydrogen atmosphere at approximately 750°C. The prime difficulty with producing clean surfaces by using chemical reaction is that it is not always possible to identify the surface contamination—without such information one can only guess as to what might turn out to be the most efficacious chemical purification procedure to choose.

3.2.2.8. High-field Evaporation or Field Desorption

When electric fields in the range 100–600 million V cm^{-1} are established in the vicinity of a point specimen, adsorbed layers on that specimen can be removed [26, 75, 76]. If the field is increased, layers of the specimen itself can be evaporated. This rather specialized method of producing clean surfaces is achieved in a field-emission microscope (see Section 3.3.5.1) with the potential of the tip positive. The method has been applied with a considerable measure of success to silicon and germanium [77] and to tungsten [78]. It has proved possible to remove atoms of tungsten from its own lattice at liquid helium temperatures at a field strength of $5 \cdot 7 \times 10^8$ V cm^{-1}. In this way, loosely held atoms situated at surface promontories were preferentially desorbed, so yielding regular surface structures. Many of the clean atomic planes produced in this manner had high Miller indices. The area of the clean surface generated by field desorption seldom exceeds $1 \cdot 5 \times 10^{-10}$ cm^2.

3.2.3. Direct study of adsorption

It is possible to determine the amount of gas adsorbed by a solid in two ways: volumetrically and gravimetrically.

In the volumetric method, the pressure of a known quantity of gas both in the absence and in the presence of the solid is measured. If an adsorption isotherm is to be constructed, several successive quantities of gas are admitted into the apparatus, the volumes of the various compartments having been previously calibrated. When the surface area, or, more correctly, when the adsorbability, of the solid is large—as is often the case with many supported catalysts—ordinary gas pressures (from a few torr to 1 atm) can be employed. But when the adsorbability or surface area is low (a few m^2g^{-1} or less), it is necessary to work with gas pressures less than a few torr and to minimize dead-space volumes, otherwise adsorption may not be detectable. The precise design of the apparatus used in the volumetric determination of adsorbed amounts will depend chiefly on the type of solid adsorbent being studied and also on the method chosen for cleaning or preparing the surface for study. Figure 1 indicates the type of arrangement required for the volumetric study of the surface properties, including physical adsorption of inert gas, and chemisorption (both adsorption and desorption) of hydrocarbons, on nickel oxide

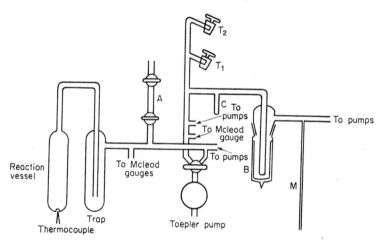

FIG. 1. Typical glass apparatus used for the volumetric study of adsorption, desorption and gas analysis. T_1 and T_2 are special taps; A, the "doser", is a known volume between two taps; and B is a cold trap.

[79]. Figure 2 illustrates schematically the layout of the volumetric apparatus used by Redhead [7] in his elegant and meticulous studies of the chemisorption of carbon monoxide on polycrystalline tungsten. Since this particular study represents well the precision and power of modern volumetric techniques, it will be described in main outline here. The sample itself consists of a 25-μm diameter (60 cm long) tungsten wire wound into a spiral of 4 cm length. (The long sample was used to minimize the fraction of wire surface which was

cooled by conduction to the support wires.) No diffusion pump or mechanical backing-pump was employed, so as to avoid contamination. The pressure in the system was reduced from atmospheric to about 2×10^{-2} torr by an adsorption pump consisting of 100 ml of Linde 13X molecular sieve (this being an artificial zeolite) contained in a metal tube immersed in liquid nitrogen. A

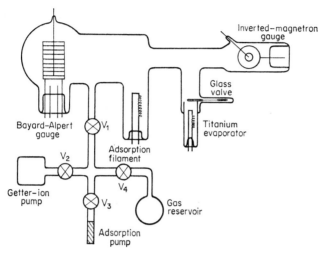

FIG. 2. Layout of volumetric apparatus used by Redhead [7].

getter–ion pump [80] (consisting primarily of a freshly sputtered titanium surface which "getters" residual gas) was started once the pressure reached 2×10^{-2} torr and valve V_3 was closed. The system was next baked at 500°C, using a large demountable furnace, for 18 h with the getter–ion pump on at a pressure of roughly 10^{-5} torr. Valve V_1 was then closed and titanium evaporated from the titanium source. After rigorous outgassing of all metal parts of the system, the pressure fell to a limiting value of 7×10^{-11} torr in about 8 h, when the Bayard-Alpert gauge (an inverted ion gauge [81] which is a device that enables pressures in the range 10^{-5}–10^{-11} torr to be measured) was operating, and to 4×10^{-11} torr with the inverted-magnetron gauge operating alone. (A full description of the operation of the inverted-magnetron gauge is given in Ref. 14.) As a matter of interest, the composition of the residual gas, as determined by mass spectrometry, was found to be 90% helium, 9% hydrogen and 1% carbon monoxide.

The tungsten samples were initially cleaned and annealed by heating to 2500°K for at least 16 h. Thereafter, they were cleaned by repeated flashing at 2200°K for 1 sec at intervals of 10 sec. The wire surface was assumed to be in a reproducible condition when the pressure peaks produced by the intermittent flashing had decreased to a constant amplitude.

Spectroscopically pure carbon monoxide at atmospheric pressure was admitted into the space between valves V_1, V_2, V_3 and V_4. V_1 was then opened and adjusted so as to give a pressure of about 5×10^{-9} mm in the system. The metal parts of the system and the glass walls adsorbed carbon monoxide readily, and Redhead found that it was necessary to allow the gas to leak into

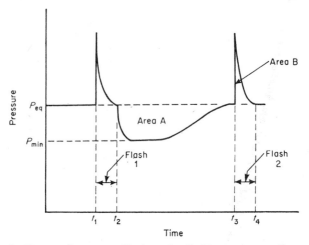

FIG. 3. Change of pressure with time after flashing a tungsten filament [7].

the system at a constant rate for at least 48 h before any measurements were made. Figure 3 shows schematically the variation in pressure p with a constant input leak, when the filament is flashed (the time-scale having been exaggerated to reveal the consequence of the flash). Redhead allowed the filament to saturate, and the pressure reached a steady value p_{eq}. The filament was then flashed at 2300°K from t_1 to t_2. After p_{eq} had again been attained, the filament was flashed again from t_3 to t_4. Now, provided that all sources and pumps in the system (other than the tungsten filament) are constant in speed, it can be shown that:

$$\int_{t_2}^{t_3} (p_{eq} - p) \, \mathrm{d}t = \int_{t_3}^{t_4} (p - p_{eq}) \, \mathrm{d}t \qquad (1)$$

In other words, area A=area B. Equality of areas A and B indicated that adsorption and re-emission rates of the electrodes and the glass envelope had come to equilibrium. Redhead took no measurements until this "area rule" was obeyed. Upon reaching this juncture, Redhead could proceed to measure the sticking probability (see p. 34) and the adsorption of carbon monoxide on the tungsten filament, the principle of the method being that a known quantity of the gas was admitted from the precalibrated volume of the system (which in this case was 1·96 litres).

3.2.3.1. *The Gravimetric Method*

Very precise measurements of adsorbed amounts can be made by recording the weight of the adsorbent during the course of adsorption. Pioneer work, together with much development of technique, has been carried out by Gulbransen [82], Rhodin [83] and Gregg [84]. In recent years great advances have been made in the construction of highly sensitive microbalances capable of recording mass changes of as little as 10^{-8} g and of following a complete adsorption isotherm when the total change in sample mass is in the region of 10^{-5} g.

A large variety of microbalance types can be adapted for adsorption studies [82, 83, 85], but by far the most commonly used, at present, are the beam types, which may be pivotal [84, 86], knife-edge [87], or simply gravity operated [82, 83, 88, 89]. The helical spring balances, which now tend to be constructed from robust copper–beryllium alloys rather than the fragile silica helices introduced by McBain and Bakr [90], are also widely used, particularly in physical adsorption studies. The actual change in mass of the solid during interaction is monitored by following the deflection of the beam, or the extension of the helix, using either a micrometer eyepiece [91] or by utilizing the principle of the optical lever in conjunction with photo-tubes [92].

A typical example of the experimental arrangement employed in the gravimetric study of adsorption is illustrated in Fig. 4. This is a schematic diagram

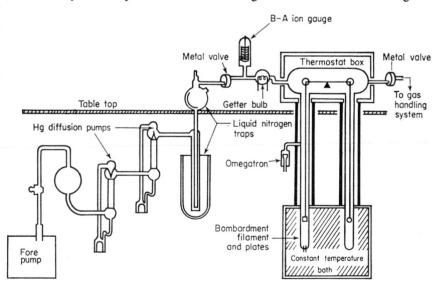

FIG. 4. Typical experimental arrangement used by Wolsky and Zdanuk [88] in the gravimetric study of adsorption.

of the vacuum microbalance system used by Wolsky and Zdanuk [88] to study the adsorption of oxygen on, and the sputtering of, orientated single crystals of germanium and silicon. Two mercury diffusion pumps in a double liquid-nitrogen trapped system and ultrahigh vacuum valves were employed. To overcome the undesirable influence of buoyancy forces, both limbs of the balance (housing the hangdown wires) were heated or cooled to the same extent during measurements. The partial pressure of the various gases present in the system was measured using an omegatron, which is a small-volume, bakeable, mass analyser suitable for the determination of the composition of a gas in the pressure range 10^{-5}–10^{-10} torr [93]. Unless rather elaborate precautions are taken to diminish the unavoidable temperature gradients along the sample or counterweight, very serious disturbances, arising from thermomolecular flow, will be encountered. Quantitative estimates of the magnitude of these spurious forces, together with a discussion of means of eliminating them, have been given by Thomas and Poulis [94, 95].

The balance used by Wolsky and Zdanuk is a gravity-operated one, and its design follows closely that developed by Gulbransen [96]. A very sensitive, yet robust, pivotal type of beam balance has been used by Czanderna and Honig [86, 97]. Figures 5 and 6 give the essential constructional details of the actual vacuum microbalance and of the microbalance system, respectively. A more sophisticated type of pivotal microbalance has been constructed by

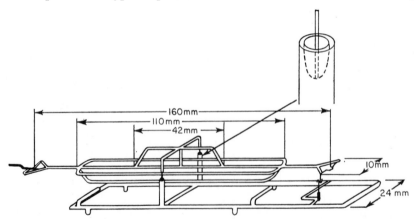

FIG. 5. The pivotal vacuum microbalance designed by Czanderna and Honig [86].

Czanderna recently [98, 99]. To illustrate the versatility of the gravimetric method, Czanderna, in studying [99] the adsorption of oxygen on silver powder, was able, by making the appropriate alterations to the ambient pressure and temperature, to measure: (a) the saturation uptake of gas as a function of pressure and temperature; (b) the rate of adsorption and desorp-

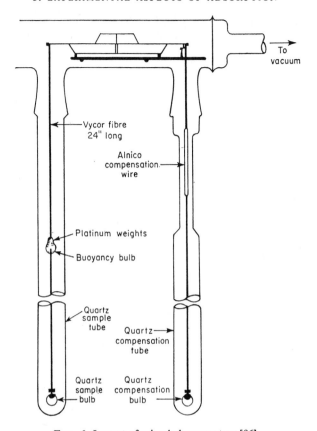

FIG. 6. Layout of microbalance system [86].

tion; (c) the thermally induced desorption of adsorbed species; (d) the surface area; and (e) long-term mass changes in the sample. The heat of adsorption can be evaluated from (a), and the activation energy of adsorption from (b).

In very recent times the sensitivity of ultrasensitive vacuum microbalances has been increased by one or two orders of magnitude. Warner and Stockbridge [100, 101] and others [102, 103] have shown that, by using a resonant quartz crystal as the "balance" (by making use of the fact that the resonating frequency of such a crystal is highly dependent upon the pressure exerted upon it), it is in practice possible to devise a system capable of detecting the formation of a fraction of a per cent of an adsorbed monolayer on a solid surface. This technique has yet to be fully exploited.

The greatest advantage that the gravimetric method has over the volumetric one is that any amount of dead-space volume can be tolerated. Unless the dead-space is small, volumetric methods become quite insensitive as the pressure increases, particularly when the surface areas are small; but the gravimetric method is potentially capable of being just as sensitive at 1 atm pressure as at 10^{-10} torr. In practice, however, convection forces and buoyancy corrections limit the highest pressure to about a third of an atmosphere. Thermomolecular corrections [104–108] to the pressure readings must, as always, be applied at low pressures when the temperature of the sample is different from that of the pressure-recording device.

The foregoing gives a general idea of the types of experimental procedure required for an adsorptive study of solid surfaces. We shall next examine the type of information that can be collected using such procedures.

3.2.4. Surface areas from physical adsorption

From what has been said in the previous chapter (Sections 2.3.1–2.3.4), it follows that, under certain highly exceptional circumstances, application of the Langmuir model for adsorption will enable an estimate of v_m, the volume of adsorbate occupying one monolayer, to be obtained. So rarely is the Langmuir equation strictly applicable to adsorption systems that we may conclude that surface areas are seldom reliably estimated by plotting v/p against $-v$, or p/v against p, and extracting the value of v_m from the slope and intercept:

$$\frac{p}{v} = \frac{1}{pv_m} + \frac{p}{v_m} \tag{2}$$

3.2.4.1. *Surface Areas from the BET Equation*

The Brunauer-Emmett-Teller equation discussed theoretically in Chapter 2 offers a much better opportunity for estimating the surface area of a solid than the Langmuir, and indeed most other, equations. The value of v_m is derived, using the BET multilayer theory, either by plotting $p/v(p_0-p)$ against p or p/p_0 (see equation (3)), or by plotting $1/v(1-x)$ against $(1-x)/x$ where $x=p/p_0$ (see equation (4)):

$$\frac{p}{v(p_0 - p)} = \frac{1}{v_m c} + \frac{(c-1)}{v_m c}\frac{p}{p_0} \tag{3}$$

$$\frac{1}{v(1 - x)} = \frac{1}{v_m} + \frac{1}{v_m c}\frac{(1-x)}{x} \tag{4}$$

Whatever the faults of the BET theory may be—and they are many and varied (see Section 2.3.8)—it has become well established [109] as a valuable and analytical method of locating point B, which marks the change from adsorption predominantly in the first layer to adsorption predominantly in the second (see Fig. 7). It has been found, however, that only type II isotherms (Section 2.3.8) having well-defined "knee-bends" or plateaux yield reliable values of v_m. (This is tantamount to saying that only when c values are high are v_m values sensible.) In other words, even though a perfectly respectable linear BET plot is obtained, the value of v_m tends to be erroneous, if the adsorption isotherm itself has a very poorly defined point B.

For most systems the BET equation is applicable in the p/p_0 range of 0·05–0·3. However, many systems have been reported to give rise to BET

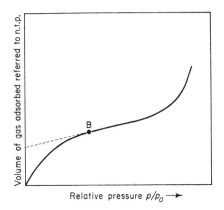

plots which are convex to the pressure axis at p/p_0 greater than about 0·2. And other systems are known where the BET is strictly applicable down to values of the relative pressure of less than 1×10^{-3}. Rather than guess which range of p/p_0 to accept, and rather than assume that the applicable range is always 0·05–0·3, it is more systematic to test the internal consistency of the results. After v_m and c have been determined from the linear plot, using what appears to be a reasonable p/p_0 range, it is then prudent to see whether equation (5), which gives the value of p/p_0 when $v=v_m$, is applicable.

$$\left(\frac{p}{p_0}\right)_{v=v_m} = \frac{\sqrt{c}-1}{c-1} \tag{5}$$

If the value of c is very large, then equation (3) becomes:

$$\frac{p}{v(p_0-p)} = \frac{1}{v_m}\frac{p}{p_0} \tag{6}$$

which reduces further to

$$v_m = v(1-x) \tag{6a}$$

Hence, it follows that a BET plot which passes through the origin has a slope equal to the reciprocal of the monolayer volume. Conversely, if only one adsorption point is available, and the heat of adsorption is high so that c is big (see equation (58), Chapter 2), it is safe to estimate the surface area from the slope of the line joining the origin and the single point on the BET diagram. Brunauer *et al.* [*110*] have shown that for nitrogen at $p/p_0 = 0.3$, the v_m obtained by the "single point" method is within 5% of the value obtained from the full BET plot.

Halasz and Schay [*111*, *112*] have introduced a more accurate modification of the "single point" method. They transformed the BET equation into:

$$v_m = v(1 - x)\left\{1 + \frac{1 - x}{c\,x}\right\} \tag{7}$$

This equation is the same as equation (4), but rearranged. The last term in equation (7) may be regarded as a correction factor which converts the approximate value of v_m obtained by the "single point" method to the v_m value derived from the full BET equation. Halasz and Schay have tabulated [*111*, *112*] the factor $\left\{1 + \frac{1 - x}{cx}\right\}$ for various values of c and x.

Before setting out to determine the monolayer capacity v_m of any solid it is useful to know in advance the approximate value of the surface area (say to within an order of magnitude). Generally, if the area is several square metres per gram, then it is customary to use nitrogen gas at liquid-nitrogen temperatures ($-195.8°C$ at 760 torr). The BET plot gives very good results for this adsorbate on the majority of solids. The reason for such success is not at all obscure. At liquid-nitrogen or liquid-air temperatures with all solids so far studied, nitrogen yields big values for c (because of the magnitude of the heat of adsorption), so that, according to equation (5), the lower the value of p/p_0 at which the "knee-bend" occurs the sharper this will be. In other words, nitrogen as adsorbate tends always to yield type II isotherms.

If the surface area is more in the region of square centimetres than square metres per gram, it is futile to use nitrogen at liquid-air temperatures as the means of producing a BET plot if a volumetric apparatus is being employed for measuring the adsorbed amount. The value of p_0 for nitrogen at these temperatures is in the vicinity of 760 torr. Now, in order to reach p/p_0 values which are required by the BET theory, adsorption at relatively high values of nitrogen gas pressure will have to be recorded and, even when dead-space volumes are kept to a minimum, adsorption measurements will prove difficult, if not impossible (with the gravimetric method, on the other hand, it is generally possible to record the adsorption data at any desired p/p_0 range). Since the volumetric method is more convenient, the majority of attempts to determine the monolayer capacity of solids with small surface areas have been made using adsorbates the saturated vapour pressures of which have been small enough to enable useful values of p/p_0 to be reached with conventional volumetric apparatus. Krypton has been extensively used [113–119] to determine small surface areas: its p_0 value at $-195\cdot8°C$ is $2\cdot0$ torr. Of late, however, the arguments for using xenon [120–124] in place of krypton have become most convincing. Xenon possesses the advantage over krypton of having a higher heat of adsorption and a lower p_0 value. The higher heat of adsorption, leading to higher c values, means that point B can occur at relative pressures p/p_0 of $0\cdot01$ or less [124]. The lower p_0 value means that, in a volumetric apparatus, dead-space corrections will be smaller than for krypton.

So far, we have not mentioned how the monolayer capacity v_m is converted to surface area. All that is required in order to do this is a knowledge of the effective area that each molecule of the adsorbate occupies in the monolayer, the number of molecules in the monolayer being calculable from v_m. It is not, however, an easy matter to evaluate the effective cross-sectional area of a molecule in the adsorbed state. One could, for example, assume hexagonal close-packing of the adsorbate molecules in the monolayer and thus calculate the cross-sectional area Σ from the molecular weight M and molar density $\bar{\rho}$ of the adsorbate. (It would be more reasonable to equate $\bar{\rho}$ with the density of the solid adsorbate, but the liquid density has often been taken [113, 125].) For hexagonal close-packing we have that:

$$\Sigma = 1\cdot09 \left(\frac{M}{N\bar{\rho}}\right)^{2/3} \tag{8}$$

where N is Avogadro's number.

Several values for the cross-sectional area of a variety of molecules have been quoted in the literature [126, 113, 127–130]. In very many instances, the values quoted for molecules relatively less frequently used in BET determinations—molecules such as neon, argon, nitrous oxide, ethylene, methane, ethane and butane—than nitrogen or krypton, have been obtained by taking the nitrogen value of 16 or $16\cdot2$ Å² as correct (or $19\cdot5$ Å² for krypton), and "standardizing" the value of the gas molecule in question, so as to make the respective v_m's lead to the same surface area.

Recent attempts to determine the relative values of the cross-sectional area of krypton and xenon in the adsorbed phase have led to a significant advance in our understanding of the physically adsorbed state, and have led us to re-examine the justification for assigning unique areas to physically adsorbed molecules. Brennan, Graham and Hayes [124], on the basis of an exhaustive comparison of the point B, and BET monolayer volumes obtained with krypton and with xenon on a large number of evaporated metal surfaces, concluded that the effective areas of these two molecules, on metal surfaces at least, are virtually the same. (Previous data [121, 131] had indicated that the ratio Σ_{Xe}/Σ_{Kr} was close to $1\cdot3$.) Following on from earlier suggestions [120] that the packing of the adsorbate molecules in a monolayer is dependent upon the adsorption sites at the surface, Brennan *et al.* [124] recalled that field-emission microscopy (see Section 3.3.5.1) has amply demonstrated the importance of the co-ordination number of a substrate site in the adsorption of the inert gases [132, 133]. The relatively large energy of activation for surface diffusion observed for krypton and xenon on tungsten [134] further supports this view. The calculations [124] of the interatomic energy (energy of adsorption, see Section 2.2.1) when these inert gases are situated at a metal surface vindicate the above results. So we are left with the conclusion that, unless the dependence of cross-sectional area of an adsorbate molecule upon its co-ordination at a solid surface is taken into consideration, estimates of the surface area from measurements of the monolayer capacity will be subject to yet a further error.

3.2.4.2. Surface Areas from the Hüttig Equation

Hüttig [135] arrived at an equation which, like the BET equation, purports to describe multilayer adsorption on a uniform surface. Whereas the BET theory assumes that molecules in the $(i+1)$th layer prevent evaporation of the underlying molecules in the multilayer, the Hüttig theory assumes that the evaporation of an ith layer molecule is completely unimpaired by the presence of molecules in the $(i+1)$th layer. The final equation can be written [136] in the nomenclature adopted for the BET theory as:

$$\frac{v}{v_m} = \left(1 + \frac{p}{p_0}\right) \frac{c \dfrac{p}{p_0}}{1 + c \dfrac{p}{p_0}} \tag{9}$$

or

$$\frac{v}{v_m} = (1 + x) \left(\frac{c\,x}{1 + c\,x}\right) \tag{9a}$$

using x to signify the relative pressure p/p_0.

The Hüttig equation has been remarkably successful in accounting for the shape of (multilayer) adsorption isotherms, and surface areas have been successfully evaluated using it at p/p_0 values up to 0·35. However, the equation rests on questionable theoretical principles [109, 137]. The principle of microscopic reversibility, which is as sacrosanct as the second law of thermodynamics, is violated in the derivation of the equation, for it is supposed that the equilibrium condition in a given layer is obtainable by equating two processes, evaporation and condensation, which are not the reverse of one another.

Notwithstanding these serious failings, v_m values deduced from the Hüttig equation agree with BET v_m values to within about 20% when the values of c are high.

3.2.4.3. *The Harkins-Jura Relative Method*

A fresh approach [138, 139] to surface area measurements by gas adsorption methods was provided by Harkins and Jura, who pursued the analogy between adsorbed layers on solids and insoluble surface films on liquids, first drawn by Gregg [140, 141]. Harkins and Jura proposed that the two-dimensional surface pressure Π of an assumed liquid condensed film, formed from the gas and held at a surface, can be represented by the empirical equation:

$$\Pi = b - a\Sigma \tag{10}$$

where a and b are constants, and Σ is the average area occupied by an adsorbed molecule. Using the Gibbs adsorption isotherm, equation (10) was transformed into a more tractable form suitable for describing adsorption isotherms. The Gibbs equation, which is widely applicable to the phenomenon of adsorption at liquid surfaces [142], may be written as:

$$\left(\frac{\partial \Pi}{\partial \ln p}\right)_T = RTc \tag{11}$$

where a change $\delta\Pi$ in energy occurs when the pressure is changed by an amount δp; and c is the surface concentration in moles per specific surface area S_g given by:

$$c = \frac{n}{S_g} = \frac{v}{\bar{V}S_g} = \frac{X}{MS_g} \tag{12}$$

where the quantity of gas adsorbed may be expressed in terms of volume, v or mass per unit mass of adsorbent X ($\bar{V}$, M and n being, respectively, the gaseous molal volume, the molecular weight and the number of moles of the adsorbate). Transformation was achieved by using the following identities:

$$\left(\frac{\partial \ln p}{\partial v}\right)_T \equiv \left(\frac{\partial \ln p}{\partial \Pi}\right)_T \left(\frac{\partial \Pi}{\partial v}\right)_T \equiv \left(\frac{\partial \ln p}{\partial \Pi}\right)_T \left(\frac{\partial \Pi}{\partial \Sigma}\right)_T \left(\frac{\partial \Sigma}{\partial v}\right)_T \tag{13}$$

and equating the excess surface pressure Π to the difference between the surface energy σ_s of the adsorbent and the surface energy σ when a volume v of gas is adsorbed,

$$\Pi = \sigma_s - \sigma \tag{14}$$

From equations (10) and (14) a small change in Π can be related to changes in σ and Σ

$$\delta\Pi = -\delta\sigma = -a\,\delta\Sigma \tag{15}$$

Expressing Σ in terms of the specific surface area and adsorbed volume, we have:

$$\Sigma = \bar{V}\, S_g/vn \tag{16}$$

By utilizing equations (15) and (16), equation (13) leads to:

$$\left(\frac{\partial \ln p}{\partial v}\right)_T = \frac{a(\bar{V}\, S_g)^2}{RTnv^3} \tag{17}$$

Integration of this equation yields:

$$\ln p = \frac{-a(\bar{V}S_g)^2}{2RTnv^2} + \text{Const.} \tag{18}$$

Hence a plot of $\ln p$ against $1/v^2$ ought to be linear if the Harkins-Jura treatment is valid, and the negative slope m will yield a numerical value for the surface area, provided that the constant k, defined by the following equation, is determined:

$$S_g = \frac{1}{\bar{V}}\left(\frac{RTn}{a}\right)^{1/2} m^{1/2} = k\, m^{1/2} \tag{19}$$

Jacobs and Tompkins [143] have emphasized that the constant k should be determined for each adsorbate–adsorbent pair, and so it should. But Harkins and Jura tacitly assumed that k was a function solely of the temperature and of the nature of the adsorbate and independent of the nature of the solid surface. Harkins and Jura claimed that they had obtained a reliable value of k by calibration using an independent surface-area method, this being the method based on the measurement of the heat of wetting. (This is why the method is a relative one.) To summarize, therefore, the surface area of a solid (anatase, i.e. TiO_2) was first determined by the heat of wetting method [138, 144], so that an independent value of S could be fed into equation (19). Secondly, by plotting $\ln p$ against $1/v^2$ (equation (18)) and measuring the negative slope, a value of m was evaluated which could also be fed into equation (19). Hence the value of k, applicable to all solids in the opinion of Harkins and Jura, is obtainable.

As mentioned previously, the Harkins-Jura method has found considerable use. However, it is an empirical equation which is now widely accepted as being inferior to another empirical method, that of Brunauer, Emmett and Teller.

3.2.4.4. Dubinin's [145] Application of Polanyi's Potential Theory of Adsorption

Polanyi [146] considered that planes of equipotential energy exist above solids with homogeneous surfaces. Each potential plane encloses a volume ϕ_i between the ith plane of potential ϵ_i and the surface where the value of ϵ is a maximum. Both ϵ and ϕ can be expressed in terms of the gas pressure, temperature and number of moles of gas adsorbed. The equation

$$\epsilon = f(\phi) \tag{20}$$

is essentially an isotherm equation, since ϵ was assumed not to change with temperature. The adsorption potential is defined as the work of compression on the gas. At pressures considerably less than the critical pressure the gas may be assumed ideal and the liquid incompressible. Therefore, the work of compression of one mole of an ideal gas from an equilibrium pressure p to the s.v.p. is

$$\epsilon_i = \int_p^{p^0} V \mathrm{d}p = \int_p^{p^0} \frac{nRT}{p} \, \mathrm{d}p = RT\ln\frac{p^0}{p} \tag{21}$$

where the work of creating a liquid surface has been neglected in comparison with ϵ_i. The volume in the adsorption space is then

$$\phi_i = \frac{m}{\rho_T} \tag{22}$$

where m is the weight of adsorbed vapour and ρ_T is the density of the liquid at temperature T. To each pair of values m and p on an adsorption isotherm there is a corresponding pair of values of ϕ_i and ϵ_i on the ϕ–ϵ curve. Since any variation in ϵ is independent of any temperature variation, this curve will represent, for any adsorbate–adsorbent system, the extent of adsorption at any relative pressure and any temperature well below that corresponding to the critical point. Hence, the ϕ–ϵ curve is fully representative of the system and is termed a characteristic curve. Various refinements for non-ideal behaviour and variation in packing density extend the validity of the characteristic curve, but need not be discussed here, as it is the basic application of the simple theory which we wish to discuss in relation to surface area determination.

Dubinin [145] related the volume adsorbed in small pores to the adsorption potential. The ratio of the volume adsorbed v at a given potential ϵ to the total adsorbed volume v_0 was considered to be an exponential function of the adsorption potential

$$\frac{v}{v_0} = \exp\left(\frac{-K\epsilon^2}{\beta^2}\right) \tag{23}$$

where β is a coefficient characterizing the polarizability of the adsorbate and K is a constant. Substituting equation (23) into equation (21),

$$\ln\frac{v}{v_0} = -K\left(\frac{nRT}{\beta}\right)^2 \left(\ln\frac{p^0}{p}\right)^2 = -B\left(\ln\frac{p^0}{p}\right)^2 \tag{24}$$

The potential theory is of value in determining surface areas provided that we can, from a plot of $\ln(v/v_0)$ against $(\ln p/p_0)^2$, extract a meaningful monolayer volume (v_m). We shall illustrate how this may be done by referring to the work of Marsh et al. [147–149] applied Dubinin's extension of the Polanyi potential theory to the determination of surface areas of carbons and compared their results with values deduced from the BET equation. It was noted that the BET surface areas of some carbons and graphitized carbons are impossibly high and are greater than the theoretical maximum surface area attributed to graphite, viz. 2680 m²g⁻¹, deduced from considerations of its theoretical density and crystallite dimensions.

The success in predicting authentic monolayer volumes for the microporous carbons which Marsh et al. [147–149] studied is attributable to their interpretation of Foster's [150] and Everett's [151] ideas concerning the

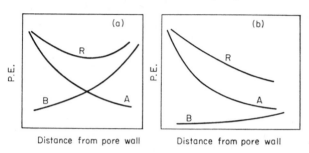

FIG. 8. Potential energy diagrams for adsorption in small (a) and in large (b) diameter pores. A and B refer to the adsorption potential and repulsive interactions respectively. R is the resultant of A and B.

variation in the adsorption potential as small and large micropores become filled progressively. The adsorption potential is constant across the diameter of a very narrow micropore, but passes through a minimum in wider micropores at the centre of the pore. Superimposed on the adsorption potential are repulsive interaction forces from the adsorbate layer. Figure 8 shows how these repulsive forces produce a resultant potential energy curve which, for small pores, passes through a minimum at a radius less than that corresponding to the pore radius. For larger pores the resultant potential does not reach a minimum before the adsorption reaches a stage at which the relative pressure is unity. Whether or not capillary condensation occurs in a pore of a given diameter therefore depends on the resultant potential energy profile. Spontaneous filling of a large-diameter pore resulting in capillary condensation would occur if the adsorption potential was high and the forces of repulsive interaction low. Such a situation would correspond to strong adsorbate-adsorbent affinity and weak interaction between adsorbate layers. In the work of Marsh et al. [147–149], the processes occurring during the adsorption of nitrogen and carbon dioxide on an activated carbon were considered to be portrayed by the course which the potential energy plot (equation (24)) follows. Figure 9 compares nitrogen and carbon-dioxide isotherms with the corresponding Dubinin plots. The section AB of the nitrogen isotherm corresponds to the filling of small-diameter micropores at low relative pressure. The greater the initial slope of the potential energy curve over the region AB, the more appreciable in extent is the filling of micropores. As the remainder of the surface is covered in a less spontaneous and more progressive manner, the slope of the potential energy curve diminishes so that the two sections AB and BC correspond to an interpretation, due to Pierce [152], that small micropores are separately filled and the remainder of the surface covered by a monolayer of adsorbate. In the wider pores, capillary condensation occurs and accounts for the final rise of the potential energy curve to D. Marsh *et al.* extrapolated their Dubinin plots for nitrogen to D', and they state that the volume corresponding to D' is v_m. They found that the value of v_m calculated from the BET equation agrees quite well with the value predicted by D'. However, such agreement does not necessarily verify the correct value for v_m deduced from the two methods. When a comparison is made of a carbon dioxide isotherm with a potential energy curve, quite a different state of affairs exists. There is only a single linear portion EF to the potential energy curve which extrapolates to the point F'. This probably indicates that the predominant adsorption process which occurs in this range is a gradual covering of the surface to a complete monolayer with the absence of capillary condensation in narrow micropores at low relative pressures. This portion of the potential energy curve does not fit the BET equation, and the application of equation (24) to predict monolayer capacities and hence surface areas is therefore restricted to those adsorbate–adsorbent systems in which there is no capillary condensation at low relative pressures. Provided

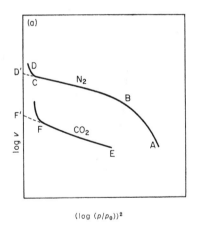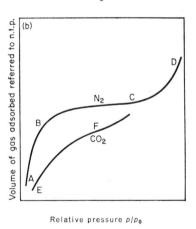

FIG. 9. Comparison of (a) adsorption data plotted from Dubinin's equation and (b) observed adsorption isotherms.

that, for a given adsorbent, an appropriate adsorbate can be found, the potential energy equation does have the advantage that the point at which the surface is covered by a monolayer is very much more distinct, since subsequent adsorption processes are reflected in a sharp rise in the curve in contrast to a normal isotherm. The adsorption data obtained with carbon dioxide

for a number of porous materials substantiate the view that the potential
energy equation of Dubinin predicts the true monolayer capacity rather than
a volume capacity apparently obtained from nitrogen isotherms.

3.2.5. Gas adsorption at high temperatures as a method of determining surface areas

Halsey and his collaborators [153–163] have showed that the adsorption of
inert gases may be measured accurately, using a volumetric technique, at
temperatures much higher than the boiling-point of the adsorbate, although
the actual extent of adsorption at 1 atm does not exceed more than about
2% of a monolayer. From their results, they formulated a theory which relates
the potential energy of an adsorbed molecule to its distance from the surface.
One of the corollaries of this theory is that it enables the surface area of the
adsorbent to be calculated. Barker and Everett [164, 165] transcribed Halsey's
imperfect gas theory into a more conventional gas adsorption theory, and
reduced the ideas to a simple approach which could be extended if necessary
to a more accurate and sophisticated model.

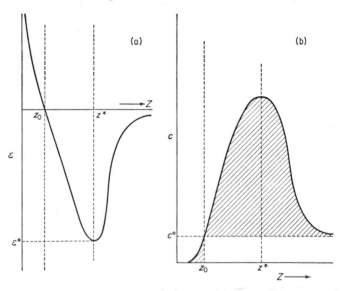

FIG. 10. (a) Potential energy as a function from surface. (b) Concentration as a function
from surface.

Consider, for simplicity, a plane adsorbing surface, the nuclei of the adsorbent surface atoms defining the xy-plane. The variation of the potential
energy ϵ of an atom with distance z from the surface is shown by the curve
in Fig. 10. Suppose also that the system is sufficiently dilute that there is no
interaction with other atoms in the gas phase and the potential energy is a

function only of the coordinates x, y, z. The average concentration c of atoms at any point in the adsorption field can then be calculated from the Boltzmann distribution law to give

$$c = c^0 \exp\left(-\frac{\epsilon}{kT}\right) \tag{25}$$

where c^0 is the concentration in the bulk gas phase at a distance far enough from the surface for ϵ to be zero. Figure 10 also shows the corresponding concentration as a function of z. The shaded area is a measure of the number of molecules adsorbed per unit surface area. Gibbs defined adsorption as the number of molecules in excess of that in the bulk gas if the latter concentration were maintained up to a surface separating the solid and gas phases. Considering the dividing surface to be at a coordinate z_0 at which ϵ is zero, the total number n of molecules adsorbed is calculated by integrating the excess concentration $(c-c^0)$ from z_0 to infinity and adding the number of molecules in the shaded area of Fig. 10 between z_0 and the ordinate,

$$\mathrm{d}n = \mathrm{d}x\mathrm{d}y \int_{z_0}^{\infty}(c - c^0)\,\mathrm{d}z + c^0\mathrm{d}x\mathrm{d}y \int_{0}^{z_0} \exp\left(-\frac{\epsilon}{kT}\right)\mathrm{d}z \tag{26}$$

The last term in equation (26) is, however, nearly always negligible in comparison with the first term and conveniently may be neglected. Hence we may write

$$\mathrm{d}n = \mathrm{d}x\mathrm{d}y \int_{z_0}^{\infty}(c - c^0)\,\mathrm{d}z \tag{27}$$

Substituting equation (25) into equation (27) we have, on integrating over the plane xy,

$$n = Sc^0 \int_{z_0}^{\infty} \left[\exp\left(-\frac{\epsilon}{kT}\right) - 1 \right]\mathrm{d}z \tag{28}$$

In terms of the pressure p of the bulk gas, assumed ideal, this becomes,

$$n = Sp \left\{ \frac{1}{RT} \int_{z_0}^{\infty} \left[\exp\left(-\frac{\epsilon}{kT}\right) - 1 \right]\mathrm{d}z \right. \tag{29}$$

Equation (29) is in the form of Henry's law, the amount adsorbed at a given temperature being directly proportional to the pressure. Provided that ϵ is known as a function of z, S can be calculated from equation (29) once the experimental quantity n/p has been found.

Whether or not the surface is plane, the same general argument applies. For a non-planar surface, both terms in equation (26) are replaced by triple integrals and ϵ should then be known for all points over the entire surface. For a heterogeneous surface, ϵ will be a function of x and y as well as z.

Steele and Halsey [154] assumed a hard-sphere interaction for ϵ, but an alternative is to assume a Lennard-Jones type potential [166] (see p. 18). The resulting values for the surface area are very dependent on the form of the potential equation used and the method of calculation.

A more refined treatment considers the adsorbed layer to be confined to a region close to the surface, and to a first approximation is regarded as a two-dimensional gas moving in a plane at a distance z^* from the solid surface. The equation of state for this two-dimensional gas is written

$$\Pi = RTc\,(1 + B^*c\,\ldots) \tag{30}$$

where Π is the surface pressure, c is the surface concentration n/S, and B^* is a two-dimensional second virial coefficient. To obtain an isotherm corresponding to equation (30) the Gibbs equation, equation (11), is written in the form

$$\left(\frac{\partial \ln p^*}{\partial \Pi}\right)_{c,\,T} = \frac{1}{RTc} \tag{31}$$

where p^* is the fugacity of the bulk gas. The isotherm is obtained by differentiating equation (30) and then substituting in equation (31). Assuming that the equation of state for the bulk gas is

$$v = \frac{RT}{p}\,(1 + Bp\,\ldots) \tag{32}$$

where B is the second virial coefficient for the bulk gas, and utilizing the exact thermodynamic relation

$$\int d\ln\frac{p}{p^*} = \int\left(\frac{V}{RT} - \frac{1}{p}\right)dp \tag{33}$$

the fugacity p^* may be expressed in terms of the bulk gas pressure. The resulting p, c isotherm is

$$\ln p + Bp = \ln c + 2B^*c + C_I \tag{34}$$

where C_I is an integration constant whose value may be arranged such that Henry's law is obeyed at small values of c. Under these circumstances equation (34) may be written in the form,

$$\ln c = Bp - 2B^*c + \ln k_H p \tag{35}$$

where k_H is the Henry's law constant of equation (29)

$$k_H = \frac{1}{RT}\int_{z_0}^{\infty}\left[\exp\left(-\frac{\epsilon}{kT}\right) - 1\right]dz \tag{36}$$

When $c\,(=n/S)$ is small, equation (35) may be written in the approximate form

$$\frac{n}{p} = k_H S\,(1 - 2B^*k_H p + Bp) \tag{37}$$

so that a plot of n/p against p would have an intercept

$$k_1 = k_H S \tag{38}$$

and a slope

$$k_2 = - k_1 \left(\frac{2B^* k_1}{S} - B \right) \tag{39}$$

whence

$$\frac{S}{B^*} = - \left(\frac{2k_1^2}{k_2 - k_1 B} \right) \tag{40}$$

The surface area may then be calculated provided an estimate for B^* can be obtained. Barker and Everett [165] calculated B^* by assuming that the forces between adsorbed molecules are the same as those between gas molecules and that the adsorbate moves only in a two-dimensional plane. In terms of the distance of approach r_0, at which the potential energy ϵ of two adsorbed atoms is zero, they showed that

$$B^* = n \pi r_0^2 f \left(\frac{kT}{\epsilon_{min}} \right) \tag{41}$$

ϵ_{min} being the minimum potential energy the value of which can be deduced from the virial coefficient of the bulk gas [167]. The function ϵ_{min} in equation (41) may be expressed in terms of the Γ-function. It may also be expressed by a simpler interpolation formula [165]. Corrections for deviations in the behaviour of the adsorbed phase from a two-dimensional gas and for inter-actions between bulk gas and adsorbed gas may be applied, but both correc-tion terms are small in comparison with the value of B^*. Perturbation of the intermolecular potential caused by the presence of the solid surface may also be taken into account, but will reduce the calculated surface areas by less than 8 %. Barker and Everett [165] also considered the formal statistical mechanics of adsorption and, by an entirely independent method, arrived at a formula which differed from that of equation (40) only by the second term in the de-nominator. Since the second virial coefficient for the bulk gas is in many cases quite negligible, the two approaches yield similar results. This in itself may indicate that the method may have a more reliable fundamental basis than the BET method, but extensive and precise experiments are still required to assess fully its reliability. For a homogeneous carbon black, the calculated areas agree quite well with those of Halsey *et al.* and also with areas estimated by the BET equation. However, large discrepancies are noted for heterogene-ous carbon surfaces. These discrepancies need not detract from the funda-mental, though laborious, nature of the method. It may mean that either the BET theory or the theory of Halsey and of Barker and Everett, or both theories, may be inapplicable to heterogeneous surfaces.

3.2.6. Less-commonly used methods, based on physical adsorption, of measuring surface areas

We shall briefly consider three methods: the use of radioactive isotopes for the direct measurement of the amount adsorbed; the application of a composite adsorption equation which can be extracted from several independent approaches to the theory of physical adsorption; and an elegant method based on measurement of work-function changes occurring during adsorption.

A moment's reflection will show that, by using a gas consisting of a radioactive isotope as an adsorbate, pressure measurements would be refined in view of the sensitivity of present-day counting equipment. Consequently, difficulties arising during the measurement of small surface areas will be alleviated. Aylmore and Jepson [168], realizing this, simplified the conventional BET method using ordinary krypton gas, by counting the β-radiation of ^{85}Kr-traced krypton in a thermostatic constant-volume cell to determine the pressure. This method, although satisfactory, still suffers from the disadvantage (which becomes acute with very small surface areas) that adsorbed amounts have to be computed by subtracting the quantity of gas present at equilibrium from the quantity initially present.

In the recent technique devised by Clarke [169] for measuring surface areas, the amount of krypton adsorbed by the solid is measured directly by counting the γ-radiation from the ^{85}Kr-traced krypton. This simplifies the calculations, renders dead-space volumes of no consequence (just as with the gravimetric method), increases the rapidity of the determination and, most important of all, enables very small specific areas (in the region of 10 cm^{2}g^{-1}) to be determined. In both this procedure and that of Aylmore and Jepson, the BET equation is the medium for arriving finally at monolayer capacity and, hence, at the surface area.

Slabaugh and Stump [170] have recently shown that, by taking Pierce's equation [171, 172] (equation (42)), which was synthesized from the independent work of Frenkel [173], Halsey [174] and Hill [175], it is possible to obtain sensible estimates of surface areas of a variety of solids from gas-adsorption data. The Frenkel-Halsey-Hill equation may be written as:

$$\theta^s = \frac{K}{\ln\left(\dfrac{p_0}{p}\right)} = \left(\frac{v}{v_m}\right)^s \tag{42}$$

or

$$\ln\left(\frac{p}{p_0}\right) = \frac{-\alpha}{RT\theta^s}$$

where α, s and K are constants related to the properties of the adsorbate and adsorbent. Fortuitously [170], it transpires that, for nitrogen as adsorbate, the values of K for several solids are almost constant. It so happens that s is usually very close to 3. (In fact, Hill [176] has given sound theoretical reasons for believing that s ought to be 3.) Hence, the value of v_m can be estimated at any single value of p/p_0, provided s and K are known for a given adsorbate-adsorbent pair.

In 1962, Pritchard [131] introduced a simple and effective method of measuring the surface area of clean evaporated metal films from surface potential measurements (see Section 3.3.5). The method is particularly effective when the film surface areas are small (in the region of about twice the geometric area of the substratum), and it avoids the necessity of precise temperature and pressure measurements and volume calibration. Moreover, it appears to be very reliable.

The surface potentials set up when some adsorbed layers are formed on a solid arise from the unsymmetrical arrangement of electrons at the surface. If the electrons are unsymmetrically arranged at the surface we would expect the adsorbed species to possess a dipole moment: this is indeed so, and there is a direct relation between the surface potential and the dipole moment. It is comparatively easy to measure the surface potential of adsorbed xenon by the diode method (see p. 134) provided that a small tungsten cathode is installed in the centre of the bulb in which the evaporated film is prepared.

Pritchard [131] showed that, as xenon is gradually adsorbed by a clean metal surface (gold, nickel, copper), there is a linear dependence of the surface

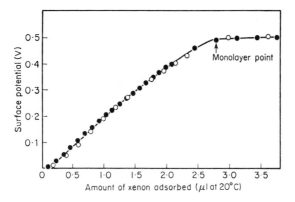

FIG. 11. Adsorption of xenon at $-183°C$ (full circles) and $-196°C$ (open circles) on a gold film deposited at $-183°C$ and sintered at $60°C$ (Pritchard [131]).

potential on coverage up to a point beyond which the surface potential changed very little, and simultaneously a small, but growing, pressure was detectable in the gas phase (see Fig. 11). Pritchard assumed that this point corresponded with completion of the monolayer, and the succeeding nearly non-polar adsorption with second-layer formation. By comparing the monolayer capacity determined in this way with that obtained using krypton gas and the conventional BET method, Pritchard's assumption was seen to be justified. Moreover, other work substantiated this view. For example, one would expect a rather sudden change in the rate of growth of the surface potential at the completion of the monolayer because the relatively large size of the xenon atoms in the first layer would diminish greatly the polarization, and the binding, of the xenon atoms in the second layer [177–179].

Since the curves for $-196°C$ and $-183°C$ are coincident (see Fig. 11), it follows that liquid air of any composition may be used during the surface potential measurements.

3.2.7. The active surface area as estimated by chemisorption

In general, we would not expect the total surface area of a catalyst to be synonymous with the so-called active surface area or active site area which represents the area at which a certain specified chemisorption, or catalysis, can take place. Experience has shown that the active surface areas range in values from minute fractions of the total area, as estimated by the BET or an equivalent method, to very nearly the total areas themselves.

Very often in catalysis research it is essential to discover what proportion of the total surface area is active in catalysis. One obvious practical advantage of having such information is that an estimate can thus be made of the time required for contaminants in the reactant gases to poison the catalyst surface. Another important item of information, especially in the assessment and selection of supported metal catalysts, is the fraction of the total area occupied by the support material or by the metal itself. Although physical techniques such as electron diffraction [180] and ion scattering [181] may be used to some extent in assessing surface composition—and there are hopes that these and other physical techniques may become even more applicable in future (see later)—investigation of the chemisorption properties of the catalysts under study have proved remarkably successful in analysing surface composition. The work of Brunauer and Emmett [182] on iron catalysts for ammonia synthesis, of Hall et al. [183] on Fischer-Tropsch catalysts (see Section 3.3.7.1), of Bridges et al. [184] on chromia–alumina catalysts used in the petroleum industry, and of Scholten and van Montfoort [185] on supported palladium catalysts, illustrate well the power and scope of this technique.

Bridges et al. [184] analysed the surface composition of a series of chromia–alumina catalysts by investigating the relative aptitudes of these catalysts to chemisorb oxygen and carbon monoxide at low temperatures. Their results showed that the amount of oxygen chemisorbed at $-195°C$ provided a reasonable measure of the chromia area present on the mixed-oxide surface. In line with some earlier work carried out by Eischens and Selwood [186]

using magnetic susceptibility measurements (see Section 3.3.4), Bridges *et al.*
[184] concluded that, at low chromia concentrations (<2% of Cr in the
mixed oxide), a two-dimensional dispersion of surface chromium ions, in a
4+ or 5+ oxidation state, existed on the chromia–alumina catalyst; and that
the percentage of the total surface covered with chromia (and, therefore, the
percentage of the surface accessible to oxygen chemisorption at low temperature) increased smoothly from zero to 8% as the weight percentage of
chromium in the catalyst increased to 2%.

MacIver and his co-workers [187, 188] have made excellent use of this
technique in the study of silica–alumina catalysts. They have made progress
in the elucidation of the mechanism of catalytic cracking (see Section 8.5), by
investigating the chemisorption of 1-butene and 1-octene on a range of
silica–alumina catalysts. It was, for example, established that 1-butene was
strongly chemisorbed on a small proportion of the catalyst at temperatures
as low as −78°C and that this adsorption was very sensitive to pretreatment of
the catalyst. They have also examined by the chemisorptive technique the
differences between the surfaces of η- and γ-alumina, and have assessed the
evidence that has led alumina to be regarded as an acid-type catalyst [189].

Scholten and van Montfoort [185], in their search for a reliable method of
separating the free palladium surface area from the total surface area of the
palladium-on-alumina, utilized the principles worked out earlier by Emmett
[182, 190] and Schuit and van Reijen [191]. The total surface area Scholten
and van Montfoort found by the conventional BET method using methane
at −196°C as the adsorbate. They could not use hydrogen chemisorption to
measure the metallic part of the area, as Schuit and van Reijen [191] had done
successfully with nickel supported on silica, since this gas would be readily
dissolved in the palladium. They found that the extent of chemisorption of
carbon monoxide at room temperature served as a reliable criterion for
determining the free palladium area.

For evaporated metal films, unlike supported metal catalysts, it often
transpires that the active site area is very nearly equal to, if not exactly the
same as, the total surface area. Beeck and Ritchie [192] measured the BET
monolayer capacity of evaporated nickel films using three different gases as
adsorbates, krypton, methane and n-butane; they also measured the volume
of hydrogen chemisorbed at −196°C and 0·1 torr pressure on the nickel film.
By making reasonable assumptions concerning the effective cross-sectional
area of each of these gas molecules in the adsorbed state, they concluded that
all the metal atoms at the film surface were joined to hydrogen atoms in a 1:1
ratio. Likewise, Trapnell [193, 194] and Lanyon and Trapnell [195] have
shown that when oxygen is chemisorbed on tungsten, rhodium or molybdenum
films, at liquid-air temperatures and at a very fast rate, a 1:1 ratio of oxygen:
surface metal atoms obtains. Furthermore, Beeck [196] produced cogent

evidence which related, directly, the following four properties of evaporated nickel films: (i), the BET area of the film (using krypton at $-196°C$); (ii) the amount of fast hydrogen chemisorption on the film at $-196°C$; (iii) the amount of carbon monoxide chemisorbed at $23°C$; and (iv) the catalytic activity of the evaporated nickel film in the hydrogenation of ethylene.

3.2.7.1. *Determination of Active Surfaces Using Radioactive Isotopes*

It is interesting to recall that one of the first applications of radioisotopes to chemical problems was in the determination of the surface area of a solid by isotopic exchange. Paneth [197] showed that by allowing a precipitate of lead sulphate to reach exchange equilibrium with a saturated solution of lead sulphate containing thorium B, it was possible to evaluate the surface area of the solid sulphate. Since this early application, much use has been made of radioisotopes in surface chemistry (see Section 3.3.7.1), and a few direct attempts have been made to analyse the chemical composition of catalyst surfaces by utilizing the exchange, or the adsorption, of radioactively labelled species from the vapour and liquid phases. Three examples of such analyses will be outlined here: two of these involve exchange between the solid and liquid phases.

Many samples of alumina catalysts are known to contain *inter alia* a certain amount of adsorbed chloride ions, which are introduced, in all probability, during the production of the alumina by precipitation from aluminium chloride solution. If it is assumed that the entire alumina surface is covered by an adsorbed layer of chloride ions, it is possible to evaluate this area in a manner analogous to that used by Paneth. Upon allowing a ^{36}Cl-labelled solution of a chloride to exchange with the surface layer on the precipitated alumina we would have, at exchange equilibrium,

$$\frac{c_0 - c}{x} = \frac{c}{y} \tag{43}$$

where c_0 and c are, respectively, the original and equilibrium count rates of the solution, y is the total chloride content of the solution, and x is the estimated chloride content of the precipitate which is available for exchange. Since y is known beforehand, and c_0 and c are measurable, x can be found. The surface area occupied by the adsorbed chloride can be estimated if we know the effective cross-sectional area of this ion. Chisnall *et al.* [198], who have made this estimate, took the area of each ion to be $11·4$ Å^2 and thus estimated the area of the alumina. They compared the area estimated in this way with that obtained from the BET theory using benzene as adsorbate. The agreement was very good, thus confirming that the entire alumina surface is covered with chloride ions.

The isotopic exchange method for measuring the surface areas of supported transition metal sulphides used by Lukens *et al.* [*199*], is similar, in principle, to that adopted by Chisnall *et al.* [*198*]. Lukens *et al.* [*199*] arrived at the total number N_s of normal sulphur atoms (^{32}S) in the surface of the transition metal sulphide that could undergo isotopic exchange with radioactive sulphur atoms (^{35}S) in H_2S (labelled with ^{35}S) dissolved in toluene. By carrying out calibration experiments in which both the number N_s and the BET area (using krypton and nitrogen) of unsupported mixed sulphides of nickel, molybdenum and tungsten were determined, it was possible to assign an effective cross-sectional area for each exchangeable surface sulphur atom. (It is to be noted that this is the reverse of what Chisnall *et al.* [*198*] did.) The technique could then be used to assay the sulphide surface area of other supported transition metal sulphides. Table 3.1 illustrates the usefulness

TABLE 3.1 [*199*]

% metal sulphide on the Al_2O_3	Total BET area (m^2g^{-1})	"Sulphide-ion" area $(m^2g^{-1}$ of catalyst)
20	290	35
34	250	27
100	23	23

of the technique. Lukens *et al.* [*199*] examined three catalysts composed of mixtures of NiS, MoS_2 and WS_2 supported on alumina. The catalysts each displayed approximately equal catalytic activity for the hydrogenation of α-methylnaphthalene. But they each had widely varying amounts of total metal sulphide on the support and their surface areas differed from one another. However, the isotopic exchange method revealed that all three catalysts had comparable sulphide ion surface area per gram of catalyst, a fact which gives some indication as to why the catalytic activity was so uniform.

The third example of the use of radioisotopes in analysing surface composition is taken from the work of MacIver, Emmett and Frank [*200, 201*]. For many years there appeared to be an exception to the widely accepted notion that at least one of the reactants in a catalysed reaction must be chemisorbed on the surface of the catalyst. This exception seemed to apply to isobutane on a silica–alumina cracking catalyst, because conventional (volumetric) techniques of recording direct adsorption failed to reveal evidence of any chemisorption of the isobutane at temperatures known to be effective catalytically. By using [2-^{14}C] isobutane instead of the ordinary alkane, MacIver, Emmett and Frank [*200*] succeeded in showing that the chemisorption of the butane at 150°C corresponded to about 0·003 ml at n.t.p. per gram of catalyst, this being equivalent to a coverage of less than a fraction 1×10^{-4} of the surface. The fact that a few hundredths of a millimole of a poison such as quinoline suffices [*202*] to eliminate almost all of the catalytic cracking activity of the silica–alumina catalysts, is clearly consonant with the observation that such a small fraction of the catalyst surface participates in the reaction.

3.2.8. Information from heats of adsorption

It is obvious that the strength of an adsorption bond is reflected in the magnitude of the heat of adsorption: the higher the heat of adsorption (that is, the greater the amount of heat liberated per mole adsorbed) the stronger is the bond. Knowledge of the magnitude of the heat of adsorption of the reactants, products and of the intermediates of a heterogeneously catalysed reaction elucidates the nature of the catalysis. If Q $(=-\Delta H)$ is very large for a given species, it is clear that catalysis will be hindered precisely because that species is too tenaciously held at the surface. If, on the other hand, Q is very low, the species in question may not be adsorbed on the catalyst surface for a time long enough to facilitate reaction (see Section 2.4.3). Successful catalysis evidently requires a situation in which the heat of adsorption is not too large to cause desorption difficulties and not too small to hamper rearrangement in the adsorbed phase. It is therefore important to know the magnitude of the heat of adsorption.

Experimentally, the heat of adsorption at a given surface coverage (see Chapter 2) may be determined either by direct calorimetric measurement or, indirectly, by applying the Clausius-Clapeyron equation to a series of adsorption isotherms. We shall sketch out the principles on which these two methods are based, but omit the finer aspects of the necessary experimental technique. For completeness it ought to be mentioned that gas–solid chromatography has recently proved very useful in the determination of heats of adsorption, particularly when the interaction is physical. The various procedures employed [203–208], which will not be described here (but see Section 3.3.10), are all ultimately related to the slope of the initial linear portion of the adsorption isotherm of the gas on the solid used in the chromatographic column. We may therefore regard the gas-chromatographic method as a special form of the isosteric (or Clausius-Clapeyron) method, which is described below.

In the calorimetric method the apparatus requirements are dictated largely by the state of subdivision of the solid. Two extreme types of calorimeter are currently employed. One is based on the familiar designs of Garner [209] and of Beebe [210], and is suitable for use with powdered specimens. The other rests heavily on the design features first introduced by Beeck [196]. This type of calorimeter is suitable for adsorption studies on evaporated films. A diagram of some of the salient features of a calorimeter used recently by Stone and his co-workers [211] in their studies on granulated catalysts is shown in Fig. 12. The adsorption calorimeter consisted essentially of two concentric glass tubes joined at the bottom by a Dewar seal and mounted inside a jacket

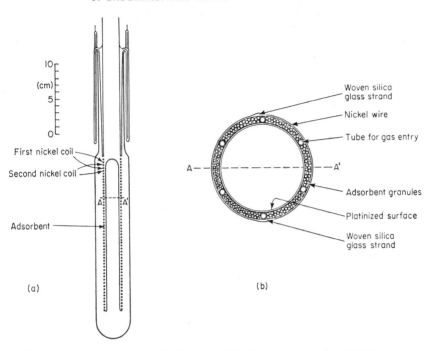

FIG. 12. (a) Diagram of the calorimeter used by Stone and co-workers [211] for measuring heats of adsorption on granulated catalysts. (b) Enlarged cross-section of central portion AA′.

which could be evacuated to 10^{-6} torr. The concentric glass tubes were thin-walled; and the annular space between the tubes was about 2 mm wide. On the outside wall of the outer tube were interwound fine-gauge nickel wire coils of a high temperature-coefficient. One of these coils served as a resistance thermometer, the other was used in the determination of the water equivalent. The whole assembly was mounted vertically and was totally immersed in water in a large Dewar vessel. The procedure of measurement is much the same as in any other calorimetric experiment [212–214]. The calorimeter could be outgassed with impunity at 500°C and was sufficiently sensitive to measure heat liberations down in the range of a few millicalories.

The Beeck type of calorimeter has been used both by Brennan *et al.* [213, 214] and by Stone [212] for measuring the heats of adsorption of oxygen and hydrogen on a variety of evaporated metal films. The calorimeter consisted of a very thin-walled glass tube (wound with platinum wire which acted as a resistance thermometer). The calorimeter tube was mounted vertically within an outer jacket through which a rapid stream of cold water could be passed when a film was being evaporated on to the inner wall of the calorimeter tube, and which could be evacuated to avoid collapse during baking and to reduce heat losses during a heat measurement. Great care must be exercised in fashioning the calorimeter tube. Unless the Pyrex tubing, which is thinned down by etching with hydrofluoric acid, has a uniform wall thickness, results tend to be very unreliable.

Both the adsorption calorimeters discussed above are used under adiabatic conditions. However, it must not be thought that the heat of adsorption so obtained corresponds to the "reversible" heat of adsorption. For most chemisorptions, and especially for low-temperature chemisorptions, the equilibrium pressures are small, so that reversible compression, during adsorption, is difficult to achieve by standard techniques. It is, in fact, customary to introduce the adsorbate gas irreversibly. Ehrlich has recently shown [215], with the aid of a thermodynamic argument, that the difference between the reversible and irreversible heats of adsorption measured calorimetrically is likely to be within the limit of the expected experimental error in the determination of the heat itself. (The difference is of the order of $2RT$.)

When the Clausius-Clapeyron equation is applied at a particular fractional coverage θ, or at a particular adsorbed volume v ($\theta = v/v_m$), the isosteric heat of adsorption Q_{iso} is obtained from:

$$Q_{iso} = RT^2 \left(\frac{\partial \ln p}{\partial T}\right)_\theta \tag{44}$$

The underlying assumption in the use of the Clausius-Clapeyron equation in the present context is that the partial molar volume of the adsorbate gas far exceeds that of the adsorbate in the adsorbed state, and that the perfect gas approximation is justified. The irreversible heat of adsorption Q_{irr}, determined from adsorption calorimetry, is related to the isosteric heat as follows:

$$Q_{iso} = Q_{irr} + RT \tag{45}$$

The difference between the two heats is again too small to warrant distinction most of the time.

Figure 13 shows, schematically, adsorption isotherms (which must be proved to be reversible in the thermodynamic sense) drawn at two temperatures T_1 and T_2 ($T_1 > T_2$). If Q'_{iso} represents the isosteric heat of adsorption at a fractional coverage $\theta = \theta'$, then clearly, from equation (44)

$$\ln\left(\frac{p_1}{p_2}\right) = \frac{Q'_{iso}}{R}\left(\frac{1}{T_2} - \frac{1}{T_1}\right) \tag{46}$$

The dependence of the isosteric heat on coverage could be readily evaluated by drawing a series of horizontal lines across Fig. 13, reading off corresponding values of pressures at the respective temperatures, and substituting in the last equation. And a reliable estimate of the isosteric heat would be obtained from a series of adsorption isosteres at several temperatures.

It is interesting to interject at this point that we should expect the heat of adsorption for a particular adsorbate–adsorbent system to vary with temperature, but not because, as discussed in Chapter 2, the actual type of adsorption bond alters with temperature (cf. adsorption isobars of Section 2.2.2). The isosteric heat of adsorption should itself be a function of temperature, because

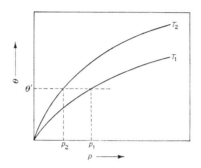

FIG. 13. Schematic diagram of adsorption isotherms at two temperatures T_1 and T_2 where $T_1 > T_2$.

of the difference ΔC_p in the partial molar heat capacities of the adsorbate in the gas phase and the adsorbate in the adsorbed state at constant pressure and composition. Kirchhoff's law gives:

$$Q_{\text{iso}} = Q_0 + \int_0^T \Delta C_p dT \qquad (47)$$

where Q_0 is the heat of adsorption at absolute zero.

That the variation of Q_{iso} with increasing temperature may turn out to be complex can be judged from the following qualitative arguments. Suppose a diatomic molecule is dissociatively adsorbed on a uniform surface. In the gas phase, each molecule has three degrees of translational freedom plus two degrees of rotational freedom, so that $C_p \sim 7R/2$, plus a small contribution from vibration, which will increase to R at higher temperatures. At low temperatures, we may envisage the two adatoms to be localized and vibrationally unexcited. They have, therefore, no translational, no vibrational and no rotational degrees of freedom. Under these circumstances ΔC_p (in going from the adsorbed state to the gas phase) amounts to $7R/2$. As the temperature is increased, however, the surface vibrations begin to come into play, and the translational motion of the adatoms will steadily increase. At the highest temperatures, each of the adsorbed atoms will have a heat capacity made up of $2R/2$ from translation, and $2R/2$ from vibration, thus making $C_p \sim 6R/2$. Hence, in the extreme case when the diatomic molecule in the gas phase has not, and the adatoms have been vibrationally excited, ΔC_p will amount to

$-5R/2$. It follows, therefore, that for no other reason than the change in heat capacities, a diminution in the isosteric heat of about 5 kcal mole^{-1} can be expected on raising the temperature from room temperature to the region of 1000°K. It does not appear at all certain that this fact, which has been emphasized by Ehrlich [215], has been fully realized in the field of desorption spectrometry, a topic which we shall consider next.

3.2.8.1. *Desorption Spectra or Flash-desorption*

From Fig. 14 of Chapter 2 it can be seen that, provided that the temperature coefficient of the rate of desorption of a species from a surface is measured, a rough estimate of the heat of adsorption of that species can be made. The activation energy of desorption E_d is related to the heat of adsorption Q and the activation energy of adsorption E_a as follows:

$$E_d = Q + E_a \qquad (48)$$

If the adsorption process is non-activated ($E_a = 0$), then the activation energy of desorption equals the heat of adsorption. If the adsorption is activated, however, the activation energy of desorption sets an upper limit to the magnitude of the heat of adsorption.

If we imagine a surface which is energetically heterogeneous in the sense that there are two or more distinct types of adsorption site—with different heats of adsorption—we should expect two or more separate desorption "bursts" when the system was rapidly heated. The burst occurring at the lower or lowest temperature would arise from the desorption of species from the adsorption sites characterized by the smallest heat of adsorption. Subse-

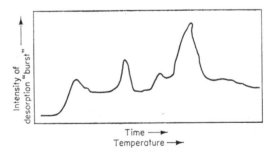

FIG. 14. Illustration of a typical desorption spectrum.

quent bursts would take place at temperatures commensurate with the heats of adsorption characteristic of the respective sites. The last burst, that is the one occurring at the highest temperature, would correspond to the sites characterized by the highest heat of adsorption. Figure 14 illustrates the type of desorption spectrum to be expected. (The height of the peaks would be determined largely by the number of species involved in the corresponding adsorption type, and their resolution would depend on the differences between the heats of adsorption of the various distinct types of adsorption sites—but see later.)

The idea of a desorption spectrometer has been known for well over a decade [216], and the principle of desorption spectrometry is much older. Only very recently has this technique been put to good advantage in the study of heterogeneous catalysis. For most of the time it has been used to confirm the existence of distinct types of bonding when simple molecules such as nitrogen and carbon monoxide are adsorbed on clean metal surfaces [217–222]. For example, Redhead [7, 222] demonstrated that carbon monoxide is chemisorbed in two distinct ways by the {110} faces of tungsten. One type of chemisorption involves the bridge-type of bonding (a), the heat of adsorption of this phase being strongly dependent on the precise structure of the surface

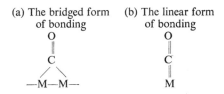

but falling in the range 65–90 kcal mole^{-1}. The other type of bonding which starts to become appreciable when the surface coverage exceeds about one-third, involves the linear form of bonding (b), and takes place at the gaps in the bridged layer. The heat of adsorption in this instance is not structure-sensitive, and is about 28 kcal mole^{-1}.

The application of the desorption-spectrometer, or of flash-desorption as it is sometimes designated, to catalyst studies involved abandoning the customary metal filaments (as adsorbents), the stringent requirements of ultrahigh vacua, and the use of refined ionization gauges and mass spectrometers (to monitor the composition of the desorption "bursts"). Instead, recourse to conditions much more similar to those ordinarily employed in catalytic systems had to be made. Amenomiya and Cvetanovic [223–226] have made significant progress in their study of alkene adsorption on alumina catalysts using the flash-desorption method. We shall consider in some detail how they set out to study the ethylene–alumina system [223].

The amount of ethylene adsorbed on the alumina was first determined in a conventional volumetric apparatus. Following this, the catalyst was evacuated, and then a stream of helium at a fixed flow-rate was passed through the powdered catalyst into a thermal conductivity cell, and finally through a liquid-nitrogen trap. When the catalyst temperature was raised at a uniform rate (by means of a programmed controller) during the flushing with helium, ethylene desorbed from the catalyst and was carried into the conductivity cell by means of which the concentration of the ethylene in the helium could be determined (using a recorder). Since the flow rate of helium was fixed, the deflection of the recorder due to the presence of the ethylene was proportional to the rate of its desorption from the alumina. The rate of desorption would continue to increase with increasing temperature, at first, but eventually it would begin to decrease as a result of depletion of adsorbed gas. A peak would, therefore, be recorded. Amenomiya and Cvetanovic took two flash-desorption chromatograms; the first after the alumina had been evacuated for 10 min at room temperature, and the second, on another sample, after evacuation for 60 min at 100°C. Figure 15(a), (b), (c), (d) summarizes their results. After several repeat determinations, using varying values for β, the speed of raising the temperature (in deg C min^{-1}), it was possible to tabulate sets of values for T_{M_1} and T_{M_2}, the temperatures at which the first and second

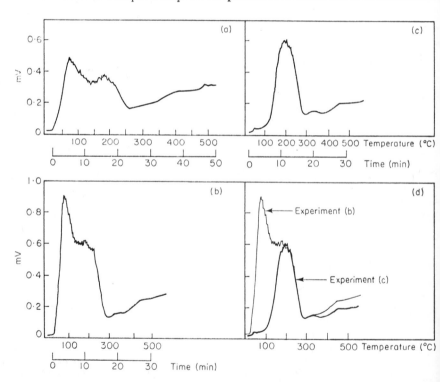

FIG. 15. Flash desorption chromatograms. (a) Evacuated before flashing for 10 min at room temperature with $\beta = 10·08°$ min^{-1}. (b) Evacuated before flashing for 10 min at room temperature with $\beta = 16·03°$ min^{-1}. (c) Evacuated before flashing for 60 min at 100° with $\beta = 15·90°$ min^{-1}. (d) Superposition of desorptograms from (b) and (c). (Amenomiya and Cvetanovic [223].)

peak maxima appeared for differing values of β. Table 3.2 includes a selection of some of the results obtained. We shall now recall the interpretation of Amenomiya and Cvetanovic [223].

TABLE 3.2. Adsorption and flash-desorption of ethylene

Expt no.	Adsorption			Flash-desorption		
	Temp. (°C)	Pressure (mm)	Amt of gas adsd (ml)n.t.p.	β (deg C min^{-1})	T_{M_1} (°C)	Amt of gas desorbed (ml)n.t.p.
22	24·9	6·13	0·218	10·08	74	—
23	24·8	6·34	0·227	20·70	79	—
24	24·7	6·37	0·239	5·05	68	—
26	25·2	5·28	0·273	5·14	72	—
27	26·2	5·63	0·259	21·05	81	—
28	26·5	5·38	0·251	10·40	74	0·091
29	25·8	5·96	0·266	16·03	77	0·095
46	24·2	9·00	0·344	16·05	78	0·105

Alumina evacuated for 10 min at room temperature prior to flash desorption. Flow rate of He $45\cdot7\pm0\cdot3$ ml min^{-1}.

The following equation holds† for a catalyst the temperature T of which is being raised:

$$\frac{\mathrm{d}^2\theta}{\mathrm{d}t^2} = \frac{\partial^2\theta}{\partial t^2} + \left(\frac{\partial^2\theta}{\partial t\,\partial T} + \frac{\partial^2\theta}{\partial T\,\partial t}\right)\frac{\mathrm{d}T}{\mathrm{d}t} + \frac{\partial^2\theta}{\partial T^2}\left(\frac{\mathrm{d}T}{\mathrm{d}t}\right)^2 + \frac{\partial\theta}{\partial T}\frac{\mathrm{d}^2T}{\mathrm{d}t^2} \tag{49}$$

where θ is the fractional coverage of the surface and t is the time. If it is arranged, as was done, for the temperature to be increased linearly with time, so that

$$T = T_0 + \beta t$$

where T and T_0 are, respectively, the temperature at time t and the initial temperature, and if it is further assumed that $\partial\theta/\partial t \gg \partial\theta/\partial T$ throughout the experiment, the above equation (49) simplifies to:

$$\frac{\mathrm{d}^2\theta}{\mathrm{d}t^2} = \frac{\partial^2\theta}{\partial t^2} + \beta\frac{\partial^2\theta}{\partial T\partial t} \tag{50}$$

(In a fast flow of helium, the rate of readsorption will be negligible, so that

† If $\theta = f(t, T)$ is any continuous function of the two variables t and T, then equation (49) is true provided, (i) that the partial derivative $\partial^2 f/\partial t\partial T$ is also a continuous function of these two variables, and (ii) that the term

$$\left(\frac{\partial f}{\partial T}\right)\left(\frac{\mathrm{d}^2T}{\mathrm{d}t dT}\right)\left(\frac{\mathrm{d}T}{\mathrm{d}t}\right)$$

is small in comparison with the other terms in the equation.

$\partial\theta/\partial t \gg \partial\theta/\partial T$ is amply justified.) We can write for the rate of desorption (compare equations (32) and (86) of Chapter 2):

$$r_d = -v_m \frac{d\theta}{dt} = k_0\theta \exp - \frac{E_d}{RT} \tag{51}$$

where k_0 is a desorption constant, E_d, as explained above (equation (48)), is the activation energy of desorption, and v_m is the monolayer capacity (when $\theta=1$). At $T=T_M$, we have $d^2\theta/dt^2=0$, so that the last two equations yield:

$$\frac{E_d \beta v_m}{RT_M^2 k_0} = \exp - \frac{E_d}{RT_M} \tag{52}$$

which can be expressed as:

$$2\log_{10}T_M - \log_{10}\beta = \frac{E_d}{2.30_3 RT_M} + \log_{10}\frac{E_d v_m}{Rk_0} \tag{53}$$

A plot of the l.h.s. of the last equation against $1/T_M$ should be linear if the interpretation is valid, and from this line both E_d and k_0/v_m are obtainable. Figure 16 shows that equation (53) is validated, and the values of E_d and k_0/v_m derived from this plot turned out to be 26·8 kcal mole^{-1} and $1\cdot6\times10^{15}$ sec^{-1}, respectively.

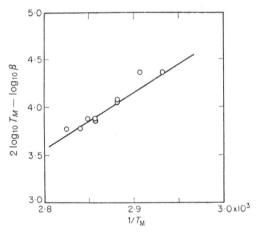

Fig. 16. Experimental verification of equation (53). (Amenomiya and Cvetanovic [223].)

26·8 kcal is therefore an upper limit for the heat of adsorption which gives rise to the first peak (T_{M_1}). But what of the pre-exponential factor? If we assume that the ethylene is non-dissociatively adsorbed on alumina, and if the adsorption is localized, the absolute rate theory gives (see equation (86), Chapter 2)

$$r_d = c_{\text{ads}} \left(\frac{kT}{h}\right) \frac{f_{\pm}}{f_{\text{ads}}} \exp \left(- \frac{E_d}{RT}\right) \qquad (54)$$

where c_{ads} is the concentration of the adsorbed ethylene, and f_{ads} and $f_{\pm}$ are the partition functions of the adsorbed species and the transition complex (with the term $kT/h\nu$ factorized out), respectively. If we now assumed that the freedom of both the activated complex and the adsorbed molecule is so small that $f_{\pm} \approx f_{\text{ads}} \sim 1$, we have

$$r_d = c_{\text{ads}} \frac{kT}{h} \exp \left(- \frac{E_d}{RT}\right) \qquad (55)$$

where kT/h now becomes synonymous with k_0/v_m in equation (51). But k_0/v_m has already been found to be $1 \cdot 6 \times 10^1$ sec^{-1}[5]. The fact that the calculated value of kT/h at $350°K$ is $0 \cdot 73 \times 10^{13}$ sec^{-1} and is in reasonable agreement with the experimental quantity is encouraging, and the slight disparity indicates that, in all probability, the freedom of the activated complex is greater than that of the adsorbed ethylene (thus making $f_{\pm}/f_{\text{ads}}$ greater than unity).

For the second peak, a tabulation of T_{M_s} values as a function of β was obtained, from which the respective E_d and k_0/v_m values were again calculable. They turned out to be $36 \cdot 4$ kcal mole^{-1} and $1 \cdot 6 \times 10^{15}$ sec^{-1}, respectively.

The existence of two different sites for the adsorption of ethylene on alumina was thus established by Amenomiya and Cvetanovic. From the total quantity of gas desorbed it was found that these occupy $2 \cdot 8\%$ of the total surface, 60% of which belong to sites having an upper limit to the heat of adsorption at $26 \cdot 8$ kcal mole^{-1}, and 40% of which belong to sites having an upper limit of $36 \cdot 4$ kcal mole^{-1}. The same technique has been used to demonstrate that there are two distinct types of site for the adsorption of *trans*-but-2-ene on alumina [224]. The total active sites occupy $2 \cdot 9\%$ of the total surface, 62% of which belong to sites I, and 38% to sites II, in very close agreement with the distribution of ethylene on alumina. The activation energies of butene desorption were found to be $12 \cdot 1$ and $16 \cdot 2$ kcal mole^{-1} for sites I and II, respectively. Almost exactly the same pattern of behaviour is displayed by propylene on alumina as by ethylene on alumina [225]. But, with ammonia on alumina, it has been found [226] that the activation energy of desorption of ammonia increases from 7 to 18 kcal mole^{-1} as the surface coverage decreases from 29 to $1 \cdot 5\%$.

We shall now consider other types of information obtainable from measurements of ΔH.

3.2.8.2. *Absolute Magnitude of the Heat of Adsorption*

It is possible to draw quite firm conclusions about the nature of the adsorbed link, and its relevance to catalysis, from the absolute magnitude of the heat of adsorption. A few examples illustrating this statement will now be cited.

Gale *et al.* [*211*], in their calorimetric study of the adsorption of isopropanol and acetone on a nickel oxide catalyst at room temperatures, found that the heat of adsorption of isopropanol on an outgassed surface was 16 kcal mole^{-1}, and that of acetone, also on an outgassed surface, was 11 kcal mole^{-1}. Straightaway, the low heats of adsorption imply, but do not necessarily confirm [*79*], that dissociative chemisorptions such as those represented (for isopropanol) by (a), (b) and (c) below can be discounted. This

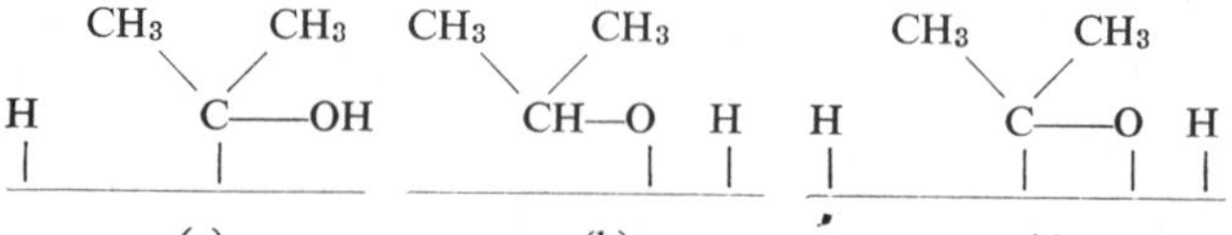

$$
\begin{array}{ccccccc}
CH_3 & CH_3 & CH_3 & CH_3 & & CH_3 & CH_3 \\
\diagdown & \diagup & \diagdown & \diagup & & \diagdown & \diagup \\
H \quad C\!-\!\!-OH & & CH\!-\!\!O \quad H & H & H & C\!-\!\!-O \quad H \\
\end{array}
$$

(a)   (b)   (c)

conclusion was supported by the experimental fact that the heat of adsorption was essentially independent of coverage, a state of affairs most exceptional for dissociative chemisorptions (see Section 2.3.5). The heat of adsorption of isopropanol is 5·5 kcal mole^{-1} higher than the heat of liquefaction of the adsorbate (10·5 kcal mole^{-1}), a difference which is rather more than we would expect if the first layer of isopropanol on the nickel oxide were held by ordinary dispersion forces. There is, however, little doubt that the true reason for the discrepancy is the considerable influence that hydrogen bonding has in the adsorption process (d).

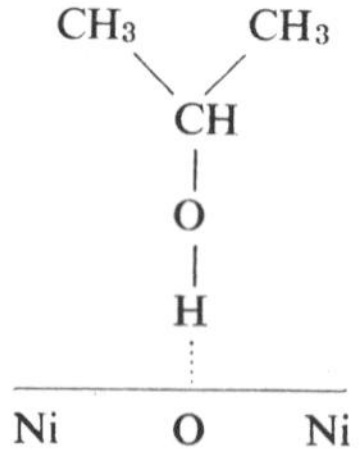

(d)

The low value of the heat of adsorption of acetone and its essential constancy with coverage also indicate that there is no chemisorption involving, say, the opening of the **ketonic** bond (e). It is interesting to note that the

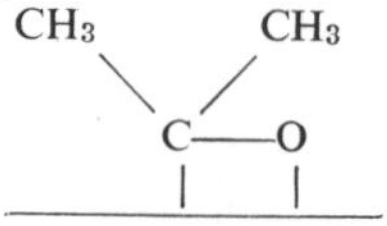

(e)

difference between the heat of adsorption and the heat of liquefaction is now only 3·7 kcal, a fact which is understandable because no hydrogen bonding can be involved in the physical adsorption of acetone.

These two examples of the use of the heat of adsorption to decide upon the actual type of adsorption recall some of the pragmatic approaches outlined when we first discussed the basic differences between physical adsorption and chemisorption (Section 2.1.1). Using an approach similar in many respects to that adopted by Gale *et al.* [*211*], Hsieh [*227, 228*] showed that, when ammonia is adsorbed by silica–alumina cracking catalysts, protons at the catalyst surface combine with the ammonia to form NH_4^+ ions in the adsorbed phase; that is, salt formation takes place when the basic ammonia reacts with the acid sites [*229–231*] on the catalyst surface. Hsieh [*227*] also concludes from the shape of ΔH versus θ plots that Lewis acid sites (which accept electrons) combine with the ammonia, and that surface hydroxyl groups form hydrogen bonds with the ammonia.

3.2.8.3. *The Value of Heats of Adsorption in Elucidating the Nature of Intermediates on a Catalyst Surface*

In this section, we shall recall the extremely elegant work of Stone [*232*], who, from the classical concepts of thermochemistry, with the aid of accurate heats of adsorption, proved the existence of a CO_3 complex on the surfaces of metallic oxides when these substances were used as catalysts for the conversion of carbon monoxide to carbon dioxide (see Section 8.1.1).

The CO_3 complex is envisaged as an entity quite distinct from a carbonate ion which has often been postulated [*233–235*] as an intermediate in the heterogeneously catalysed oxidation of carbon monoxide. The CO_3 complex is thought to be formed by interaction of CO_2 and adsorbed oxygen, whereas Garner's carbonate complex is thought of as being formed by the interaction of CO_2 and oxide ions of the oxide lattice. The first clue to the existence of the CO_3 complex was uncovered when it was found that carbon dioxide could be adsorbed on cuprous oxide (an efficient catalyst for the reaction:

$$CO(g) + \tfrac{1}{2} O_2(g) \rightleftharpoons CO_2(g))$$

only if oxygen had been preadsorbed or if oxygen and carbon dioxide were admitted together. Stone and his co-workers [*236–240*] then set out to measure, for a series of metal oxides, (i) the heat of adsorption of carbon monoxide on an outgassed surface and on a preadsorbed film of oxygen, (ii) the heat of adsorption of oxygen on an outgassed surface, on a preadsorbed film of carbon monoxide, and on a preadsorbed film of carbon dioxide, and (iii) the heat of adsorption of carbon dioxide on an outgassed surface (wherever this was possible) and on a preadsorbed oxygen film. From such quantitative data the heat of formation of the CO_3 complex could be computed. Stone [*232*], in fact, also computed the heat of formation of a postulated CO_2 adsorbed complex to see which of the two heats agreed best with the observed values.

TABLE 3.3. Experimental values of the heats of adsorption of carbon monoxide, carbon dioxide, and oxygen in kcal mole^{-1} on surfaces of copper, nickel and cobalt oxide in different states of pretreatment [232]

Oxide	Adsorption of carbon monoxide		Adsorption of carbon dioxide		Adsorption of oxygen		
	Outgassed surface	Presorbed oxygen	Outgassed surface	Presorbed oxygen	Outgassed surface	Presorbed CO	Presorbed CO$_2$
	a	b	c	d	e	f	g
Cu$_2$O	20	49	—	21	55	100	—
NiO	26	88	28	37	43	100	71
CoO	20	52	22	22	59	95	74

Table 3.3 lists the experimental values of the heats of adsorption of carbon monoxide, carbon dioxide, and oxygen on surfaces of copper, nickel and cobalt oxides, the designations a, b, c, d, e, f and g referring to the respective states of pretreatment. If the heat of combustion of carbon monoxide (67 kcal mole^{-1} to form carbon dioxide) is designated by h, it follows that the heat of adsorption of carbon monoxide on a surface carrying preadsorbed oxygen is $h-(e/2)+d$ on the basis of formation of the CO$_3$ complex, and $h-(e/2)+c$ on the basis of the CO$_2$ complex. Likewise, the heat of adsorption of oxygen on a surface carrying preadsorbed carbon monoxide is $h-a+d+(e/2)$ on the basis of the formation of the CO$_3$ complex, and $2(h-a+c)$ on the basis of the CO$_2$ complex. Table 3.4 lists the various calculated and experimental heats of adsorption for the models of CO$_2$(ads) and CO$_3$(ads). The evidence is distinctly in favour of the CO$_3$ complex. Furthermore, as shown in Table 3.5, the four

TABLE 3.4. Comparison of calculated and experimental heats of adsorption for the models of CO$_2$(ads) and CO$_3$(ads) formation [232]†

Oxide	Heat of adsorption of CO on a surface carrying presorbed oxygen (kcal mole^{-1})			Heat of adsorption of oxygen on a surface carrying presorbed carbon monoxide (kcal mole^{-1})		
	Calculated		Observed	Calculated		Observed
	On the basis of formation of CO$_2$(ads)	On the basis of formation of CO$_3$(ads)		On the basis of formation of CO$_2$(ads)	On the basis of formation of CO$_3$(ads)	
	$h-\dfrac{e}{2}+c$	$h-\dfrac{e}{2}+d$	b	$2(h-a+c)$	$h-a+d+\dfrac{e}{2}$	f
Cu$_2$O	59‡	60	49	134	96	100
NiO	73	82	88	138	100	100
CoO	59	59	52	138	99	95

† h, the heat of the reaction CO(g) + $\frac{1}{2}$O$_2$(g) = CO$_2$(g) is taken as 67 kcal mole^{-1}.
‡ Using an assumed value of 20 kcal mole^{-1} for c.

TABLE 3.5. Heats of formation of the CO_3 complex (kcal) $CO(g) + O_2(g) = CO_{3(ads)}$ [232]

Oxide	$a + f$	$e + b$	$\frac{e}{2} + h + d$	$h + c + \frac{g}{2}$	Mean
Cu_2O	120	104	116	—	113
NiO	126	131	126	131	129
CoO	115	111	119	126	118

independent ways available for the computation of the heat of formation of the CO_3 complex from gaseous carbon monoxide and gaseous oxygen each lead to entirely consistent and sensible results.

Stone [232] has demonstrated the relevance of data such as those presented in this section in discussing the mechanism of the sustained catalysis of carbon monoxide oxidation on metal oxides.

3.2.8.4. A General Rule for the Heat of Adsorption of Simple Molecules on Metals

An empirical relationship is better than no relationship at all; and the one that we shall adumbrate here is that which—for reasons yet to be fully understood—connects the heat of adsorption and the heat of formation of the corresponding compound.

Brennan, Hayward and Trapnell [213] noted that there exists a close similarity, in a number of instances, between the heat of adsorption of oxygen on evaporated metal films and the heat of formation of the bulk oxide. They pointed out that this was contrary to expectation because there are a number of factors associated with the surface which might be expected to affect profoundly the heat of formation of a single layer of oxide and render it quite different from the heat for the bulk oxide. Be this as it may, there is a substantial body of evidence [241, 242] which shows that such a close similarity may be more than fortuitous. Such a similarity might have evoked little interest were it not for the fact that evidence has recently come to light which suggests that, in the catalysed decomposition of formic acid, for instance [243, 244], there appears to be a definite correlation between the catalytic activity of the metal and the heat of formation of the metal formate. Roberts [245] has consequently drawn attention to the important part that the heat of adsorption vis-à-vis the heat of formation may play in the understanding of catalysis. Roberts [245] showed that the heat of formation of many metal oxides, nitrides, hydrides and sulphides varied smoothly with the heat of adsorption of the corresponding simple diatomic gases on the clean metal surfaces. More recently, Tanaka and Tamaru [246] have formulated a general rule according to which the initial heats of chemisorption of gases such as oxygen, ethylene, nitrogen, hydrogen and ammonia on various metal surfaces are empirically expressed by the equation:

$$Q = a \{(- \Delta H_f^0) + 37\} + 20 \text{ kcal mole}^{-1}$$

Q is the initial heat of adsorption, $-\Delta H_f^0$ is the heat of formation of the highest oxide per metal atom, and a is a constant dependent on the nature of the gas. Consequently, if a heat of adsorption of a particular gas or a metal surface is known (from which a is computed for that gas), all the values of the heats of chemisorption of the gas on metal surfaces can be estimated.

3.2.9. Information from entropies of adsorption

For both physical adsorption and chemisorption, two extreme models may be distinguished depending upon whether the adsorbed species have very restricted or totally unrestricted mobility. According to the concept of immobile adsorption, site adsorption or localized adsorption—all these terms are synonymous—adsorbed molecules are bound to definite adsorption sites on the surface. In this model, the number of molecules that can be adsorbed in a completely filled monolayer is determined by the number of surface sites, and, in the first instance, by the structure of the solid surface. According to the concept of mobile adsorption (or non-localized adsorption) the adsorbed species are thought to behave as a two-dimensional gas, moving freely over the surface. In this model, the number of molecules that can be adsorbed in a monolayer is determined by the dimensions of the adsorbed molecules themselves. In practice, the behaviour of any adsorbed species may be anywhere between the two extremes of immobile and mobile adsorption. If the energy barriers between adsorption sites are very low and the thermal energy is large, there will be a strong tendency for mobile adsorption to obtain (cf. Section 2.4.3); whereas, if the energy of activation for surface migration (E_m in equation (106) of Chapter 2) is much larger than the thermal energy ($E_m \gg 10RT$), the properties of the system approach those of the model of immobile adsorption.

Since the mobility of an adsorbed species is directly related to the entropy of that species, knowledge of the entropy change accompanying adsorption elucidates the nature of the process. This approach has been pioneered by Kemball [247–249], Everett [250–252] and de Boer [253–257] and their co-workers.

In order to evaluate entropies of adsorption from experimental data, the following procedure is employed [254]. From two adsorption isotherms at temperatures T_1 and T_2 the corresponding values for the equilibrium pressures p_1 and p_2 are selected for a given amount adsorbed. The difference in Gibbs free energy ΔG_1 between the three-dimensional gas standard at temperature

T_1 and the adsorbed species in equilibrium with gas at a pressure p_1 is given by:

$$\Delta G_1 = - RT_1 \ln \frac{p^0}{p_1} \qquad (56)$$

where p^0 is the standard pressure (760 torr). Similarly, ΔG_2 for the same amount adsorbed at temperature T_2 is given by:

$$\Delta G_2 = - RT_2 \ln \frac{p^0}{p_2} \qquad (57)$$

But

$$\Delta G_1 = \Delta H - T_1 \Delta S$$

and $\qquad (58)$

$$\Delta G_2 = \Delta H - T_2 \Delta S$$

where ΔH is the differential heat of adsorption and ΔS is the differential entropy of adsorption, both of which are calculable, at a variety of adsorbed amounts, from a series of equations such as (56), (57) and (58).

The next step is to convert ΔS values corresponding to various amounts adsorbed to the respective ΔS^0 values, ΔS^0 being the difference in differential molar entropy between the three-dimensional gas in its standard state and in its adsorbed standard state. The standard state for the model of immobile adsorption is different from the standard state for that of mobile adsorption. For the immobile adsorption picture, it is convenient [253] to choose the standard adsorbed state as that corresponding to half coverage of the surfaces, that is, at $\theta = \frac{1}{2}$. For mobile adsorption it is convenient [253] to use a standard state which is similar to the standard of the normal three-dimensional gases. Kemball and Rideal [247] defined a standard state by dividing the standard volume of the three-dimensional gas (1 atm pressure) by an arbitrary chosen thickness of the adsorbed film of 6 Å. The resulting area per molecule, designated A_0, is $22 \cdot 53\ T\text{Å}^2$. After noting the difference in standard adsorbed states, we may proceed to calculate the values of ΔS^0. For the model of immobile or site adsorption we have the change in differential molar entropy ΔS_i^0 being given by [254]:

$$- \Delta S_i^0 = - \Delta S - R \ln \frac{\theta}{1 - \theta} \qquad (59)$$

And, for the mobile adsorption, the change ΔS_m^0 is given by:

$$- \Delta S_m^0 = - \Delta S - R \ln \frac{A_0}{A} \qquad (60)$$

where A refers to the area per molecule at the amount adsorbed under consideration. It is clear that, to evaluate ΔS^0 figures, it is necessary to know not only the amount adsorbed, but also the maximum amount which can be adsorbed (so as to yield θ values) and the actual surface area of the solid (so as to yield A values). From previous sections in this chapter, we know how to arrive at values of the monolayer capacity and the surface area, so that the *experimental* values of ΔS_i^0 and ΔS_m^0 can be obtained.

The final step in the method of assessing the mobility of the adsorbate from entropy data is to compare experimental values of ΔS^0 with those calculated theoretically from partition functions (see Section 2.4.1) on the basis of an assumed model. Thus, if a nonlinear molecule is adsorbed in an immobile manner, there will be a loss of three degrees of translational freedom, three degrees of rotational freedom, and a small amount of vibrational freedom—often the vibrational freedom is retained on adsorption. From the partition functions, the entropy changes ΔS_i^0 associated with the loss of translational and rotational freedom can be calculated in terms of the atomic masses and moments of inertia (see Ref. *249* for full details of such calculations). If the theoretical value of ΔS_i^0 agrees very closely with the experimental value derived as explained above, it may be concluded that the adsorbate is immobile. Likewise, if the theoretical and experimental values of ΔS_m^0 agree well, the adsorbate may be regarded as mobile.

Although entropy computations of the type we have just described have yielded much useful information about adsorbate freedom, this approach to the study of mobility has been largely supplanted by more reliable direct techniques, such as those of field-emission microscopy, field-ion microscopy and infrared spectrophotometry. It turns out that the entropy approach is, in the final analysis, insensitive. It cannot reliably be used to distinguish between the possibilities of immobile and mobile layers for certain adsorptions simply because the effective areas of molecules on surfaces, or the surface areas of the solids themselves, are known to an inadequate degree of precision, thus affecting significantly the magnitudes of ΔS_i^0 and ΔS_m^0 extracted from equations such as (59) and (60). This difficulty has been emphasized by Everett [*252*]. Another difficulty with using entropy data is that it is tacitly assumed that the solid does not undergo any entropy change during the adsorption, but that all the entropy change emanates from the species being adsorbed. This objection has been raised previously [*258*], and has been considered briefly in Section 2.2.3.

3.3. The Study of Chemical and Physical Changes Accompanying Adsorption on Catalysts

Quite apart from the utilization of thermodynamic data, such as heats and entropies of adsorption, it is possible to investigate the intermediaries of a catalytic reaction on catalyst surfaces by means of a direct molecular study. Recently, one of the most successful approaches has involved the application with suitable modification of well-known techniques such as infrared spectroscopy. Also, some attempts have been made to use the electronic spectra of adsorbed molecules. Other spectroscopic techniques have been harnessed, notably nuclear magnetic resonance and electron spin resonance. Yet other techniques, which involve following the changes in certain properties of the solid catalyst as adsorption occurs and reaction proceeds, have been useful in investigating the properties of the adsorbed state. Into the latter category falls work-function and electrical-resistance measurements. We shall now summarize the more important aspects of these modern techniques and consider, in particular, the usefulness of the various methods of probing into the mechanisms of catalysis.

3.3.1. The infrared spectra of adsorbed species

The technique of recording the spectra of adsorbed molecules on metal surfaces represents one of the most noteworthy achievements of surface chemistry. Much of the credit for the development of this technique goes to Eischens [259–261] and Terenin [262, 263] and their co-workers. The former, in particular, has studied several important gas–metal systems of great interest in the field of catalysis.

The overwhelming bulk of the work carried out [264] on the infrared spectra of adsorbed species has been accomplished on metal adsorbents prepared in a rather special way. It would appear, at first sight, impracticable to record the adsorption spectra of adsorbed species in view of their inevitably small proportion compared with the adsorbents themselves. However, if the adsorbents have very large areas, it is possible to detect the absorption spectra of species with quite low spectral extinction coefficients. A moment's reflection will reveal that supported metal adsorbents, used so widely in catalysis, are, in general, ideal for the study of the spectra of the adsorbed state because the proportion of adsorbed species is enhanced.

The experimental aspects of the infrared of the adsorbed state differ from those of conventional infrared spectroscopy only in so far as rather special optical cells and sample preparations are required. It is the sample and not the instrument that usually determines the precision of the technique. If adsorption on metals is being studied, it is most convenient to support minute particles (approximately 50–100 Å in diameter) of the metal in question on to small, nonporous silica particles, the diameter of which falls in the range 150–200 Å. (Small particle size is important here, because it reduces the amount of radiation which is lost by scattering.) An alternative procedure is to use porous glass as a support for the metals on an evaporated film [265]. If adsorption is being studied on silica-gel or silica–alumina catalysts, it is possible to use massive samples of these materials [266, 267]. Figure 17 shows the type of arrangement of the catalyst sample within the optical cell used by Griffiths [268]. When powdered catalysts are pressed into discs [268] it is not always necessary to mount the cell vertically.

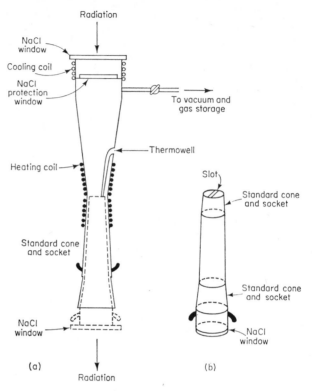

Fig. 17. High-temperature infrared cell used by Griffiths [268].

One of the most helpful aspects of the infrared of the adsorbed state is that we can often conclude, unequivocally, that chemisorption has taken place, this being one of the techniques, mentioned briefly in Section 2.1.1, which give direct experimental guidance as to the distinction between physical adsorption and chemisorption. Observation of a new absorption band in the infrared region of the spectrum can always be taken as an indication of chemisorption. However, the converse is not true. If, during an adsorbent-adsorbate interaction, no new bands are observed, it must not be inferred that chemisorption is absent: for other reasons, which are outside the scope of this chapter, the new bond formed between the adsorbent and adsorbate may not be active in the infrared. While on this topic, it is as well to point out one further loophole through which the unguarded may fall. In a particular catalysis study a new absorption band may be noticed; it may then be safely assumed that a reaction intermediate has been formed on the surface. However, the observation of this new band does not prove that the corresponding reaction intermediate is the vital intermediate in the course of the catalytic reaction. It may well be that the crucial reaction intermediate is not detectable spectroscopically.

When ethylene is adsorbed on porous glass it is found [269, 270] that the adsorbed phase has an infrared absorption spectrum closely akin to that of liquid ethylene. Here is an example where we can safely conclude, using an experimental criterion, that ethylene has been physically adsorbed.

Just as flash-desorption has revealed two or more distinct types of chemisorption on catalyst surfaces, so also has infrared spectroscopy. Pliskin and Eischens [271], working with silica-supported and alumina-supported platinum catalysts, have shown that a weakly bonded form of hydrogen is retained at the metal surface as well as a strongly bonded form. Moreover, Pickering and Eckstrom [265], working with evaporated rhodium films, found a total of no less than eighteen new absorption bands (in the range 2193–1316 cm^{-1}). The suggestion is that the bands may correspond to chemisorption of hydrogen on different crystal faces of a highly polycrystalline material. (If this result is reliable it is a salutary reminder of the importance of crystallographic structure in deciding the properties of surfaces.)

So far as chemisorption of carbon monoxide on metals is concerned, Eischens [261] has shown convincingly that there are two major types of bonding, the bridged form and the linear form (see Section 3.2.8.1). Whereas for palladium, platinum, rhodium and nickel the two forms of bonding have been observed, for copper and iron only the single-bonded (linear) species has been detected. Recently, Yates and Garland [272] have concluded that there are five distinct species of carbon monoxide adsorbed on nickel (see Fig. 18 and Table 3.6). It was established that the five absorption bands

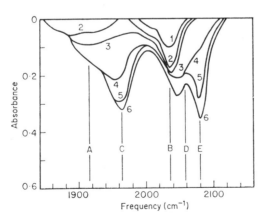

FIG. 18. Infrared spectrum of CO adsorbed on nickel at various pressures. Curve 1, 1.7×10^{-4} torr; curve 2, 2.0×10^{-4} torr; curve 3, 3.5×10^{-4} torr; curve 4, 2.3×10^{-2} torr; curve 5, 1.2 torr; curve 6, 1.2 torr after 12 h. Five distinct surface species are formed according to Yates and Garland [272].

TABLE 3.6. Assignment of infrared bands to surface species for carbon monoxide chemisorbed on a supported nickel catalyst [272]

Band	Frequency (cm⁻¹)	Structure	Relative strength of absorption
A	1915	O | C ╱ ╲ —Ni—Ni—	Strong
B	2035	O | C | —Ni—	Strong
C	1963	O O O | | | C C C |╱╲| —Ni Ni—	Medium
D	2057	O | C | —Ni—	Medium
E	2082	O | C | —Ni—	Weak

represented five distinct surface species, by observing independent variations in the intensities during changes in the experimental conditions.

The effect of the support or carrier material on the infrared spectra of carbon monoxide adsorbed on platinum [261] and on nickel [273] has been found to be significant, thus clarifying another mystery associated with heterogeneous catalysis. Eischens and Pliskin [261] found that the ratio of the concentrations of bridged to linear forms was much larger when platinum was supported on silica than when it was supported on alumina. O'Neill and Yates [273] showed that not only the relative numbers of the two surface species, but also the strength of the two types of adsorption, varied with the nature of the support. These workers further revealed that the silica support has comparatively little effect on the adsorption of carbon monoxide on nickel, but that alumina and titania exert a marked effect.

Infrared spectroscopy has been most helpful in deciphering the mechanism of formic acid decomposition, the catalysis of which has for so long been a subject of debate. Using silica-supported nickel catalysts for the decomposition, several investigators [243, 244, 274–278] have shown that (i) a formate ion is formed on the surface upon adsorption of the carboxylic acid; (ii) this ion is the reactive intermediate in the catalytic decomposition of the acid; (iii) the decomposition proceeds from a formate-covered metal surface, since as the pressure rises (owing to the production of carbon dioxide and hydrogen) the optical density of the band corresponding to the formate ion (1575 cm^{-1}) falls concomitantly, and completely vanishes when no further pressure increase occurs. In view of what has already been said in Section 3.2.8.4 concerning the correlation of the thermodynamic properties of bulk and surface "compounds", it is pertinent to note that the heat of adsorption of formic acid at monolayer coverage is 18 kcal mole^{-1} and the heat of formation of one-half of a mole of nickel formate is 13 kcal (see also p. 315).

So far as chemisorption of hydrocarbons on metals is concerned, one of the most important factors appears to be the pretreatment of the catalyst surface. In a series of studies on the chemisorption of olefins and paraffins carried out by Eischens and Pliskin [261] it emerged that both associative and dissociative adsorption could occur, depending on the nature of the catalyst pretreatment. The associative type of adsorption—in which the two carbon atoms of ethylene, say, are linked to the metal surface—is favoured when the olefin is adsorbed on nickel which has a preadsorbed layer of hydrogen. If the hydrogen is removed from the catalyst surface, by outgassing, prior to admitting the olefin, dissociative chemisorption predominates. Paraffins cannot, of course, be chemisorbed associatively, so that, not surprisingly, they are not adsorbed by nickel containing preadsorbed hydrogen. Acetylene displays unique behaviour. Irrespective of the pretreatment of the silica-supported nickel, acetylene produces the spectrum assigned to chemisorbed ethyl radicals (see Chapter 8), which are often represented by *CH$_2$CH$_3$. From this experimental fact, we can infer that extensive self-hydrogenation of adsorbed acetylene occurs with the simultaneous formation of surface carbide.

When acetylene is adsorbed on an oxide such as alumina, the molecule retains its triple-bond character. Moreover, it can be chemisorbed in two configurations [279, 280]. That part of the acetylene which is rapidly adsorbed establishes an end-on configuration with respect to the surface; the remaining part of the acetylene is held weakly and parallel to the surface. When ethylene is adsorbed on alumina, the process is slow [280], and a complex reaction involving only ethylene and ethyl groups (by self-hydrogenation) takes place.

3.3.1.1. *Far Infrared Spectra of Adsorbed Molecules*

In the infrared region of the electromagnetic spectrum it is the vibrational-rotational spectra that occur. In the far infrared region, rotational transitions only of the molecules occur. The question may well be asked, therefore, whether the far infrared region is an accessible one for the study of adsorbed molecules. Yates [281] has recently succeeded in exploring the far infrared spectra of methane, ammonia, sulphur dioxide and a number of other gases (in the range 300–30 cm⁻¹) adsorbed on commercial silica samples (Cabosil), but no bands due to adsorbed molecules could be detected. One can only assume either that the expected transition bands lie outside the explored range, or that the bands due to adsorbed molecules may be present in the region examined, but are too weak to be observed. In view of this result, any attempts to study the rotation of adsorbed molecules would still be better made via ordinary infrared measurements. For example, the contours of the absorption bands can be calculated [269, 270], for a certain model of molecular rotation, and the resulting calculated band can then be compared with the experimentally observed shape. It was in this way that Sheppard and Yates [269] concluded that, when methane was adsorbed on silica, the molecule could be pictured as having three of its hydrogens contacting the surface with rotation taking place only around the axis perpendicular to the surface, that is, with one degree of rotational freedom.

3.3.2. The visible and ultraviolet spectra of adsorbed species

Whereas infrared studies have been largely devoted to the study of the adsorption of relatively simple molecules such as carbon monoxide or ethane, ultraviolet spectra, being due to electronic transitions, can be most valuable in the characterization of much more complicated species (occurring often as intermediates in catalysis).

Experimentally, the difficulties associated with this technique are very similar to those associated with the use of infrared spectra, except that the

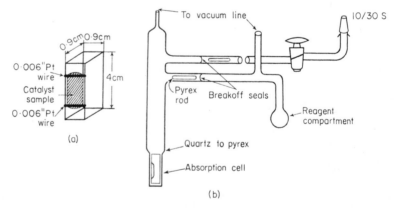

FIG. 19. Cell used by Leftin [282] for spectral studies of catalysts in the visible and ultraviolet regions.

problems involved in preparing suitably transparent samples are now more severe (since light scattering varies inversely as the fourth power of the wavelength). A typical example of an optical cell suitable for the spectral studies in the visible and ultraviolet ranges is shown in Fig. 19. Leftin [264, 282], who designed this cell, has also fashioned another [283] which is suitable for spectral studies in both the infrared and ultraviolet regions.

We shall now cite a few examples to illustrate the usefulness of visible and ultraviolet spectroscopy in catalysis research. On the basis of much reliable and independent work [284–286], it is now generally believed that many hydrocarbons, but particularly arylalkanes, produce carbonium ions when they are chemisorbed on to catalyst surfaces. Before we can ever hope to understand how the carbonium ion fits into the scheme of catalysis, it is essential that we first endeavour to characterize the precise nature of the ion

TABLE 3.7. Compounds used in spectral studies of adsorption on catalysts [264]

Hydrocarbon	Carbonium ion	Principal absorption band(s) (mμ)
Triphenylmethane	Triphenylcarbonium ion	4040; 4320
1,1-Diphenylethane	Methyldiphenylcarbonium ion	4230
Cumene	Dimethylphenylcarbonium ion	3950

and, hence, to say something about the nature of the surface site at which the carbonium ion is held. It is in this context, and for this purpose, that visible and ultraviolet spectroscopy plays a central role in the study of catalysis.

Consider the adsorption of a series of phenyl analogues of isobutane on to silica–alumina catalysts such as those used in catalytic cracking. Leftin [282] and his associates set out to investigate whether these arylalkanes yielded carbonium ions (see Table 3.7) on the catalyst surface. (Phenyl derivatives were chosen because phenyl groups are effective in stabilizing carbonium ions, as evidenced by the indisputable presence of carbonium ions in sulphuric acid solutions of carbinols [287], and because the spectra of these carbonium ions have been well characterized [288, 289].) When the catalyst was exposed to the vapour of triphenylmethane, a yellow colour slowly developed on the surface (see the absorption spectra contained in Fig. 20). The comparisons made possible by Fig. 20 leave no doubt concerning the existence of the triphenylcarbonium ion.

Leftin and Hall [290] have studied the electronic spectra of 1,1-diphenylethylene and of α-methylstyrene adsorbed on silica–alumina catalysts. They demonstrated that carbonium ions and some intermediates involving radical ions were formed. Carbonium ions are thought to be held in the vicinity of

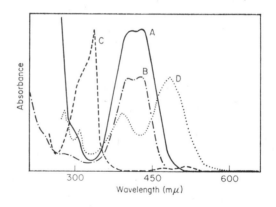

FIG. 20. Absorption spectra of carbonium ions. Curve A, triphenylmethane (ϕ_3CH) chemisorbed on silica–alumina; curve B, ϕ_3COH in conc. H_2SO_4; curve C, ϕ_3C^0 in ether; curve D, ϕ_3C^- in ether (Leftin [282]).

the catalyst surface by electrostatic forces, and they may be pictured as constituting the positively charged half of an electrical double layer [264].

Like infrared spectroscopy, ultraviolet spectroscopy can be utilized to decide between physical adsorption and chemisorption. For a system in which physical adsorption is involved, the electronic spectrum of the adsorbed molecule is very similar to that of the free molecule [291].

3.3.3. Magnetic resonance spectroscopy

In recent years it has been possible to study the chemical and physical properties of adsorbed phases, and the surface structure of solids, using relatively new branches of spectroscopy which lie in the radio-frequency region of the electromagnetic spectrum, generally beyond 1 Mc/s. Magnetic resonance techniques utilize the fact that atomic nuclei and electrons possess magnetic dipole moments and spin angular momenta and that, upon applying a magnetic field, a number of energy levels are produced by Zeeman splitting of the quantum states of the magnetic moment.

3.3.3.1. *Nuclear Magnetic Resonance*

Nuclear magnetic resonance (n.m.r.) is concerned with the resonant absorption of energy by atomic nuclei when they are placed in a strong magnetic field and radiation of radio-frequency [292–295]. If a nucleus has a spin number I (and therefore spin angular momentum $Ih/2\pi = I\hbar$), there will be $2I+1$ different orientations when that nucleus is placed in a constant magnetic field, corresponding to $2I+1$ equally spaced energy levels (see Fig. 21).

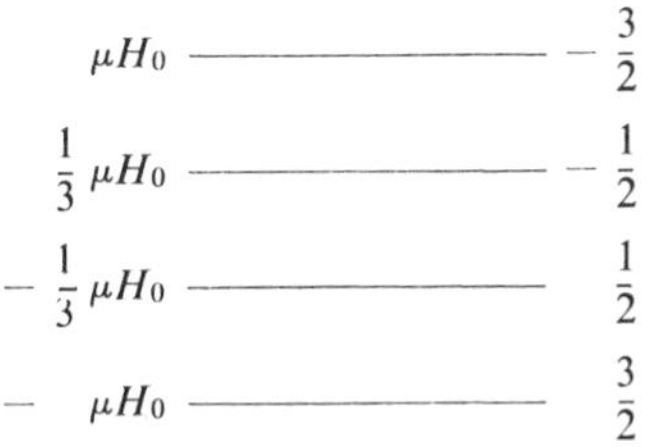

$$\mu H_0 \qquad\qquad -\frac{3}{2}$$

$$\frac{1}{3}\mu H_0 \qquad\qquad -\frac{1}{2}$$

$$-\frac{1}{3}\mu H_0 \qquad\qquad \frac{1}{2}$$

$$-\ \mu H_0 \qquad\qquad \frac{3}{2}$$

FIG. 21. When a nucleus has, for example, a spin number $I=3/2$, there are four equally spaced energy levels. μ stands for the magnetic moment of the nucleus and H_0 the magnetic field in which it is placed. The interval of energy between successive levels is $\mu H_0/I$.

N.M.R. is confined to those nuclei which possess a finite magnetic moment and spin angular momentum: over a hundred stable or long-lived radioactive nuclei fall into this category. Widely occurring nuclei such as ^{12}C, ^{16}O and ^{32}S do not give rise to n.m.r., since they have zero magnetic moment.

For transitions to be induced between adjacent energy levels, the frequency ν of the electromagnetic radiation must satisfy the equation:

$$h\nu = \frac{\mu H_0}{I} \qquad\qquad (61)$$

and the angular frequency ω of the radiation is thus

$$\omega = 2\pi\nu = \frac{\mu}{I\hbar} H_0 = \gamma H_0 \qquad\qquad (62)$$

where γ, the ratio of the magnetic moment of the nucleus to the spin angular momentum, is called the magnetogyric (or gyromagnetic) ratio. The value of γ for a nucleus possessing a non-zero nuclear magnetic moment characterizes that nucleus. Now the usefulness of the n.m.r. technique in chemistry generally, and in heterogeneous catalysis in particular, rests in the fact that the widths, splittings, and shifts of the magnetic resonance of atomic nuclei depend in a sensitive manner on the particular magnetic and electronic environment of the nuclei. The precise value of H_0, the magnetic field which the nucleus experiences, is seldom, if ever, the same as the applied magnetic field. In general we can say that

$$H_0 = H + H_{local} \tag{63}$$

where H and H_{local} are the applied and local field *at* the nuclei under consideration. In a sense, the nucleus giving rise to n.m.r. can be regarded as a probe by which we can ascertain details of the electronic and nuclear structure of the substance under investigation.

The thermal motions of the lattice of atoms in which nuclei giving rise to n.m.r. are embedded can be interpreted and understood if, along with absorption line shape and position, we measure a property known as the spin-lattice relaxation time. To appreciate what is conveyed by this term, we must first consider what happens when continued absorption of electromagnetic energy occurs during a magnetic resonance experiment. The absorption will tend ultimately to equalize the populations of the energy levels, raising the effective temperature of the nuclear assembly and leading to a condition known as saturation. However, the tendency towards equalization is opposed by interactions between the thermal motions of the lattice of atoms and the resonant nuclei themselves; the interactions tend to restore the temperature of the nuclear assembly to that of the lattice and to sustain the appropriate difference in population. When, after saturation, a nuclear assembly is allowed to return to equilibrium with the lattice, it does so exponentially, and the time-constant is, by definition, the spin-lattice relaxation time. In practice, the relaxation time can range from 10^{-5} to 10^5 sec.

The schematic arrangement of the apparatus used by Andrew and his co-workers [294] to study n.m.r. in solids is shown in Fig. 22. The solid is mounted in a small coil (about a dozen turns) between the poles of a permanent magnet—the field strength employed [296] is usually 6000–10,000 gauss. When current at the resonant frequency energizes the coil, additional loss in the coil is caused by the absorption of energy by the nuclei. The field of the magnet is swept in cycles through the resonant condition; and the balance of the bridge, into which the coil is connected, is displaced at each traverse of the resonant condition. The output of the bridge, after amplification, is displayed on an oscillograph or other recording devices. With a field of 10,000 gauss, equation (61) enables us to compute that, for protons, which have a spin angular momentum of $\hbar/2$, the resonance frequency is 42·6 Mc/s.

Line-width measurements of the resonant protons in methane adsorbed on titanium dioxide have yielded information concerning the rotation and translation of the adsorbed hydrocarbon molecule [297]; spin-lattice relaxa-

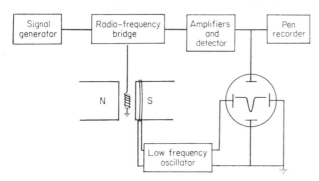

FIG. 22. Schematic arrangement of apparatus used by Andrew and co-workers [294] for detecting nuclear magnetic resonance.

tion times of the protons in adsorbed water have illustrated how n.m.r. can be used to record the "freezing" of molecules in the fine pores of an adsorbent [298–300]. While it is likely that n.m.r. will be much used in the future to study the defect centres [301] in solids exhibiting catalytic activity, it now appears [302] that *high-resolution* n.m.r. will be of comparatively little value in the catalyst field because the broad lines observed in solids tend to obscure the shifts of the resonance line (the so-called chemical shift [292]). If more sophisticated techniques, such as the mechanical spinning of the sample during resonance measurement [303, 304], can be developed, it is likely that n.m.r. will become a powerful tool for the study of solid catalysts. In the relatively new field of Ziegler catalyst systems (see Section 5.3.4.2), conventional n.m.r. measurements on ^{27}Al by DiCarlo and Swift [305] have already gone a long way towards supporting the views of Cossee and his associates [306–308] concerning the mechanism of the stereospecific polymerization of α-olefins on mixtures of a transition metal compound and a metal alkyl derived from a strongly electropositive metal.

3.3.3.2. *Electron Spin Resonance*

The principles of electron spin resonance (e.s.r.), also known as electron paramagnetic resonance (e.p.r.), are in many respects similar to those of n.m.r. The prime differences emanate from the facts that magnetic moments of the electrons are over a thousand times as large as nuclear moments, and that electrons, owing to their rapid motions, usually possess a contribution to their magnetic moment from the orbital angular momentum as well as from the spin angular momentum [293]. The condition for resonance is

$$h\nu = \Delta E = g\beta H \qquad (64)$$

where β is the Bohr magneton $\left(\beta = \dfrac{e\,h}{2mc}\right)$, H the magnetic field, g is known as the spectroscopic splitting factor, ΔE is the separation between adjacent Zeeman levels, and ν is the resonance frequency (h, e, m and c are fundamental constants being, respectively, Planck's constant, the charge and the mass of the electron, and the velocity of light).

Using equipment which is in principle similar to that employed in n.m.r., the resonance condition for a particular sample is conveniently located by operating the e.s.r. spectrometer at a fixed frequency and subjecting the field H to continuous, cyclical, variation. This procedure also permits the definitive measurements of line width (expressed in gauss), and the nature of the fine structure of the resonance line, to be made. Equation (64) shows that resonance for unpaired electrons (with $g=2$) would occur at about 10,000 Mc/s if a magnetic field of several kilogauss were employed. Microwave generators, known as klystrons, operating either in the "K-band", which refers to frequencies in the range 18,000–26,500 Mc/s, or in the "X-band" (8200–12,400 Mc/s) are almost always used in commercial e.s.r. spectrometers.

The great advantage of e.s.r. is that it is an extremely sensitive method of detecting unpaired electrons, and it has been extensively used for this purpose ever since the technique was discovered by Zavoisky [309]. It has not yet found great applicability in the catalysis field because many catalysts do not contain unpaired electrons, but it has already demonstrated that when certain hydrocarbons are adsorbed on industrially important supported catalyst free radicals are produced. Thus, Rooney and Pink [310] reported that well-defined e.s.r. spectra were obtained when polynuclear aromatic hydrocarbons, such as anthracene and perylene, were adsorbed on a silica–alumina catalyst. These workers [311] and others [312–314] have established, using e.s.r., that: (i) formation of positive radical ions from polynuclear hydrocarbons occurs on the Lewis acid sites, where there is a deficiency of electrons, at the surfaces of silica–alumina catalysts; (ii) aromatic hydrocarbon radicals react with oxygen at silica–alumina surfaces (see also Chapter 8).

E.S.R. has also furnished valuable information about the effects of different supports, the influence of various promoters and the extent and nature of sintering of industrial catalysts. For example, MacIver and his colleagues [315–318] have shown that chromia–alumina catalysts, after reduction at high temperature, show two distinct phases, a dispersed phase which predominates at low concentrations of chromium and a bulk phase which is prevalent at the higher concentrations; the dispersed phase consists of isolated Cr^{3+} ions not coupled electronically, and the bulk phase of clusters of Cr^{3+} ions with strong exchange coupling of $3d$-electrons. Some doubts exist [319, 320] about the interpretation of the e.s.r. spectra of supported chromia catalysts, and, in particular, evidence for the existence of Cr^{4+} and Cr^{5+} ions has been critically debated.

The significance of e.s.r. measurements in the general study of the electronic structure of solid catalysts has been emphasized recently by Stone [302], who draws attention to the fact that, once the problem of calculating how, precisely, the e.s.r. of a magnetic ion is affected by the symmetry of the crystal field around it, the technique will acquire fundamental importance.

3.3.4. Magnetic susceptibility

Substances which possess unpaired electrons have a positive value of magnetic susceptibility and are, by definition, paramagnetic. Paramagnetic susceptibilities χ are of the order of a few units times 10^{-6} c.g.s., and are independent of field strength except at very high fields and low temperatures, when magnetic saturation is said to occur. For very many paramagnetic substances, the susceptibility may be represented by the Curie-Weiss law [321, 322]:

$$\chi = \frac{C}{(T + \varDelta)} \tag{65}$$

where T is the absolute temperature and both C and $\varDelta$ are constants. A plot of the reciprocal of χ versus T enables C to be determined from the slope and $\varDelta$ from the intercept at $T=0$. The relationship between the atomic, ionic or molecular moment μ and the magnetic susceptibility per mole, χ_M, of substance is given by:

$$\mu = 2 \cdot 84 \, (\chi_M \, (T + \varDelta))^{1/2} \tag{66}$$

or
$$\mu = 2 \cdot 84 \sqrt{C} \tag{67}$$

Some typical values of μ, expressed in Bohr magnetons, are: Cr^{3+}, 3·8; O_2, 2·8; Fe^{3+}, 5·9 and Cu^{2+}, 1·7.

The magnetic moment is not always easily related to the oxidation state: for transition metals, however, μ in Bohr magnetons is $(n \, (n+2))^{1/2}$ where n is the number of unpaired electrons per atom [321]. The value of $\varDelta$ yields information about the environment and the crystal field in which the atom giving rise to para-magnetism is placed; but difficulties are encountered in relating $\varDelta$ quantitatively to the crystal field.

Enough has now been said to indicate the theoretical principles involved in calculating the number of unpaired electrons from magnetic susceptibility measurements. Since chemisorption often involves unpaired electrons in solids, and occasionally in adsorbates, it is obvious why magnetic susceptibilities are useful in computing the changes in electronic structure of catalytic systems.

Selwood and his co-workers [323] showed that it was possible to follow the course of an adsorption such as that of molecular oxygen on graphite by measuring the paramagnetic susceptibility of the oxygen. The adsorption of other paramagnetic gases such as nitric oxide could also be followed. This technique (that is, measurement of the susceptibility of the adsorbate) is obviously limited in its applicability, since the majority of gases adsorbed on heterogeneous catalysts are diamagnetic. Besides, if the magnetic properties of the adsorbate need to be studied, it is best to do so using magnetic resonance (see previous section).

Morris and Selwood [324] were the first to demonstrate that the magnetic properties of the adsorbent could be conveniently followed during adsorption, and that it was relatively easy to draw at least semi-quantitative conclusions about changes in the electronic structure of the adsorbent. Dilke, Eley and Maxted [325] also noted the usefulness of the technique.

The experimental determination of the magnetic susceptibility of a solid is quite straightforward if the substance exhibits only paramagnetism (and not ferromagnetism). The susceptibility is obtained by using either the Gouy method, which entails the use of a uniform magnetic field, or the Faraday method, which utilizes a non-uniform field. When ferromagnetic substances, such as iron, nickel or cobalt, are being studied, it is sometimes best to measure the specific magnetization, χH (where H is the field strength), the point being that, for ferromagnetics, the susceptibility is large and dependent upon field strength. It is better, however, to measure the saturation magnetization, that is, the magnetization at infinite field strength and zero temperature. The saturation moment per atom is equal to the number of unpaired electrons. Complications in measurement arise when ferromagnetic materials are being studied in a fine state of subdivision. With particle sizes in the region of 100–300 Å, the magnetic behaviour of ferromagnetic substances differs from that of ferromagnetic substances in bulk [322, 326] in that, with the small particles, magnetization is dependent on field strength and temperature in a manner similar to that of paramagnetic substances (see equation (65)). This phenomenon has been variously called collective paramagnetism [322], superparamagnetism [327, 328] and subdomain behaviour. The numerous experimental aspects of susceptibility measurements have been adequately described by Selwood [322, 329, 330], Fensham [331], Gray [332] and Stone [333, 334]. Since supported catalysts have the catalytic metal or metal oxide almost invariably in a fine state of subdivision, a low-frequency a.c. permeameter method of measuring susceptibility is employed [329]. Since a measurement of a change in bulk property due to a surface effect is being made, the need to use finely divided solids (or inert supports) which have large surface-to-volume ratios is compelling.

The nickel–hydrogen system, and its relevance to the catalytic hydrogenation of olefins, is one system that has been comprehensively studied by magnetic susceptibility measurements. A decrease in magnetization of approximately one Bohr magneton on adsorption is observed that can be due either to covalent bonding (H—Ni) or to positive ion formation of the type depicted by H$^{\pm}$–Ni (see Ref. *323*). The decrease does not distinguish between these two possibilities, but merely denotes the pairing of the *d*-band electrons of the metal. Other evidence, however, notably the thermochemical calculations described in Section 2.2.1.1, rules out the formation of H$^+$ or H$_2^+$ ions at the surface, and the susceptibility decrease therefore implies covalent bond formation. On nickel surfaces in general Selwood and his associates have shown that: (i) on the surface of the hydrogenation catalyst there are three or four times as many sites available to hydrogen as to ethylene; (ii) when hydrogen sulphide is adsorbed, dissociation takes place resulting in the formation of two adatoms of hydrogen and a divalent adatom of sulphur; (iii) the bonding of carbon monoxide is dependent on nickel particle size and on surface coverage; (iv) when carbon dioxide is chemisorbed on nickel, two bonds are formed for every molecule adsorbed (as shown below), but the maximum volume which may be chemisorbed at room temperature is only about one-eighth the volume of hydrogen which may be chemisorbed on the same surface.

Magnetic susceptibility measurements have been carried out during adsorption on numerous other solids—chiefly supported metals. Some recent Russian work has dealt with the correlation of the variation in magnetic susceptibility of platinum supported on silica gel, quartz, and sugar charcoal with the dilution of the metal on the support material. Bylina, Evdokimov and Kobozev [*335*] report that as the dilution increases the catalytic activity of the platinum in the hydrogenation of 1-hexene also increases monotonically. There is an implication here that the catalytic activity depends on the particle size of the metal, a fact which features eminently in Kobozev's theory of ensembles [*336*] and which has been noted in other heterogeneous systems [*337, 338*]. In contradistinction, Strel'nikova and Lebedev [*339*] have reported that the catalytic activity of platinum supported by silica gel which has been impregnated with mononuclear and binuclear coordination complexes does not increase steadily with increasing dilution of the metal on the support: the catalytic activity in the hydrogenation of cyclohexene goes through maxima and minima on dilution of the platinum (see also pp. 306, 444).

A magnetic study of the non-stoichiometry of cuprous oxide, and its relevance to catalysis on oxide surfaces has been made by O'Keefe and Stone [334]. This work focuses attention on the importance of the defect solid state in heterogeneous catalysis (see Chapter 5). The paramagnetism of the surfaces of noble metals, transition metals, and of various metal alloys has also been studied magnetically with a view to elucidating the mechanism of ortho–para hydrogen conversion by metal surfaces [340].

3.3.5. Work-function measurements

In previous sections (see Sections 2.2.1.1, 2.2.3 and 3.2.6) we have had occasion to refer, in passing, to the work function of a solid. This property is usually symbolized by $e\phi$ and is defined as the energy required to remove an electron from the highest occupied level inside the solid into a vacuum outside its surface. If an adsorbed layer possesses some ionic character (or ionicity, see Section 2.2.1.1), the value of the work function of the adsorbent surface will obviously be modified by the presence of the adsorbate. If the negative end of the dipole in the adsorbed layer points away from the surface, then the potential barrier through which the electrons escaping from the solid must pass will be increased, so that $e\phi$ is, in turn, increased. Conversely, an electropositive adsorbed layer, that is, one which has the positive end of the dipole pointing away from the surface, will decrease the value of the electron work function of the solid.

If the dipolar adsorbed layer is regarded as a parallel plate condenser, we may relate [341] the apparent dipole moment μ to the potential drop ΔV across the layer (also called the surface potential difference). The Helmholtz equation gives:

$$\Delta V = 4\pi c_s \theta \mu \tag{68}$$

where c_s is the number of surface sites per cm^2 and θ is the fractional coverage of the surface. Since $e\Delta V$ is the change in energy required to remove an electron from the solid after the formation of the adsorbed layer, it is synonymous with $e\Delta\phi$, the change in work function brought about by the adsorbed layer. The greatest significance of equation (68) is that it enables us to calculate the dipole moment of the adsorption bond from measurement of the surface potential. To illustrate this point we recall the work of Tompkins and his associates [342], who found that the dipole moment of the chemisorbed bond of carbon monoxide to iron was practically the same as that of the corresponding bond between a carbonyl group and the iron atom in the free molecule of some iron carbonyls. It was therefore inferred that the surface link can best be represented by:

These investigators also showed that, in general, the dipole moment of the adsorption bond alters as the coverage alters, a point of considerable significance in the theory of chemisorption, for it had often been assumed, in default of experimental evidence, that the dipole moment in the adsorbed layer at monolayer coverage can be used for calculations of dipole–dipole interaction on sparsely covered surfaces. In point of fact, as $\theta \to 1$, $\mu \to -0 \cdot 066$ debye and, as $\theta \to 0$, $\mu \to -0 \cdot 33$ debye for the copper+hydrogen system [341].

Work-function measurements have been extensively used in surface chemistry recently, notably by Tompkins and his school and by Roberts [343–346]. Many of the studies have aimed at elucidating the nature of chemisorption itself and also the mechanisms of processes such as sulphidation and oxidation. However, other studies have involved the use of mixed adsorbates, an approach which impinges directly upon problems in catalysis. Siddiqi and Tompkins [347] have increased our knowledge of the configuration of intermediaries present at the catalyst surface by measuring the surface potential of several mixtures of adsorbates on a series of metal films. Their method of detecting intermediaries during the course of a simple heterogeneously catalysed process is complementary to the spectroscopic methods described in Sections 3.3.1 and 3.3.2. Although it is not yet possible to describe unambiguously the mechanism of the oxidation of carbon monoxide by oxygen (see Chapter 8) on metal surfaces, Siddiqi and Tompkins [347] have provided some evidence for the existence of bridge structures and other configurations which have been suggested by infrared spectroscopy. On nickel, for example, the following structures have been formulated [347]

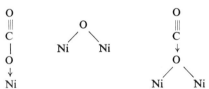

3.3.5.1. Methods of Measuring Work Functions, including Field-emission Microscopy [348, 349]

There are two main divisions into which the various methods of measuring work functions fall: those based on electron emission and those based on the principles of condenser action. In the first of these divisions we shall briefly consider the following four methods: the space-charge-limited diode, the photoelectric method, the thermionic saturation-current method and the field-emission microscope.

It is well known that the current from an emitter filament—the cathode of a diode valve—is controlled by the negative space charge near the filament surface when small voltages are applied between it and an anode. Under these circumstances, the current collected by the anode is dependent upon the applied voltage and the magnitude of the work function of the anode: it is largely independent of the work function of the cathode. Hence, in measuring work-function changes using this principle, the solid under study, which may be in the form of a filament or an evaporated film, is made the anode of the space-charge-limited diode. This procedure has yielded reliable results in the hands of Mignolet [350, 351], Pritchard and Tompkins [341, 348, 352] and Pethica [353]. Figure 23 shows the film vessel used by Pritchard in an improved method of measuring the surface potentials of gases adsorbed on

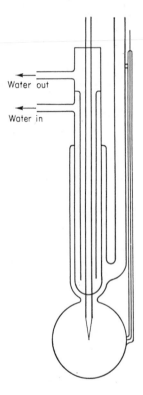

FIG. 23. Film used by Pritchard [352] for measuring the surface potentials of gases adsorbed on evaporated films of metals.

evaporated films of copper and gold. The bulb had a diameter of about 4·5 cm, and the metal to be evaporated was placed on the "V" of the cathode as a small loop of spectrographically pure wire. Following the practice of Mignolet [351], a stream of water served to thermostat the cathode leads when the bulb was immersed in different-temperature baths, thereby avoiding drifts due to thermal equilibration. A typical circuit required for measuring the current collected by the anode is shown in Fig. 24.

Although widely used and of unquestionable value, this particular method has distinct disadvantages [347]. In the first place, it is limited to pressures below 10^{-3} torr during actual measurement of the surface potential. Secondly,

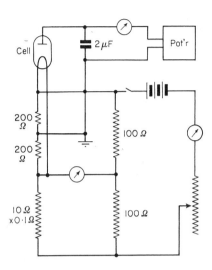

FIG. 24. Typical circuit required for measuring the current collected by the anode (Pritchard [352]).

with certain adsorbates, particularly oxygen, there is the possibility of chemical attack on the hot filament, which is almost invariably made of tungsten. Such attack vitiates the results. Lastly, some of the surface reactions being studied (with, say, mixed adsorbates) are influenced to an unknown degree by the electron concentration inevitably present. All of these difficulties are avoided if methods based on condenser action (Section 3.3.5.2) are used, and removed to a greater or lesser extent by using other electron emission methods.

The equation

$$h\nu_0 = e\phi \tag{69}$$

forms the basis of the photoelectric method of measuring changes in work function [354, 355], where ν_0 is the threshold frequency below which no electron emission takes place when the surface in question is irradiated with electromagnetic radiation. Strictly speaking, we would expect equation (69) to hold only at $0°K$, because, at normal temperatures, the threshold is blurred because of the thermally stimulated emission of those electrons whose energies are above the Fermi level. The consequence of this fact, and its influence on the details of measurement, have been discussed by Hayward and Trapnell [349]. The method suffers from two disadvantages: firstly, photocurrents as low as 10^{-14} A may have to be measured accurately; secondly, for large values of $e\phi$ (>5 V), the threshold frequency lies far in the ultraviolet region of the spectrum, which makes experimentation difficult.

The thermionic saturation-current method [356] makes use of an equation which can be derived theoretically [349] and which relates the saturation current per unit area of surface at a particular temperature to the work function of the solid. The method is not widely used in the study of systems of interest in catalysis for the simple reason that the temperatures required to produce a measurable current are often so high that desorption from the adsorbent surface takes place in the course of measurement.

In field emission, which is the principle governing the action of one of the most useful tools for studying the adsorbed phase, the field-emission microscope, the total current emitted by a solid surface is essentially independent of the temperature up to about 300°K, and is given by the renowned Fowler-Nordheim equation [357, 358]:

$$\frac{i}{V^2} = S \exp \left(- \frac{B(e\phi)^{3/2}}{KV} \right) \tag{70}$$

in which i is the total current, V is potential applied between the surface and the anode, and S, B and K are constants which, separately, incorporate such properties as the surface area, the average work function of the surface, and a geometrical factor embracing the radii of curvature of the cathode and the anode. Fundamentally, the importance of equation (70) is that it yields $(e\phi)^{3/2}/V$ if $\ln(i/V^2)$ is plotted against $1/V$. However, since the constant K cannot be evaluated a priori, equation (70) is normally used in conjunction with the following equation:

$$\Delta\phi = \phi' \left[\left(\frac{s_{\text{ads}}}{s'} \right)^{2/3} - 1 \right] \tag{71}$$

where $e\phi'$ refers to the work function of the clean surface, $e\phi$ to the surface on which adsorption has taken place, and where s' and s_{ads} are, respectively, the slopes of the above-mentioned plots before and after adsorption.

Field-emission microscopy was introduced by Müller in 1937 [359] and was somewhat neglected in subsequent years. Latterly, Gomer [358], Becker [360] and others [4, 361–365] have used it extensively not only for the purposes of measuring work-function changes but to observe the surface diffusion of adsorbed species. The solid under study is in the form of a minute tip (see Fig. 25), the radius of which is several microns or less. Most of the solids

studied to date have been the refractory metals such as tungsten and molybdenum. Recently, however, Melmed and Gomer [365–367] have demonstrated that, through our increased knowledge of the mechanism of crystal
whisker growth [368], it has become possible to study almost any conducting
material by this method. (In one paper [365] recently, Melmed and Gomer
examined the electron emission from the surfaces of twelve different substances.) Catalysed decompositions are also within reach [363].

Since the radius of the tip (Fig. 25), compared with the radius of the
spherical anode employed, is extremely small, field strengths of the order of
10^7–10^8 V cm^{-1} are easily attained. A high field strength stimulates cold
emission from the tip, the resulting fluorescent pattern displayed on the

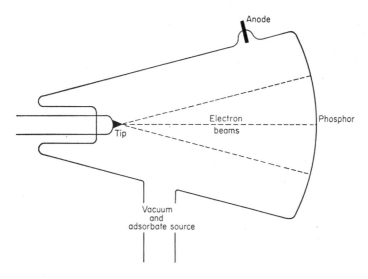

FIG. 25. Field-emission microscope (schematic).

phosphor gives a magnified image of the emission from the solid. Moreover,
as the intensity of the fluorescence at any point depends on the number of
impinging electrons at that point, it is a measure of the work function of the
corresponding point on the solid surface. Figure 26 portrays a typical field-
emission micrograph.

To summarize, therefore, field-emission microscopy can yield the work
function of particular crystallographic planes, and the details of adsorption,
surface migration and desorption of various adsorbates on, at or from those
planes. Field desorption, a closely allied phenomenon, has been developed
very recently by Gomer and Swanson [369–370]. As yet it has had limited application, but it holds the rich promise of being able to offer a direct determination

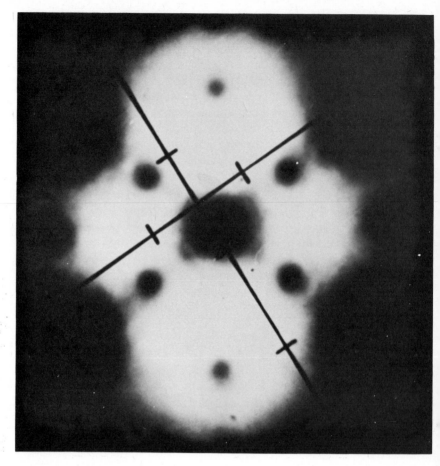

FIG. 26. Field-emission micrograph of a "clean" tungsten surface, [110] orientated. (By kind permission of Dr J. P. Jones.)

of potential-energy curves in adsorption (see Section 2.2) and, therefore, of affording a deeper insight into catalysed reactions.

3.3.5.2. Condenser Method of Measuring Work Function

The electrostatic potential difference V_{12} which is set up when two materials possessing dissimilar work functions ($e\phi_1$ and $e\phi_2$,' say) are connected electrically is also known as the contact potential difference. The origin of the contact potential difference is symbolized [349, 371] in Fig. 27. If we regard the two materials as the plates of a condenser, it can be appreciated that the application of a compensating potential V_c, equal to V_{12} in magnitude but opposite in sign, will remove the field between the two materials. Clearly, V_{12} is equal to $\phi_2-\phi_1$. Hence, by measuring the change in the magnitude of the compensating potential during the course of adsorption on one of the materials—the other being "inert" in the sense that, even if adsorption does take place upon it, the work function is not altered—the work function change

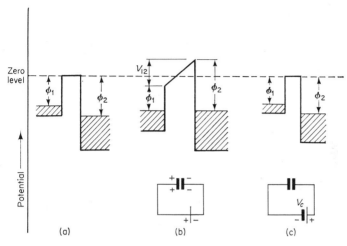

FIG. 27. The application of a compensating potential V_C removes the contact potential difference. $V_{12}=\phi_2-\phi_1$.

can be followed. The procedure, in essence, is to apply a continuously varying compensating potential and to note the potential at which the electrical field between the material and the "inert" reference electrode disappears.

One way in which the disappearance of the electrical field can be detected is by the use of a vibrating condenser cell such as that developed by Mignolet [372]. Roberts et al. [345, 373] have made accurate measurements of work function changes accompanying the adsorption of oxygen on evaporated films of molybdenum, chromium and other metals using a condenser cell similar to that depicted in Fig. 28. The hollow glass tuning-fork is rigidly fixed, at the upper end, to a metal block, whereas the two free ends can be made to vibrate by means of an electromagnet. The metal film to be investigated is evaporated on the vibrating electrode which is joined to the inside wall of one limb of the tuning-fork. Provisions are made, during evaporation, to prevent any condensation of film on to the non-vibrating electrode which is fixed to the rigid part of the cell. As long as an electric field exists between the surfaces of the two electrodes, current will be produced in a circuit connecting the electrodes on changing the electrical capacity of the cell. The electrical capacity is changed by subjecting the vibrating electrode to an oscillation. If, by means

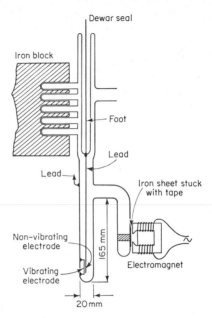

FIG. 28. Design of vibrating condenser cell, used by Mignolet [*372*] to measure changes of work function.

of a "backing-off" potentiometer [*345*] a compensating potential applied to the cell nullifies the contact potential difference, no a.c. signal will be produced when the cell is set vibrating. Modern cells of the Mignolet type usually consist of gold foil as the non-vibrating electrode, and have a separation distance between the two electrodes of about 0·5 mm.

Tompkins and his co-workers [*374*] have recently described an elegant method of measuring the compensating potential by the steady condenser method [*375*]. The principle of the method is to arrange for the compensating potential to be regulated by the current which flows between the electrode under study and the "inert" electrode. In the final (ideal) case, no current will flow, since the compensating potential should equal the contact potential difference between the electrodes.

3.3.5.3. *The Field-ion Microscope*

Field-ion microscopy is considered at this juncture not because it is directly related to the measurement of work function but because it bears striking resemblance to the technique of field-emission microscopy discussed in Section 3.3.5.1. Field-ion microscopy is of enormous significance in catalysis not because of what it has already accomplished but because it permits of a direct study, under favourable circumstances, of catalysed reactions in atomic detail.

Magnifications of one million times are quite easily achieved with this technique, and its full potentialities have yet to be realized.

The specimen geometry in field-ion microscopy is identical to that used in field-emission microscopy, and the general experimental set-up is very similar in both techniques. If a small amount of helium gas is admitted into a vacuum chamber such as that shown in Fig. 25, and a large positive potential is applied

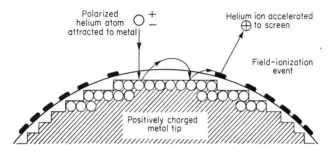

FIG. 29. Illustration of how the electric field at a metal tip may ionize helium atoms [378].

to the fine specimen tip, the helium atoms will be attracted towards the tip by polarization forces. If the tip has a sufficiently small radius, the electric field at its tip will be large enough to ionize the helium gas, as illustrated in Fig. 29 [376–379]. The ions will then be accelerated radially outwards and impinge upon the phosphor, where they will form a projection image of the end of the tip. Since the probability of ionization of the helium atoms depends very sensitively upon the local atomic structure of the surface of the end of the tip, the image on the phosphor reflects the fine details of this atomic structure. If it is ensured that the temperature of the tip is very low (less than $50°K$)—so as to keep the thermal velocity of the gas small—the resolution of the projection image depends only on the radius of the helium ions (or of other ions such as neon, argon or hydrogen, which can also be used [367, 376, 378]), and a helium field-ion micrograph is capable of such magnification that individual atoms and individual atom vacancies can be "seen". Figure 30 illustrates the type of image produced in field-ion microscopy. From this type of photograph it has been possible to see "directly" such features as grain boundaries—the extent of disorder across the boundary is surprisingly low and extends for only a few atom distances [378, 381]—and vacancy clusters produced by radiation damage. The central screw dislocation (see Section 5.2.2) running up the axis of a crystal whisker has also been "seen" by Melmed [367].

The current trend with field-ion microscopy is to use it in association with field-emission microscopy [377], so that the atomic details of the surfaces giving rise to electron emission prior to or after adsorption can be properly characterized. It is when the changes brought about by mixed adsorbates in

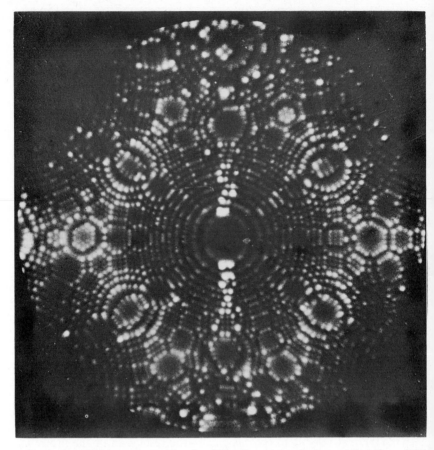

FIG. 30. Field-ion micrograph of a "clean" tungsten single crystal surface, [110] orientated (10,000 V and 2×10^{-3} torr of helium). (By kind permission of **Dr D. W. Bassett**.)

ion microscopy are measured that we can expect advances in our knowledge of catalysed reactions.

3.3.6. Electrical measurements

The measurement of electrical conductance and related phenomena such as thermoelectric power and Hall coefficients has been of considerable value in the study of solid surfaces. Most of the information has been restricted to the study of chemisorption as such, and comparatively little work has been done on the electrical properties of catalysts.

Several investigators [382–391] have devised simple conductance cells which have enabled the conductance changes of thin metal films to be recorded during the chemisorption of simple adsorbates. Suhrmann and his collaborators [384] have used essentially the same approach to follow the decomposition of formic acid on evaporated nickel film, and from their results were able to prove that the products of decomposition must have been water vapour and carbon monoxide rather than hydrogen and carbon dioxide.

The electrical conductance of a solid catalyst will be altered during acts of chemisorption and desorption if the electrons of the adsorbate molecules become part of the conduction electrons of the catalyst itself, or if the conduction electrons of the catalyst become part of the electron shells of the adsorbate molecules. It follows, therefore, that if this change of conductance is to be brought within the realm of the measurable, the catalyst must have a very large proportion of its constituent atoms situated at the solid surface. If this condition is not met, the conduction electrons at the surface will not constitute a significant proportion of the total number of conduction electrons in the solid. Realizing this fact, Clark and Benets [392] set about studying the electrical properties of vanadium pentoxide catalysts by preparing the oxide in the form of very thin films (thickness 220 Å and 1600 Å) on quartz particles which acted as supports. Following a suggestion by Mott [393], they measured the conductance using high-frequency a.c. so as to "short out" the boundaries between conducting grains. By measuring the temperature coefficient of conductance, the thermoelectric power and its

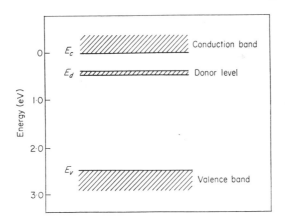

FIG. 31. Energy levels of V_2O_5. The donor level is O_v^{2-}. (See also Section 5.2.4.)

temperature coefficient, Clark and Benets [392] were able to formulate a picture of the defect structure of the vanadium pentoxide catalyst. Figure 31 reproduces one of their conclusions: it shows that the oxygen vacancies in the crystal lattice form electron donor levels 0·42 eV below the conduction band. The oxygen vacancies are in sufficient concentration to be a good source of electrons for the catalyst. The nature of the defects arising from non-stoichiometry in chromic oxide [394] and molybdena [395] re-forming catalysts have also been elucidated by electrical measurements. From measurements of the extremely low conductances (10^{-15}–10^{-10} mho cm^{-1}) of porous silica catalysts [396] it has been possible to prove that, upon adsorption of methanol, CH_3OH^{2+} and CH_3O^- ions are formed as a result of autoprotolysis of adjacent adsorbed alcohol molecules.

Hall coefficients and thermoelectric voltage measurements play an important part in the study of the surface properties of semiconductors. These two properties yield information regarding the nature of the current carriers in semiconductors in addition to the concentration of defects which can also be derived [397] from measurements of the temperature coefficient of electronic conduction and thermionic emission. (When an electric current passes down a rectangular conductor in the presence of a transverse magnetic field, a potential difference is generated between the sides of the conducting plate. The Hall coefficient R is defined by the relation $R = Ed/iH$, where E is the generated potential difference, i the current and d the thickness of the plate parallel to the magnetic field strength H. The sign of the potential difference is determined by the sign of the current carriers. Similarly, the sign of the thermoelectric power or the Peltier coefficient, determined by the measurement of the e.m.f. in a metal–semiconductor thermoelectric circuit [398] depends on the sign of the charge carriers.) Weisz and co-workers [399] showed that chromia–alumina catalysts changed from p-type to n-type semiconductors during the catalysis of the dehydrogenation of cyclohexane.

3.3.7. Investigation of catalysis and chemisorption by isotopic techniques

Some of the advantages gained by studying chemisorption on catalyst surfaces using radioactively labelled compounds have been cited in Sections 3.2.6 and 3.2.7.1; in particular, the work of Emmett and his colleagues [200, 201, 400], who estimated the active surface area of catalyst using ^{14}C-labelled gases, has been discussed. Here we shall deal with some of the recent trends in the study, by radiochemical methods, of catalysis and catalyst poisoning, and the use of stable isotopes and isotopic exchange.

3.3.7.1. *Use of Radioactive Isotopes*

To illustrate the power of the radiochemical approach and the far-reaching implications of the results obtained we shall pay particular attention to the elegant researches of Thomson and his school at Glasgow [401–407]. This school has enlarged our knowledge of catalyst poisoning, and has again affirmed that the fraction of chemisorbed species participating in catalysis is often very small.

We shall deal first with the question of catalyst poisoning. Campbell and

Thomson [402], by using tritiated hydrogen and the mercury isotope [203]Hg (half-life 47 days) were able to answer whether, when well-known "poisonous" molecules such as those of mercury arrive at a nickel catalyst surface, adsorption of the poison and concomitant displacement of previously adsorbed species takes place. They found that mercury readily displaced hydrogen from nickel surfaces at 20°C, but that the displacement was incomplete in spite of multilayer mercury adsorption. With films covered to extents less than 50% of their adsorptive capacity, hydrogen corresponding to 7% of that capacity could not be displaced. One explicit consequence of the arrival of the mercury is that the nickel catalyst is poisoned for the hydrogenation of cyclopropane. The question we must now ask is how the mercury exerts its influence. It is extremely unlikely that the poisoning is a result of alloy formation [403], and it is much more reasonable to ascribe the loss of catalytic activity to the fact that mercury atoms are preferentially chemisorbed at those very sites which hitherto retained hydrogen in a state free to participate in catalysis. This view is augmented by the results of Bond and Sheridan [408], who demonstrated that the kinetics of cyclopropane hydrogenation on nickel and other metals were dictated by the reaction between chemisorbed hydrogen and either gasphase or physically adsorbed cyclopropane. The failure of poisoned nickel to facilitate the hydrogenation of cyclopropane may therefore be regarded as the result of a drastic depletion in the supply of chemisorbed hydrogen caused by the presence of the mercury.

In a study of the adsorption and hydrogenation of [14C]crotonic and [14C]vinylacetic acid on thiophen-poisoned palladium [405], results quite different from those appertaining to mercury-poisoned nickel were obtained. The extent of adsorption of the acids was not at all affected by the presence of thiophen, although hydrogenation rates were markedly diminished. It was concluded that only a small fraction of the adsorbed molecules of acid takes part in the catalytic hydrogenation, and that the catalyst surface must be heterogeneous. In a later study [407], dealing with the direct observation of [14C]ethylene on the surface of nickel films during hydrogenation, it was again established that only a fraction of the chemisorbed species participated in the reaction. This proves beyond dispute that the ethylene–nickel surface system is heterogeneous, and the difference in reactivity of the adsorbed species almost certainly arises from adsorption on different planes exposed in the nickel film.

What the results of Thomson and his co-workers reveal, with refreshing clarity, is that attempts to relate heats of adsorption, conductance and magnetic changes in films, surface potentials, adsorption isotherms, absolute rate calculations and other such properties or concepts, to the process of catalysis, can be valid only *if they apply to that fraction of adsorbed molecules active in the catalytic process.*

Although Thomson's work has all been carried out since 1957, it must not be supposed that the radiochemical technique was, until recently, ignored in catalysis research. Emmett and Roginskii, with their respective teams of investigators, fully appreciated the value of using radioactively labelled compounds to decide, *inter alia*, which particular reaction mechanism was the more likely to be correct in a given system. Consider the classic example of the highly popular synthesis of hydrocarbons from hydrogen and carbon monoxide, a reaction for which Fischer and Tropsch [409, 410] found efficient catalysts nearly forty years ago. From the beginning, there has been doubt about the way in which the iron and cobalt catalysts used in this reaction succeeded to produce copious quantities of liquid-phase hydrocarbons. One possibility was that hydrocarbons might be formed through a process involving metallic carbide as an intermediate; for example, Fe_2C could be formed which, upon reaction with hydrogen, would yield the hydrocarbon ethylene (one of the products of reaction):

$$2Fe(s) + 2CO(g) \rightleftharpoons Fe_2C(s) + CO_2(g)$$

followed by:

$$2Fe_2C(s) + 2H_2(g) \rightleftharpoons C_2H_4(g) + 4Fe(s)$$

Another possibility [411, 412] was that carbon monoxide and hydrogen formed, initially, carbon–hydrogen–oxygen complexes on the catalyst surface. These complexes would then act as nuclei for building up by a chain-like process the higher hydrocarbons as successive monoxide molecules were added to the initial complex. By employing radioactive (^{14}C-labelled) carbon monoxide, Emmett [412–418] concluded that the first of these possibilities was mistaken. The procedure adopted was as follows. A sample of iron catalyst was partially converted to Fe_2C by exposure to radioactive carbon monoxide, and the fraction of the surface consisting of radioactive carbon was assayed. The iron catalyst containing the radioactive iron carbide was then exposed to a mixture of non-radioactive carbon monoxide and hydrogen, and the gas mixture circulated over the catalyst. Emmett found, upon determining the quantity of radioactive hydrocarbons in the product, that only about 10% of the radioactivity had been transferred from the catalyst surface to the gas phase. Hence, carbide formation cannot be one of the principal pathways through which hydrocarbon products are formed (see Chapter 8). Later researches with [14]C [412, 419, 420], utilizing the same ratiocinative arguments, led to the conclusion that primary and secondary alcohols combined were, in all probability, involved as the principal intermediaries in the formation of higher hydrocarbons from mixtures of hydrogen and carbon monoxide passing over Fischer-Tropsch catalysts. The use of ^{14}C in catalysis research continues to increase [421].

3.3.7.2. *Use of Stable Isotopes*

In our discussion of the possible reasons for the decrease in heat of adsorption with increasing coverage (Section 2.3.5), we learned of the ingenious method devised by Roginskii of proving the existence of surface heterogeneity by using the two stable isotopes light hydrogen and deuterium. It will be recalled that Roginskii simply measured the ease of desorption of hydrogen and of deuterium to probe the energy contours of a metal surface. Another, quite distinct use of hydrogen and deuterium in the investigation of catalyst surfaces is to make use of the exchange reaction:

$$H_2 + D_2 \rightleftharpoons 2HD$$

one of the simplest examples of a catalysed reaction. This heterogeneously catalysed reaction is known to involve the chemisorption of hydrogen atoms [422], so that the catalytic activity of a surface and its dependence upon the availability of unpaired electrons at the surface can be gauged from the ease with which the exchange proceeds [423–426]. The efficacy of a solid surface in catalysing the para-hydrogen conversion:

$$p\text{-}H_2 \rightleftharpoons o\text{-}H_2$$

is also a measure of the availability of electrons at the surface, and this reaction is also used to gain information about the performance of a given surface as a catalyst in another reaction [422, 427], because here again we have chemisorption of hydrogen atoms.

But there is another, and far more extensive, use for stable isotopes in the field of catalysis. The study of the exchange of both deuterium and oxygen-18 between chemisorbed species and other species taking part in a catalysed reaction has enhanced our knowledge of the mechanisms of several important reactions. In particular, this study has yielded information concerning the existence and the reactivity of radicals adsorbed on the surfaces of catalysts [428–430]. The study of exchange reactions in heterogeneous catalysis has received much attention from Kemball, Bond, Winter and their co-workers.

It is an established fact [422] that hydrogen atoms in saturated hydrocarbons exchange with molecular deuterium homogeneously only above about 600°C. Hence, in following the course of deuterium exchange in heterogeneous systems (at temperatures which seldom exceed 300°C [430]) we need have no fears that a significant quantity of the deuterium has been transferred from one species to another in the gas phase. Likewise, at temperatures where ^{18}O exchange is being followed in heterogeneous systems, it has been proved beforehand that no homogeneous transfers occur. In illustrating the theory, scope and practice of exchange reactions in heterogeneous catalysis we shall concentrate on deuterium exchange and draw very largely from a seminal paper by Kemball [428].

If, in the course of deuterium exchange with a saturated hydrocarbon C_nH_m there is a percentage x_i of the isotopic species $C_nH_{m-i}D_i$ present at a time t, then the extent of exchange can be represented by a function f defined as:

$$f = x_1 + 2x_2 + 3x_{3+} \ldots . + mx_m \qquad (72)$$

If all the isotopic species (the deuterated hydrocarbons) react at the same rate, the exchange reaction will follow the first-order equation [431]:

$$- \log_{10} (f - f_\infty) = \frac{k_f t}{2 \cdot 303 f_\infty} + C \qquad (73)$$

where C is a constant, f_∞ is the equilibrium value of f, and k_f is a rate constant equivalent to the number of deuterium atoms entering 100 molecules of the hydrocarbon in unit time. (When the hydrocarbon contains more than one kind of carbon–hydrogen bond, equation (73) requires slight modification [432].) If it happens that the course of the exchange reaction follows equation (73), we may deduce that all the hydrogen atoms in the molecule are equally accessible for exchange with deuterium: if the equation does not hold, we have a useful indication of differences in reactivity between the various hydrogen atoms of the molecule.

The rate of disappearance of the light hydrocarbon C_nH_m may also be represented by a constant k (expressed in terms of percentage per unit time) where:

$$- \log (x_0 - x_\infty) = \frac{kt}{2 \cdot 303 (100 - x_\infty)} + C^1 \qquad (74)$$

in which C^1 is another constant, x_0 is the percentage of C_nH_m present at time t, and x_∞ is the equilibrium percentage of this species. The ratio k_f/k (usually denoted M) of the two constants is the mean number of deuterium atoms entering each hydrocarbon molecule in the initial part of the exchange reaction, where equation (74) is most likely to be valid.

Where only one of the hydrogen atoms in a hydrocarbon molecule is replaced by deuterium on each occasion that the molecule reacts on the catalyst, the exchange reaction is called a *simple step-wise exchange*. Deuterated species containing two or more deuterium atoms would be formed only by successive reactions of the first products. The way in which we would know whether such a step-wise exchange prevails is firstly from the value of M and secondly from the nature of the initial product: M would be unity, and only monodeuterated species such as $C_nH_{m-1}D$ would be produced initially.

Where more than one of the hydrogen atoms in a hydrocarbon molecule are replaced by deuterium atoms on each occasion that the molecule reacts on the catalyst the process is referred to as *multiple exchange*. This type of exchange is characterized (i) a value of M greater than unity and (ii) the appearance, initially, of products containing more than one deuterium atom.

he actual experimental method of following the course of an exchange reaction is usually based on the use of the mass spectrometer [433–435], although other techniques such as infrared spectroscopy [436] can be used. The experimental set-up in which the mass spectrometer is directly linked to the reaction vessel is very useful [428] in that analyses—of the increase in the deuterium content of the hydrocarbon, or the increase in the hydrogen content of the deuterium or, better, both increases—can be made at any stage in the reaction, but especially at the initial stage. Two rather important experimental details are worth noting, even if we are not considering here the precise arrangement of the apparatus. Firstly, the amounts of reactants and products which are bled away into the spectrometer can be made extremely small and, secondly, it is desirable to employ high ratios of deuterium to exchanging molecule so as to minimize the influence of isotopic dilution of the deuterium on the rate of production of the more highly exchanged species. (It is difficult to determine the initial product distribution accurately if the rate of their production is decreasing rapidly.)

One of the most encouraging results of the use of deuterium exchange reactions for the study of catalysis has been that, in large measure, there is consistency between the patterns of catalysis observed with evaporated metal films on the one hand and with the corresponding bulk metal on the other. Thus, we can say [422] with some degree of confidence that, whatever their physical form, the noble metals of group VIII (rhodium, palladium, iridium and platinum) will tend to favour multiple exchange, whereas metals in groups IVA, VA and VIA will have little tendency to favour multiple exchange. The mechanism of the multiple exchange which appears to be most commonly obeyed for hydrocarbons other than methane is that which involves adjacent carbon atoms, and is termed [437] α–β exchange. In such an exchange, loss of hydrogen atoms takes place from each of two neighbouring carbon atoms of the hydrocarbon.

Enough has been said to indicate how exchange of deuterium at the surfaces of catalysts reveals the various pathways of the heterogeneous reactions. It ought to be stated that use of oxygen-18 has provided equally valuable information, especially concerning the part played by lattice oxygen in the processes of chemisorption on, and catalysis at, various oxide surfaces. Often, the catalytic exchange of oxygen-18 between, say, gaseous oxygen and water vapour on the surfaces of solid oxides has been studied largely for its own sake [438–442], but, more recently, two reactions readily catalysed by metallic oxides—the decomposition of nitrous oxide [443] and the oxidation of carbon monoxide by gaseous oxygen [443–445]—have received particular attention.

Just as in deuterium exchange, mass spectrometry turns out to be the most efficacious technique for following the course of oxygen-18 exchange. And it is imperative to certify prior to the actual exchange experiments that neither reactants nor products suffer a too-rapid direct oxygen-18 exchange with the catalyst surface under the conditions of the reaction.

Apart from emphasizing the importance of participation of lattice oxygen in the decomposition of nitrous oxide and the oxidation of carbon monoxide (see Chapter 8 for fuller details of the mechanism of the latter reaction) on oxide surfaces, the oxygen-18 technique has demonstrated, just as did the radiochemical (Section 3.3.7.1) and flash-desorption (Section 3.2.8.1) techniques, that the fraction of total surface sites which takes part in the catalytic reaction is very small [*443, 446*].

3.3.8. Low-energy electron diffraction

In a recent discussion of the impact of the low-energy (or slow) electron diffraction technique upon the study of heterogeneous catalysis, Germer and Macrae [*447*] stated: "Studies of catalysis upon polycrystalline surfaces may some day be outmoded". In view of the optimism expressed in the previous section concerning the conclusions of Kemball [*428*] and Bond [*422*]—that there was a large measure of consistency between the patterns of catalysis observed with evaporated films and with polycrystalline solids—it is incumbent upon us to comment on the influence of the physical state of a solid on its catalytic behaviour. In no aspect of our subject is this topic more obviously relevant than in the discussion of the results of recent electron diffraction studies. Some of the facts that have emerged are of fundamental importance.

It has been known for a long time that the arrangement of foreign atoms adsorbed as a single monolayer upon a crystal surface can be determined by the diffraction of very low-energy electrons (having speeds and wavelengths corresponding to potential differences of the order of 50–100 V rather than the customary 50 kV), and many investigators [*448–451*] have endeavoured to develop a suitable apparatus which could take full advantage of this fact. Germer and his associates [*452, 453*] have recently developed the ideas of Ehrenberg [*451*] to the point where a comparatively simple apparatus can be built that enables the low-energy electrons carrying diffraction-pattern information to be accelerated through a large potential difference so that the entire patterns can be seen on a fluorescent screen. With this new technique, Germer *et al.*, Farnsworth [*454*] and Lander [*455, 456*] have determined the arrangements of various kinds of atoms and molecules upon a variety of crystal surfaces. Figure 32 shows the experimental arrangement used by Germer *et al.* [*457*]. The electrons produced by an electron gun are arranged to strike the crystal at normal incidence, and the electrons of the resulting diffraction pattern move in a field-free space until they reach a fine-mesh grid. Between this grid and a second grid is a decelerating field which removes

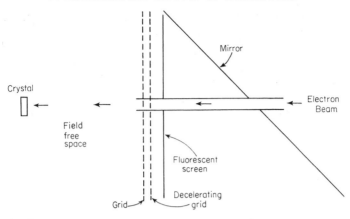

FIG. 32. Scale drawing of the experimental arrangement used by Germer *et al.* [*457*] to study the diffraction of low-energy electrons.

most of the slow electrons. Beyond the second grid, the electrons are accelerated by 1–4 kV to a fluorescent screen, on which the pattern can be observed and photographed directly. Figure 33 shows a more complete sketch of the diffraction tube, which is itself part of an ultra-high-vacuum unit. (From our discussion in Section 3.2.2, we are aware that a cleaned crystal surface will become covered by a monolayer of foreign species in less than 3 min if the residual pressure is greater than about 10^{-9} torr.)

Some of the most remarkable results obtained with low-energy electron diffraction have emerged from studies of adsorption on the {100}, {110} and {111} faces (stringently cleaned) of nickel single crystals [*458–466*]. In the

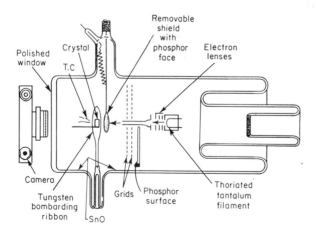

FIG. 33. Sketch of the entire electron diffraction tube (see Ref. *452*).

first place, the spacing between the outer and the next layer of nickel atoms is increased about 5% over that in the bulk. Secondly, when adsorption of hydrogen or oxygen takes place on the relatively open {110} nickel faces, there is a concomitant rearrangement of the surface atoms of nickel. Such reconstructive rearrangement, or reconstitution as it is called [461], does not take place when the same gas molecules are chemisorbed on the more densely packed {111} and {100} planes. So pronounced is the rearrangement, and so unmistakable is the change in electron diffraction pattern following re-arrangement, that the course of the adsorption of hydrogen on the {110} faces of nickel can be followed, indirectly, by following the change in arrange-ment of the nickel atoms. (The hydrogen atoms on a crystal surface do not scatter electrons sufficiently to modify the diffraction pattern directly.) Thirdly, the reconstitution of the {110} faces during hydrogen chemisorption occurs at a fixed pressure at a given temperature; in other words, a type of surface phase-transition occurs. The new structure formed by the adsorp-tion of hydrogen on this surface is composed of both nickel and hydrogen atoms, and it is referred to as the (2×1) structure, since its repeat distance is twice that of the underlying nickel surface structure in the [10] crystallographic direction and equal to that of the nickel in the [01] direction. (Since the scattering of the incident electrons by hydrogen is so small, it is not possible to determine the location of the hydrogen atoms in this structure.) It is not yet known whether, as seems likely [461], at sufficiently low temperatures hydro-gen will be adsorbed without a concomitant reconstitution of the {110} planes, but it is known [463] that when carbon monoxide is chemisorbed on the {110} faces at ordinary temperatures no reconstitution occurs. Park and Farnsworth [465] have shown that, on a clean {110} face, the carbon mon-oxide is adsorbed with one molecule to each atom of the clean nickel surface, the two-dimensional surface unit cell of the monoxide structure having the same dimensions as the nickel substrate.

Conclusions as unexpected as those drawn from the work on nickel have also been drawn from recent work on other surfaces, especially those of the elemental semiconductors silicon and germanium [456, 467, 468]. For instance, the surfaces of silicon and germanium single crystals involve a surface lattice which is different from that of the underlying material. Moreover, ordered structures and two-dimensional "superstructures", reconstructive first-order transitions and order-disorder phenomena are found to be common [456]. But we are, as yet, still in the embryonic stages in the development of the field of low-energy electron diffraction. Exciting experiments still await execution [447]. With rare exception [465], no studies have been carried out to ascertain what happens to the crystal surface when a second gas is put on the surface under conditions likely to lead to catalysis. It has, however, already become overwhelmingly evident that enormous differences of behaviour can be expected from one crystal face to another—it is for this reason that it is now felt, in some quarters, that studies on polycrystalline materials are likely to prove sterile in view of the bewildering complexity of the various crystallographic planes. If it is agonizingly difficult to explain major differences in the surface behaviour of the {110} and {100} planes of nickel, how much more difficult is it going to be to account for the behaviour of a polycrystalline solid or a badly defined evaporated film?

As a last word on low-energy electron diffraction, it is pertinent to emphasize what may turn out to be one of the most surprising consequences of the technique, namely that it has been demonstrated experimentally that surface coverage (as of chemisorbed hydrogen on {110} nickel) may change discontinuously with pressure. In Chapter 2 (Sections 2.3–2.3.8.1) it was tacitly assumed that any adsorption isotherm must represent surface coverage as a continuous function of the equilibrium pressure (for chemisorptions). Lander [456] has pointed out the necessity of considering adsorption isotherms such as those formulated by Fowler and Guggenheim [469], largely for academic interest, but which, unlike the isotherms of Langmuir, Freundlich, Temkin, etc., predict discontinuities in coverage. Fowler and Guggenheim's formulation led to an equation relating the gas-phase pressure p to the fraction coverage θ:

$$p = p' \frac{\theta}{(1 - \theta)} \exp \left(\frac{2 \theta w}{kT} \right) \tag{75}$$

where

$$p' = \frac{(2\pi m)^{3/2} (kT)^{3/2} j_G \exp \left(-\frac{\epsilon_0}{kT} \right)}{h^3 j_s} \tag{76}$$

ϵ_0 is the interaction energy between an adsorbed species and a surface site; $2 w/z$ is the interaction energy between an adsorbed unit and nearest adsorbed neighbours (z being the lateral coordination number); and j_G and j_s are the internal energy partition functions for species in the gas phase and in the surface phase respectively. When w is small (or repulsive in effect), the Fowler and Guggenheim isotherm (equation (75)) takes the Langmuir form. But a discontinuity appears when w is strong (and attractive) [470].

3.3.9. Optical and electron microscopy

Far too little use has been made of optical microscopy in the general study of catalysis. Of late, however, there has been an increased awareness of the utility of this simple, direct method, which is particularly well suited for the identification of separate phases in mixed catalysts [302]. One very striking illustration of the value of optical microscopy on a pragmatic level in industrial catalysis is to be found in the recent work of van Zoonen [*471*], who was able to measure, microscopically, the distribution of coke (deposited during catalytic hydroisomerization) in the body of a 3-mm catalyst pellet. Optical microscopy can also reveal the extent to which the internal surface of a catalyst has been effectively used. (See also p. 206.) But the most far-reaching consequence of the use of optical microscopy has emerged from the work of Gwathmey and Cunningham [*3, 472–475*] on the studies of changes in surface structure during the course of a catalysed reaction. For reasons which are still largely obscure, all the faces of pure (99·999%) copper crystals rearrange to produce facets parallel to the close-packed {111} planes when a single crystal of the copper functions as a catalyst for the reaction of hydrogen and oxygen at 400°C. By using spherical single crystals of copper, Gwathmey and Cunningham were also able to establish that the catalytic activity varied from face to face and that there was a slow formation of copper powder (similar to a previously reported [*476*] effect) during catalysis.

The application of electron microscopy has steadily increased ever since it was realized [*46, 477–482*] that replication techniques could reveal considerable topographical details of bulk catalysts and evaporated films. More information can be gleaned from transmission electron microscopy, as this technique not only permits selected-area electron diffraction patterns to be taken (for identification of the particular crystallographic face) but it also enables the structural imperfections of a solid to be assayed directly. Moreover, the resolution of transmission electron microscopy exceeds that of the replication technique [*44–46, 483*]. To illustrate the power of transmission electron microscopy, we shall cite the results of Sanders and his co-workers, previously adumbrated in Section 3.2.2.2. The prime contributions made by these investigators is first their experimental demonstration to show that, as expected from previous work [*463, 475, 484*], it is not sufficient to know the overall orientation of the surface—the detailed structure must be known—and, secondly, their analysis of the nature and influence of the structure of (and defects in) evaporated metal films.

For example, in the transmission electron micrograph shown in Fig. 34, small nickel crystals of width in the range (200–1000 Å) are visible: so also are the gaps between the crystals. (The nickel film was deposited at 273°K and, although it was thought, not unnaturally, that the film was formed on a glass substrate, it was proved by selected-area electron diffraction that the nickel film grew on a nickel oxide substrate [*44*].) Figure 35 shows a similar (transmission) micrograph after the nickel film had been heated in vacuum at 670°K for 60 min. It is apparent from the increased crystal width, the decreased dimensions of the gaps, as well as from the presence of the stacking faults (see, for example, Ref. *485*), that extensive crystal growth has occurred on heating. A carbon–platinum replica (see Fig. 36) of the surface after

FIG. 34. Transmission electron micrograph of an evaporated nickel film. The width of the nickel crystallites falls in the range 200–1000 Å. × 200,000. (By kind permission of Dr J. V. Sanders [44].)

FIG. 35. Transmission electron micrograph (× 200,000) after the film shown in Fig. 34 had been heated at 670°K for 60 min. Note the presence of stacking faults [44].

heating shows clearly the elimination of the surface asperities and the production of a comparatively featureless surface.

A second example is apposite. Bagg, Jaeger and Sanders [46] used evaporated silver films, deposited on to glass and mica, as catalysts for the decomposition of formic acid. In addition to following the detailed kinetics of the decomposition, they examined the films by transmission electron microscopy, and thereby evaluated the concentrations of various structural defects. They concluded that variations in the concentration of points and lines of emergence of dislocations constituting grain boundaries do not seem to influence the kinetics of decomposition, and that the activation energy of the reaction is low only when a large proportion of the surface is parallel to {111} planes. (See also pp. 261, 266.)

Allpress and Sanders [486] have recently completed an elegant demonstration of how, with the aid of transmission electron microscopy, minute particles of gold, condensed on a silver surface in a vacuum, can be used to decorate or reveal atomic steps and other microtopographical features on the

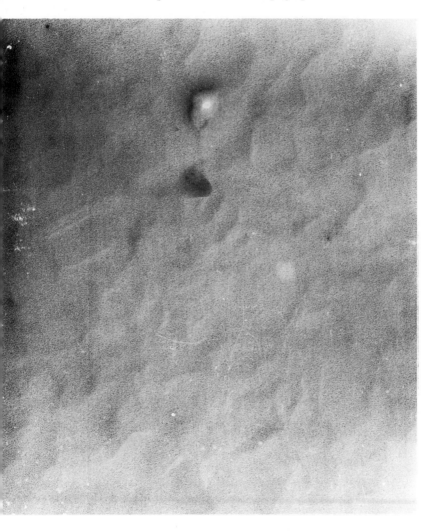

FIG. 36. Replica electron micrograph of the surface shown in Fig. 35 (× 160,000). Note the elimination of surface asperities [44].

FIG. 37. Electron micrograph (× 35,000) of a silver crystal which has been etched to simple planes and subsequently "decorated" with gold. Twin planes (tp), slip steps (ss) and internal (i) and external (e) edges are readily apparent [44].

silver surface. Their technique is an extension of the method pioneered by Bassett *et al.* [43, 487]. Figure 37 illustrates a typical, thermally etched surface decorated with gold: twin boundaries, a slip step (of monatomic dimensions), and internal and external edges (see Fig. 38) stand out clearly.

3.3.9.1. *Diffraction Techniques*

The ability to analyse the chemical composition of a minute portion of a catalyst surface by selected-area electron diffraction is one of the most noteworthy features of the electron microscopic approach. Lippens and de Boer [488, 489], in their elegant studies of the surface composition of alumina catalysts, were able to identify unambiguously microcrystals of boehmite, nordstrandite, beyerite and other hydrated aluminas in their samples. X-Ray diffraction [490] is also of value as a means of phase identification in catalysis, but it is not as successful as in other fields of chemistry, largely because the finely divided materials used as catalysts yield diffraction patterns which are unavoidably diffuse, and consequently less informative.

X-Ray spectroscopy as a spectrochemical technique for identifying the

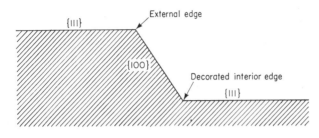

FIG. 38. Internal and external edges at the junction of simple planes [44].

various elements present in a solid [491] has not been put to the fullest use, and neither has X-ray fluorescence spectroscopy. However, increasing use is currently being made of the X-ray microscopic technique pioneered by Berg [492] and Barrett [493] over thirty years ago. For example, Newkirk [494] has used the Berg-Barrett technique to locate areas of misfit, such as the immediate environment of a dislocation, in certain solids. Lang [495] has developed the technique of X-ray transmission microscopy to the state where it can safely be used [496] to determine the concentration and distribution of dislocations, and their relevant Burgers vectors, in most solids. Being able to ascertain the degree of crystallographic misorientation, stacking fault and strain in a solid is a prerequisite [497] to an understanding of the role of lattice defects in heterogeneous catalysis (see Chapter 5).

3.3.9.2. Electron-probe Microanalysis

One extremely powerful new technique of surface analysis which may well facilitate the identification of separate phases and trace impurities in industrial catalysts is that of electron-probe microanalysis [498–500]. Basically, the technique entails X-ray spectrochemical analysis of areas between 0·1 and 3μm diameter on the surface of a solid specimen. A very narrow but powerful beam of electrons is made to strike a given microscopic area of the specimen. The electrons generate the characteristic X-ray spectra of the chemical elements contained in the area irradiated to a depth of about 1–3μm below the surface, and the emitted X-rays are analysed according to wavelength and intensity. This permits both qualitative and quantitative chemical analysis of the micron-sized volumes. The commercial instruments now available [*500*] are fitted with an optical microscope to assist in the selection of the exact area to be analysed. The limit of detectability of the electron-probe microanalyser is, at present, about 100–500 ppm w/w, but this limit should not be interpreted too severely, as it must be remembered that a local concentration (say in 1 μm^3) of 100 ppm may correspond to an overall average concentration within the entire specimen of 1 ppm or less. The technique has been successfully used (i) to identify unknown phases or particles in a parent solid [*501, 502*] and (ii) to determine the local concentration of materials present in very low-average total concentrations [*500, 503, 504*].

3.3.10. Gas chromatography and other techniques

3.3.10.1. *Gas Chromatography*

It has already been stated (see Section 3.2.8) that gas–solid chromatography has recently proved most useful in the evaluation of heats of adsorption. Nowadays both gas–solid and gas–liquid chromatography have found increasing use in the direct study of catalysis.

The accurate measurement of the evolution of small quantities of product, or of mixtures of chemically similar products, has hitherto been carried out by specialized techniques such as mass spectrometry (see, for example, the work of Kemball [*429, 433, 505–508*] and Bond [*422*]) and refined pressure gauges which respond to a particular gaseous product [*509, 510*]. At present, however, there is a preference for chromatographic analysis to be used [*511–526*] to follow the course of catalysed reactions. Quite apart from the rapidity with which the analysis of the products can be carried out, the technique also enables kinetic orders, activation energies and heats of adsorption to be readily determined. Gas chromatography has already proved effective in the screening of a large series of catalysts to ascertain the nature and extent of products in particular reactions.

Emmett and his co-workers [*412, 511–513*] have demonstrated well the value of using a small catalytic reactor in a circuit containing an analytical chromatographic column: this constitutes the so-called microcatalytic technique, latterly developed by Hall [*515*] and by Owens and Amberg [*519*]. Details of this technique are illustrated in Fig. 39, which shows the apparatus used by Kokes, Tobin and Emmett [*511*] in their study of the catalytic cracking of hydrocarbons. The hydrocarbon in question was swept over the catalyst, through a chromatographic column and out through an analyser in a stream of helium (the carrier gas). A flow-type Geiger counter could be put in series with the analyser, which was a thermal conductivity cell, whenever radioactive species such as ^{14}C-labelled compounds were being studied. Typical results obtained in the catalytic cracking of trimethyl pentane are shown in Fig. 40. Hall *et al.* [*515*] have used their semiautomatic microcatalytic

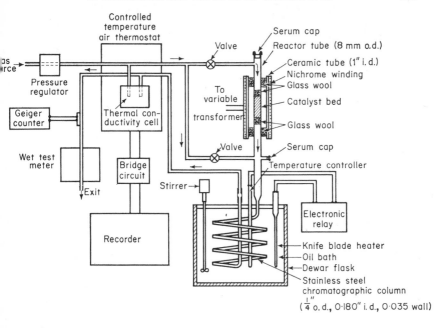

FIG. 39. Diagram of micro-catalytic (chromatographic) apparatus used by Kokes, Tobin and Emmett [511] in their study of the cracking of hydrocarbons.

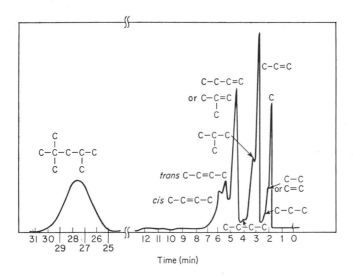

FIG. 40. Product distribution obtained by the cracking of 2,2,4-trimethylpentane over a silica–alumina catalyst in a micro-reactor [512].

reactor to investigate the polymerization of propylene and the cracking of 2,3-dimethylbutane.

Cremer and Roselius [526] pointed out that gas chromatography could, under certain conditions, yield information concerning the surface area, particle size and porosity of a solid catalyst. Suppose t_0 is the transit time of a carrier gas through a chromatographic column and t is the corresponding time for a particular gas (the test substance) to be swept through the column by the carrier gas. The retardation time $\Delta t = t - t_0$ is characteristic of the adsorbing power of the solid catalyst contained in the chromatographic column. Now, if we compare two catalysts 1 and 2 and assume that: (i) the concentration of the test substance is so small that no effective blocking of the adsorbing surface occurs (i.e. assume that the linear portion of the adsorption isotherm is being obeyed); (ii) only the adsorbability of the active centres and not their number is different; and (iii) the retention of the test substance is due only to adsorption; then the heats of desorption on the two catalysts Q_1 and Q_2, are, according to Cremer, related to the times Δt_1 and Δt_2 as follows [527]:

$$\Delta Q = RT \ln \frac{\Delta t_1}{\Delta t_2} \tag{77}$$

It is to be emphasized [526] that, even if a nonlinear adsorption isotherm is being obeyed, and even if the number of active sites fluctuates, the parameter Δt is always characteristic of the particular catalyst filling the column, and varies sensibly with the smallest alteration in the condition of the catalyst. It is for this reason that such properties as porosity and particle size can be assayed by gas chromatography.

Gas–solid chromatography offers a very convenient method for measuring the amounts of reactants adsorbed on a catalyst surface *during the actual reaction*. Let us first consider some general comments. It is important to realize that, for a certain pressure of a given reactant, the amount adsorbed on a catalyst at equilibrium, as determined from direct adsorption measurements, may bear little or no relation to the actual adsorbed amount of the given reactant, at the same partial pressure, during the catalysed reaction. Thus, in the catalytic decomposition of germane (GeH_4) on germanium, the surface is saturated with a monolayer of chemisorbed hydrogen during the decomposition reaction, whereas the maximum coverage attained in separate adsorption equilibrium measurements of hydrogen on germanium is less than 60% [528–532]. The surface coverages relevant in the actual course of the catalysis will be dependent upon the mechanism of the reaction and, in particular, on the rate-determining step; this fact has now been incontrovertibly established by Tamaru [532–540], chiefly in the study of the synthesis and decomposition of ammonia on metal catalysts. (In contrast to the approach adopted by a host of predecessors studying [541] this reaction, Tamaru set about measuring the amount adsorbed during catalysis. He found, incidentally, that, during the synthesis of ammonia on doubly promoted iron catalysts, the adsorption of nitrogen was considerably accelerated by the presence of the hydrogen, and that its rate of adsorption is almost ten times faster than the production of ammonia (see Chapter 8).) Tamaru was one of the first investigators to appreciate [533] that adsorption measurements during surface catalysis can be carried out using the principle of gas chromatography.

In essence, the method proposed by Tamaru entails using a chromatographic column charged with the catalyst under study. The test gas—comprising the reactants of the catalysed reaction, for example—is injected into the carrier gas (which is itself not influenced by the catalyst), and is allowed to pass through the column under well-defined conditions of temperature and pressure. The effluent gas is then analysed to determine how much of the reactants have been consumed; the retention time is a direct measure of the strength of adsorption on the catalyst. This is basically the method used by Bassett and Habgood [516–518] to study various reactions, such as the isomerization of cyclopropane to propylene on the recently discovered [542–547] zeolitic catalysts such as Molecular Sieves X and Y. Habgood has shown [518] that complications can arise in the chromatographic method for studying surface reactions if the adsorption and desorption steps are slow relative to the rate of surface reaction.

3.3.10.2. *Other Techniques*

It has been the aim of this chapter to present a comprehensive picture of the experimental aspects of the study of catalysis and adsorption. Almost all of the procedures and techniques currently used have been discussed to a greater or lesser extent, but one or two specialized approaches, such as the study of exo-electron emission of catalysts [548–550], X-ray K-absorption edges [551–553] and differential thermal analysis [554–556] have been omitted.

References

1. D. Graham, *J. phys. Chem.* **61**, 49 (1957).
2. D. W. L. Griffiths, W. J. Thomas and P. L. Walker, Jr, *Carbon* **1**, 515 (1964).
3. A. T. Gwathmey and R. E. Cunningham, *Adv. Catalysis* **X**, 57 (1958).
4. G. Ehrlich, *Adv. Catalysis* **XIV**, 255 (1963).
5. J. M. Thomas, Ph.D., Thesis, University of Wales (1957).
6. J. P. Hobson and R. A. Armstrong, *J. phys. Chem.* **67**, 2000 (1963).
7. P. A. Redhead, *Trans. Faraday Soc.* **57**, 641 (1961).
8. P. A. Redhead and E. V. Kornelsen, *Vakuum-Techn.* **10**, 31 (1961).
9. E. V. Kornelsen, Paper presented at the 7th National Vacuum Symposium of the American Vacuum Society, Cleveland, Ohio (1960).

10. A. Venema, Trans. 8th Vacuum Symp. and 2nd Int. Congr., Vol I, p. 1. Pergamon Press, Oxford (1962).
11. P. A. Redhead, *Can. J. Phys.* **36,** 255 (1958).
12. P. A. Redhead, *Can. J. Phys.* **37,** 1260 (1959).
13. M. Onchi, Trans. 8th Vacuum Symp. and 2nd Int. Congr., Vol. I, p. 544. Pergamon Press, Oxford (1962).
14. J. P. Hobson and P. A. Redhead, *Can. J. Phys.* **36,** 271 (1958).
15. F. J. Norton, Trans. 8th Vacuum Symp. and 2nd Int. Congr., Vol. I, p. 8. Pergamon Press, Oxford (1962).
16. F. G. Ciapetta and C. J. Plank, *in* "Catalysis", ed. by P. H. Emmett, Vol. I., p. 315. Reinhold, New York (1954).
17. R. W. Roberts, *Brit. J. appl. Phys.* **14,** 537 (1963).
18. F. G. Allen, J. Eisinger, H. D. Hagstrum and J. T. Law, *J. appl. Phys.* **30,** 1563 (1959).
19. B. M. W. Trapnell, "Chemisorption". Butterworths, London (1955).
20. M. W. Roberts and K. W. Sykes, *Proc. R. Soc.* **A242,** 534 (1957).
21. H. E. Farnsworth, R. E. Schlier, T. H. George and R. M. Burger, *J. appl. Phys.* **29,** 1150 (1958).
22. R. E. Schlier and H. E. Farnsworth, *J. chem. Phys.* **30,** 917 (1959).
23. H. E. Farnsworth, "The Surface Chemistry of Metals and Semi-conductors", p. 21. Wiley, New York (1960).
24. L. H. Germer and C. D. Hartman, *J. appl. Phys.* **31,** 2085 (1960).
25. R. H. Good and E. W. Müller, *Handb. Phys.* **21,** 176 (1956).
26. E. W. Müller, *in* "Advances in Electronics and Electron Physics", ed. by L. Marton, Vol. 13, p. 87. Academic Press, New York (1960).
27. J. Eisinger, *J. chem. Phys.* **29,** 1154 (1958).
28. T. A. Vanderslice and N. R. Whetten, *J. chem. Phys.* **37,** 535 (1962).
29. H. D. Hagstrum, *J. appl. Phys.* **32,** 1015, 1020 (1961).
30. R. W. Roberts, *Trans. Faraday Soc.* **58,** 1159 (1962).
31. R. W. Roberts, *Ann. N.Y. Acad. Sci.* **101** (III), 766 (1963).
32. J. A. Dillon, Jr, Trans. 8th Vacuum Symp. and 2nd Int. Congr., Vol. I, p. 113. Pergamon Press, Oxford (1962).
33. J. A. Dillon, Jr, Proc. Conf. Ultrapurification of Semi-conductor Materials, Boston, p. 604. Macmillan, New York (1961).
34. O. Beeck, A. E. Smith, and A. Wheeler, *Proc. R. Soc.* **A177,** 62 (1940).
35. F. C. Tompkins and R. V. Culver, *Adv. Catalysis* **11,** 68 (1959).
36. R. Suhrmann, *Adv. Catalysis,* **7,** 303 (1955).
37. C. A. Neugebauer, J. B. Newkirk and D. A. Vermilyea (eds.), "Structure and Properties of Thin Films". Wiley, New York (1959).
38. M. W. Roberts, *Trans. Faraday Soc.* **59,** 698 (1963).
39. J. A. Allen, *Rev. pure appl. Chem.* **4,** 133 (1954).
40. J. S. Anderson, *Disc. Faraday Soc.* **8,** 362 (1950).
41. G. Somarjai, *J. Phys. Chem. Solids* **24,** 175 (1963).
42. D. W. Pashley, *Adv. Phys.* **5,** 173 (1956).
43. D. W. Pashley, *Phil. Mag.* **4,** 316 (1959).
44. J. R. Anderson, B. G. Baker and J. V. Sanders, *J. Catalysis* **1,** 443 (1962).
45. J. V. Sanders, private communication (1964).
46. J. Bagg, H. Jaeger and J. V. Sanders, *J. Catalysis* **2,** 449 (1963).
47. D. R. Palmer, S. R. Morrison and C. E. Dauenbaugh, *Phys. Rev. Letters* **6,** 170 (1961).

48. D. Haneman, *J. phys. chem. Solids* **11**, 205 (1959).
49. H. E. Farnsworth, *Ann. N.Y. Acad. Sci.* **101** (III), 658 (1963).
50. R. G. Lye, *Phys. Rev.* **99**, 1647 (1955).
51. P. J. Bryant, Nat. Symp. Vacuum Technology Trans., p. 311. Macmillan, New York (1962).
52. O. Stern, *Naturwissenschaften* **17**, 391 (1929).
53. R. H. Savage, *J. appl. Phys.* **19**, 1 (1948).
54. M. Green, J. A. Kafalas and P. H. Robinson, "Semiconductor Surface Physics", p. 349. University of Pennsylvania Press (1957).
55. M. Green, *J. Phys. Chem. Solids* **14**, 77 (1960).
56. K. H. Maxwell and M. Green, *J. Phys. Chem. Solids* **14**, 94 (1960).
57. A. J. Rosenberg, P. H. Robinson and H. C. Gatos, *J. appl. Phys.* **29**, 771 (1958).
58. R. E. Schlier, *J. appl. Phys.* **29**, 1162 (1958).
59. J. R. Young, *J. appl. Phys.* **30**, 1671 (1959).
60. J. A. Becker, E. J. Becker and R. G. Brandes, *J. appl. Phys.* **32**, 411 (1961).
61. J. A. Dillon, Jr, and H. E. Farnsworth, *J. appl. Phys.* **29**, 1195 (1958).
62. F. G. Allen, T. M. Buck and J. T. Law, *J. appl. Phys.* **31**, 979 (1960).
63. J. T. Law, *J. Phys. Chem. Solids* **14**, 9 (1960).
64. W. T. Sproull, *Phys. Rev.* **43**, 516 (1933).
65. E. D. Hondros and A. J. W. Moore, *Acta metal.* **8**, 647 (1960).
66. A. J. W. Moore, *Acta metal.* **10**, 579 (1962).
67. H. E. Farnsworth and H. H. Madden, Jr, *J. appl. Phys.* **32**, 1933 (1961).
68. M. J. Duell and A. J. B. Robertson, *Trans. Faraday Soc.* **57**, 1416 (1961).
69. K. H. Kingdon and I. Langmuir, *Phys. Rev.* **22**, 148 (1923).
70. A. J. W. Moore, *Am. Scient.* **48**, 109 (1960).
71. J. H. Leck, *in* "Chemisorption", ed. by W. E. Garner, p. 162. Butterworths, London (1957).
72. J. A. Dillon, Jr, and H. E. Farnsworth, *J. appl. Phys.* **28**, 174 (1957).
73. J. A. Dillon, Jr, and R. M. Oman, *J. appl. Phys.* **31**, 26 (1960).
74. J. V. Laukonis and R. V. Coleman, *J. appl. Phys.* **32**, 242 (1961).
75. E. W. Müller, *Phys. Rev.* **102**, 618 (1956).
76. E. W. Müller, *J. appl. Phys.* **28**, 1 (1957).
77. F. G. Allen, *J. Phys. Chem. Solids* **8**, 119 (1959).
78. E. C. Cooper and E. W. Müller, *Rev. sci. Instr.* **29**, 309 (1958).
79. W. J. Thomas, *Trans. Faraday Soc.* **53**, 1224 (1957).
80. L. D. Hale, *Rev. sci. Instr.* **29**, 367 (1958).
81. R. T. Bayard and D. Alpert, *Rev. sci. Instr.* **21**, 571 (1950).
82. E. A. Gulbransen, *Adv. Catalysis* **5**, 120 (1953).
83. T. N. Rhodin, *Adv. Catalysis* **5**, 40 (1953).
84. S. J. Gregg, *J. chem. Soc.* 561 (1946).
85. J. M. Thomas and B. R. Williams, *Q. Rev.* **19**, 231 (1965).
86. A. W. Czanderna and J. M. Honig, *Analyt. Chem.* **29**, 1206 (1957).
87. W. D. Machin, Ph.D., Thesis, Rensselaer Polytechnic Institute, N.Y. (1961).
88. S. P. Wolsky and E. J. Zdanuk, "Vacuum Microbalance Techniques", Vol. 1, p. 35. Plenum Press, New York (1961).
89. J. M. Thomas and B. R. Williams, *in* "Vacuum Microbalance Techniques", Vol. 4, p. 204. Plenum Press, New York (1965).
90. J. W. McBain and A. M. Baker, *J. Am. chem. Soc.* **48**, 690 (1926).
91. S. J. Gregg, D. W. Aylmore and Jepson, *J. Inst. Metals* **88**, 205 (1959–60).

92. R. V. Jones, *J. sci. Instr.* **38**, 37 (1961).
93. L. R. McNary, National Research Council (Canada), Report No. 4259 (ERA-311) (1956).
94. J. M. Thomas and J. A. Poulis, *in* "Vacuum Microbalance Techniques", Vol. 3, p. 15. Plenum Press, New York (1963).
95. J. A. Poulis and J. M. Thomas, *J. sci. Instr.* **40**, 95 (1963).
96. E. A. Gulbransen, *Rev. sci. Instr.* **15**, 201 (1944).
97. A. W. Czanderna and J. M. Honig, *J. phys. Chem.* **63**, 620 (1959).
98. A. W. Czanderna, *in* "Vacuum Microbalance Techniques", Vol. 4. Plenum Press, New York (1965).
99. A. W. Czanderna, *J. phys. Chem.* **68**, 27 (1964).
100. A. W. Warner and C. D. Stockbridge, *in* "Vacuum Microbalance Techniques", Vol. 2, p. 93. Plenum Press, New York (1962).
101. A. W. Warner and C. D. Stockbridge, *in* "Vacuum Microbalance Techniques", Vol. 3, p. 55. Plenum Press, New York (1963).
102. I. Haller and P. White, *Rev. sci. Instr.* **34**, 677 (1963).
103. K. H. Behrndt and R. W. Love, *7th Vacuum Symp. Trans.*, p. 87. Pergamon Press, London (1960).
104. S. C. Liang, *J. phys. Chem.* **57**, 910 (1953).
105. M. J. Bennett and F. C. Tompkins, *Trans. Faraday Soc.* **53**, 185 (1957).
106. G. A. Miller, *J. phys. Chem.* **67**, 1359 (1963).
107. T. Takaishi and Y. Sensui, *Trans. Faraday Soc.* **59**, 2503 (1963).
108. H. H. Podgurski and F. N. Davis, *J. phys. Chem.* **65**, 1343 (1961).
109. D. M. Young and A. D. Crowell, "Physical Adsorption of Gases". Butterworths, London (1962).
110. S. Brunauer, P. H. Emmett and E. Teller, *J. Am. chem. Soc.* **60**, 309 (1938).
111. I. Halasz and G. Schay, *Z. anorg. allg. Chem.* **287**, 242 (1956).
112. I. Halasz, G. Schay and K. Wencke, *Z. anorg. allg. Chem.* **287**, 253 (1956).
113. R. A. Beebe, J. B. Beckwith and J. M. Honig, *J. Am. chem. Soc.* **67**, 1554 (1945).
114. O. Beeck, *Adv. Catalysis* **2**, 151 (1950).
115. D. F. Klemperer and F. S. Stone, *Proc. R. Soc.* **A243**, 375 (1957).
116. R. A. Pierotti and G. D. Halsey, *J. phys. Chem.* **63**, 680 (1959).
117. M. W. Roberts, *Trans. Faraday Soc.* **56**, 128 (1960).
118. J. M. Thomas, *Nature, Lond.* **189**, 134 (1961).
119. J. M. Haynes, *J. phys. Chem.* **66**, 182 (1962).
120. J. R. Anderson and B. G. Baker, *J. phys. Chem.* **66**, 482 (1962).
121. W. A. Cannon, *Nature, Lond.* **197**, 1000 (1963).
122. J. H. Singleton and G. D. Halsey, Jr, *J. phys. Chem.* **58**, 330 (1954).
123. V. Ponec and Z. Knor, *Coll. Czech. chem. Commun.* **27**, 1091 (1962).
124. D. Brennan, M. J. Graham and F. H. Hayes, *Nature, Lond.* **199**, 1152 (1963).
125. P. H. Emmett and S. Brunauer, *J. Am. chem. Soc.* **59**, 1553 (1937).
126. P. H. Emmett, *in* "Catalysis", ed. by P. H. Emmett, Vol. I, p. 31. Reinhold, New York (1954).
127. P. H. Emmett, T. W. De Witt and R. T. Davies, *J. Phys. Colloid Chem.* **57**, 1232 (1947).
128. H. K. Livingston, *J. Colloid Sci.* **4**, 447 (1949).
129. P. H. Emmett and M. Cineo, *J. Phys. Colloid Chem.* **51**, 1248 (1947).
130. S. Ross, *J. Am. chem. Soc.* **70**, 3830 (1948).
131. J. Pritchard, *Nature Lond.* **194**, 38 (1962).

132. G. Ehrlich and F. G. Hudda, *J. chem. Phys.* **30**, 493 (1959).
133. W. J. M. Rootsaert, L. L. van Reijen and W. M. H. Sachtler, *J. Catalysis* **1**, 416 (1962).
134. R. Gomer, *Disc. Faraday Soc.* **28**, 23 (1959).
135. G. F. Hüttig, *Mh. Chem.* **78**, 177 (1948).
136. S. Ross, *J. phys. Chem.* **53**, 383 (1949).
137. T. L. Hill, *J. Am. chem. Soc.* **72**, 5347 (1950).
138. G. Jura and W. D. Harkins, *J. chem. Phys.* **11**, 430 (1943).
139. G. Jura and W. D. Harkins, *J. Am. chem. Soc.* **68**, 1941 (1946).
140. S. J. Gregg, *J. chem. Soc.* 696 (1942).
141. S. J. Gregg and F. A. P. Maggs, *Trans. Faraday Soc.* **44**, 123 (1948).
142. N. K. Adam, "Physics and Chemistry of Surfaces". Oxford University Press, London (1941).
143. P. M. W. Jacobs and F. C. Tompkins, *in* "Chemistry of the Solid State", ed. by W. E. Garner, Ch. 4. Butterworths, London (1955).
144. W. D. Harkins and G. Jura, *J. Am. chem. Soc.* **66**, 1366 (1944).
145. M. M. Dubinin, *Chem. Rev.* **60**, 235 (1960).
146. M. Polanyi and F. Goldman, *Z. phys. Chem.* **A132**, 313 (1928).
147. H. Marsh and W. F. K. Wynne-Jones, *Carbon* **1**, 269 (1964).
148. T. G. Lamond and H. Marsh, *Carbon* **1**, 281 (1964).
149. T. G. Lamond and H. Marsh, *Carbon* **1**, 293 (1964).
150. A. G. Foster, *J. chem. Soc.* 1806 (1952).
151. D. H. Everett, "Industrial Carbon and Graphite", p. 272. Society of Chemical Industry, London (1958).
152. C. Pierce, *J. phys. Chem.* **57**, 149 (1953).
153. W. A. Steele and G. D. Halsey, *J. chem. Phys.* **22**, 979 (1954).
154. W. A. Steele and G. D. Halsey, *J. phys. Chem.* **59**, 57 (1955).
155. P. I. Freeman and G. D. Halsey, *J. phys. Chem.* **59**, 181 (1955).
156. G. Constabaris and G. D. Halsey, *J. chem. Phys.* **27**, 1433 (1957).
157. P. I. Freeman, *J. phys. Chem.* **62**, 73 (1958).
158. P. I. Freeman, *J. phys. Chem.* **62**, 729 (1958).
159. G. Constabaris, J. H. Singleton and G. D. Halsey, *J. phys. Chem.* **63**, 1350 (1959).
160. J. R. Sams, G. Constabaris and G. D. Halsey, *J. phys. Chem.* **64**, 1689 (1960).
161. J. R. Sams, *Molec. Phys.* **9**, 17 (1965).
162. R. Wolfe and J. R. Sams, *J. phys. Chem.* **69**, 1129 (1965).
163. J. R. Sams, *J. chem. Phys.* **43**, 2243 (1965).
164. D. H. Everett, "Powders in Industry", p. 98. Society of Chemical Industry, London (1961).
165. J. A. Barker and D. H. Everett, *Trans. Faraday Soc.* **58**, 1608 (1962).
166. J. E. Lennard-Jones, *Proc. R. Soc.* **A106**, 463 (1924).
167. J. O. Hirschfelder, C. F. Curtiss and R. B. Bird, "Molecular Theory of Gases and Liquids". Wiley, New York (1954).
168. D. W. Aylmore and W. B. Jepson, *J. sci. Instr.* **38**, 156 (1961).
169. J. T. Clarke, *J. phys. Chem.* **68**, 884 (1964).
170. W. H. Slaughbaugh and A. D. Stump, *J. phys. Chem.* **68**, 1251 (1964).
171. C. Pierce, *J. phys. Chem.* **63**, 1076 (1959).
172. C. Pierce, *J. phys. Chem.* **64**, 1184 (1960).
173. J. Frenkel, "Kinetic Theory of Liquids". Oxford University Press, London (1946).

174. G. D. Halsey, Jr, *J. chem. Phys.* **16**, 931 (1948).
175. T. L. Hill, *Adv. Catalysis* **4**, 211 (1952).
176. T. L. Hill, *J. chem. Phys.* **17**, 668 (1949).
177. J. C. P. Mignolet, *J. chem. Phys.* **21**, 1298 (1953).
178. J. Pritchard and F. C. Tompkins, *Trans. Faraday Soc.* **56**, 540 (1960).
179. J. C. P. Mignolet, *Recl. Trav. chim. Belg. Pays-Bas* **74**, 685 (1955).
180. S. Ymaguchi, *J. chem. Phys.* **27**, 1114 (1957).
181. S. Rubin, T. O. Passell and L. E. Bailey, *Analyt. Chem.* **29**, 736 (1957).
182. S. Brunauer and P. H. Emmett, *J. Am. chem. Soc.* **62**, 1732 (1940).
183. W. K. Hall, W. H. Tarn and R. B. Anderson, *J. Am. chem. Soc.* **72**, 5436 (1950).
184. J. M. Bridges, D. S. MacIver, H. H. Tobin, Proc. Int. Congr. Catalysis, 2nd Congr., Paris, 1960, p. 2161. Technip Press, Paris (1961).
185. J. J. F. Scholten and A. van Mortfoort, *J. Catalysis* **1**, 85 (1962).
186. R. P. Eischens and P. W. Selwood, *J. Am. chem. Soc.* **69**, 1590 (1947).
187. D. S. MacIver, R. C. Zabor and P. H. Emmett, *J. phys. Chem.* **63**, 484 (1959).
188. D. S. MacIver, H. H. Tobin and R. T. Barth, *J. Catalysis* **2**, 485 (1963).
189. H. Pines and W. O. Haag, *J. Am. chem. Soc.* **82**, 2471 (1960).
190. P. H. Emmett and S. Brunauer, *J. Am. chem. Soc.* **59**, 310 (1937).
191. G. C. A. Schuit and L. L. van Reijen, *Adv. Catalysis* **10**, 243 (1958).
192. O. Beeck and A. W. Ritchie, *Disc. Faraday Soc.* **8**, 159 (1950).
193. B. M. W. Trapnell, *Proc. R. Soc.* **A206**, 39 (1951).
194. B. M. W. Trapnell, "Chemisorption", p. 180. Butterworths, London (1955).
195. M. A. H. Lanyon and B. M. W. Trapnell, *Proc. R. Soc.* **A227**, 387 (1955).
196. O. Beeck, *Adv. Catalysis* **2**, 151 (1950).
197. F. Paneth and V. Vorwerk, *Z. Phys. Chem.* **101**, 445, 480 (1922).
198. K. T. Chisnall, J. W. Lucas and K. S. W. Sing, *Chemy Ind.* 1517 (1959).
199. H. R. Lukens, Jr, R. G. Meisenheimer and J. N. Wilson, *J. phys. Chem.* **66**, 469 (1962).
200. D. S. MacIver, P. H. Emmett and H. S. Frank, *J. phys. Chem.* **62**, 935 (1958).
201. P. H. Emmett, *in* "New Approaches to the Study of Catalysis" (Thirty-Sixth Annual Priestley Lectures). The Pennsylvania State University (1962).
202. G. A. Mills, E. R. Boedeker and A. G. Oblad, *J. Am. chem. Soc.* **72**, 1554 (1950).
203. S. A. Grune and H. Pust, *J. phys. Chem.* **62**, 55 (1958).
204. H. W. Habgood and J. F. Hanlan, *Can. J. Chem.* **37**, 843 (1959).
205. P. E. Eberly, *J. phys. Chem.* **65**, 68 (1961).
206. S. Ross, J. K. Saelens and J. P. Oliver, *J. phys. Chem.* **66**, 696 (1962).
207. R. L. Gale and R. A. Beebe, *J. phys. Chem.* **68**, 555 (1964).
208. R. S. Hansen, J. A. Murphy and T. C. McGee, *Trans. Faraday Soc.* **60**, 597 (1964).
209. W. E. Garner and F. J. Veal, *J. chem. Soc.* 1436 (1935).
210. C. H. Amberg, W. B. Spencer and R. A. Beebe, *Can. J. Chem.* **33**, 305 (1955).
211. R. L. Gale, J. Haber and F. S. Stone, *J. Catalysis* **1**, 32 (1962).
212. D. F. Klemperer and F. S. Stone, *Proc. R. Soc.* **A243**, 375 (1957).
213. D. Brennan, D. O. Hayward and B. M. W. Trapnell, *Proc. R. Soc.* **A256**, 81 (1960).
214. D. Brennan and F. H. Hayes, *Trans. Faraday Soc.* **60**, 589 (1964).
215. G. Ehrlich, *J. Chem. Phys.* **36**, 1499 (1962).
216. H. D. Hagstrum, *Rev. sci. Instr.* **24**, 1122 (1953).
217. T. W. Hickmott and G. Ehrlich, *J. Phys. Chem. Solids* **5**, 49 (1958).

218. G. Ehrlich, *J. phys. Chem.* **60**, 1388 (1956).
219. J. Eisinger, *J. chem. Phys.* **27**, 1206 (1957).
220. G. Ehrlich, *J. chem. Phys.* **34**, 39 (1961).
221. G. Ehrlich, *J. chem. Phys.* **36**, 1171 (1962).
222. P. A. Redhead, Paper presented at 20th Annu. Conf. Physical Electronics. M.I.T., March (1960).
223. Y. Amenomiya and R. J. Cvetanovic, *J. phys. Chem.* **67**, 144 (1963).
224. Y. Amenomiya and R. J. Cvetanovic, *J. phys. Chem.* **67**, 2046 (1963).
225. Y. Amenomiya and R. J. Cvetanovic, *J. phys. Chem.* **67**, 2705 (1963).
226. Y. Amenomiya, J. H. B. Chevier and R. J. Cvetanovic, *J. phys. Chem.* **68**, 52 (1964).
227. P. Y. Hsieh, *J. Catalysis* **2**, 211 (1963).
228. P. Y. Hsieh, *J. phys. Chem.* **68**, 1068 (1964).
229. A. G. Oblad, T. H. Milliken, Jr, G. A. Mills, *Adv. Catalysis* **3**, 199 (1951).
230. H. Benesi, *J. phys. Chem.* **61**, 970 (1957).
231. V. C. F. Holm, G. C. Bailey and A. Clark, *J. phys. Chem.* **63**, 129 (1959).
232. F. S. Stone, *Adv. Catalysis* **13**, 1 (1962).
233. W. E. Garner and J. Maggs, *Trans. Faraday Soc.* **32**, 1744 (1936).
234. W. E. Garner, *J. chem. Soc.* 1239 (1947).
235. W. E. Garner and T. Ward, *J. chem. Soc.* 857 (1939).
236. W. E. Garner, T. J. Gray and F. S. Stone, *Disc. Faraday Soc.* **8**, 246 (1950).
237. W. E. Garner, F. S. Stone and P. F. Tiley, *Proc. R. Soc.* **A211**, 472 (1952).
238. R. M. Dell and F. S. Stone, *Trans. Faraday Soc.* **50**, 501 (1954).
239. R. Rudham and F. S. Stone, *in* "Chemisorption", ed. by W. E. Garner, p. 205. Butterworths, London (1957).
240. F. S. Stone, R. Rudham and R. L. Gale, *Z. Elektrochem.* **63**, 129 (1959).
241. A. Hickling, *Trans. Faraday Soc.* **41**, 338 (1947).
242. A. Hickling and D. Taylor, *Trans. Faraday Soc.* **44**, 262 (1948).
243. J. Fahrenfort, L. L. van Reijen and W. M. H. Sachtler, *in* "The Mechanism of Heterogeneous Catalysis", ed. by J. H. de Boer, p. 23. Elsevier, Amsterdam (1960).
244. W. M. H. Sachtler and L. L. van Reijen, *Shokubai* **4**, 147 (1962).
245. M. W. Roberts, *Nature, Lond.* **188**, 1020 (1960).
246. K. Tanaka and K. Tamaru, *J. Catalysis* **2**, 366 (1963).
247. C. Kemball and E. K. Rideal, *Proc. R. Soc.* **A187**, 53 (1946).
248. C. Kemball, *Proc. R. Soc.* **A190**, 117 (1947).
249. C. Kemball, *Adv. Catalysis* **2**, 233 (1950).
250. D. H. Everett, *Trans. Faraday Soc.* **46**, 453, 942, and 957 (1950).
251. D. H. Everett and D. M. Young, *Trans. Faraday Soc.* **48**, 1164 (1952).
252. D. H. Everett, *Proc. chem. Soc.* 28 (1957).
253. J. H. de Boer, "The Dynamical Character of Adsorption". The Clarendon Press, Oxford (1952).
254. J. H. de Boer and S. Kruyer, *Proc. K. ned. Akad. Wet.* **B56**, 67, 236, 415 (1953).
255. J. H. de Boer and S. Kruyer, *Proc. K. ned. Akad. Wet.* **B57**, 92 (1954).
256. J. H. de Boer and S. Kruyer, *Proc. K. ned. Akad. Wet.* **B58**, 61 (1955).
257. J. H. de Boer and S. Kruyer, *Trans. Faraday Soc.* **54**, 540 (1958).
258. J. M. Thomas, *J. chem. Educ.* **38**, 138 (1961).
259. R. P. Eischens, W. A. Pliskin and S. A. Francis, *J. chem. Phys.* **22**, 1786 (1954).
260. R. P. Eischens, S. A. Francis and W. A. Pliskin, *J. phys. Chem.* **60**, 194 (1956).

261. R. P. Eischens and W. A. Pliskin, *Adv. Catalysis* **10**, 1 (1958).
262. N. G. Yaroslavskii and A. N. Terenin, *Dokl. Akad. Nauk SSSR* **66**, 885 (1949).
263. A. N. Terenin, *Mikrochim. Acta* **2**, 467 (1955).
264. H. F. Leftin and M. C. Robson, Jr, *Adv. Catalysis* **14**, 115 (1963).
265. H. L. Pickering and H. C. Eckstrom, *J. phys. Chem.* **63**, 512 (1959).
266. L. H. Little, *J. phys. Chem.* **63**, 1616 (1959).
267. M. Falman and D. J. C. Yates, *Proc. R. Soc.* **A246**, 32 (1958).
268. D. W. L. Griffiths, Ph.D. Thesis, Universty of Wales (1964).
269. N. Sheppard and D. J. C. Yates, *Proc. R. Soc.* **A238**, 69 (1956).
270. D. J. C. Yates, *in* "Chemisorption", ed. by W. E. Garner, p. 93. Butterworths, London (1957).
271. W. A. Pliskin and R. P. Eischens, *Z. phys. Chem.* **24**, 11 (1960).
272. J. T. Yates, Jr, and C. W. Garland, *J. phys. Chem.* **65**, 617 (1961).
273. C. E. O'Neill and D. J. C. Yates, *J. phys. Chem.* **65**, 901 (1961).
274. K. Hirota, K. Fueki, Y. Nakai and T. K. Shindo, *Bull. chem. Soc. Japan* **31**, 780 (1958).
275. J. Fahrenfort and H. F. Hazebrock, *Z. phys. Chem.* **20**, 105 (1959).
276. J. Fahrenfort, L. L. van Reijen and W. M. H. Sachtler, *Z. Electrochem.* **64**, 216 (1960).
277. J. K. A. Clarke and A. D. E. Pullin, *Trans. Faraday Soc.* **56**, 534 (1960).
278. R. P. Eischens and W. A. Pliskin, Proc. 2nd Int. Congr. Catalysis, Vol. 1, p. 789. Editions-Technip, Paris (1961).
279. D. J. C. Yates and P. J. Lucchesi, *J. chem. Phys.* **35**, 243 (1961).
280. D. J. C. Yates and P. J. Lucchesi, *J. phys. Chem.* **67**, 1197 (1963).
281. D. J. C. Yates, *J. chem. Phys.* **40**, 1157 (1964).
282. H. P. Leftin, *J. phys. Chem.* **64**, 1714 (1960).
283. H. P. Leftin, *Rev. sci. Instr.* **32**, 1418 (1961).
284. M. W. Tamele, *Disc. Faraday Soc.* **8**, 270 (1950).
285. C. L. Thomas, *Ind. Engng Chem.* **41**, 2565 (1949).
286. B. S. Greensfelder, H. H. Voge and G. M. Good, *Ind. Engng Chem.* **41**, 2573 (1949).
287. V. Gold, B. M. V. Haines and F. L. Tye, *J. chem. Soc.* 2167 (1952).
288. A. G. Evans, *J. appl. Chem.* **1**, 240 (1951).
289. N. C. Deno, J. J. Jurazelski and A. Schriesheim, *J. org. Chem.* **19**, 155 (1954).
290. H. P. Leftin and W. K. Hall, *J. phys. Chem.* **66**, 1457 (1962).
291. A. Ron, M. Folman and O. Schnepp, *J. chem. Phys.* **36**, 2449 (1962).
292. E. R. Andrew, "Nuclear Magnetic Resonance". Cambridge University Press (1955).
293. D. E. O'Reilly, *Adv. Catalysis* **12**, 31 (1960).
294. E. R. Andrew, *Brit. J. appl. Phys.* **10**, 431 (1959).
295. J. N. Schoolery, Proc. 1st Int. Instrument Congr. and Exposition, Philadelphia, 1954. Paper No. 54–19–2. Instrument Society of America.
296. E. R. Andrew, A. Bradbury and R. G. Eades, *Nature, Lond.* **182**, 1659 (1958).
297. N. Fuschillo and C. A. Renton, *Bull. Am. phys. Soc.* **2**, 226 (1957).
298. H. A. Resing, J. K. Thompson and J. J. Krebs, *J. phys. Chem.* **68**, 1621 (1964).
299. J. R. Zimmerman and J. Lasater, *J. phys. Chem.* **62**, 1157 (1958).
300. W. K. Hall, H. P. Leftin, F. J. Cheselske and D. E. O'Reilly, *J. Catalysis* **2**, 506 (1963).
301. G. D. Watkins and R. V. Pound, *Phys. Rev.* **89**, 658 (1953).

302. F. S. Stone, *Chemy Ind.* 1810 (1963).
303. E. R. Andrew, *Disc. Faraday Soc.* **34**, 38 (1962).
304. E. R. Andrew and S. Clough, Remarks at Conf. on Nuclear Magnetic Resonance, Bangor (1961).
305. E. N. DiCarlo and H. E. Swift, *J. phys. Chem.* **68**, 551 (1964).
306. P. Cossee, *J. Catalysis* **3**, 80 (1964).
307. E. J. Arlman, *J. Catalysis* **3**, 89 (1964).
308. E. J. Arlman and P. Cossee, *J. Catalysis* **3**, 99 (1964).
309. E. Zavoisky, *Fiz. Zh.* **9**, 211 (1945).
310. J. J. Rooney and R. C. Pink, *Proc. chem. Soc.* 70 (1961).
311. J. J. Rooney and R. C. Pink, *Trans. Faraday Soc.* **58**, 1632 (1962).
312. D. M. Brouwer, *J. Catalysis* **1**, 372 (1962).
313. G. B. Pariiski, G. M. Zhidomirov and V. B. Kazanskii, *Zh. strukt. Khim.* **4**, 364 (1963).
314. A. N. Terenin, V. A. Barachevskii, E. I. Kotov and V. Kholmogorov, *Spectrochim. Acta* **19**, 1797 (1963).
315. D. E. O'Reilly and D. S. MacIver, *J. phys. Chem.* **66**, 276 (1962).
316. C. P. Poole, W. L. Kehl and D. S. MacIver, *J. Catalysis* **1**, 407 (1962).
317. D. E. O'Reilly and C. P. Poole, *J. phys. Chem.* **67**, 1762 (1963).
318. C. P. Poole, *J. phys. Chem.* **67**, 1297 (1963).
319. P. Cossee and L. L. van Reijen, Proc. 2nd Int. Congr. Catalysis, 1960, p. 1679. Technip Press, Paris (1961).
320. L. L. van Reijen and P. Cossee, Remarks at Symp. Katalyse, Eindhoven, August 1962.
321. P. W. Selwood, *Adv. Catalysis* **3**, 27 (1951).
322. P. W. Selwood, "Adsorption and Collective Paramagnetism", p. 10. Academic Press, New York (1962).
323. P. W. Selwood, "Magnetochemistry", p. 279. Interscience, New York (1956).
324. H. Morris and P. W. Selwood, *J. Am. chem. Soc.* **65**, 2245 (1943).
325. M. H. Dilke, D. D. Eley and E. B. Maxted, *Nature, Lond.* **161**, 804 (1948).
326. J. W. Geus, A. P. P. Nobel and P. Zwietering, *J. Catalysis* **1**, 8 (1962).
327. N. I. Kobozev, *Russ. J. phys. Chem.* **33**, 641 (1959).
328. C. P. Bean and J. D. Livingston, *J. appl. Phys.* **30**, 1208 (1959).
329. I. E. Den Besten, P. G. Fox and P. W. Selwood, *J. phys. Chem.* **66**, 450 (1962).
330. C. R. Abaledo and P. W. Selwood, *J. chem. Phys.* **37**, 2709 (1962).
331. P. J. Fensham, Ph.D. Thesis, Bristol (1952).
332. T. J. Gray, *in* "Chemistry of the Solid State", ed. by W. E. Garner, p. 153. Butterworths, London (1955).
333. T. J. Jennings and F. S. Stone, *Adv. Catalysis* **9**, 441 (1957).
334. M. O'Keeffe and F. S. Stone, *Proc. R. Soc.* **A267**, 501 (1962).
335. E. A. Bylina, V. B. Evdokimov and N. I. Kobozev, *Russ. J. phys. Chem.* **36**, 1392 (1962).
336. N. I. Kobozev, *Acta phys.-chim. URSS* **21**, 294 (1946).
337. H. Amariglio, Ph.D. Thesis, Nancy (1962).
338. J. M. Thomas, *in* "The Chemistry and Physics of Carbon", ed. by P. L. Walker, Vol. 1, Ch. 3. Dekker, New York (1965).
339. Zh. V. Strel'nikova and V. P. Lebedev, *Russ. J. Phys. Chem.* **36**, 842 (1962).
340. J. T. Kummer, *J. phys. Chem.* **66**, 1715 (1962).
341. J. Pritchard and F. C. Tompkins, *Trans. Faraday Soc.* **56**, 540 (1960).

342. F. C. Tompkins, "Superficial Chemistry and Solid Imperfections", p. 143. Inaugural Lectures, Imperial College of Science and Technology, London (1960).
343. C. M. Quinn and M. W. Roberts, *Nature, Lond.* **200**, 648 (1963).
344. J. G. Little, C. M. Quinn and M. W. Roberts, *J. Catalysis* **3**, 57 (1964).
345. C. M. Quinn and M. W. Roberts, *Trans. Faraday Soc.* **60**, 899 (1964).
346. J. M. Saleh, B. R. Wells and M. W. Roberts, *Trans. Faraday Soc.* **60**, 1865 (1964).
347. M. M. Siddiqi and F. C. Tompkins, *Proc. R. Soc.* **A268**, 452 (1962).
348. R. Culver and F. C. Tompkins, *Adv. Catalysis* **11**, 68 (1959).
349. D. O. Hayward and B. M. W. Trapnell, "Chemisorption", p. 38. Butterworths, London (1964).
350. J. C. P. Mignolet, *J. chem. Phys.* **20**, 341 (1952).
351. J. C. P. Mignolet, *Rec. Trav. chim. Pays-Bas Belg.* **74**, 685 (1955).
352. J. Pritchard, *Trans. Faraday Soc.* **59**, 437 (1963).
353. P. L. Jones and B. A. Pethica, *Proc. R. Soc.* **A256**, 454 (1960).
354. R. Suhrmann and W. M. H. Sachtler, Proc. Int. Symp. Reactivity of Solids, Gothenburg, p. 601, 1952 (1954).
355. J. S. Anderson and D. F. Klemperer, *Proc. R. Soc.* **A258**, 350 (1960).
356. A. L. Reimann, "Thermionic Emission", Ch. 3. Chapman and Hall, London (1934).
357. R. H. Fowler and L. W. Nordheim, *Proc. R. Soc.* **A119**, 173 (1928).
358. R. Gomer, "Field Emission and Field Ionization", Chs. 2, 4. Harvard University Press, Cambridge, Massachusetts (1961).
359. E. W. Müller, *Z. Phys.* **106**, 541 (1937).
360. J. A. Becker, *Adv. Catalysis* **7**, 136 (1955).
361. R. Klein, *J. chem. Phys.* **31**, 1306 (1959).
362 A. A. Holscher, *J. chem. Phys.* **41**, 579 (1964).
363. J. M. Saleh, M. W. Roberts and C. Kemball, *J. Catalysis* **3**, 189 (1963).
364. W. J. M. Rootsaert, L. L. van Reijen and W. M. H. Sachtler, *J. Catalysis* **1**, 416 (1962).
365. A. J. Melmed and R. Gomer, *J. chem. Phys.* **34**, 1802 (1961).
366. A. J. Melmed, *J. chem. Phys.* **36**, 1101 (1962).
367. A. J. Melmed, *J. chem. Phys.* **38**, 607 (1963).
368. J. J. Gilman (ed.), "The Art and Science of Growing Crystals". Wiley, New York (1963).
369. L. R. Gomer and L. W. Swanson, *J. chem. Phys.* **38**, 1613 (1963).
370. L. W. Swanson and R. Gomer, *J. chem. Phys.* **39**, 2813 (1963).
371. A. Eberhagen, *Fortschr. Phys.* **8**, 245 (1960).
372. J. C. P. Mignolet, *Disc. Faraday Soc.* **8**, 326 (1950).
373. C. M. Quinn and M. W. Roberts, *Proc. chem. Soc.* 246 (1962).
374. T. A. Delchar, A. Eberhagen and F. C. Tompkins, *J. sci. Instr.* **40**, 105 (1963).
375. A. Eberhagen, R. Jaeckel and F. Strier, *Z. angew. Phys.* **11**, 131 (1959).
376. E. W. Müller, 4th Int. Conf. Electron Miscroscopy. Springer-Verlag, Berlin (1958).
377. E. W. Müller, Int. Conf. Crystal Lattice Defects, 1962; see *J. phys. Soc. Japan* **18**, 1 (1963) (Supplement II).
378. D. G. Brandon, *Endeavour* **23**, 90 (1964).
379. D. G. Brandon, M. Wald, M. J. Southon and B. Ralph, *J. phys. Soc. Japan* **18**, 324 (1963) (Supplement II).

380. G. Ehrlich and F. G. Hudda, *J. chem. Phys.* **36**, 3233 (1962).
381. E. W. Müller, S. Wakamura, O. Nishikawa and S. B. McLane, *J. appl. Phys.* **36**, 2496 (1965).
382. R. Suhrmann, *Z. Elektrochem.* **56**, 351 (1952).
383. R. Suhrmann and K. Schultz, *Z. phys. Chem.* **1**, 69 (1954).
384. R. Suhrmann and G. Wedler, *Adv. Catalysis* **9**, 223 (1957).
385. R. Suhrmann, G. Wedler and G. Schumicki, *in* "Structure and Properties of Thin Films", ed. by C. A. Neugebauer, P. Newkirk and F. Vermilyea, p. 268. Wiley, New York (1958).
386. V. Ponec and Z. Knor, *Coll. Czech. chem. Commun.* **25**, 2913 (1960).
387. V. Ponec and Z. Knor, Proc. 2nd Int. Congr. Catalysis, 1960, Vol. 1, p. 195. Editions-Technip, Paris (1961).
388. W. M. H. Sachtler and G. J. H. Dorgelo, *Z. phys. Chem. Frankf. Ausg.* **25**, 69 (1960).
389. D. Brennan and J. M. Jackson, *Proc. chem. Soc.* 375 (1963).
390. M. Onchi, American Vacuum Society Symp., Boston (1963).
391. R. Suhrmann, *Adv. Catalysis* **7**, 303 (1955).
392. H. Clark and D. J. Benets, *Adv. Catalysis* **9**, 204 (1957).
393. N. F. Mott, *in* "Semi-conducting Materials", ed. by H. K. Henisch, p. 6. Academic Press, New York (1951).
394. S. W. Weller and S. E. Voltz, *Adv. Catalysis* **9**, 215 (1957).
395. G. S. John, M. J. Den Herder, R. J. Mikovsky and R. F. Waters, *Adv. Catalysis* **9**, 252 (1957).
396. S. Levy and M. Folman, *J. phys. Chem.* **67**, 1278 (1963).
397. A. L. G. Rees, "Chemistry of the Defect Solid State", p. 79. Methuen, London (1954).
398. C. A. Klein, *Rev. mod. Phys.* **34**, 56 (1962).
399. P. B. Weisz, C. D. Prater and K. D. Rittenhouse, *J. chem. Phys.* **21**, 2236 (1953).
400. J. W. Hightower and P. H. Emmett, Proc. 3rd Int. Congr. Catalysis, Amsterdam (1964).
401. S. J. Thomson and A. Walton, *Trans. Faraday Soc.* **53**, 821 (1957).
402. K. C. Campbell and S. J. Thomson, *Trans. Faraday Soc.* **55**, 306 (1959).
403. K. C. Campbell and S. J. Thomson, *Trans. Faraday Soc.* **55**, 985 (1959).
404. K. C. Campbell and S. J. Thomson, *Trans. Faraday Soc.* **57**, 279 (1961).
405. S. Affrossman, D. Cormack and S. J. Thomson, *J. chem. Soc.* 3217 (1962).
406. S. Affrossman and S. J. Thomson, *J. chem. Soc.* 2024 (1962).
407. S. J. Thomson and J. L. Wishlade, *Trans. Faraday Soc.* **58**, 1170 (1962).
408. G. C. Bond and J. Sheridan, *Trans. Faraday Soc.* **48**, 713 (1952).
409. F. Fischer and H. Tropsch, *Gesamm. Abh. Kennt. Kohle* **10**, 313 (1932).
410. F. Fischer and H. Tropsch, *Brennstoff-Chem.* **7**, 97 (1926).
411. O. C. Elvins and A. W. Nash, *Nature, Lond.* **118**, 154 (1926).
412. P. H. Emmett, "New Approaches to the Study of Catalysis", p. 124. Thirty-Sixth Priestley Lecture, Pennsylvania State University (1962).
413. J. T. Kummer, T. W. Dewitt and P. H. Emmett, *J. Am. chem. Soc.* **70**, 3632 (1948).
414. J. T. Kummer, H. H. Podgurski, W. B. Spencer and P. H. Emmett, *J. Am. chem. Soc.* **73**, 564 (1951).
415. J. T. Kummer and P. H. Emmett, *J. Am. chem. Soc.* **75**, 5177 (1953).
416. W. K. Hall, R. J. Kokes and P. H. Emmett, *J. Am. chem. Soc.* **79**, 2983 (1957).

417. W. K. Hall, R. J. Kokes, and P. H. Emmett, *J. Am. chem. Soc.* **82,** 1027 (1960).
418. P. H. Emmett and J. T. Kummer, Proc. 3rd World Pet. Congr. Section IV, p. 15 (1951).
419. S. Z. Roginskii, Proc. 1st UNESCO Congr., p. 1 (1958).
420. O. A. Golovina, M. M. Sacharov, S. Z. Roginskii, and E. S. Dakakins, *Russ. J. Phys. Chem.* **33,** 471 (1959).
421. W. A. Van Hook, Ph.D. Thesis, The Johns Hopkins University (1961).
422. G. C. Bond, "Catalysis by Metals". Academic Press, London (1962).
423. D. S. MacIver and H. H. Tobin, *J. phys. Chem.* 451 (1960).
424. D. S. MacIver and H. H. Tobin, *J. phys. Chem.* **65,** 1665 (1961).
425. D. A. Dowden, N. Mackenzie and B. M. W. Trapnell, *Proc. R. Soc.* A237, 245 (1953).
426. K. S. De, M. J. Rossiter and F. S. Stone, Proc. 3rd Int. Congr. Catalysis, Amsterdam (1964). Paper 1.28.
427. D. R. Ashmead, D. D. Eley and R. Rudham, *J. Catalysis* **3,** 280 (1964).
428. C. Kemball, *Bull. Soc. chim. Pays-Bas Belg.* **67,** 373 (1958).
429. C. Kemball, *Proc. chem. Soc.* 264 (1960).
430. C. Kemball, *Adv. Catalysis* **11,** 223 (1959).
431. G. M. Harris, *Trans. Faraday Soc.* **47,** 716 (1951).
432. J. R. Anderson and C. Kemball, *Proc. R. Soc.* **A233,** 361 (1954).
433. C. Kemball, *Proc. R. Soc.* **A207,** 539 (1951).
434. W. J. Dunning, *Q. Rev.* **9,** 23 (1955).
435. A. J. B. Robertson, "Mass Spectrometry". Methuen, London (1954).
436. K. Morikawa, N. R. Trenner and H. S. Taylor, *J. Am. chem. Soc.* 1103 (1937).
437. H. C. Rowlinson, R. L. Burwell, Jr, and R. H. Tuxworth, *J. phys. Chem.* **59,** 225 (1955).
438. N. Morita and T. Titani, *Bull. chem. Soc. Japan* **13,** 357 (1938).
439. N. Morita and T. Titani, *Bull. chem. Soc. Japan* **17,** 217 (1942).
440. N. Morita, *J. chem. Soc. Japan* **63,** 659 (1942).
441. N. Morita, *Bull. chem. Soc. Japan* **15,** 1 (1940).
442. S. M. Kanpacheva and A. M. Rozen, *Dokl. Akad. Nauk SSSR* **68,** 1057 (1949).
443. E. R. S. Winter, *Adv. Catalysis* **10,** 196 (1958).
444. G. Ya. Turouskii and F. M. Vainshtein, *Dokl. Akad. Nauk SSSR* **78,** 1173 (1951).
445. A. Kanome and T. Chitani, *J. chem. Soc. Japan* **63,** 36 (1942).
446. Y. L. Sandler and W. M. Hickain, Proc. 3rd Int. Congr. Catalysis, Amsterdam (1964). Paper I.4.
447. L. H. Germer and A. U. MacRae, *The Robert A. Welch Foundation Research Bull.* No. 11, 5 (1961).
448. C. J. Davisson and L. H. Germer, *Phys. Rev.* **30,** 705 (1927).
449. C. J. Davisson, *J. Franklin Inst.* **205,** 597 (1928).
450. H. E. Farnsworth, *Nature, Lond.* **123,** 941 (1929).
451. W. Ehrenberg, *Phil. Mag.* **18,** 878 (1934).
452. E. J. Scheibner, L. H. Germer and C. D. Hartman, *Rev. sci. Instr.* **31,** 112 (1960).
453. L. H. Germer and C. D. Hartman, *Rev. sci. Instr.* **31,** 784 (1960).
454. H. E. Farnsworth, R. E. Schlier, T. H. George and R. M. Burger, *J. appl. Phys.* **29,** 1150 (1958).

455. J. J. Lander and J. Morrison, *J. chem. Phys.* **37**, 729 (1962).
456. J. J. Lander, *Surface Sci.* **1**, 125 (1964).
457. L. H. Germer, E. J. Scheibner and C. D. Hartman, *Phil. Mag.* **5**, 222 (1960).
458. L. H. Germer and C. D. Hartman, *J. appl. Phys.* **31**, 2085 (1960).
459. L. H. Germer, A. U. MacRae and C. D. Hartman, *J. appl. Phys.* **32**, 2432 (1961).
460. L. H. Germer and A. U. MacRae, *J. appl. Phys.* **33**, 2923 (1962).
461. L. H. Germer and A. U. MacRae, *J. chem. Phys.* **37**, 1382 (1962).
462. L. H. Germer and A. U. MacRae, *J. chem. Phys.* **36**, 1555 (1962).
463. L. H. Germer and A. U. MacRae, *Proc. natn. Acad. Sci., U.S.A.* **48**, 997 (1962).
464. A. U. MacRae and L. H. Germer, *Phys. Rev. Letters* **8**, 489 (1962).
465. R. L. Park and H. E. Farnsworth, *J. chem. Phys.* **40**, 2534 (1964).
466. A. U. MacRae, *Science, N.Y.* **139**, 379 (1963).
467. J. J. Lander, J. Morrison and F. Unterwald, *Rev. sci. Instr.* **33**, 784 (1962).
468. M. Green and R. Seiwatz, *J. chem. Phys.* **37**, 358 (1962).
469. R. H. Fowler and E. A. Guggenheim, "Statistical Mechanics". Cambridge University Press (1956).
470. A. R. Miller, "The Adsorption of Gases on Solids". Cambridge University Press (1949).
471. D. van Zoonen, Proc. 3rd Int. Congr. Catalysis, Amsterdam (1964). Paper II.9.
472. A. T. Gwathmey and A. F. Benton, *J. phys. Chem.* **44**, 35 (1940).
473. H. Leidheiser and A. T. Gwathmey, *J. Am. chem. Soc.* **70**, 1200 (1948).
474. R. E. Cunningham and A. T. Gwathmey, *J. Am. chem. Soc.* **76**, 391 (1954).
475. R. Y. Meelheim, R. E. Cunningham, K. R. Lawless, S. Azim, R. H. Kean and A. T. Gwathmey, Proc. 2nd Int. Congr. Catalysis, Vol. 1, p. 2005. Editions-Technip, Paris (1961).
476. W. E. Garner, T. J. Gray and F. S. Stone, *Proc. R. Soc.* **A197**, 294 (1949).
477. R. G. Picard and O. S. Duffendack, *J. appl. Phys.* **14**, 291 (1943).
478. H. Levinstein, *J. appl. Phys.* **20**, 306 (1949).
479. W. M. H. Sachtler, G. J. H. Dorgelo and W. van der Knaap, *J. Chim. phys.* **51**, 491 (1954).
480. G. Garton and J. Turkevich, *J. chem. Phys.* **51**, 516 (1954).
481. S. Z. Roginskii, I. Tretyakov and A. B. Shekhter, *Zh. fiz. Khim.* **29**, 1921 (1955).
482. A. B. Shekhter and I. Tretyakov, "Soviet Research in Catalysis", Vol. **1**, p. 201. Consultants Bureau, Inc., New York (1960).
483. R. L. Moss, M. J. Duell and D. H. Thomas, *Trans. Faraday Soc.* **59**, 216 (1963).
484. G. A. Bassett, *Nature, Lond.* **198**, 468 (1963).
485. M. J. Whelan, P. B. Hirsch, R. W. Horne and W. Bollman, *Proc. R. Soc.* **A240**, 524 (1957).
486. J. G. Allpress and J. V. Sanders, *Phil. Mag.* **9**, 645 (1964).
487. G. A. Bassett, J. W. Menter and D. W. Pashley, *Disc. Faraday Soc.* **28**, 7 (1959).
488. B. C. Lippens and J. H. de Boer, *J. Catalysis* **3**, 44 (1964).
489. B. C. Lippens, Thesis, University of Delft, The Netherlands (1961).
490. H. P. Klug and L. E. Alexander, "X-ray Diffraction Procedures". Wiley, New York (1954).
491. G. von Hevesy, "Chemical Analysis by X-rays and its Applications". McGraw-Hill, New York (1932).

178   INTRODUCTION TO HETEROGENEOUS CATALYSIS

492. W. Berg, *Naturwissenschaften* **19**, 391 (1931).
493. C. S. Barnett, *Phys. Rev.* **38**, 832 (1931).
494. J. B. Newkirk, *Phys. Rev.* **110**, 1465 (1958).
495. A. R. Lang, *J. appl. Phys.* **28**, 497 (1957).
496. A. E. Jenkinson, *Philips tech. Rev.* **23**, 82 (1962).
497. E. M. Hofer and H. E. Hinterman, *Trans. Faraday Soc.* **60**, 1457 (1964).
498. J. Hillier, U.S. Patent, 2, 418, 029 (1947).
499. R. Castaing, Thesis, University of Paris (1951).
500. L. S. Birks, "Electron Probe Microanalysis". Interscience, New York (1963).
501. G. C. Wood and D. A. Melford, *J. Iron St. Inst.* **198**, 142 (1961).
502. E. W. White, J. M. Thomas and P. L. Walker, Jr, Unpublished results.
503. D. B. Whittry, Semiconductor Conference, American Institute of Mechanical Engineers, Boston, August 1959.
504. J. W. Menter, *J. R. Inst. Chem.* **86**, 415 (1962).
505. E. Crawford, M. W. Roberts and C. Kemball, *Trans. Faraday Soc.* **58**, 1761 (1962).
506. C. Kemball and J. J. Rooney, *Proc. R. Soc.* **A264**, 567 (1961).
507. C. Kemball and W. R. Patterson, *Proc. R. Soc.* **A270**, 219 (1962).
508. J. Erkelens, C. Kemball and A. K. Galwey, *Trans. Faraday Soc.* **59**, 1181 (1963).
509. J. M. Saleh, C. Kemball and M. W. Roberts, *Trans. Faraday Soc.* **57**, 1771 (1961).
510. J. M. Saleh, M. W. Roberts and C. Kemball, *Trans. Faraday Soc.* **58**, 1642 (1962).
511. R. J. Kokes, H. Tobin and P. H. Emmett, *J. Am. chem. Soc.* **77**, 5860 (1955).
512. W. K. Hall and P. H. Emmett, *J. Am. chem. Soc.* **79**, 2091 (1957).
513. P. H. Emmett, *Adv. Catalysis* **9**, 645 (1957).
514. K. Tamaru, *Nature, Lond.* **183**, 319 (1959).
515. W. K. Hall, D. S. MacIver and H. P. Weber, *Ind. Engng Chem.* **52**, 421 (1960).
516. D. W. Bassett and H. W. Habgood, *J. phys. Chem.* **64**, 769 (1960).
517. D. W. Bassett and H. W. Habgood, *Chemy Can.* **13**, 50 (1960).
518. H. W. Habgood, *Annu. Rev. Phys. Chem.* **13**, 259 (1962).
519. P. J. Owens and C. H. Amberg, "Solid Surfaces and the Gas-Solid Interface", p. 182. No. 33, Advances in Chemistry Series. American Chemical Society (1961).
520. P. W. Darby and C. Kemball, *Trans. Faraday Soc.* **55**, 83 (1959).
521. F. E. Shepherd, J. J. Rooney and C. Kemball, *J. Catalysis* **1**, 379 (1962).
522. A. K. Galwey, *Proc. R. Soc.* **A271**, 218 (1963).
523. A. K. Galwey, *Trans. Faraday Soc.* **59**, 503 (1963).
524. R. L. Wilson and C. Kemball, *J. Catalysis* **3**, 426 (1964).
525. G. C. Bond, J. J. Phillipson, P. B. Wells and J. M. Winterbottom, *Trans. Faraday Soc.* **60**, 1847 (1964).
526. E. Cremer and L. Roselius, *Adv. Catalysis* **9**, 659 (1957).
527. E. Cremer and R. Mueller, *Mikrochem. Acta* **37**, 553 (1951).
528. K. Tamaru, M. Boudart and H. S. Taylor, *J. phys. Chem.* **59**, 801 (1955).
529. P. J. Fensham, K. Tamaru, M. Boudart and H. S. Taylor, *J. phys. Chem.* **59**, 806 (1955).
530. K. Tamaru, *J. phys. Chem.* **61**, 647 (1957).
531. K. Tamaru and M. Boudart, *Adv. Catalysis* **9**, 699 (1957).
532. K. Tamaru, *Bull. chem. Soc. Japan* **31**, 669 (1958).

533. K. Tamaru, *Nature, Lond.* **183,** 319 (1959).
534. K. Tamaru, *Bull. chem. Soc. Japan* **33,** 430 (1960).
535. K. Tamaru, *Bull. Fac. Engng Yokohama natn. Univ.* **8,** 81 (1959).
536. K. Tamaru, *Trans. Faraday Soc.* **55,** 824 (1959).
537. K. Tamaru, *Trans. Faraday Soc.* **57,** 1410 (1961).
538. K. Tamaru, Proc. 2nd Int. Congr. Catalysis, Paris, 1960, p. 325. Editions-Technip, Paris (1961).
539. K. Tamaru, *Trans. Faraday Soc.* **59,** 979 (1963).
540. K. Tamaru, Proc. 3rd Int. Congr. Catalysis, Amsterdam (1964). Paper I.39.
541. H. S. Taylor, *Annu. Rev. Phys. Chem.* **12,** 134 (1961).
542. J. A. Rabo, P. E. Pickert, D. N. Stamines and J. E. Boyle, Proc. 2nd Int. Congr. Catalysis, Paris (1960), p. 2055. Editions-Technip, Paris (1961).
543. V. J. Friletti, P. B. Weisz and R. L. Golden, *J. Catalysis* **1,** 301 (1962).
544. C. J. Norton, *Chemy Ind.* 258 (1962).
545. G. M. Schwab and R. Sieb, *Z. Naturf.* **18a,** 164 (1963).
546. P. E. Pickert, J. A. Rabo, E. Dempsey and V. Schomaker, Proc. 3rd Int. Congr. Catalysis, Amsterdam (1964). Paper I.43.
547. J. Turkevich, F. Nozaki and D. N. Stamines, Proc. 3rd Int. Congr. Catalysis, Amsterdam (1964). Paper No. I.33.
548. J. Kramer, *Z. Phys.* **125,** 739 (1949).
549. O. Haxel, F. G. Houtermans and K. Sieger, *Z. Phys.* **130,** 109 (1951).
550. N. I. Kobozev, I. V. Krilova and A. S. Shashkov, Proc. 3rd Int. Congr. Catalysis, Amsterdam (1964). Paper I.35.
551. R. A. van Nordstrand, *Adv. Catalysis* **12,** 149 (1960).
552. P. H. Lewis, *J. phys. Chem.* **66,** 105 (1962).
553. L. M. Naphtali and L. M. Polinski, *J. phys. Chem.* **67,** 369 (1963).
554. Y. Trambouze, T. H. The, M. Perrin and M. V. Matheiu, *J. Chim. phys.* **51,** 425 (1954).
555. A. A. Balandin and T. V. Rode, *Problemy Kinet. Katal.* **5,** 135 (1948).
556. S. K. Bhattacharyya, V. S. Ramachandran and J. C. Gosh, *Adv. Catalysis* **2,** 114 (1957).

CHAPTER 4

The Significance of Pore Structure and Surface Area in Heterogeneous Catalysis

4.1. The Importance of Pore Structure and Surface Area

The accessibility of a catalyst surface to reacting gases is of considerable importance in the selection of a solid material which is to function as an active catalyst for heterogeneous gas reactions. For a given catalyst, the greater the amount of surface available to the reacting gas the better is the conversion to products. Few catalyst preparations have surfaces which are energetically homogeneous in the sense that all adsorption sites are equivalent and the same amount of energy is exchanged between each molecule of adsorbate and adsorbent site. If such a catalyst could be prepared, then its activity would be directly proportional to the surface area exposed to the adsorbing gas. However, as discussed in Chapter 2, catalysts have some *a priori* heterogeneity and also heterogeneity induced by interaction between adsorbed species. Irrespective of the cause of such heterogeneity, the effect is to render certain areas of the catalyst surface more active than others. In such an event, the activity of a catalyst is not directly proportional to the surface area, but rather is dependent on the way in which the activity is distributed over the available surface area. Despite the existence of energetically heterogeneous surfaces, there are many catalysts which display an activity proportional to surface area and any *in situ* heterogeneity is only a small fraction of the total chemically active surface. One of the earliest applications of surface area measurement was the prediction of catalyst poisoning. If, on continued use, the activity decreases more rapidly than surface area, then poisoning may be suspected, whereas, if a decrease in surface area is concomitant with a decrease in activity, then thermal deactivation is indicated. Another application provides a method of assessing the efficacy of catalyst supports and promoters. A support or promoter may either increase the surface area available for adsorption and reaction or it may increase the catalyst activity per unit surface area. Hence, surface area measurement is an important expedient in predicting catalyst performance and determining the role which the catalyst surface plays in any heterogeneous gas reaction. It should be emphasized, however, that often only a small fraction of the surface area determined by physical techniques is chemically active.

Surface area is by no means the only physical property which determines the extent of adsorption and reaction. Equally important, especially for nonmetallic catalysts, is the pore structure, which, although contributing to the total surface area, must be regarded as a separate factor. This is because, in a given catalyst preparation, the distribution of pore sizes may be such that some of the catalyst is completely inaccessible to large reactant molecules and, furthermore, may restrict the rate of conversion to products by impeding the diffusion of reactant in the internal pore structure. Accordingly, it is an advantage to know something about the detailed pore structure of a catalyst. Commercial catalyst preparations always have a high internal surface area. If this were not so the external surface, being quite small, would quickly become poisoned and the catalyst rapidly lose activity. To be able to predict the correct pore size necessary to achieve a given activity requires a pore model of the catalyst. Commercial catalysts do not have simple pore structures, so that one aspect of the general problem of reaction rates and selectivity in catalyst pores is the selection of an appropriate model from physical data concerning pore volume and surface area. The choice of the correct model depends on how close the experimental adsorption data can be fitted to a general geometric configuration of the pore structure in such a way that the surface area and pore volume are adequately described. Once an appropriate model has been selected to characterize the porous material, the remaining problem of how the chemical kinetics are affected by diffusion in the pore structure can be tackled. Furthermore, an accurate prediction can be made of any likely improvement in activity or catalytic selectivity resulting from a judicious choice of pellet or pore size. Other useful deductions can also be made from pore size determinations. For example, whether or not steam deteriorates a cracking catalyst can be ascertained by the fact that a considerable increase in pore radius accompanies a loss in surface area if steam deactivation is prevalent.

Having emphasized the utility of surface area and pore size determination, let us review the experimental methods of studying surface area and pore volumes by methods which are particularly pertinent to an estimation of the internal surface area of porous solids. Before the last section, dealing with diffusion in catalyst pores, a discussion is given of the problems of characterizing pore structures and the selection of suitable models.

4.2. Experimental Methods of Estimating Surface Areas

Experimental techniques which involve the estimation of the surface areas of solids, both porous and non-porous, by application of low-temperature gas adsorption methods were discussed in Chapters 2 and 3. Such methods occupy a significant position in the chronological order of events which gave the study of heterogeneous catalysis such impetus. The techniques outlined included that due to Langmuir, the celebrated method of Brunauer, Emmett and Teller and the semi-empirical approach of Harkins and Jura. Each one of these methods is admirably suited to the estimation of the surface areas of solids: their respective merits and attendant disadvantages have already been compared. Also discussed in Chapter 3 are the more recent application of Polanyi's potential theory of adsorption to the estimation of surface areas, and the high-temperature method due to Barker and Everett. Both are important developments in an attempt to overcome some of the discrepancies existing between the various low-temperature gas adsorption theories. In the following sections other methods of estimating surface areas are outlined. In the gas permeability method (Section 4.2.1), conditions are similar to those existing under conditions of forced flow through a packed bed of catalyst. If a knowledge of the external surface area (not including the contribution to the total area from pores) of a solid is required, other techniques (Sections 4.2.2–4.2.4) are applied.

4.2.1. Gas permeability methods for estimating surface areas

If a gas flows through a packed bed of finely divided solids a pressure drop results across the bed. Measurement of this pressure drop ΔP at a known gas flow rate will enable the external surface area of the solid to be assessed. Forced flow conditions (see p. 215) are used and so Poiseuille's formula [1] for the average velocity u_l over the cross-section of a conduit of length L is applicable in the modified form

$$u_l = \frac{(d')^2}{K\eta} \frac{\Delta P}{L} \tag{1}$$

The numerical factor of 32 in Poiseuille's equation has been replaced by a constant K, η is the viscosity of the gas and d' is an equivalent diameter of the void spaces between the particles. This void space is measured by the ratio of the volume of voids to the total surface of the material in the bed. If ψ_b is the void space per unit volume and S_v the surface area of the material per unit volume, then

$$d' = \frac{\psi_b}{(1 - \psi_b)S_v} \tag{2}$$

Now the average velocity as measured over the whole cross-sectional area of the bed will be ψ_b times the average velocity $(u_l)_i$ in the interstices between the particles, so

$$u_l = (u_l)_i \psi_b \tag{3}$$

Both equations (2) and (3) may be used to calculate the pressure drop in fixed beds of solids. Substituting these two latter equations into equation (1) and assuming that the length L' of the flow passages will be directly proportional to the length or thickness L of the bed,

$$u_l = \frac{1}{K'} \frac{\psi_b}{(1 - \psi_b)^2} \frac{1}{\eta S_v^2} \frac{\Delta P}{L'} \tag{4}$$

from which S_v can be calculated. The constant K' is known as Kozeny's constant [2], the commonly accepted value of which is 5, although Coulson [3] has shown that it is a function of the porosity, particle shape and size range. It should be noted that this equation allows only an estimate of the external surface area of the solid.

Barrer and Grove [4] measured the time lag in reaching the steady flow state when a gas streams through a packed column of solid material. The diffusion equation applicable to the transient flow of a gas under molecular streaming conditions is

$$\frac{\partial c}{\partial t} = D \frac{\partial^2 c}{\partial x^2} \tag{5}$$

where c is the molar concentration at a time t and position x and D the diffusion coefficient. The boundary conditions applicable are

$$c = c_1 \text{ at } x = 0 \text{ for all } t \tag{6}$$

$$c = 0, 0 < x < L, \text{ at } t = 0 \tag{7}$$

$$c \approx 0, x = L \text{ for all } t \tag{8}$$

Barrer and Grove's experimental conditions [5] approximated the above boundary conditions closely. The solution to this equation has been given by several authors [6–8] and the value of D for a particle containing cylindrical capillaries is

$$D = \frac{4}{3} r \sqrt{\frac{RT}{\pi M}} \tag{9}$$

where r is the average radius of the capillaries constituting the pore structure of the material. The theory and experiments of Barrer and Grove showed, in particular, that with the above boundary conditions the rate of diffusion of gas at the downstream end of the porous material increases with time until a final steady rate is attained. A typical curve of the downstream pressure against time is shown in Fig. 1. The curve approaches a straight line asymptotically, provided always that the pressure at the downstream end is negligible in comparison with the pressure at the upstream end. The asymptote may be extrapolated back to the time axis to give an intercept τ which is a measure

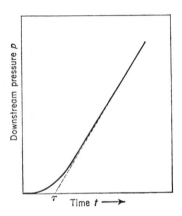

FIG. 1. Time required to reach steady state when gas is forced through a tube packed with porous particles.

of the time required to reach steady flow conditions. This value is related to the length L of the column of porous material by the equation

$$\tau = \frac{L^2}{6D} \tag{10}$$

Thus a very simple method is available for measuring the diffusion coefficient and the mean pore radius of the porous solid.

By utilizing Arnell's equation [8, 9], typical of those based on equation (4) for streamline flow and combined with an equation for molecular streaming, it is possible to deduce the porosity ψ and hence the internal surface area per unit volume S_v of the solid. Arnell's equation for the permeability K is

$$K = \frac{\psi^3 P}{K'\eta S_v{}^2} + \frac{8\delta}{3} \left(\frac{2RT}{\pi M}\right)^{1/2} \frac{k_0}{K'} \frac{\psi^2}{S_v} \tag{11}$$

where P is the mean pressure across the bed, K' is Kozeny's constant, k_0 is a shape factor and δ is a constant near unity. The permeability K is easily calculated from ΔP, the experimental pressure drop, and G the mass flow rate per unit area. Thus

$$K = \frac{A_c \Delta P}{GL} \tag{12}$$

where A_c is the cross-sectional area of the bed and L its length. The first term in equation (11) gives the Poiseuille permeability K_P defined as

$$K_P = \frac{n_P \pi r^4}{8\eta} = \frac{\psi^3}{K'\eta S_v{}^2} \tag{13}$$

where n_P is the number of pores per unit surface area. K_P is a measure of the contribution to the total permeability from forced flow. The second term gives the Knudsen permeability

$$K_K = \frac{4}{3}\delta \left(\frac{RT}{\pi M}\right)^{1/2} n_P \pi r^3 = \frac{8}{3}\delta \left(\frac{2RT}{\pi M}\right)^{1/2} \frac{k_0}{K'} \frac{\psi^2}{S_v} \tag{14}$$

which is a measure of the contribution to the total permeability from molecular diffusion when the mean free path of the gas molecules is considerably greater than the mean diameter of the pores. At moderate pressures, a plot of the total permeability K against the mean pressure P gives a straight line. From equations (11) and (12), the slope of such a line gives K_P and the intercept K_K. The ratio of the slope to the intercept will give a value for ψ/S_v, and since, for a cylindrical set of pores,

$$\frac{\psi}{S_v} = \frac{r}{2} \tag{15}$$

the values of r and of n_P are therefore known. The internal surface area is then $2\pi n_P r$ and this should be multiplied by a tortuosity factor to allow for the the tortuous nature of the pores. The resulting value may be compared with that obtained from Barrer's method based on the measurement of diffusion coefficients. Surface areas calculated from Barrer's method are lower by a factor of 4 or 5 when compared with values obtained from Arnell's equation. This is accounted for by the fact that blind pores are not measured by forced flow methods, whereas, in Barrer's method, all pores including cavities and blind channels are filled during the approach to a steady state. Hence, this steady-state forced flow method gives only a rough estimate of the internal surface area of the porous solid.

Returning again to Barrier's time lag method, if, during the tortuous passage which the gas takes through the honeycomb of pores, the gas is adsorbed, the rate of increase of concentration in the gas phase at the downstream end of the column is modified. If the adsorption isotherm is linear and obeys Henry's law (see p. 33), it is found that the time required to reach steady flow conditions is

$$\tau = \frac{L^2}{6D_1} = L^2 \frac{\left(1 + \dfrac{2k_H}{r}\right)}{6D} \tag{16}$$

where k_H is Henry's law adsorption constant and D_1 the modified value of the diffusion coefficient. The technique adopted is to determine the time lag τ_i for a gas which is not adsorbed (e.g. He or Ne) and compare it with the time lag τ_1 observed using an adsorbate. From equations (9) and (10) it follows that the time lag τ_0 predicted for the adsorbate, if adsorption were absent, is

$$\frac{\tau_i}{\tau_0} = \sqrt{\frac{M_i}{M}} \tag{17}$$

where M_i is the molecular weight of the inert gas and M the molecular weight of adsorbate gas. From equation (16) the ratio is given by

$$\frac{\tau_1}{\tau_0} = \frac{D_0}{D_1} = \left(1 + \frac{k_H}{r}\right) \tag{18}$$

and D, may thus be found. The calculation of the mean pore radius and surface area then follows the same procedure as before.

The results of Barrer and Grove [4, 5] have been interpreted as consistent with the assessment of the total surface area, including the area of closed-end capillaries. Thus, the term in equation (11) for Knudsen flow may be considered a measure of the collisions of the molecules with the periphery of the particles, and therefore probably characteristic of the external area of each particle. On the other hand, the term for viscous flow may be considered to measure the surface roughness of all pores which are accessible to the molecules used as a probe. Kraus and Ross [10] express some doubt as to whether the surface area of closed-end capillaries are measured. Although excellent agreement with BET surface areas has been obtained with nitrogen as the permeating gas (BET surface areas determined using N_2 at $-196°C$), agreement has been less satisfactory when measurements using helium are compared with the same surface areas. The areas obtained using helium by the transient flow method are as much as 20% larger than those obtained by nitrogen adsorption. It has been suggested that helium is able to penetrate into some narrow capillaries inaccessible to nitrogen.

4.2.2. Estimation of external surface areas by sedimentation

If, in a sample of fine solids of known density ρ_s, the particles are allowed to settle in a suitable liquid of density ρ_l, then it is possible to calculate the average particle diameter d. Let the drag force on a spherical particle be F; then, equating this drag force to the accelerating force of gravity,

$$F = \frac{1}{6}\pi d^3 (\rho_s - \rho_l) g \tag{19}$$

Provided that the Reynolds number of the freely falling sphere is less than 0·2, the drag force R' exerted on the particle per unit area is [11]

$$R' = \frac{12\eta u_t}{d} \tag{20}$$

where u_t is the terminal falling velocity. Since the projected area of the particle is $\pi d^2/4$, then the total drag force F on the particle is

$$F = \frac{12\eta u_t}{d} \frac{\pi d^2}{4} = 3\pi d\eta u_t \tag{21}$$

a formula which was given by Stokes [12]. Equating the two expressions for F, the diameter of the sphere is therefore

$$d = \sqrt{\frac{18\eta u_t}{g(\rho_s - \rho_l)}} \tag{22}$$

For non-spherical particles, provided that the Reynolds number is low, the same formula applies where d is now the diameter of the circle having the same area as the projected area of the particle. Equation (22) is applicable to particle diameters greater than 1000 Å.

If the particle size lies in the range 100–1000 Å, the settling rate is best determined by means of experiments in a centrifuge. The rate of separation of suspended particles from a liquid is much greater in a centrifuge than by gravitational settling, but the settling velocity of the particles does not approach a terminal value, because the accelerating force increases as the particles approach the walls of the container. The equation of motion for the particle may quite easily be written down. Consider an element of liquid depth dy and width dh. Let the radius of the inner surface of the thin element of liquid be x at a distance y from the bottom of the container. Equating the accelerating force to the net restoring force on the particle

$$m \frac{d^2h}{dt^2} = m (x + h) \omega^2 - F \tag{23}$$

where m is the average mass of the particles, ω the angular velocity of the centrifuge and F is the resisting drag force on the particle. The accelerating force is negligible compared with the centrifugal force and the drag force, so the left-hand side of equation (23) may be set equal to zero. Writing the mass in terms of the density of the solid relative to that of the liquid we have, from equation (23)

$$(\rho_s - \rho_l)\frac{\pi d^3}{6}(x + h)\omega^2 = F = 3\pi d\eta \frac{dh}{dt} \tag{24}$$

since the conditions will be those of streamline flow and the drag force will be given by equation (21). Integrating and rearranging,

$$d = \frac{6}{\omega}\sqrt{\left(\frac{\eta}{2t(\rho_s - \rho_l)}\ln\frac{x + h_2}{x + h_1}\right)} \tag{25}$$

where t is the time taken for the particle to settle from a distance h_1 to h_2 from the surface in a radial direction.

Sedimentation is used extensively to measure the particle size and the size distribution of many finely divided materials. Measurement of particle size will, however, yield information only about the external area. For porous substances, therefore, incomplete data are obtained when particle diameter measurements are made. One useful application, however, is the assessment of the average particle diameter of fluidized cracking catalysts [13]. It is also important to determine the size distribution of solids in designing catalytic fluidized beds, but the use of sedimentation is, for the most part, restricted to this utility. If it is required to know the proportion of total surface area which is external, as, for example, in the application of some of the equations developed in Section 4.5.5 for the prediction of reaction rates in porous solids, then sedimentation is a rapid and reliable method for giving average particle size.

4.2.3. X-Ray examination of powders [14]

Crystallites of most solid materials will behave as three-dimensional gratings towards X-rays. The X-ray pattern obtained for a powder will depend on the size of the crystallites present in the solid and the diffraction lines will be broadened. For a thin parallel plate with N_a atomic planes distance d apart, diffracted rays are in phase if the Bragg relation

$$n\lambda = 2d\sin\alpha \tag{26}$$

is obeyed, where n is an integer, α the angle of diffraction and λ the wavelength of the X-rays. The deviation necessary to obtain the rays reflected from the first and last plane in phase is then given by

$$\delta(2\alpha) = \frac{\lambda}{2N_a d\cos\alpha} \tag{27}$$

Since ($N_a d$) is the thickness of the crystal, equation (27) predicts that the broadening of the diffraction angle is inversely proportional to the thickness of the crystal. (It is important to distinguish between line broadening due to crystallite size and that due to lattice distortion). The thickness of the crystal is a measure of particle size.

A superior application of the use of X-rays in the examination of powders is the determination of particle size and surface area by the low-angle scattering of X-rays. The usual diffraction pattern of the material is formed by the scattered rays at relatively wide angles to the incident beam. In addition to this there is intense scattering at very small angles of about $\frac{1}{2}°$ from the incident beam. The technique is to measure the intensity at small angles to the direct beam (which is obtained by X-rays rendered monochromatic by reflection from a crystal). The intensity can be measured either photographically or by means of a scanning Geiger-Müller tube. The low-angle scattering intensity of an assembly of particles which have a distribution of sizes characterized by a linear dimension R is given by

$$I(\alpha) = C \int_0^\infty m(R)R^3 \exp\left(\frac{-\kappa^2 R^2}{5}\right) dR \qquad (28)$$

where, for small angles,

$$\kappa = 4\pi\alpha/\lambda \qquad (29)$$

$m(R)$ is the weight fraction of particles of dimension R, and C is a constant for the material. To compute the contribution of various sizes to the total low angle intensity, an analytical distribution function having two adjustable parameters is selected and the parameters varied until the height and breadth of the experimental curve of $I(\alpha)$ against κ is matched. Once the size distribution function is found, the calculation of the surface area is a relatively easy matter, although some assumption must be made about the geometry of the particles. Both the internal and the external areas of the particle are measured by the X-ray technique, since it is the ultimate particle diameter of the minutest crystallites which is found, and not the size of agglomerates which may have a porosity arising from the packing together of the smallest particles. The results of X-ray measurements agree very well with the BET gas adsorption method for many solids.

4.2.4. Optical and electron microscopic determination of particle size [15]

The limit of resolution using the optical microscope is about $0\cdot2\mu$ using light of about $0\cdot5\mu$ wavelength [16]. However, particles as small as 20 Å may be resolved by means of the electron microscope, which operates by the same principle as the optical microscope except that electron magnets instead of glass lenses are used for focusing and collimating the beam. A reasonable estimate of surface area is obtained only if the material is non-porous or if the roughness factor is near unity. On the other hand, the method is capable of yielding information about the number and size distribution of large pores. This latter information is extremely useful in predicting the rates of reaction in porous solids, as it is the large pore openings which provide access to the internal surface for reactant molecules.

There are some restrictions in applicability of the electron microscope. The solid specimen must be able to tolerate high vacuum without any change in structure. Furthermore, electron bombardment should not affect the material in any way. Since the area of the specimen which is examined is only about 10^{-7} mm^2, the best results are achieved with homogeneous solids. If inhomogeneous powders are used, a non-representative sample is likely to be examined and unreliable results obtained.

4.3. Methods of Assessing Pore Volume and Diameter

4.3.1. The Kelvin equation

One of the problems which is associated with catalysis by porous materials is the estimation of a mean pore diameter or, better still, a pore size distribution. Simple methods of determining total pore volume are inadequate, since knowledge of at least a mean pore radius is essential if it is desired to predict the effect of pore size on reaction rates. A suitable method of estimating pore radii is based on the fact that capillary condensation occurs in narrow pores at pressures less than the saturated vapour pressure of the adsorbate. A simple equation relating the lowering of the vapour pressure above a cylindrical column of liquid contained in a capillary of radius r may be obtained by equating the work done in enlarging a spherical drop of liquid to the work done in adding molecules to the interior of the drop.

If σ is the surface tension, then the work done in enlarging the surface area of the drop is

$$(\mu_0 - \mu)\, \delta n = \sigma \delta S = 8\pi r \sigma \delta r \qquad (30)$$

where μ_0 is the chemical potential of the vapour in equilibrium with a small liquid drop, μ is the chemical potential of the vapour over a plane liquid surface, δn is the increase in number of moles and δS the corresponding increase in surface area of the drop. The increase in volume of the drop will be given by

$$\bar{V}\, \delta n = 4\pi r^2\, \delta r \qquad (31)$$

where $\bar{V}$ is the molar volume of the liquid. Hence, from equations (30) and (31),

$$(\mu_0 - \mu) = \frac{2\sigma \bar{V}}{r} \qquad (32)$$

From the standard thermodynamic relations

$$\mu_0 = \mu^0 + RT \ln p_0 \qquad (33)$$

$$\mu = \mu^0 + RT \ln p \qquad (34)$$

where p_0 is the saturated vapour pressure and μ^0 is the standard chemical potential at unit pressure, substitution gives the Kelvin equation

$$\ln \frac{p_0}{p} = \frac{2\sigma \bar{V}}{rRT} \tag{35}$$

This equation may be applied to the liquid inside a capillary. If the angle of wetting between solid and liquid is α, then the component of the surface tension is $\sigma \cos \alpha$, and equation (35) is modified by this factor.

Equation (35) indicates that the smaller the radius of the capillary the greater is the lowering of the vapour pressure. Thus, in capillaries of very small diameter, vapour will condense to liquid at pressures considerably less than the normal vapour pressure. The equilibrium adsorption pressure p_a in the region of capillary condensation is greater than the corresponding desorption pressure p_a, since when desorption occurs from completely filled capillaries the wetting angle is zero. Thus, it is necessary to take the adsorption process to a relative pressure of unity and follow the desorption branch of the hysteresis loop of the isotherm (see Section 4.4.1) when there is nothing arbitrary about assigning a wetting angle to equation (35). A plot of v as a function of r can therefore be constructed by applying equation (35) to a given volume on the desorption loop of an experimental isotherm. The resulting curve gives the volume of gas necessary to fill all pores up to a radius r. From the v–r curve a plot of dv/dr as a function of r gives a pore size distribution curve. In effect, this latter curve, usually shaped like a Gaussian error curve, describes the extent to which pores of a given radius contribute to the total internal volume. Almost without exception, there is a clearly defined maximum which depicts the mean value of the radius of all accessible capillaries less than 300 Å.

It is as well to examine in more detail the premises upon which equation (35) is founded. A plot of v against r for a given adsorbent should be identical at different temperatures, which is normally the case, but often it differs for different adsorbates. As the v–r curve should be unique for a given adsorbent (since equation (35) imputes nothing concerning adsorbate–adsorbent interaction), it is probable that values of σ and $\bar{V}$ are in error when such behaviour is noted, although it has been suggested that differences in orientation of adsorbed molecules may give rise to various packing factors. If authentic values of r are to be expected, it is important to have dependable values for molar volume and surface tension. A second source of error lies in the hypothesis that all pores are cylindrical, so that a hemispherical meniscus is formed. This gives rise to the term $2/r$ in equation (35). For pores having other shapes, the factor $2/r$ may be replaced by a more appropriate term depending on the geometry and classification by which the pore structure is characterized. This problem is discussed generally in Section 4.4.3.

4.3.2. Pore size distribution by gas adsorption

Since multilayer adsorption usually accompanies capillary condensation in the pores of solids, the Kelvin equation will not give the correct radius, since the pore radius will have been effectively reduced by the thickness of the adsorbed multilayer. The maximum radius which will be filled by capillary condensation at a pressure p is therefore

$$r = t + \frac{2\sigma \bar{V}}{RT\ln \dfrac{p_0}{p}} \tag{36}$$

The thickness of the adsorbed layer may be calculated from the BET equation by evaluating the number of molecules adsorbed at a given pressure p, dividing by the number adsorbed in a monolayer and multiplying by the adsorbate molecular diameter. Hence, both r and t are known functions of the equilibrium pressure p. Other procedures for calculating the thickness of the adsorbed layer have been suggested by Wheeler [17] and by Shull [18]. To calculate the distribution function describing the contribution to the total volume of cylindrical pores having radii between r and $(r+\delta r)$ an isotherm equation in terms of such a distribution function must be written. Suppose $L(r)\delta r$ is the function which defines the length of pores with radii between these limits. The total pore volume unfilled is the total volume of all the cylindrical shells whose radii are between $(r-t)$ and infinity. If the total volume of pores occupied by adsorbate at the saturation pressure is v_s and v is the pore volume occupied by adsorbate at an equilibrium pressure p, then (v_s-v) is the volume of unfilled pores and so

$$v_s - v = \int_0^\infty \pi r^2 L(r)\mathrm{d}r - v = \int_{r-t}^\infty \pi(r - t)^2 L(r)\mathrm{d}r \tag{37}$$

Equation (37) is an integral equation in which the only unknown is the distribution function $L(r)$. Wheeler and Shull [18] assumed either a Maxwellian type distribution

$$L(r) = A_M r \exp\left(-\frac{r}{r_0}\right) \tag{38}$$

or a Gaussian type distribution

$$L(r) = A_G \exp\left\{-\left(\frac{Br_0}{r - r_0}\right)^2\right\} \tag{39}$$

where r_0 is the most probable pore radius, A_M and A_G are constants giving the frequency of occurrence of pore sizes and B is a factor which determines the sharpness of the Gaussian distribution curve. Either of equations (38) or (39) can be inserted in equation (37) and integrated. A family of curves for given r and r_0 is then calculated as a function of r, and the resulting curves are matched against the experimental curve of $(v_s - v)$ as a function of r. The mean pore radius is then given by

$$\bar{r} = \frac{2v_s}{S} = \frac{2v_s}{\displaystyle\int_0^\infty 2\pi r\, L(r)\mathrm{d}r} \tag{40}$$

Barrett, Joyner and Halenda [19] proposed solving equation (37) by means of a numerical stepwise integration. Differentiating equation (37) with respect to r and writing the equation in finite difference form, gives

$$\Delta v = \pi(r - t)^2\, L(r)\, \Delta r + 2\pi \Delta t \int_r^\infty (r - t)\, L(r)\mathrm{d}r \tag{41}$$

Thus, the volume Δv of gas desorbed when the pressure is lowered by Δp is accounted for by (i) those pores whose radii are between r and $(r+\Delta r)$ and which are empty except for a thickness t of adsorbate, and (ii) an additional amount desorbed from pores with radii greater than r as a result of the thickness of the adsorbed layer decreasing by an amount Δt. To solve equation (41), the volume of gas desorbed is plotted as a function of r by application of the Kelvin equation, and the graph is divided up into a large number of segments at equal intervals of r as shown in Fig. 2. For each segment, the

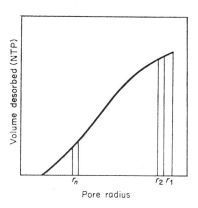

FIG. 2. Volume of gas desorbed in increments as a function of capillary radius.

values of r and t will be the mean values for the volume Δv of gas desorbed over that particular interval of r. At some pressure near the saturation pressure there will be a final segment which is a measure of the volume of all pores greater than the value for the radius (say r_2) corresponding to the penultimate segment. An arbitrary radius (say r_1) is then chosen for an imaginary large pore which will contain the volume of all pores greater than r_2. The distribution function for the largest (imaginary) pore is then

$$L(r_1) = \frac{\Delta v_1}{\pi (r_1 - t_1)^2 \Delta r_n} \tag{42}$$

which follows from equation (41), since, for the last segment, the value of the integral will be zero, there being no pores with a radius greater than r_1. Hence the numerical value of $L(r_1)$ is determined. Equation (41) is then solved for $L(r_2)$ by writing $(r_1-t_1)L(r_1)\Delta r_1$ in place of the integral. This value of $L(r_2)$ then enables an estimation to be made for $L(r_3)$ by accumulating the values of $(r-t)L(r)\Delta r$ for each segment. Thus, for any increment n,

$$L(r_n) = \frac{\Delta v_n}{\pi(r_n - t_n)^2 \Delta r_n} - \frac{2\pi\Delta t_n}{(r_n - t_n)^2\Delta r_n} \sum_{m=n-1}^{1} (r_m - t_m) L(r_m)\Delta r_m \tag{43}$$

Wheeler [17] suggested the following alternative method of solution to equation (37). If this latter equation is differentiated with respect to r we have

$$\frac{dv}{dr} = v'(r) = \frac{d}{dr}\int_0^r \pi r^2 L(r)dr - \frac{d}{dr}\left\{\int_r^\infty 2t\pi rL(r)dr\right\} - \frac{d}{dr}\left\{t^2\int_r^\infty \pi L(r)dr\right\} \tag{44}$$

Since the surface area accommodated by all pores larger than r is given by

$$S(r) = \int_r^\infty 2\pi r L(r)dr \tag{45}$$

equation (44) may be rewritten

$$v'(r) = \frac{d}{dr}\int_r^\infty \pi r^2 L(r)dr - \frac{d}{dr}\{tS\} \tag{46}$$

if the last term in t^2 is neglected in comparison with the other terms. Hence

$$v'(r) = \frac{(r - 2t)}{2}\frac{dS}{dr} - S\frac{dt}{dr} \tag{47}$$

Equation (47) is a simple linear first-order differential equation the solution of which may be written in terms of $v'(r)$ and $t'(r)=dt/dr$. Multiplying equation (47) through by the integrating factor

$$I = \exp\left\{-\int_0^r \frac{2t'(r)dr}{(r - 2t)}\right\}$$

gives

$$\frac{2}{(r - 2t)} v'(r \exp)\left\{-\int_0^r \frac{2t'(r)dr}{r - 2t}\right\} = \frac{d}{dr}\left(S \exp\left\{-\int_0^r \frac{2t'(r)dr}{r - 2t}\right\}\right) \tag{48}$$

Hence the solution is

$$S(r) = \left[\exp\left\{ -\int_0^r \frac{2t'(r)dr}{r-2t} \right\} \right]^{-1} \left[\int_r^{\infty} \exp\left\{ -\int_0^r \frac{2t'(r)dr}{r-2t} \right\} \frac{2}{r-2t} \frac{dv}{dr} \, dr \right] \quad (49)$$

The integral inside parentheses is simply the area underneath a curve of $2t^1/(r-2t)$ as a function of τ and therefore may be evaluated. The integral of the complete function contained in the large brackets may be calculated by finding the area under a curve of $2I/(r-2t)$ against v, where I is the exponential function (the integrating factor) containing the first integral. By such a procedure (and in a relatively short time) the value of $S(r)$ may be found for given values of r. The resulting cumulative distribution function is based on surface area and provides information about the amount of surface contained in pores of a given size range. In this sense it differs from other methods involving non-cumulative functions which give the surface area or pore volume at a particular value of r. A refinement of the method is to take account of the last term neglected from equation (44) by computing its value from the approximate answer obtained by means of equation (49) and subtracting it from the value of v. The whole computation for S is then repeated using the corrected value of v in equation (49).

4.3.3. The pressure porosimeter

A more direct approach to pore size distribution for micropores is to measure the volume of liquid (which does not wet the adsorbent) forced under pressure into the capillaries. The effect of interfacial surface tension is to oppose the entry of liquid into the capillary. The force tending to impede the entry of a liquid into a narrow cylindrical channel of radius r is $2\pi r \sigma \cos\alpha$ where σ is the surface tension and α the contact angle between liquid and solid. If a pressure P is imparted to the mercury, the force which tends to drive mercury into the cylindrical pores is $\pi r^2 P$. Equating these two forces gives

$$P = \frac{-2\pi\sigma\cos\alpha}{r} \quad (50)$$

as the pressure necessary to force mercury into a pore of radius r. For mercury, a contact angle of $140°$ and a surface tension of 480 dynes cm^{-1} are typical figures, so equation (50) indicates that a pressure of 10,000 lb in^{-2} must be applied to fill pores of 100 Å. A pressure greater than this is, under ordinary circumstances, impracticable, so the pressure porosimeter suffers from the disadvantage that capillaries with diameters less than about 100 Å remain unfilled and therefore escape detection. Nevertheless, since very large pores are accounted for, the porosimeter is especially useful in investigating the pore size distribution of porous materials containing pores up to 100,000 Å radius. The slope to a plot of the volume of liquid absorbed by the solid as a function of P will give a value for dv/dP at a particular p and hence for a particular r. Now, since the volume of all pores larger than r is given by

$$v = \int_r^\infty \pi r^2 L(r) dr \tag{51}$$

then

$$\frac{dv}{dr} = \pi r^2 L(r) = v(r) \tag{52}$$

Utilizing equation (50), the volume distribution function becomes

$$v(r) = \frac{dvdP}{dPdr} = -\frac{P}{r}\frac{dv}{dP} \tag{53}$$

When the right-hand side of equation (53) is plotted as a function of r, the resulting distribution curve gives the volume of pores which have a given radius.

4.3.4. Density measurement

For porous solids, the pellet density ρ_p differs from the true density ρ. This is simply because the material contains void space. If the fraction of pore space in the material is ψ, then

$$1 - \frac{\rho_p}{\rho} = \psi \tag{54}$$

and the total pore volume v_p of the catalyst pellets is ψ/ρ_p. The pellet density is easily measured with a pyknometer using an inert non-penetrating liquid. The true density is measured by expanding a gas which is not adsorbed (such as helium) into a vessel containing a known weight of the material. If the volume of the container and the volume of the expansion vessel are accurately known, then the density may be calculated by observing the pressure before and after expansion. If p_1 is the initial pressure of gas in a volume V_1 and p_2 the pressure after expansion into the container of volume V_2 then

$$\rho = \frac{m}{V_1 + V_2 - \dfrac{p_1 V_1}{p_2}} \tag{55}$$

where m is the mass of material in the container.

The method is suitable only for determining an average pore radius and the measurement of porosity must be combined with surface area data. Assuming that the pores are cylindrical, then the ratio of the experimental pore volume per gram V_g to specific surface area S_g is given by

$$\frac{V_g}{S_g} = \frac{\dfrac{v_p}{m}}{S_g} = \frac{\pi r^2 L}{2\pi r L} = \frac{\bar{r}}{2} \tag{56}$$

where L is the pore length and $\bar{r}$ is the average pore radius. It is fortuitous that two factors not included in equation (56) tend to cancel. If the surface is rough the experimental value of $\bar{r}$ would be too low, while if the pores of the material intersect the value would be too high. Despite this disadvantage, average pore radii calculated by this method agree quite well with those obtained from other methods.

4.4. Pore Structure of Adsorbents and Catalysts

4.4.1. Hysteresis and the shapes of capillaries

Above relative pressures of about 0·2, porous adsorbents desorb a larger quantity of vapour at a given relative pressure than that corresponding to adsorption. Typical hysteresis loops are shown in Figs. 3 and 7 (types A and E). Several explanations have been proposed to account for such hysteresis. Zsigmondy [20] assumed that during adsorption the vapour does not completely wet the walls of the adsorbent capillaries because an impurity, such as air, may be permanently adsorbed on the walls. Raising the pressure displaces any impurities until, at the saturated vapour pressure of the adsorbate, complete wetting takes place. On desorption, the angle of contact in the Kelvin equation is thus zero. Hence, for a given volume adsorbed, the pressure p_a on adsorption is greater than that on desorption p_d. Such an explanation is not acceptable for completely reversible hysteresis phenomena. McBain [21] assumed that pores are shaped like ink bottles with a narrow neck and a larger diameter body. On adsorption, the neck will fill at relatively low pressures but the body will not fill until the pressure, as given by the Kelvin equation, is

$$p_a = p^0 \exp - \frac{2\sigma\bar{V}}{r_b RT} \tag{57}$$

where r_b is the radius of the body of the capillary. The pore will not empty on desorption until the pressure is reduced to such an extent that the liquid in the neck is unstable, i.e. when

$$p_d = p^0 \exp - \frac{2\sigma\bar{V}}{r_n RT} \tag{58}$$

where r_n is the radius of the neck of the pore. Since $r_n < r_b$ then $p_a > p_d$ and a given volume is desorbed at a pressure lower than that at which it is adsorbed. Now the bulk of the liquid in the body is in equilibrium with the vapour, so in this theory equilibrium corresponds to the adsorption branch of the hysteresis loop.

Experiments [22] in which the adsorption of vapours on glass was studied by observing the change in colour of interference fringes indicated that the pore size distribution of the glass was discontinuous. The radii of the pores deduced from the Kelvin equation showed consistent values on desorption, but on adsorption the calculated radii were twice as large. Prior to these experiments, Cohan [23] provided a satisfactory explanation for such an anomaly and deduced a relation between p_a and p_d for open-ended cylindrical capillaries. Cohan supposed that an annular ring of liquid of length L is formed in the capillary of radius r, but that a meniscus is not formed until the pore is full. If, previous to condensation, the vapour had occupied the pore volume $\pi r^2 L$, then the change in surface on formation of an annular ring of width δr is $2\pi L \delta r$ and the consequent change in surface energy $2\pi \sigma L \delta r$. This must be equal to the free energy of formation of the liquid. If μ represents the chemical potential of the vapour at a pressure p_a and μ_0 the corresponding potential at the saturated vapour pressure p_0, then

$$(\mu - \mu_0)\delta n = -2\pi L \sigma \delta r \tag{59}$$

Noting that the number of moles transferred is

$$\delta n = \frac{2\pi L r \delta r}{\bar{V}} \tag{60}$$

and utilizing the standard relations given in equations (33) and (34)

$$\frac{2\pi L r \delta r}{\bar{V}} RT \ln \frac{p_a}{p_0} = -2\pi L \sigma \delta r \tag{61}$$

whence

$$p_a = p_0 \exp - \frac{\sigma \bar{V}}{rRT} \tag{62}$$

Only when the pore is full is a meniscus formed, and then the pressure p_d on desorption will be given by the Kelvin equation (35). Hence, by combination of the Kelvin equation with equation (62)

$$\left(\frac{p_a}{p_0}\right)^2 = \left(\frac{p_d}{p_0}\right) \tag{63}$$

Therefore, in terms of the Kelvin equation, the radius corresponding to p_a is twice that corresponding to p_d.

Implicit in the above equations is the cylindrical shape of the capillaries. Barrer et al. [24] have examined the influence of various shapes of capillaries on the form of isotherms. de Boer [25] has discussed the reverse problem and shown what conclusions concerning the shapes of pores may be derived from the form of the adsorption–desorption isotherm.

Five types of hysteresis loop may be distinguished and examples are shown in Figs. 3–7. The types of capillary shape that could be responsible for each loop are also sketched.

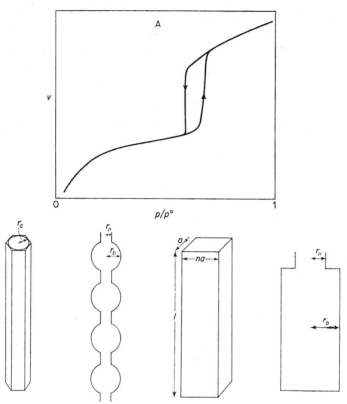

FIG. 3. Capillary shapes responsible for type A hysteresis.

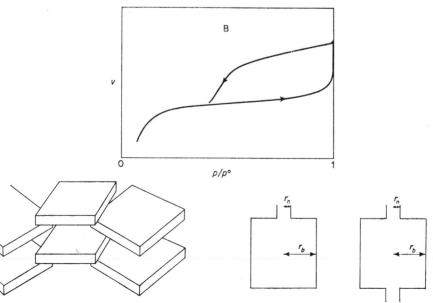

FIG. 4. Capillary shapes responsible for type B hysteresis.

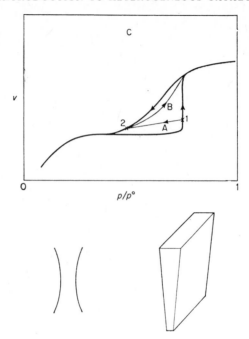

FIG. 5. Capillary shapes responsible for type C hysteresis, and the phenomenon of scanning.

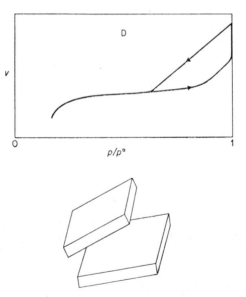

FIG. 6. Capillary shapes responsible for type D hysteresis

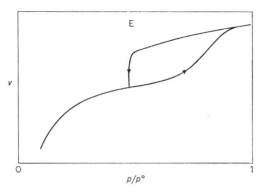

FIG. 7. Type E hysteresis.

4.4.1.1. *Type A Hysteresis Loops*

Those sorption isotherms in which either the adsorption or desorption branches are steep at intermediate relative pressures are characteristic of tubular-shaped capillaries open at both ends. The cross-section need not necessarily be circular, for a capillary with a polygonal cross-section will give rise to condensation along the inside edges until a cylindrically shaped meniscus is formed, after which, as shown by equation (62) the capillary will fill at a relative pressure corresponding to an effective radius $2r_c$, where r_c is the radius of the cylinder formed by the condensate forming a continuum along the edges of the polygon. The relative pressure on desorption will correspond to the radius r_c according to the Kelvin equation. Thus the relation between p_a and p_d is given by equation (63).

If the open capillaries are composite, with slightly widened bodies of radius r_b, each of these cavities having a narrower neck of radius r_n, then the narrow parts will fill at a relative pressure corresponding to $2r_n$. Consequently, a spherical meniscus is formed at both ends of the wider spherical body. If $r_b \leqslant 2r_n$, then the vapour pressure is saturated or supersaturated with respect to the radius r_b and so the composite pore is completely filled at a relative pressure corresponding to $2r_n$. The pore is emptied at a relative pressure corresponding to the radius r_n at the open ends. If the pore has only one narrow neck and is accessible only through the wider body, then it will empty at a pressure corresponding to r_b. Hence, equation (63) is obeyed. If, on the other hand, the open tubular capillary has a rectangular cross-section of dimensions a and na (where $n > 1$) and a length $L \gg a$, although a type A hysteresis loop is formed, equation (63) is not obeyed. Capillary condensation starts along the inside edges until an inside meniscus with a radius $a/2$ is formed. The pores then fill at a pressure corresponding to an effective radius a. Emptying of such pores occurs when the effective radius of the inscribed cylinder is $na/(n+1)$. For square capillaries, equation (63) is obeyed, but in all cases for $n>1$ we have

$$\left(\frac{p_a}{p_0}\right)^2 < \left(\frac{p_d}{p_0}\right) \qquad (64)$$

If n is very large, then the adsorption and desorption branches coincide and no hysteresis is observed.

The ink-bottle-shaped capillaries, assumed by McBain [21] to explain the phenomenon of hysteresis, also exhibit type A hysteresis. If one end of the capillary is closed and the radius of the body is r_b and that of the neck is r_n then, provided that $r_n<r_b<2r_n$, the pore fills at a pressure corresponding to r_b, since, in this case, a meniscus forms only at the top of the body and is spheroidal in shape and not cylindrical. Consequently, equations (30) and (31) apply to the change in surface energy and the number of moles transferred from vapour phase to liquid. The resulting relation is therefore identical with the Kelvin equation (35) and not equation (62). The pore will empty at a pressure corresponding to r_n as usual. The pressure on adsorption is therefore less than that on desorption and the inequality (64) is obeyed. If the radius of the body of the ink-bottle capillary is greater than the diameter of the neck, then the short necks are filled at a relative pressure corresponding to $2r_n$, but the whole pore will fill only at a relative pressure corresponding to r_b. In this latter case, the steep rise of the hysteresis loop is delayed until the whole pore is filled. Emptying takes place at a pressure corresponding to r_n. A wide hysteresis loop is produced and the inequality

$$\left(\frac{p_a}{p_0}\right)^2 > \left(\frac{p_d}{p_0}\right) \qquad (65)$$

holds because, for this case, $r_b>2r_n$.

4.4.1.2. *Type B Hysteresis Loops*

If a pore is formed by two parallel plates, a meniscus cannot form until the vapour pressure of the adsorbate is raised to the saturation vapour pressure. On desorption, the pore is emptied at a relative pressure corresponding to the width of the capillary, since this will be the effective radius of curvature of the meniscus. Thus the isotherm has a steep adsorption branch at a relative pressure of unity and a sloping desorption branch at intermediate relative pressures. Such an isotherm is said to show type B hysteresis.

Graphite oxide, montmorillonites and aluminium hydroxides give rise to type B hysteresis loops. These materials are of a crystalline nature and the packing together of these particles in laminae leads to the formation of open slit-shaped capillaries with parallel walls. Such structures have been confirmed by examining the shapes of the capillaries using an optical birefringence technique [26].

Another type of capillary which may lead to type B hysteresis is an ink-bottle pore with a body which is so wide, *ca* 1000 Å, that the adsorption branch practically coincides with a relative pressure of unity.

4.4.1.3. *Type C Hysteresis Loops*

Barrer *et al.* [24] predicted that type C hysteresis may be characteristic of materials with spheroidal pores all with a circular cavity radius but with various-sized entrances. Also open and closed ink bottles with a heterogeneous distribution of neck radii would cause such a hysteresis loop.

A heterogeneous distribution of capillary dimensions should exhibit the phenomenon of scanning, which is the ability of a system to follow a course across a hysteresis loop. Consider point 1 on the adsorption branch of the type C hysteresis loop shown in Fig. 5, and further suppose that this loop characterizes an assemblage of ink-bottle pores with various-sized necks. If the pressure is reduced, any capillaries which are partly filled will empty, but those which are completely full will retain the condensate until the pressure is lowered to p_2 the equilibrium pressure in the neck of the pore. Hence the hysteresis loop is crossed along path A. At no point on the desorption branch are there any partially filled pores. At point 2, a proportion of pores are empty and the remainder are full. If the pressure is now increased, those pores which are empty will fill again and the path B is followed. The adsorption branch is regained only at the saturation pressure because at point 2 some of the pores which had radii smaller than r_n have large bodies and remain full. Therefore, there are fewer large pores to fill than under normal circumstances and, for a given volume adsorbed, the equilibrium pressure will be lower than on the original adsorption branch.

4.4.1.4. *Type D Hysteresis Loops*

A heterogeneous assembly of capillaries with large body radii and a sufficiently varying range of narrow short necks will cause the adsorption branch of the type C loop to be displaced to higher relative pressures. On account of the wide size range of narrow necks the desorption branch will have a sloping character. Hysteresis loops of this type are rare, but have been observed when water is adsorbed on gibbsite partly converted to boehmite [27].

4.4.1.5. *Type E Hysteresis Loops*

This type of hysteresis loop, in contrast to type C, has a sloping adsorption branch and a steep desorption branch, both at intermediate relative pressures. It may arise from the same types of open capillaries responsible for type A hysteresis when the effective radii of the bodies of the pores are heterogeneously distributed but the effective radii of the narrow entrances are all of equal size. Similarly, a distribution of various-sized spheroidal cavities with the same entrance diameter will give a type E hysteresis loop. Scanning behaviour can be expected from both types of pore.

Ink-bottle pores with large bodies of varying effective radii and small narrow necks start filling at an effective cylindrical meniscus radius of $2r_n$ and continue filling until the whole body is full. They will empty at a relative pressure corresponding to r_n. Scanning behaviour will be absent for this type of pore, since on desorption the pores will empty spontaneously.

Type E hysteresis loops are well known and have been noted for the sorption of benzene by ferric oxide gel [28], for the sorption of water by silica gel [29] and also for nitrogen adsorption on many silica–magnesia and silica–alumina cracking catalysts [30].

4.4.1.6. *Hysteresis Loops with Two Sloping Branches*

The five types of hysteresis loop discussed above all have one vertical adsorption or desorption branch because the dimensions of the cross-sections of either the wide or the narrow part of the pores are all of equal size and are predominantly of one shape. If such conditions are not fulfilled and there are pores whose bodies and necks have cross-sections covering a wide range of radii, hysteresis loops with sloping adsorption and desorption branches will result.

4.4.1.7. *Absence of Hysteresis*

Tubular type capillaries closed at one end will not exhibit hysteresis, because, in this case, there will be no delay in the formation of a meniscus on adsorption. The Kelvin equation will therefore apply to both adsorption and desorption and p_a will equal p_d.

The absence of hysteresis is also apparent for an assembly of open capillaries which are sufficiently narrow to preclude the adsorption of a layer more than four molecular diameters thick. For such narrow capillaries, the thickness t of the adsorbed layer is not negligible in comparison with the capillary radius. In such a case Cohan's equation (62) should be written

$$p_a = p_0 \exp\left(-\frac{\sigma \bar{V}}{(r-t)RT}\right) \qquad (66)$$

Since p_a is given by equation (58), the relation between p_a and p_d will now be

$$\left(\frac{p_a}{p_0}\right)^{2\left(1-\frac{t}{r}\right)} = \frac{p_d}{p_0} \qquad (67)$$

From equation (67), p_a is equal to p_d when

$$r = 2t \qquad (68)$$

Since the smallest possible value of t is one molecular diameter, equation (67) indicates that hysteresis does not occur in open-ended capillaries with diameters less than four molecular diameters, or, if the pores are not cylindrical, with cross-sections which will contain an imaginary inscribed cylinder four molecular diameters in width. For open-ended capillaries it also follows that hysteresis will not commence until pores of radii $r > 2t$ are filled.

4.4.2. Surface areas calculated from hysteresis loops

If gas is desorbed in decremental volumes, then the volume desorbed from pores having radii between r and $(r + \Delta r)$ is approximately

$$\Delta v_p = \pi r_k^2 L(r_k) \Delta r_k \qquad (69)$$

where r_k is the Kelvin radius calculated from equation (36) corresponding to this decrement, the thickness t of the adsorbed layer having been taken into account. Similarly, a quantity ΔS can be defined which measures the surface area of these pores. It follows that the surface area ΔS of pores with radii between r and $r + \Delta r$ is given by

$$\frac{\Delta S}{2 \Delta v_p} = \frac{2\pi r_k L(r_k) \Delta r_k}{2\pi r_k^2 L(r_k) \Delta r_k} = \frac{1}{r_k} \qquad (70)$$

The cumulative surface area S_{cum} can then be found by summing all the decrements into which the desorption has been divided.

When the capillaries are truly cylindrical the cumulative surface area will be equal to the real internal surface S_{real} of the assembly of capillaries. If, during a pore size analysis, no values of r are counted lower than the beginning of the hysteresis loop (where the radius is, say, r') the cumulative surface area S_{cum} found for open-ended polygonal capillaries would be

$$S_{\text{cum}} = S(r') = \int_{r'}^{\infty} 2\pi r_i L(r_i) \mathrm{d}r_i \qquad (71)$$

where r_i is the radius of the cylinder inscribed in a polygon of n sides. In this case, the cumulative surface area will be less than that found from BET measurements, since pores smaller than r' are not counted. In terms of the radius of the inscribed cylinder, the true geometrical surface area enclosed by all the polygonal pores is

$$S_{\text{real}} = \int_0^{\infty} 2\pi r_i L(r_i) \tan\left(\frac{\pi}{n}\right) \mathrm{d}r_i \qquad (72)$$

Now we can write

$$S_{\text{BET}} \approx \int_0^{\infty} 2\pi r_i L(r_i) \, \mathrm{d}r_i \qquad (73)$$

since the BET area will include pores less than a radius corresponding to r' where hysteresis commences. Thus, from equations (72) and (73),

$$\frac{S_{\text{BET}}}{S_{\text{real}}} = \frac{\pi}{n} \cot\left(\frac{\pi}{n}\right) \qquad (74)$$

On the other hand, if a pore size analysis is continued to lower relative pressure than that corresponding to r', then the additional volume desorbed is divided by a Kelvin radius (according to equation (70)) smaller than the inscribed circle. Consequently, the cumulative surface area is larger than the true area. For a very wide distribution of pore sizes, however,

$$S_{\text{cum}} \approx S_{\text{BET}} \tag{75}$$

since, at the closing point of the loop, $S_{\text{cum}} < S_{\text{BET}}$, and, at lower relative pressures, $S_{\text{cum}} > S_{\text{BET}}$, the two effects being roughly compensating. Hence, cumulative surface areas, calculated from a pore size analysis, are less than the true geometric area by the factor given in equation (74). For triangular-shaped capillaries this amounts to 0·6 and for hexagonal capillaries 0·9.

If some of the pore volume is composed of capillary intersections, then there will be some pore space without a corresponding surface area. Steggerda [27] shows that the average pore radius calculated from equation (70) should be corrected by a factor

$$\bar{r} = \frac{2V_p}{S} \frac{1 - \psi}{1 - \frac{1}{2}\psi} \tag{76}$$

where ψ is the porosity of the material (defined earlier by equation (54)), v_p is the pore volume and S the surface area. Hence, the values of ΔS, calculated for an assembly of intersecting pores, will be too high, and therefore $S_{\text{cum}} > S_{\text{BET}}$.

4.4.3. Pore geometry and suitable models for characterizing porous structures

4.4.3.1. *Geometrical Models of Pores*

Only recently has careful consideration been given to pore models other than simple cases of non-intersecting capillaries and parallel-sided fissures [24, 29, 31–33]. Electron microscope methods are potentially capable of yielding information concerning the basic geometry of many porous materials, so that the seemingly complex relation between pore structure and adsorptive properties can be unravelled. As pointed out in Sections 4.4.2 and 4.4.3, the experimental parameters that are used to describe the pore structure are the total pore volume v_p, the surface area S, and the pore size distribution function. It has been shown through equations (40) and (56), how, for cylindrical capillaries, the mean pore radius is related to the pore volume and the surface area. For other geometric shapes, the relation between v_p and S is given by

$$\bar{r} = \frac{1}{\gamma} \frac{2v_p}{S} \tag{77}$$

where γ is a factor characteristic of the particular pore geometry. Some indication of the pore geometry could therefore be deduced if an independent measurement of $\bar{r}$ is made, thus leading to an evaluation of γ. A knowledge of γ, however, does not always lead to a unique geometrical specification. Everett [*32*] tabulated values of γ for simple uniform structures in which only a single geometric parameter r defines the system. These values are given in Table 4.1. More complicated uniform structures are characterized by two

TABLE 4.1

Basic structure	Length parameter	Numerical value of γ
Non-intersecting cylindrical capillaries	Capillary radius	1
Parallel-sided fissures	Distance apart of walls	1
Non-intersecting close-packed cylindrical rods	Radius of rod	0·104
Cubic packing of spheres	Radius of sphere	0·613
Orthorhombic packing of spheres	Radius of sphere	0·433
Rhombohedral packing of spheres	Radius of sphere	0·229

parameters D (or r) and R. Now γ depends upon the ratio R/D (or R/r) as well as on the basic type of geometrical structure. Table 4.1 gives the relationship between γ and the structural parameters for various types of geometry. Figure 8 shows the structural type which the parameters define. Unless R/D

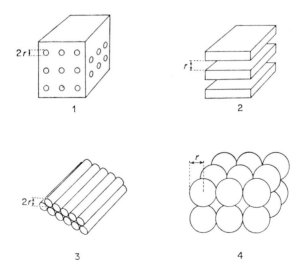

Fig. 8. Four single-parameter geometric models proposed for the structure of porous solids.

is known, no reliable estimate of the pore size parameter can be obtained since, for values of R/D of practical significance, γ lies in the range 1–3. A moment's consideration, however, shows that R/D determines the porosity of the structure so that, if the pore volume per unit weight is independently known, R/D may be determined from the relations in Table 4.1 and a plot of $v_p/m(=V_g)$ versus R/D shown in Fig. 9 for the various structures. Even

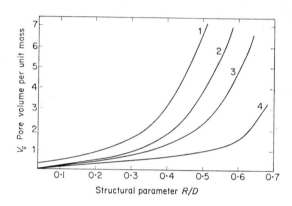

Fig. 9. Pore volume as a function of structure parameter for four types of geometry. The numbers on each curve correspond to the structural types shown in Fig. 8.

then, a unique solution is not obtained unless one has determined, by electron-microscopic methods, the structural type to which the material belongs. The above considerations apply to an assembly of uniform non-intersecting and intersecting capillaries defined by various structural types. For a non-uniform assembly of non-intersecting cylindrical capillaries the pore size distribution may be described either by

$$\bar{r}_v = \frac{\int_0^\infty r\pi r^2 L(r)\,dr}{\int_0^\infty \pi r^2 L(r)\,dr} = \frac{\int_0^\infty rv(r)\,dr}{\int_0^\infty v(r)\,dr} \tag{78}$$

or

$$\bar{r}_L = \frac{\int_0^\infty rL(r)\,dr}{\int_0^\infty L(r)\,dr}. \tag{79}$$

Unless there is a very sharp distribution of sizes, the ratio $2v_p/S$ will differ from the mean pore size by some factor, say β, and so, from equation (76),

$$\beta_v r_v = \frac{2v_p}{S} = \frac{\displaystyle\int_0^\infty 2\pi r^2 L(r)dr}{\displaystyle\int_0^\infty 2\pi r L(r)dr} = \frac{\displaystyle\int_0^\infty v(r)dr}{\displaystyle\int_0^\infty \frac{1}{r}v(r)dr} = \frac{1}{\left(\dfrac{1}{r_v}\right)} \tag{80}$$

where $\overline{(1/r_v)}$ is the mean value of $1/r_v$. From equation (77) an alternative expression is

$$\beta_L \bar{r}_L = \frac{2V_p}{S} = \frac{\displaystyle\int_0^\infty \pi r^2 L(r)dr}{\displaystyle\int_0^\infty 2\pi r L(r)dr} = \frac{\displaystyle\int_0^\infty L(r)dr}{\displaystyle\int_0^\infty r L(r)dr} = \frac{\dfrac{\displaystyle\int_0^\infty r^2 L(r)dr}{\displaystyle\int_0^\infty L(r)dr}}{\bar{r}_L} = \frac{\overline{r^2_L}}{\bar{r}_L} \tag{81}$$

where $\overline{r^2_L}$ is the mean value of r^2_L. By substituting either a Maxwellian or a Gaussian distribution function into equations (80) and (81), Everett [31] calculated β_V and β_L as a function of the width of distribution of r values. In this way it was shown that the errors likely to arise in calculating a mean radius from $2v_p/S$ for a non-uniform structure are unlikely to exceed 10%, while there is no more than a difference of 2% between values for $\bar{r}_V$ and $\bar{r}_L$. If the system is bidisperse, or even tridisperse, there is no close relation between $2v_p/S$ and the mean pore radius which must now be carefully defined. For non-uniform two-parameter structures, the mean parameters can be defined in terms of the number of units having values within a given range and the ratio $2v_p/S$ then involves mean values of the type $\bar{R}, \bar{D}, \overline{R^2}, \overline{RD}, \overline{R^2 D}$. If D is constant there is only one way of defining the mean values and a value for β may be found. If D is variable, then in addition to defining mean values in terms of numbers of units, alternative mean values may be derived in terms of the volumes of units.

The ratio of the pore volume to the surface area will, in general, give a mean pore radius, but at least one set of independent experiments is necessary to characterize the structure of the pore geometry. Albeit these measurements provide information about pore geometry, one example will suffice to show that the ratio v_p/S, as deduced geometrically, is proportional to, but not equal to, the experimentally derived value. Consider a cylindrical pore whose radius of curvature is large compared with the diameter of an adsorbed molecule. Suppose that the area which a single molecule occupies is d^2 and that its volume is d^3. Figure 10 illustrates the geometric problem of filling the actual capillary volume with cubes and calculating the corresponding area of surface covered. Let the pore volume which is to be filled be $(\pi r^2 d)$. Since the complete geometrical surface of the pore is not covered by the squares, we can say that the locus of points representing the distance from the centre of

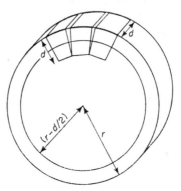

FIG. 10. The geometric problem of finding the surface area occupied by cubes filling the capillary volume.

the face of each cube to the pore wall describes a circle the radius of which is the measured average effective radius of the capillary. If the total number of molecules covering the surface which this volume contains is n, then the total surface area covered is nd^2. This may be equated to $\{2\pi(r-d/2)d\}$ the surface area of the imaginary cylinder formed by joining the centres of the squares. Hence the number of molecules covering the surface of the pore wall is $2\pi(r-d/2)/d$. Since each molecule occupies an area d^2, the area of the wall which is covered (the experimental surface area S_{exp}) is $2\pi(r-d/2)d$. Hence

$$\frac{2v_{p\ exp}}{S_{exp}} = \frac{2\pi r^2 d}{2\pi(r - d/2)d} = \frac{r}{1 - d/2r} \qquad (82)$$

Since the geometric ratio $2v_p/S$ is equal to r, then the proportionality factor relating the geometric ratio to the experimental one is $(1-d/2r)$. Everett [32] calculated this factor for several other geometrical models and his results are given in Table 4.2. Concave surfaces lead to an underestimate of the geometrical surface area, while convex surfaces produce an overestimate.

Sufficient indication has already been given to conclude that the ratio of the pore volume to surface area is of only limited value in determining the detailed pore geometry of porous media. Two points should be noted: (1) the estimation of precise values of v_p and S from experimental data is obtained by the somewhat arbitrary selection of points on the isotherm representing complete pore filling and the completion of a monolayer over the

TABLE 4.2

Basic structure	Proportionality factor $\dfrac{2v_p}{S} \times \dfrac{1}{r}$
Non-intersecting cylindrical capillaries	$1 - \dfrac{d}{2r}$
Non-intersecting cylindrical rods	$1 + \dfrac{d}{2r}$
Intersecting cylindrical capillaries	$1 - \dfrac{d}{2R} + \dfrac{9d}{2R}\left(\dfrac{1 - d/2R}{1 + 10r}\right)$
Intersecting cylindrical rods	$1 + \dfrac{d}{2R} - \dfrac{9d}{2R}\left(\dfrac{1 + d/2R}{1 + 10r}\right)$

entire surface; (2) the geometrical ratio $2v_p/S$ does not provide a unique characterization of pore structure. Despite these shortcomings, it is obvious that some significance may be given to an average pore dimension derived from pore volume and surface-area measurements. The principal defect of this approach is its failure to define exactly the pore geometry. Nevertheless, by assuming that the dimension so derived is an effective pore radius, then a pore model in which the capillaries are equivalent inscribed cylinders will not detract too much from a solution to the problem of predicting chemical reaction rates in porous catalysts.

4.4.3.2. *Practical Model of Pores*

In order to provide a practical solution to this problem, it is convenient to relate experimentally derivable quantities to the mean effective pore radius. Wheeler [34, 17] proposed a model in which the mean radius $\bar{r}$ and the length L of the pores of a catalyst pellet are determined in such a way that the sum of the surface areas of all of the pores is equal to the BET surface area and the sum of the pore volume is equal to the experimental pore volume. Suppose the external surface area (e.g. measured by the sedimentation technique, Section 4.2.2) of the catalyst is S_x and that there are n_p pores per unit external surface area. The pore volume contained by $(S_x n_p)$ pores is therefore $(S_x n_p)\pi r^2 L$. If the total volume of the catalyst pellet is V_p and its porosity ψ, then the experimental pore volume is $V_p \psi$. Now the porosity is the product of the pellet density ρ_p and the specific pore volume V_g, both measurable quantities. By equating the experimental pore volume to the pore volume of the model, one obtains

$$V_p \rho_p V_g = (S_x n_p)\pi r^2 L \tag{83}$$

If the pore walls are rough and the pores intersect, the total surface area of a single pore wall is $2\pi L r \tau (1 - \psi)$, where τ is a roughness factor and

$(1-\psi)$ represents the fraction of the pore walls which are not interrupted by inter-sections. The total surface area of the $(S_x n_p)$ pores is therefore $[(S_x n_p)2\pi r L\tau(1-\psi)]$. Equating the experimental surface area to the surface area of the model

$$V_p \rho_p S_g = (S_x n_p)\tau(1 - \psi)2\pi r L \qquad (84)$$

where $(V_p \rho_p)$ is the mass of the catalyst pellet considered. Dividing equation (83) by (84), the mean pore radius is

$$\bar{r} = \frac{2V_g}{S_g} \tau(1 - \psi) \qquad (85)$$

The number of pore openings n_p per unit external surface area would be $(\psi/\pi r^2)$ if all the pores were orientated in the same direction. Wheeler [34, 17] argued that as some pores are exactly parallel to the gas flow while some are at right-angles, a reasonable assumption is that the average orientation of pores is 45° to the direction of gas flow. In this case n_p will be $(\psi/\sqrt{2}\pi r^2)$. Substituting this latter expression for n_p into equation (83) gives

$$L = \sqrt{2}\frac{V_p}{S_x} \qquad (86)$$

so that the average pore length is $\sqrt{2}$ times the ratio of the volume to external surface area of the pellet. In terms of the pellet size d_p, this ratio is $d_p/6$ for spheres, cylinders and cubes. Thus, for most practical catalyst pellets, the average length of a pore is $(\sqrt{2}d_p/6)$. This value is the average length of the pore used in subsequent equations formulated to assess reaction rates in pores, since most commercial catalysts are pressed into pellets in the shape of spheres or cylinders and in which all external faces are exposed to the re-acting gas. By means of this model the reaction rate in a single pore can be computed and this may be related to the rate of reaction in a practical catalyst pellet of known geometrical shape and known pore volume and surface area.

4.5. Reaction Rates in Porous Catalysts

4.5.1. Mass transport of gases through pores

Since there are many conditions under which a catalytic reaction can operate, it is to be expected that these conditions may affect the mode of transport of gaseous molecules through the porous structure and, if the rate of transport is comparable with the rate of chemical reaction, influence the observed kinetics. It is therefore pertinent to consider the various modes of transport of gaseous molecules through porous media. The rate at which molecules diffuse through porous media is very much less than that associated with the translational velocity of molecules. This is because molecules collide with the pore walls and with other molecules during their passage through the catalyst structure, resulting in a completely random molecular motion. However, it is sufficient to determine the net rate of molecular transport through an imaginary plane across which the concentration gradient is known. This net rate will depend on the magnitude of the pore radius as compared with the mean free path of the molecules and on whether or not there is a total pressure difference across the pore length. We shall distinguish three types of mass transport in pores.

4.5.1.1. *Knudsen Flow*

When the mean free path of molecules is considerably greater than the pore diameter, Knudsen flow is the mode of molecular transport. Such conditions usually prevail with gas reactions carried out at moderate pressures on catalysts with small pores of 1000 Å radius or less. For example, at one atmosphere pressure the mean free path of a molecule with a diameter of about 2×10^{-8} cm is of the order of 10^{-5} cm: many catalysts (e.g. alumina cracking catalysts) have pores less than 100 Å, so that for these catalysts Knudsen flow will predominate at pressures not far removed from atmosphere.

Under Knudsen flow conditions, collisions between molecules in the gas phase are less numerous than those between molecules and pore walls. The flow is now governed by the collisions of molecules with the capillary walls, which, Knudsen assumed, have irregular molecular projections distributed over the surface. Consequently, a gaseous molecule will be reflected from the walls in a direction independent of the angle of incidence. Knudsen [35, 36] showed that their reflections obeyed the cosine law, which states that of the n molecules present the number dn which are reflected in a cone of solid angle ω is

$$dn = n \frac{\omega}{\pi} \left(\cos \alpha \right) dS \tag{87}$$

where α is the angle between the axis of the cone and the normal to the element of surface dS. By making use of the cosine law, it may be shown from geometrical considerations [37] that the excess flow of molecules in the direction x of gas flow is $B dN/2dx$ where, for a circular cross-section of radius r, B is $16\pi r^3/3$ and N denotes the molecular flux. Knudsen confirmed by experiment that the number of molecules decreased linearly with the distance along the tube. From kinetic theory, the number of molecules crossing unit area in unit time is

$$\frac{Nc\bar{u}}{4} = \frac{p}{\sqrt{(2\pi MRT)}} \tag{88}$$

where N is Avogadro's number, c is the concentration, $\bar{u}$ the mean Maxwellian velocity and p the pressure at the plane under consideration. Now, if

n decreases linearly with the length of the capillary, we can say that the rate of flow of gas through the pore is

$$\frac{dn}{dt} = \frac{1}{2}\left(\frac{16\pi r^3}{3}\right)\frac{dN}{dx} = \frac{8}{3}\,\pi r^3\,\frac{\Delta p}{\Delta x}\,\frac{1}{\sqrt{(2\pi MRT)}} \tag{89}$$

where Δp represents the difference in pressure between two planes a distance Δx apart along the pore length. From the gas laws, the pressure gradient in equation (89) may be written in terms of a concentration gradient, so that

$$\frac{dn}{dt} = \frac{8}{3}\,\pi r^3\,\sqrt{\left(\frac{RT}{2\pi M}\right)}\,\frac{\Delta c}{\Delta x} = \pi r^2\,\frac{2}{3}\,r\,\sqrt{\left(\frac{8RT}{\pi M}\right)}\frac{\Delta c}{\Delta x} \tag{90}$$

Equation (90) is in the form of Fick's law of diffusion, since πr^2 is the area presented to the diffusing molecules. The Knudsen diffusion coefficient D_K may therefore be defined as

$$D_K = \frac{2}{3}\,r\,\sqrt{\left(\frac{8RT}{\pi M}\right)} = \frac{2}{3}\,r\bar{u}. \tag{91}$$

where $\bar{u}$ is the mean Maxwellian velocity. We can therefore include Knudsen flow in a general diffusion equation by utilizing the Knudsen diffusion coefficient as given by equation (91).

4.5.1.2. *Bulk Flow of Gases*

At 1 atm pressure the mean free path is about 10^{-5} cm. In pores larger than about 10^{-4} cm, therefore, the mean free path is much smaller than the pore diameter and gaseous collisions will now be more frequent than collisions with walls. The rate of diffusion will now be independent of the pore radius.

The kinetic theory of gases shows that the net rate of flow of molecules in one direction is given by the difference between the rates of flow at two given points a distance λ (the mean free path) apart. Suppose there are two kinds of molecule, A and B. The rate at which A pass from left to right through unit area in unit time is, from kinetic theory, $n(c_A)_+\bar{u}/6$, where $n(c_A)_+$ is the number of A molecules travelling in this direction at a distance λ from the area considered. If c_A is the concentration of A a distance λ from the plane, then the number of A molecules moving from left to right in unit time is

$$n(c_A)_+\frac{\bar{u}}{6} = n\left(c_A - \lambda\frac{\partial c_A}{\partial x}\right)\frac{\bar{u}}{6} \tag{92}$$

and, by analogy, the number of A molecules moving from right to left in unit time is

$$n(c_A)_-\frac{\bar{u}}{6} = n\left(c_A + \lambda\frac{\partial c_A}{\partial x}\right)\frac{\bar{u}}{6} \tag{93}$$

the net flow of A molecules from right to left is therefore

$$\frac{\partial n_A}{\partial t} = \frac{n\bar{u}}{3}\,\lambda\,\frac{\partial c_A}{\partial x} \tag{94}$$

The same number of B molecules pass through the same area in the opposite direction, so

$$\frac{\partial c_A}{\partial x} = -\frac{\partial c_B}{\partial x} \tag{95}$$

Equation (95) shows that diffusion does not change the total pressure. The diffusion coefficient D_B for bulk flow may be defined from equation (94) as

$$D_B = \frac{1}{3}\lambda \bar{u} \tag{96}$$

4.5.1.3. Forced Flow

Both of the above diffusive processes are independent of a total pressure difference across the pore. If a pressure difference is maintained, forced flow occurs, and when the mean free path of the molecules is large compared to the pore diameter, forced flow is indistinguishable from Knudsen flow and is not affected by pressure differentials. When, however, the mean free path is small compared to the pore diameter and a pressure difference is still maintained, superimposed on bulk flow will be flow resulting from such a pressure difference. A relation giving the rate of flow of a fluid forced through a pipe was deduced experimentally by Hagen [38] and, independently, by Poiseuille [1]. Such a relation may also be applied to forced flow in narrow channels such as catalyst pores. Consider an element of fluid, length ΔL and radius a, being forced through a cylindrical pore of radius r. Let the linear velocity of the outer edge of this element be u_l. The force resulting from the shear stress at the pore wall is balanced by a force which is maintained by the pressure difference ΔP across the elementary cylinder of fluid. Now the viscosity η is the shear stress arising from unit velocity gradient, so the shearing force is

$$\eta 2\pi a \Delta L \frac{du_l}{da} = \pi a^2 \Delta P \tag{97}$$

Integrating (97) from the axis of the cylinder to any radius a with the boundary condition that the velocity u_l is zero at the cylinder wall ($a=r$),

$$u_l = -\left(\frac{\Delta P}{4\eta \Delta L}\right)(r^2 - a^2) \tag{98}$$

Assuming that this velocity is constant in an elementary annulus of radii a and $(a+\delta a)$, the volumetric velocity is given by

$$\frac{-\Delta P}{4\eta \Delta L} \int_0^r 2\pi a(r^2 - a^2)da = \frac{\pi r^4}{8\eta}\frac{\Delta P}{\Delta L} \tag{99}$$

Rewriting this equation in terms of the number of molecules flowing per unit time and replacing the pressure gradient term by a concentration gradient

$$\left(\frac{dn}{dt}\right)_P = \frac{\pi r^4}{8\eta}\frac{\Delta P}{\Delta L}c_T = \pi r^2\left(\frac{r^2}{8\eta}c_T RT\right)\frac{\Delta c_T}{\Delta L} \tag{100}$$

where c_T is the total molar concentration of all the molecules, both reactant and product, that are present. The diffusion coefficient for forced flow may therefore be written

$$D_P = \frac{r^2 c_T RT}{8\eta} \qquad (101)$$

Since the viscosity of most gases at pressures of about 1 atm is about 10^{-4} centipoise, then, for pores of 10^{-4} cm radius, D_P is of the order of 10^{-1} cm^2 sec^{-1}. In small pores, it is apparent that forced flow will compete with bulk diffusion. For rapid reactions accompanied by an increase in the number of moles, a considerable excess pressure is developed in the interior recesses of the particle. This develops because there will be insufficient driving force for the outward flow (by Poiseuille flow) of the excess molecules until the interior pressure is substantially greater than the pressure at the exterior of the catalyst. For capillaries with $r > 10^{-2}$cm, forced flow is very much more rapid than bulk flow and excess pressures would not be developed in the interior of the catalyst particles. Except in the case of reactions at high pressure, the pressure drop which must be maintained across a fixed catalyst bed to maintain flow through the packed reactor is insufficient to cause forced flow through the capillaries of the catalyst pellet, and the gas flow is diverted around the exterior of the pellets. Reactants then reach the interior surface of the catalyst either by Knudsen or bulk flow.

4.5.2. Concentration profiles in pores of definite geometry

If diffusion into the catalyst particle is fast compared to chemical reaction, it is obvious that the entire accessible internal surface of the catalyst will be available for reaction, since the reactants will have reached the interior pore structure before they have had time to react. In such a case, although only a small concentration gradient exists between the exterior and the interior of the particle, there is a finite diffusive flux of reactant molecules in and product molecules out. In a slow chemical reaction, the entire surface of the catalyst is likely to be useful.

Contrast such a slow reaction with a fast reaction in a very active catalyst. Here the reactants will have been converted into products before they have had time to diffuse more than a small distance along the pores. There will be a steep concentration gradient of both reactants and products near the periphery of the pellet, and reactant molecules will diffuse rapidly a short distance into the pore and product molecules diffuse out rapidly. Almost the entire reaction takes place on the periphery of the catalyst pellet, and the interior pore structure is not used.

Thiele [39] developed differential equations the solution of which describe the variation of concentration along the length of pores of different pellet shapes. Wheeler [34, 17] extended Thiele's treatment in such a way that the equations would be adaptable to Wheeler's general model of a practical catalyst pellet. We consider various important cases which enable some conclusions to be drawn concerning reaction rates in commercial catalysts.

4.5.2.1. A Single Cylindrical Pore

Figure 11 shows a single open-ended cylindrical pore of radius r and length $2L$. Suppose the concentration of a single reactant at the pore mouth ($L=0$),

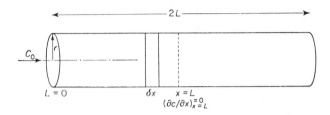

FIG. 11. Conservation of mass within a cylindrical pore.

where we fix the origin of coordinates, is c_0. If we consider an elementary section δx of this pore, a material balance demands that, in the steady state, the change in mass flow rate of the reactant in and out of the section is equal to the rate of reaction. Assuming the reaction to be first order, then the change in mass flow rate as a result of diffusion along the pore length will be

$$\pi r^2 D \left(\frac{\partial c}{\partial x}\right)_x - \pi r^2 D \left(\frac{\partial c}{\partial x}\right)_{x+\delta x} = -\pi r^2 D \frac{\partial^2 c}{\partial x^2} \delta x \qquad (102)$$

where c is the concentration at a point x along the pore, and D is the diffusion coefficient, the form of which depends on the conditions under which mass transport occurs (Knudsen, bulk or forced flow). Equating this to the reaction rate in the surface area ($2\pi r\delta x$) within the element

$$\pi r^2 D \frac{\partial^2 c}{\partial x^2} \delta x = 2\pi r\delta x k_1 c \qquad (103)$$

where k_1 is the intrinsic rate constant per unit surface area of solid. The boundary conditions are

$$c = c_0 \text{ at } x = 0 \qquad (104)$$

$$\frac{\partial c}{\partial x} = 0 \text{ at } x = L \qquad (105)$$

the second condition implying that, at the centre of the pore, there is no net flow of reactant. The solution to equation (103), a simple linear second order differential equation, is

$$c = c_0 \frac{\cosh\left\{ L \sqrt{\left(\frac{2k_1}{rD}\right)} - x \sqrt{\left(\frac{2k_1}{rD}\right)} \right\}}{\cosh\left\{ L \sqrt{\left(\frac{2k_1}{rD}\right)} \right\}} \qquad (106)$$

The rate of reaction $R_{1/2}$ in a half-pore is equal to the rate of flow of reactant into the pore, so

$$R_{1/2} = \pi r^2 D \left(\frac{\partial c}{\partial x}\right)_{x=0} = \pi r^2 D c_0 \sqrt{\left(\frac{2k_1}{rD}\right)} \tanh\left\{ L \sqrt{\left(\frac{2k_1}{rD}\right)} \right\} = \frac{\pi r^2 D c_0 h_1 \tanh h_1}{L} \quad (107)$$

where the dimensionless quantity h_1 is defined by

$$h_1 = L \sqrt{\left(\frac{2k_1}{rD}\right)} \qquad (108)$$

and referred to as the Thiele modulus.

A simple expression for the fraction of pore area available for reaction can now be obtained by dividing equation (107) by the rate $R_0 (= 2\pi r L k_1 c_0)$ which would be observed if there were no concentration gradient along the pore (i.e. the rate is fast compared to diffusion and the whole pore is available for reaction). We therefore obtain

$$f = \frac{R_{1/2}}{R_0} = \frac{\pi r^2 D c_0 \sqrt{\left(\frac{2k_1}{rD}\right)} \tanh h_1}{2\pi r L k_1 c_0} = \frac{1}{h_1} \tanh h_1 \qquad (109)$$

When $h_1 > 2$ (reaction fast, small pores) $f = \frac{1}{h_1}$ and the fraction of surface available is inversely proportional to the pore length which, to a first approximation, is itself proportional to the pellet size. The function is plotted as curve A in Fig. 12.

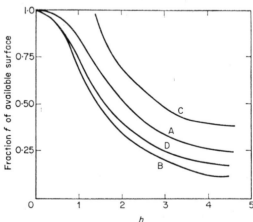

FIG. 12. Fraction of surface available for reaction as a function of the Thiele modulus. A, first order; B, second order; C, zero order; D, first order on spherical pellets.

If the kinetic order of reaction is two, equation (103) becomes

$$\pi r^2 D \frac{\partial^2 c}{\partial x^2} = 2\pi r k_2 c^2 \tag{110}$$

The solution to this equation involves elliptic integrals and is complex. However, as shown below, it is a simple matter to calculate the fraction of surface available. We first define a dimensionless parameter

$$h_2 = L \sqrt{\left(\frac{2k_2 c_0}{rD}\right)} \tag{111}$$

In order to integrate equation (110) we multiply both sides of the equation by $\partial c/\partial x$. It is convenient to choose the origin of coordinates at the pore centre. The boundary conditions are then $c=c_c$ and $\partial c/\partial x=0$ at $x=0$. A single integration then gives

$$\frac{\partial c}{\partial x} = \sqrt{\left(\frac{2k_2}{rD} \frac{2}{3} (c^3 - c_c^3)\right)} \tag{112}$$

The total amount of reaction in the half pore is then

$$R_{1/2} = 2\pi r k_2 \int_0^L c^2 dx = 2\pi r k_2 \int_{c_c}^{c_0} \frac{c^2 dc}{\sqrt{\left(\frac{2k_2}{rD} \frac{2}{3} (c^3 - c_c^3)\right)}} = \pi r^2 D \sqrt{\left(\frac{2k_2}{rD} \frac{2}{3} (c_0^3 - c_c^3)\right)} \tag{113}$$

which result could also have been obtained by equating $R_{1/2}$ to the rate of diffusion $\pi r^2 D (\partial c/\partial x)_{x=L}$ of reactant into the pore. Since the maximum possible rate of reaction that could take place is $2\pi r L k_2 c_0^2$, the fraction of surface available is

$$f = \frac{R_{1/2}}{R_0} = \frac{1}{L\sqrt{\left(\frac{2k_2 c_0}{rD}\right)}} \sqrt{\left\{\frac{2}{3}\left[1 - \left(\frac{c_c}{c_0}\right)^3\right]\right\}} = \frac{1}{h_2} \sqrt{\left\{\frac{2}{3}\left[1 - \left(\frac{c_c}{c_0}\right)^3\right]\right\}} \tag{114}$$

For fast reactions on small pores $c_c \ll c_0$ and the fraction of surface available is then

$$f = \frac{1}{h_2} \sqrt{\frac{2}{3}} \tag{115}$$

It should be noted that for this case the dimensionless parameter depends on the square root of the concentration. Curve B in Fig. 12 shows how the fraction of surface available varies with h_2.

Another important case is when the reaction is zero order. Many catalytic reactions show zero order kinetics. This is usually attributable to the active surface being completely covered by a strongly adsorbed reactant or product. The diffusion equation analogous to equation (101) for this case is

$$\pi r^2 D \partial^2 c / \partial x^2 = 2\pi r k_0 \qquad (116)$$

with the boundary conditions as before ($c = c_0$ at $x = 0$, and $\partial c / \partial x = 0$ at $x = L$). The solution is immediately written

$$\frac{c}{c_0} = 1 - \frac{2L^2 k_0}{r D c_0}\frac{x}{L} + \frac{L^2 k_0}{r D c_0}\left(\frac{x}{L}\right)^2 = 1 - h_0^2 \left\{\frac{x}{L} - \frac{1}{2}\left(\frac{x}{L}\right)^2\right\} \qquad (117)$$

where the dimensionless parameter h_0 is defined by

$$h_0 = L \Big/ \sqrt{\left(\frac{2k_0}{r D c_0}\right)} \qquad (118)$$

In any part of the pore c is always greater than zero for values of $h_0 < \sqrt{2}$, so that, under these conditions, the surface is completely available. If the surface is not completely available, the boundary conditions must be changed such that both the concentration c and the concentration gradient $\partial c / \partial x$ become zero at some point, say x_0, in the pore. Equation (116) then integrates to

$$\frac{c}{c_0} = 1 - 2\left\{\frac{h_0 x}{\sqrt{2}L} - \frac{1}{2}\left(\frac{h_0 x}{\sqrt{2}L}\right)^2\right\}, \quad x \leqslant x_0 \qquad (119)$$

$$c = 0, \qquad\qquad\qquad\qquad x \geqslant x_0 \qquad (120)$$

The point x_0 at which $\partial c / \partial x$ is zero is, from equation (119), $\sqrt{2}L/h_0$. Hence, when $h_0 > \sqrt{2}$, the fraction of surface available is

$$f = \frac{2\pi r k_0 x_0}{2\pi r k_0 L} = \frac{\sqrt{2}}{h_0} \qquad (121)$$

and the reaction rate in the half pore is

$$R_{1/2} = 2\pi r L k_0 f = 2\pi r \sqrt{(r k_0 D c_0)} \qquad (122)$$

A plot of the fraction of available surface as a function of h_0 for a zero order reaction is given by curve C in Fig. 12.

In the case of a reaction accompanied by a volume change, the solution is complicated. Thiele [39] assumed that the total concentration of reactant and product in the pore is constant and thereby simplified the calculation considerably. The number of moles of reactant passing through an infinitesimal cross-section of the pore in one direction is equal to the difference between the amount of reactant displaced out of the section by forced flow as a result of volume increase and the amount entering by diffusion. Further, in the steady state, the number of moles of reactant passing through this element must be equal to the amount of forced flow per mole of original reactant. If the reaction is

$$A \rightarrow \nu B \qquad (123)$$

then the increase in number of moles is $v-1$ so

$$D\pi r^2 \frac{\partial c_A}{\partial x} - \left(\frac{dn}{dt}\right)_P \frac{c_A}{c_T} = \frac{1}{v-1}\left(\frac{dn}{dt}\right)_P \qquad (124)$$

where the term $(dn/dt)_P$ accounts for forced Poiseville flow caused by the volume change. From equation (124)

$$\left(\frac{dn}{dt}\right)_P = \frac{(v-1)\pi r^2 D c_T}{c_T + (v-1)c_A} \frac{\partial c_A}{\partial x} \qquad (125)$$

and so the net number of moles of reactant crossing the element is

$$\frac{1}{v-1}\left(\frac{dn}{dt}\right)_P = \frac{\pi r^2 D c_T}{c_T + (v-1)c_A} \frac{\partial c_A}{\partial x} \qquad (126)$$

Considering, as before, an infinitesimal section of the pore and equating the mass remaining in the section to the extent of reaction on the pore walls,

$$\left[\frac{1}{v-1}\left(\frac{dn}{dt}\right)_{P=x}\right] - \left[\frac{1}{v-1}\left(\frac{dn}{dt}\right)_{P=x+\delta x}\right] = \frac{\partial}{\partial x}\left\{\frac{1}{v-1}\left(\frac{dn}{dt}\right)_P\right\}\delta x =$$

$$\frac{\partial}{\partial x}\left\{\left(\frac{\pi r^2 D c_T}{c_T + (v-1)c_A}\right)\frac{\partial c_A}{\partial x}\right\}\delta x = 2\pi r \delta x k_1 c \qquad (127)$$

with the boundary conditions

$$x = 0, \quad \frac{\partial c_A}{\partial x} = 0 \qquad (128)$$

$$x = 0, \quad c = c_0 \qquad (129)$$

Thiele integrated this equation analytically and obtained curves for the fraction of surface available which depends on the values of c_A/c_T and x. As c_A/c_T decreases, the fraction of available surface approaches that for a first order reaction without volume change. Wheeler [34] criticized the assumption that c_T is constant and suggested that an equation in c_T should be solved simultaneously with equation (127). From equation (126) we see that the equation for c_T should be

$$\frac{1}{v-1}\left(\frac{dn}{dt}\right)_P = \frac{1}{v-1}\pi r^2 D_P \frac{\partial c_T}{\partial x} = \frac{\pi r^2 D c_T \frac{\partial c_A}{\partial x}}{c_T + (v-1)c_A} \qquad (130)$$

A second criticism which Wheeler propounded is that the diffusive flow of molecules was neglected in comparison with forced flow in deriving equations (127) and (130). However, Thiele's original computations are sufficiently accurate to give a qualitative picture of the situation.

4.5.2.2. *The Influence of Diffusion on Adsorption and Desorption in Single Pores*

If, during a surface reaction, one of the reactants or products is adsorbed, the reaction rate is modified. If θ is the fraction of surface covered by adsorbed molecules, it is reasonable to suppose that the first-order rate constant k_1 will decrease linearly with the fraction $(1-\theta)$ of surface remaining uncovered. Thus, a dimensionless group h_θ is now defined

$$h_\theta = L\sqrt{\left(\frac{2k_1(1-\theta)}{rD}\right)} = h_\theta^0 \sqrt{(1-\theta)} \tag{131}$$

where h_θ^0 is the value of h_θ when the pore is free of adsorbed molecules. Provided that the vacant sites are evenly distributed over the surface, the activity of the pore is proportional to $h_\theta \tanh h_\theta$ (cf. equation (107)). The ratio of the activity of the pore containing adsorbate to that of the uncovered pore will be the fraction of surface available for reaction, and is therefore

$$f = \frac{h_\theta^0 \sqrt{(1-\theta)} \tanh\{h_\theta^0 \sqrt{(1-\theta)}\}}{h_\theta^0 \tanh h_\theta^0}. \tag{132}$$

When h_θ^0 is small (slow reactions in large pores), then the fraction of surface available for reaction is simply $\sqrt{(1-\theta)}$.

For an extremely active catalyst, it may be expected that only a few collisions of adsorbate molecules with the walls are sufficient to cause the pore entrance to be covered with an immobile layer of adsorbed molecules. Since a fraction of the pore length θL near the pore mouth will be covered with adsorbate, molecules must now travel a distance $(1-\theta)L$ along the pore before reaching any surface free of adsorbate. In the case of a fast reaction, diffusion to the uncovered surface may be sufficiently slow to determine the overall rate. If transport through that part of the pore covered with adsorbate is by diffusion, then in the steady state the rate of diffusion through the length θL of pore covered by adsorbate is equal to the reaction rate in the length of pore $(1-\theta)L$ remaining free of adsorbate. The rate of reaction in the length $(1-\theta)L$ will be given by an equation analogous to equation (107), so the equation determining the concentration c_L at the end of the length θL will be

$$\pi r^2 D\left(\frac{\partial c}{\partial x}\right)_{x=\theta L} = \pi r^2 D\left(\frac{c_0-c_L}{\theta L}\right) = \frac{\pi r^2 D c_L}{(1-\theta)L} h_\theta^0(1-\theta)\tanh\{h_\theta^0(1-\theta)\} \tag{133}$$

where h_θ^0 is the value of h_θ for that part of the surface free of adsorbed molecules. The rate of reaction in that part of the pore which is covered by adsorbate is thus obtained by solving for c and substituting this into the left-hand side of the equation. We therefore find that this ratio is

$$\pi r^2 D\frac{c_0 - c_L}{\theta L} = c_0 \frac{\pi r^2 D}{L} \frac{\tanh\{h_\theta^0(1-\theta)\}}{1+h_\theta^0\theta} \tag{134}$$

Since the rate in the uncovered portion is $\pi r^2 (D/L)c_0 \tanh h_\theta^0$, the ratio of the rates in the covered section of the pore to the uncovered section is

$$f = \frac{1}{1 + h_\theta^0 \theta} \frac{\tanh\{h_\theta^0(1 - \theta)\}}{\tanh h_\theta^0} \tag{135}$$

This is a measure of the extent to which the reaction rate is retarded by strong adsorption in small pores. If the rate is high and the pores are very small, the reaction is retarded by a factor $1/(1 + h_\theta^0 \theta)$.

Although Wheeler did not consider the influence of diffusion on desorption, Sykes and White [40] considered the problem when investigating the adsorption of sulphur and carbon disulphide on various active carbons. If, during a reversible adsorption, the coverage θ does not vary appreciably, the rate at which gas is desorbed from the surface may be written as $k_0(1 - c/c_\infty)$, where c is the gas concentration and c_∞ its equilibrium value. Desorption will be retarded if the gas concentration gradient required to initiate diffusion out of the pore causes c to approach c_∞. Considering an elementary section, thickness δx, of the pore, a material balance gives

$$\pi r^2 D \left(\frac{\partial c}{\partial x}\right)_x - \pi r^2 D \left(\frac{\partial c}{\partial x}\right)_{x+\delta x} = \pi r^2 D \frac{\partial^2 c}{\partial x^2} \delta x = 2\pi r k_0 \left(1 - \frac{c}{c_\infty}\right) \delta x \tag{136}$$

where the boundary conditions are the same as for a first order reaction without volume change (i.e. $c = c_0$ at $x = 0$, and $\partial c/\partial x = 0$ at $x = L$). The solution to this simple linear second order differential equation is

$$\frac{c - c_\infty}{c_0 - c_\infty} = \frac{\cosh\left(L\sqrt{\left(\frac{2k_0}{rDc_0}\right)} - x\sqrt{\left(\frac{2k_0}{rDc_0}\right)}\right)}{\cosh\left(L\sqrt{\left(\frac{2k_0}{rDc_0}\right)}\right)} \tag{137}$$

The rate of desorption from the half pore will therefore be

$$-\pi r^2 D \left(\frac{\partial c}{\partial x}\right)_{x=0} = \frac{-\pi r^2 D}{L} (c_0 - c_\infty)h_0 \tanh h_0 \tag{138}$$

The maximum rate in the pore is $2\pi r k_0 L(1 - c_0/c_\infty)$, so the fraction of surface available is, in this case,

$$f = \frac{1}{h_0} \tanh h_0 \tag{139}$$

where

$$h_0 = L\sqrt{\left(\frac{2k_0}{rDc_0}\right)} \tag{140}$$

The fraction f is a measure of the extent to which desorption is retarded by diffusion.

The effect of temperature on the rate of adsorption is worth considering at this point. If, in equation (134), $h_\theta^0(1-\theta)$ is large, then the rate of reaction is simply equal to the rate of diffusion (under a concentration gradient $(c_0/\theta L)$) into that part of the pore not covered by adsorbate. The temperature coefficient of reaction will then depend entirely on the effect of temperature on diffusion. Under Knudsen or bulk diffusion conditions, the rate of diffusion, and hence the rate of reaction, is proportional to $\sqrt{T}$. Under these conditions, an apparent activation energy (see p. 301), deduced from a plot of ln(rate) versus the reciprocal of absolute temperature, will be very small indeed, since the curve is almost parallel to the abscissa. At a sufficiently low temperature, when the rate is measurably slow, the rate constant k_1 is small enough to make h_θ^0 small and so now the fraction of surface available will be near unity, and the true activation energy is observed. On gradually raising the temperature, therefore, there will be a smooth transition from a true chemical rate to a rate determined by diffusion. If the pores are large enough, however, the temperature has to be very high before the activation energy decreases substantially.

In the case of desorption from large-diameter pores h_0 is small and, as would be expected, the surface is completely available, so that true desorption kinetics are observed and the normal activation energy is measured. In small pores, however, when h_0 is large, the rate becomes proportional to $\sqrt{k_0}$ and hence the apparent activation energy is only one-half its true value and desorption is accelerated.

4.5.3. Reaction rates in spherical particles

Figure 13 shows a spherical catalyst pellet of radius R. Consider a first-order reaction occurring within the porous structure of the pellet. A material balance over a spherical shell within the particle of radii a and $a+\delta a$ enables the differential equation for steady diffusion accompanied by reaction to be written down. In this case, however, we must make a suitable assumption about the variation in the number of pores which have the same radius as the centre of the particle is approached. The simplest assumption is that the number of pores of a given radius r decrease in proportion to the area of the spherical shell. If ψ_p is the fraction of surface which consists of pores, the change in mass flow as a result of diffusion is

$$D\frac{\partial}{\partial a}\{4\pi\psi_p a^2 c\}_a - D\frac{\partial}{\partial a}\{4\pi\psi_p a^2 c\}_{a+\delta a} = 4\pi\psi_p a\frac{\partial^2 c}{\partial a^2}\delta a + 8\pi\psi_p a\frac{\partial c}{\partial a}\delta a \quad (141)$$

neglecting second order differences. In the steady state, this may be equated to the amount of reaction in the elementary shell of surface area $(4\pi a^2\delta a)S_v$, where S_v is the specific surface area per unit volume of catalyst pellet. Hence the differential equation becomes

$$\frac{\partial^2 c}{\partial a^2} + \frac{2}{a}\frac{\partial c}{\partial a} = \frac{S_v k_1 c}{D_e} = \frac{k_v c}{D_e} \tag{142}$$

where k_1 is the rate constant per unit surface area, $k_v(=S_v k_1)$ is the rate constant per unit volume and D_e is an effective diffusivity. The two boundary conditions are

$$c = c_0 \text{ at } a = 0 \tag{143}$$

and

$$\frac{\partial c}{\partial a} = 0 \text{ at } a = R \tag{144}$$

Equation (142) can be reduced to a standard Bessel equation the solution of which is written in terms of half-order Bessel functions. Half-order Bessel functions can conveniently be expressed in terms of hyperbolic functions and the complete solution of equation (142) may be written

$$c = \frac{c_0 R}{a}\frac{\sinh\left\{a\sqrt{\left(\dfrac{k_v}{D_e}\right)}\right\}}{\sinh\left\{R\sqrt{\left(\dfrac{k_v}{D_e}\right)}\right\}} \tag{145}$$

The rate of reaction inside the pellet must balance the rate of diffusion into the particle so

$$4\pi R^2 D_e\left(\frac{\partial c}{\partial a}\right)_{a=R} = 4\pi R^2 D_e c_0 \sqrt{\left(\frac{k_v}{D_e}\right)}\left[\frac{1}{\tanh\left\{R\sqrt{\left(\dfrac{k_v}{D_e}\right)}\right\}} - \frac{1}{R\sqrt{\left(\dfrac{k_v}{D_e}\right)}}\right] \tag{146}$$

We now define a similar dimensionless quantity,

$$\phi_s = R\sqrt{\left(\frac{k_v}{D_e}\right)} \tag{147}$$

(This is the symbol adopted by Weisz [41] and others [42] for spherical geometry: Wheeler writes [34, 17] $3h_1 = \phi_s$.) If there is no resistance to diffusion, the reaction rate is $k_v c_0$ per unit internal area. Since the internal pore volume of the pellet is $\frac{4}{3}\pi R^3$, then the uninhibited rate of reaction is $\frac{4}{3}\pi R^3 k_v c_0$. The ratio of the rate in the pellet to the uninhibited rate in the absence of diffusion is a measure of the fraction f of surface available for reaction,

$$f = \frac{4\pi R D_e c_0 \phi_s}{\dfrac{k_v c_0 4\pi R^3}{3}}\left\{\frac{1}{\tanh\phi_s} - \frac{1}{\phi_s}\right\} = \frac{3}{\phi_s}\left\{\frac{1}{\tanh\phi_s} - \frac{1}{\phi_s}\right\} \tag{148}$$

This function is plotted as curve D in Fig. 12. For $\phi_s > 5$ (i.e. very active surface), as in case A, f is inversely proportional to h and hence the pellet size.

When reactants and/or products retard the reaction rate, the fraction of surface available for reaction (termed the effectiveness factor by Thiele and

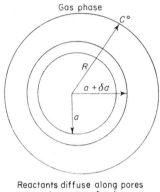

Reactants diffuse along pores
towards centre of sphere

FIG. 13. Conservation of mass within spherical porous particle.

others; see Section 9.2.4) may differ considerably from that calculated on the assumption that the reaction is first order. The magnitude of this effect has recently been calculated [43] by substituting the appropriate Langmuir-Hinshelwood kinetic expression for the reaction rate into the basic differential equation, the solution of which describes the concentration profile along the catalyst pellet. Thus, for example, $k_v c$ in equation (142) would be replaced by the appropriate expression containing terms in the denominator which allow for the retardation of reaction rate by reactants and/or products.

It should be noted that, in contrast to the problem of reaction rates in single pores, the pore size and, indeed, the pore shape have not been assumed. The effective diffusivity D_e is a quantity which is a function of the porosity and tortuosity of the pellet. It thus describes adequately the way in which species diffuse into the pellet. Various authors [44, 45] have successfully measured effective diffusivities of porous catalysts and have shown that D_e depends directly on porosity. The tortuosity factor can be regarded as a factor which allows for the tortuous route which a molecule takes when diffusing through the pore structure.

Wagner [46] and Amundson [47] considered the case of two reversible first-order reactions

$$A \underset{k_v'}{\overset{k_v}{\rightleftharpoons}} B$$

taking place in a flow reactor packed with spherical catalyst pellets. Two diffusion equations must now be written for the reactant A and the product B. By taking a material balance over an infinitesimally thin shell of a single spherical particle, two equations result

$$D\left\{\frac{\partial^2 c_A}{\partial a^2} + \frac{2}{a}\frac{\partial c_A}{\partial a}\right\} - k_v c_A + k_v' c_B = 0 \tag{149}$$

$$D\left\{\frac{\partial^2 c_B}{\partial a^2} + \frac{2}{a}\frac{\partial c_B}{\partial a}\right\} + k_v c_A = k_v' c_B = 0 \tag{150}$$

with the boundary conditions $c_A = c_A{}^0$ and $c_B = c_B{}^0$ at $a = R$. In addition to these equations, if it is desired to express the conversion of A as a function of the distance along a packed tubular reactor, a material balance must be written for the whole reactor. As shown later (Chapter 9), for a flow reactor the rate of reaction r is

$$r = -u_l \frac{\partial c_A{}^0}{\partial z} \tag{151}$$

where u_l is the linear velocity of flow through the tube and $\partial c^0/\partial z$ is the concentration gradient of A at a point z along the reactor length. The term c^0 implies the concentration of A in the bulk gas stream at a point z in the reactor. If there are m grams of catalyst packed in the tube, then the rate at which A diffuses into the pore structure of all the pellets is

$$\frac{4\pi R^2 D\left(\dfrac{\partial c_A}{\partial a}\right)_{a=R}}{\dfrac{4\pi R^3 \rho_b}{3m}} = \frac{3mD}{\rho_b R}\left(\frac{\partial c_A}{\partial a}\right)_{a=R} \tag{152}$$

where ρ_b is the density of the packed bed. Hence the two equations describing the change of concentration of A and B along the reactor length are

$$-u_l \frac{\partial c_A{}^0}{\partial z} = \frac{3mD}{\rho_b R}\left(\frac{\partial c_A}{\partial a}\right)_{a=R} \tag{153}$$

and

$$-u_l \frac{\partial c_B{}^0}{\partial z} = \frac{3mD}{\rho_b R}\left(\frac{\partial c_B}{\partial a}\right)_{a=R} \tag{154}$$

with the boundary condition $c_A{}^0 = (c_A{}^0)_i$ and $c_B{}^0 = (c_B{}^0)_i$ at $z = 0$. If the four simultaneous linear differential equations (149) (150) (152) and (153) are solved, there results

$$\ln \frac{c_A{}^0}{(c_A{}^0)_i} = \frac{mz}{R\rho_b u_l}(k_v + k_v')\frac{3}{\phi_s}\left\{\frac{1}{\tanh \phi_s} \frac{1}{\phi_s}\right\} \tag{155}$$

where ϕ_s is defined by

$$\phi_s = R\sqrt{\left(\frac{k_v + k_v'}{D}\right)} \tag{156}$$

For a very active surface with small pores, when $\phi_s > 5$,

$$\ln \frac{c_A^0}{(c_A^0)_i} = \frac{3mz(k_v + k_v')}{R\rho_b u_l h_s} = \frac{3mz}{\rho_b u_l} \frac{\sqrt{\{D(k_v + k_v')\}}}{R^2} \qquad (157)$$

so that at a given point in the reactor the conversion to product would increase with decrease in particle size.

4.5.4. Reaction rates in pores of arbitrary shape

In Section 4.4.3 we discussed several models to characterize porous structures. If the pore structure is known and can be characterized geometrically, then one has only to modify the differential equation describing the way in which the rate is influenced by diffusion. Except in the case of very simple shapes such as cylinders and spheres, the solution to the diffusion equation cannot be obtained by ordinary analytical means and numerical methods must be used. A much simpler approach is the one outlined by Wheeler [34] in which the experimental surface area and pore volume conform to the geometric surface area and pore volume of the catalyst model. Such a model can be used to describe the rate at which a reaction occurs on practical catalyst pellets. We previously showed that the number of pores n_p in a catalyst particle is given by $\psi/\sqrt{2\pi r^2}$ where ψ is the porosity and r the average pore dimension. If such a pellet is a composite of N pores of length L the rate of reaction per catalyst pellet is N times the rate per pore, the latter quantity, $R_{1/2}$, given by equation (145). Since the number of pores is $(n_p S_x)$ where n_p is the number of pores per unit external surface area S_x, then the rate R_p per pellet is

$$R_p = S_x n_p R_{1/2} = \left(\frac{S_x \psi}{\pi r^2 \sqrt{2}}\right) \frac{\pi r^2 D}{L} c_0 \, h_p \tanh h_p \qquad (158)$$

where L given by equation (86) and is $\sqrt{2}$ times the ratio of the pellet volume to the external surface area. h_p is now defined by

$$h_p = L \sqrt{\left(\frac{2k_1(1-\psi)\tau}{rD}\right)} = \frac{2V_p}{S_x} \sqrt{\left(\frac{k'_1}{rD}\right)} \qquad (159)$$

where $k_1(1-\psi)\tau$, $(=k'_1)$, is the modified value of k_1 as a result of pore inter-sections and tortuosity. The fraction of surface available for reaction is given, as before, by $(1/h_p)\tanh h_p$. The quantity measured experimentally is usually the activity per unit bulk volume of reactor. This may be obtained by multiplying equation (158) by the number of pellets per unit volume $(\rho_b/\rho_p V_p)$ where ρ_b is the bulk density of the catalyst. Thus the rate per unit bulk volume is

$$R_b = \frac{\rho_b}{\rho_p V_p} \frac{S_x \psi D}{\sqrt{(2L)}} c_0 \, h_p \tanh h_p = \frac{1}{2}\left(\frac{S_x}{V_p}\right)^2 \rho_b V_g D c_0 \, h_p \tanh h_p \qquad (160)$$

since the porosity θ is the product of the pellet density and the specific volume per gram V_g, and L is $(\sqrt{2}V_p/S_x)$. As indicated in Section 4.4.3, for almost all pellet shapes the ratio V_p/S_x is one-sixth of the pellet size. Substituting in equations (159) and (160), we may therefore write for practical pellets

$$R_b = \frac{6}{a_p}\, \rho_b V_g c_0 \sqrt{\left(\frac{k_1 D}{r}\right)}\, \tanh\left(\frac{a_p}{3}\sqrt{\left(\frac{k_1}{rD}\right)}\right) \tag{161}$$

where a_p is the pellet size.

The diffusion equation (142) for spherical catalyst pellets may be made to coincide with this practical model in the following way. The elementary spherical shell considered in Fig. 13 has a total surface area of $(4\pi a^2 \delta a \rho_p S_g)$ where $(\rho_p S_g)$ is the surface area per unit pellet volume. Thus we replace S_v in equation (142) by $(\rho_p S_g)$. If the pellet size d_p is written in place of the sphere diameter, the dimensionless group is now

$$\phi_s = \frac{d_p}{2}\sqrt{\left(\frac{k'_1 \rho_p S_g}{D_e}\right)} = \frac{3V_p}{S_x}\sqrt{\left(\frac{k'_1 \rho_p S_g}{D_e}\right)} \tag{162}$$

and in this case the shape of the pores is immaterial provided always the quantity $(\rho_p S_g)$ is a true measure of the surface area per unit volume. As indicated in Section 4.4.2, S_g is usually a fairly close assessment of the surface area in pores except in cases where there is a very narrow distribution of pore sizes.

4.5.5. Pressure and temperature gradients in porous pellets

In Section 4.5.2 it was shown that sharp partial pressure gradients occur in the pores of a pellet. Quite large total pressure gradients may also be expected. Consider a spherical catalyst pellet in which the principal mode of mass transport is by diffusion. Suppose that the reaction $A \rightarrow \nu B$ occurs in the pores of this pellet. In the steady state, the number of molecules of B formed which diffuse outward through an infinitesimally thin spherical shell at a distance a from the centre of the particle must be ν times the reaction rate within the sphere. Since the reaction rate within the sphere just balances the rate of diffusion of A into the shell,

$$4\pi a^2 D_B \nu \frac{\partial c_B}{\partial a} = 4\pi a^2 D_A \frac{\partial c_A}{\partial a} \tag{163}$$

Integrating this equation from $a=0$, where the concentration of A is $c_A{}^0$ and of B is $c_B{}^0$, to a radial position a,

$$(c_B - c_B{}^0) = \nu \frac{D_A}{D_B}(c_A{}^0 - c_A) \tag{164}$$

which implies that the concentration increase in B is $(\nu D_A/D_B)$ times the concentration decrease in A. The total concentration is therefore

$$c_T = c_A + c_B = \nu \frac{D_A}{D_B}(c_A{}^0 - c_A) + c_B{}^0 + c_A \tag{165}$$

If we consider a pellet located in the entrance to a packed tubular reactor, then $c_B{}^0 \approx 0$, since hardly any product has formed at this point in the reactor. The fractional increase in concentration at the centre of such a pellet, where c_A will have fallen to zero, will therefore be given by

$$\frac{c_T}{c_A{}^0} = \nu \, \frac{D_A}{D_B} = \sqrt{\nu} \tag{166}$$

provided that Knudsen diffusion is prevalent, so that D_A/D_B is equal to $\sqrt{(1/\nu)}$. Thus, the total pressure at the centre of an active catalyst pellet is $\sqrt{\nu}$ times that at the exterior surface. If $\nu > 1$ the pressure is greater at the centre than at the exterior surface. If $\nu < 1$ the converse is true. In some cracking reactions, for example, where ν may be as high as 4, the pressure at the centre of the pellet is twice that in the bulk gas stream. Such total pressures in the interior of a pellet may conceivably cause catalyst disintegration. As well as total pressure gradients, quite high temperature gradients may be expected. All the reactions considered in the previous sections of this chapter were assumed to occur under strictly isothermal conditions. If the catalyst pellet is a poor thermal conductor, then, for a fast exothermic reaction, the temperature in the interior of the catalyst will be quite different from that at the periphery. Consider an infinitesimally thin spherical shell of a spherical catalyst particle. In the steady state the heat flux within the shell must be balanced by the heat produced by chemical reaction, whereas the rate of the chemical reaction will just equal the rate of diffusion of reactant into the shell. Hence we may write

$$- 4\pi a^2 \kappa \, \frac{\partial T}{\partial a} = \Delta H 4\pi a^2 D_e \frac{\partial c}{\partial a} \tag{167}$$

where κ is the thermal conductivity and ΔH is the heat of reaction per mole of reactant. The negative sign indicates that the temperature is higher inside the pellet than at the periphery. This is because it is some finite distance from the pore mouth that the conversion takes place. Integrating this equation from the exterior surface, where the temperature is T_0 and the concentration c_0, to the centre of the pellet, where the temperature and concentration are T_R and c_R, respectively

$$T_R - T_0 = \frac{D \Delta H}{\kappa} \, (c_0 - c_R) \tag{168}$$

For a fast reaction the concentration of reactant will be zero at the centre of the pellet, so c_R in equation (168) is zero under these conditions. For an exothermic reaction with ΔH about 20,000 cal mole^{-1}, a thermal conductivity of 2×10^{-3} cal deg C^{-1} cm^{-2}sec^{-1}, typical of metal oxides, and a Knudsen diffusion coefficient about 10^{-1} cm sec^{-1} (corresponding to pores of about 200 Å radius), the factor $D \Delta H \kappa$ in equation (168) has a value 10^6 deg C cm^3 mole^{-1}. If the concentration of reactant gas at the exterior of the pellet is

10^{-4}mole cm^{-3}, then the difference in temperature between the centre of the pellet and the exterior surface is as much as 100°C. Equation (168) only predicts the difference in temperature between the centre of the pellet and the exterior surface. If the temperature profile is required, a heat balance must be taken over an infinitesimally thin spherical shell within the pellet. In this way, an equation analogous to the diffusion equation (142) is obtained,

$$\frac{\partial^2 T}{\partial a^2} + \frac{2}{a}\frac{\partial T}{\partial a} = -\frac{\Delta H k_v c}{\kappa} \qquad (169)$$

To calculate the temperature profile within the pellet, this equation must be solved simultaneously with equation (142) describing the concentration profile. Since the chemical rate constant k_v is an exponential function of temperature, an analytical solution presents difficulties. However, the two simultaneous equations may be solved numerically by writing both equations in difference form and obtaining an estimate for temperature and concentration in small discrete increments. The procedure is repeated until refined values are obtained. The method of solution for a similar case is outlined in Chapter 9, Section 9.4.3. Other numerical procedures have also led to solutions.

It is seen that equations (168) and (169) contain two more independent variables than the case considered for isothermal diffusion. These additional independent variables are the exponent of the Arrhenius equation (E/RT) and a heat generation function $-\Delta HDc_0/\kappa T_0$ which Weisz and Hicks [48] designate γ and β respectively. Solutions to the equations have been given by plotting the ratio of the inhibited and uninhibited reaction rates (referred to earlier as the fraction of surface available, but, for non-isothermal cases, best referred to as an effectiveness factor) as a function of the Thiele modulus for spherical pellets, $\phi_s = R\sqrt{(k_v/D_e)}$, with γ and β as fixed parameters. Figure 14 shows such a plot. β represents the maximum temperature difference that could exist in the particle relative to the temperature at the particle periphery and may thus be written $(T-T_0)_{max}/T_0$. For an exothermic reaction, β is positive and it is seen from Fig. 14 that the effectiveness factor may exceed unity. This is because the increase in rate caused by the temperature rise inside the particle more than compensates for the decrease in rate caused by the negative concentration gradient effecting a decrease in concentration towards the centre of the particle. The overall rate of reaction in the pellet interior is thus greater than it would be if the concentration and temperature were the same as at the periphery, i.e. no mass- and heat-transfer effects. The curve $\beta=0$ represents the isothermal case and is similar to those depicted in Fig. 12.

Such calculations [48] indicate that the temperature gradient is highest at the periphery of the pellet and that it flattens out considerably as the centre is approached. For fast exothermic reactions, the centre of the pellet will be

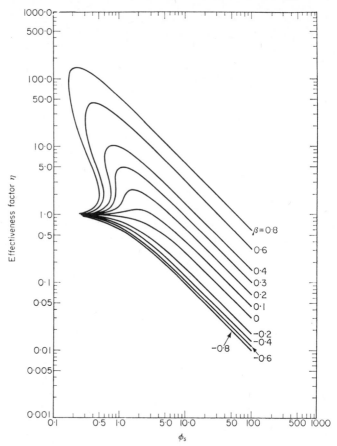

FIG. 14. The effectiveness factor η as a function of ϕ_s. $\gamma = E/RT = 20$; $\beta = -\Delta H D_{c_0}/\kappa T_0$ [49].

at a uniform high temperature with a fairly sharp decrease in temperature near the exterior surface.

4.5.6. The effect of in-pore diffusion on experimental parameters

Having examined several cases of diffusion accompanying reactions in porous catalysts, we now use the results from these simplified models to predict the effect which diffusion has on some parameters which are observed experimentally. Unless one is aware of the possible effect which diffusion has of obscuring the true chemical reaction kinetics, it is easy to obtain misleading information concerning reaction rates and to deduce an incorrect mechanism. It is also important to decide whether the experimental conditions are the correct ones for observing the true kinetics. Thus, if one wishes to study the reaction mechanism, it is advisable to choose conditions such that the chemical rate is not influenced by diffusion. On the other hand, it is a distinct advantage to be able to assess the most suitable pellet size for a commercial reaction, for an incorrect choice may lead to pecuniary disaster! Since it is expensive to install heat-exchange equipment and to supply fuel for large reactors, the fiscal success of such an industrial venture depends to a great extent on having a reactor which is of a minimum size compatible with output demands. If pellet size is overestimated, there is an unnecessary wastage of reactor space. Conversely, an underestimate of pellet size may, in the case of slow reactions, adversely affect product output.

4.5.6.1. *Effect of Diffusion in Reaction Orders*

That the observed reaction kinetics may not be representative of the true kinetics in the absence of diffusion is qualitatively evident. If reaction were so fast that the rate at which reactant is consumed depended entirely on diffusion, then the observed rate would be independent of the concentration of reactant. When the rate of reaction is comparable with the rate of diffusion, the situation is different. It is seen from the equations obtained in Section 4.5.2 that, except for first-order reactions, the rate $R_{1/2}$ in the half pore is a function of concentration which differs from the original kinetic law. Thus, for a fast second-order reaction, $R_{1/2}$ depends on $c_0^{3/2}$. For a zero-order reaction, the rate in the half pore is proportional to $c_0^{1/2}$. On the other hand, the rate in the half pore remains a linear function of concentration for fast first-order reactions. These results may be generalized by saying that, for large values of the Thiele modulus, corresponding to fast reactions in small pores, the observed order of reaction is $(n+1)/2$ where n is the true kinetic order.

In some reactions there is a pseudo-dependence of reaction rate on the partial pressure of one of the reactants. This occurs for bimolecular reactions under Knudsen diffusion conditions in porous active catalysts. If the reaction is $A+B \rightarrow C$, the diffusion of B will be $\sqrt{(M_B/M_A)}$ times slower than the diffusion of A, where the molecular weight M_B of B is greater than the molecular weight M_A of A. Hence, the concentration gradient of B is greater than that of A by this same amount. The interior of the catalyst pellet will therefore be depleted in the reactant B. However, if the concentration of B in the reactant mixture is increased, more B will be able to reach the pellet interior and the rate will increase. Thus, irrespective of the exponent of c_B in the true kinetic equation, there will be an additional dependence of reaction rate on the pressure of B due to the slower diffusion of this reactant. An example of this behaviour is the hydrogenation of ethylene on porous metallic catalysts. Normally, the rate is independent of the partial pressure of ethylene, but under conditions of Knudsen diffusion on an active catalyst the rate can turn out to be proportional to the square root of the ethylene partial pressure.

4.5.6.2. *Influence of Diffusion on Activation Energy*

When the Thiele modulus is large, the observed rate of reaction in the half pore depends, in each case, on the square root of the true reaction velocity constant. Since k_v determines the true activation energy E through the exponential function $e^{-E/RT}$, the observed activation energy for fast reactions in small pores is only one half of the true value in large pores where diffusion does not influence the rate. Between these two extremes, the observed activation energy will be less than the true value, but greater than one-half its true value. Since the parameter h contains k, then, for a reaction in pores of a given radius, its value will increase with temperature. For pores of intermediate size, the rate of a first-order reaction, for example, depends on $h \tanh h$. Over a temperature range, therefore, the observed rate is not a simple exponential function of reciprocal temperature. Assuming that in large pores the true activation energy is 20 kcal mole^{-1} and in small pores, where the rate is proportional to $\sqrt{k}$, it is 10 kcal mole^{-1}, then we can calculate a curve of rate as a function of reciprocal temperature for intermediate size pores. Assigning a value of 0·5 to h at a temperature of 500°K, we can compute $h \tanh h$ at any higher temperature, since h varies as $\sqrt{k}$, which is itself an exponential function of temperature. If h_1 is the value of h at a temperature T_1, and h_2 at T_2, then

$$h_2 = h_1 \exp\left\{\frac{E}{R}\left(\frac{1}{T_1} - \frac{1}{T_2}\right)\right\} \tag{170}$$

where E is the true activation energy for the reaction in large pores. Figure 15 shows $h \tanh h$ computed over the range 500°–700°K. The curve shows a

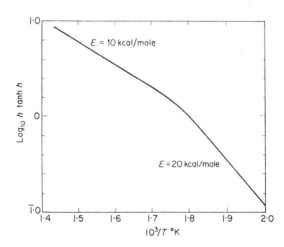

FIG. 15. An illustration of the effect of temperature on the Thiele modulus.

gradual change of slope from 20 kcal mole^{-1} to 10 kcal mole^{-1}, indicating that for intermediate-size pores the true activation energy is observed only at comparatively low temperatures where h is small and, hence, the factor tanh h is unity. At high temperatures where h is large, tanh h becomes equal to h and so the rate now depends on h^2 and is hence proportional to k and not its square root. Such a transcendence from apparent kinetics to true kinetics is therefore to be expected (a) with increasing pore size at a fixed temperature, and (b) for pores of a given size if the temperature is decreased.

4.5.6.3. *Dependence of Reaction Rates on Surface Area and Pore Volume*

Examination of equation (161) reveals that the rate per unit bulk volume of catalyst depends on the factor tanh h. For fast reactions on small pores this factor is unity, so that equation (161) now becomes

$$R_b = \frac{6}{a_p} \, \rho_b V_g c_0 \sqrt{\left(\frac{k'_1 D}{r}\right)}, \text{ for } h_1 > 2 \tag{171}$$

Since Knudsen diffusion is usually the mode of transport in small pores, the diffusion coefficient will be directly proportional to the pore radius. Hence, under Knudsen flow conditions, fast reactions become independent of pore size. Furthermore, if we write equation (171) in an alternative form by substituting $(2V_g/S_g)$ for r,

$$R_b = \frac{6}{a_p\sqrt{2}} \, \rho_b c_0 \sqrt{(k_1 D V_g S_g)}, \text{ for } h_1 > 2 \tag{172}$$

it is seen that for fast reactions the rate is no longer directly proportional to the specific surface area but to the square root of both the specific surface area and the pore volume under conditions of ordinary diffusion. Under conditions of Knudsen diffusion, when D is proportional to r, the rate per unit bulk volume is directly proportional to the pore volume and independent of surface area.

Table 4.3 shows how the observable parameters considered differ in the two extreme cases of slow reactions in large pores and fast reactions in small

TABLE 4.3

	Reaction order	Activation energy	Bulk diffusion			Knudsen diffusion		
			Pore radius	Surface area	Pore volume	Pore radius	Surface area	Pore volume
Slow reactions, large pores	n	E	r	S_g	Independent	—	—	—
Fast reactions, small pores	$\dfrac{(n+1)}{2}$	$\dfrac{E}{2}$	$r^{3/2}$	$\sqrt{S_g}$	$\sqrt{V_g}$	r	Independent	V_g

pores. From normal kinetic behaviour, observed on relatively inactive catalysts with large pores, there is a gradual transition in the functional dependence of these parameters to the extreme case of a fast reaction occurring in small pores.

4.5.6.4. *Experimental Evaluation of the Thiele Modulus*

It is possible to estimate the value of the Thiele modulus from a single catalyst-activity measurement. Equation (172) gives the rate of reaction per unit bulk volume. If we regard the measured rate (moles converted per unit volume per unit time) as a rate constant times concentration, then we may define the rate constant k_v per unit bulk volume as,

$$k_v = \frac{18}{a_p{}^2} \rho_b V_g D \, h_1 \tanh h_1 \tag{173}$$

in which the factor $(S_x/V_p)^2$ has been replaced by $36/a_p{}^2$ for catalysts which have an exterior surface to pellet volume one-sixth the pellet size. The rate constant may be measured by means of a single flow experiment. As shown in Chapter 9 (Section 9.2, equation (14)), a general equation for the first-order velocity constant of a reaction $A+B \rightarrow \nu P$ is

$$k = \frac{F}{V_R c_A{}^0} \left\{ (1 - \delta)\ln \left(\frac{1}{1 - x_A} \right) - \delta x_A \right\} \tag{174}$$

where F is the feed rate in moles per unit time, $c_A{}^0$ is the initial concentration of A at the reactor inlet, x_A is the fractional conversion of A, and V_R the reactor volume. The coefficient $\delta(=1-\nu)$ is a measure of the change in volume of the reacting system at constant total pressure and is given by the stoichiometry of equation (173). Equating this experimental rate constant to the rate constant in equation (173),

$$h_1 \tanh h_1 = \frac{a_p{}^2}{18\rho_b V_g D} \frac{F}{V_R c_A{}^0} \left\{ (1 - \delta)\ln \left(\frac{1}{1 - x_A} \right) - \delta x_A \right\} \tag{175}$$

Only experimental quantities are contained in the right-hand side of the above equation. If the physical properties of the catalyst pellet are already known, then a_p, ρ_b, V_g and D (which depends on the pore size and the total pressure conditions) do not have to be determined independently, and one single-flow experiment using a given mass of catalyst is all that is necessary to evaluate $h_1 \tanh h_1$ and therefore h_1.

4.5.7. Catalyst poisoning

Active catalysts are extremely susceptible to poisoning by foreign molecules. Particularly strong poisons are molecules with lone pairs of electrons capable of forming covalent bonds with solid surfaces. Examples are ammonia, phosphine, arsine, carbon monoxide, sulphur dioxide and hydrogen sulphide. Other poisons include hydrogen, oxygen, halogens and mercury. The surface of a porous catalyst becomes poisoned, thus inhibiting further reaction, by virtue of the foreign molecule being adsorbed in the porous structure of the catalyst and covering a fraction of its active surface. The reactant molecules must now be transported to the unpoisoned part of the surface before reaction occurs, and so poisoning increases the average distance the reactant must diffuse through the porous structure. The equations deduced in Section 4.5.2.2 apply to the poisoning of surfaces. Thus, we distinguish two types of poisoning: (a) homogeneous adsorption of poison with the foreign molecules distributed evenly over the surface and (b) selective poisoning in which an extremely active surface first becomes poisoned at the exterior surface and then progressively along the pore length. '

When homogeneous poisoning occurs, the intrinsic activity k of the pore walls decreases to $k(1-\alpha)$ where α is the total fraction of the surface covered with poison. Thus, to find the ratio F of the activity of the poisoned pore to the unpoisoned pore, $k(1-\alpha)$ replaces k in the equation for reaction rate in a half pore. The ratio of activity in the poisoned pore to that in the unpoisoned pore is then

$$F = h_\theta^0 \frac{\sqrt{(1 - \alpha)} \tanh (h_\theta^0 \sqrt{(1 - \alpha)}]}{h_\theta^0 \tanh h_\theta^0} \qquad (176)$$

where h_θ^0 is the value of the Thiele modulus for the unpoisoned pore. This equation is analogous to equation (132) for the effect which diffusion has on adsorption. For slow reactions h_θ^0 is small, so F becomes equal to $(1-\alpha)$ and the activity decreases linearly with the amount of poison added. For fast reactions in small pores h_θ^0 is large and the fraction F becomes simply $\sqrt{(1-\alpha)}$. This means that the activity of the catalyst now decreases less than linearly with increasing amount of poison added. This is because a first-order reaction does not penetrate so far towards the centre of the pellet as a slow reaction. Curves A and B in Fig. 16 show how homogeneous poisoning of a catalyst affects a slow reaction and a fast one, respectively.

Selective poisoning occurs with very active catalysts. Initially, the pore mouth adsorbs poison and then, as more poison is added, an increasing fraction of the pore length becomes covered and inaccessible to reactant molecules. A pore of length L whose surface has been poisoned such that a fraction α is covered with adsorbed molecules will have a length $(1-\alpha)L$ free of contaminant. Any reactant molecules have to diffuse through the poisoned length before reaching the active surface, and, under some circumstances, this may be a very slow process. To find the ratio of the activity in the poisoned length to that in the unpoisoned length we first equate the rate of diffusion of

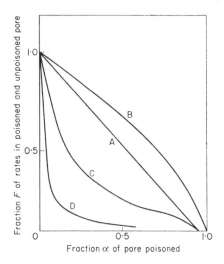

FIG. 16. Poisoning of catalyst pores: (a) the effect of homogeneous poisoning on a slow reaction; (b) the effect of homogeneous poisoning on a fast reaction; (c) selective poisoning of active catalysts; (d) selective poisoning in small pores.

reactant through the length L to the rate of reaction in the length $(1-\alpha)L$. Equation (134), Section 4.5.2, expresses the rate of diffusion of molecules through a fraction of pore length covered by an adsorbate, while the quantity $\pi r^2 D\dfrac{c_0}{L}\tanh h_\theta^0$ gives the rate of reaction in that fraction of the pore which does not contain adsorbate. Thus, an expression analogous to equation (175) gives the ratio F of the activity in the poisoned pore to the activity in the unpoisoned pore,

$$F = \frac{\pi r^2 D\left(\dfrac{c_0 - c_L}{\alpha L}\right)}{\pi r^2 D\dfrac{c_0}{L}\tanh h_\theta^0} = \frac{1}{1 + h_\theta^0 \alpha}\frac{\tanh\left\{h_\theta^0(1-\alpha)\right\}}{\tanh h_\theta^0} \qquad (177)$$

where c_L is the concentration at the end of the poisoned length and h_θ^0 is the Thiele modulus for the uncontaminated pore. When h_θ^0 is large for an active catalyst with small pores, the fraction $h_\theta^0(1-\alpha)$ will usually be sufficiently large that

$$F = \frac{1}{1 + h_\theta^0 \alpha}, \quad h_\theta^0(1-\alpha) > 2 \qquad (178)$$

Curve C in Fig. 16 shows the selective poisoning of active catalysts near the periphery of the particles, and is the function represented by equation (177). Curve D describes the effect of selective poisoning in small pores. For the latter case, the activity decreases drastically after only a small amount of poison has been added.

Reaction in poisoned porous catalysts will show quite abnormal activation energies. Since the same equations apply as in the cases discussed for the effect of diffusion on adsorption, similar general conclusions are deduced. When h_θ^0 is larger than 20 and a moderate amount of poisoning has occurred ($\alpha > 0.2$), the rate becomes equal to the rate of diffusion of reactant past the contaminated length of pore. Hence, the temperature coefficient of reaction is extremely small (only a few kcal mole^{-1}). On the other hand, at a low temperature where the rate is measurably small, the rate of reaction is the same as that in an uncontaminated pore, and then the true activation energy will be measured over a limited range of temperature.

References

1. J. Poiseuille, Inst. de France Acad. des Sci. Mem. **9**, 433 (1846).
2. J. Kozeny, *Ber. Wien Akad.* **136A**, 271 (1927).
3. J. M. Coulson, *Trans. Inst. chem. Engng* **27**, 237 (1949).
4. R. M. Barrer and D. M. Grove, *Trans. Faraday Soc.* **47**, 837 (1951).
5. R. M. Barrer and D. M. Grove, *Trans. Faraday Soc.* **47**, 826 (1951).
6. R. M. Barrer, *Disc. Faraday Soc.* **3**, 61 (1948).
7. P. Clausing, *Ann. Phys. Lpz.* **7**, 489 (1930).
8. J. C. Arnell, *Can. J. Res.* **A24**, 103 (1946).
9. P. C. Carman, and J. C. Arnell, *Can. J. Res.* **A26**, 129 (1948).
10. G. Kraus and J. W. Ross, *J. phys. Chem.* **57**, 334 (1953).
11. J. M. Coulson and J. F. Richardson, "Chemical Engineering", Vol. 2, p. 100. Pergamon Press, London (1960).
12. G. G. Stokes, *Trans. Camb. phil. Soc.* **9**, 8 (1851).
13. W. B. Innes and K. D. Ashley, *Proc. Am. Petr. Inst.* **27**, III, 9 (1947).
14. J. Biscoe and B. E. Warren, *J. appl. Phys.* **13**, 364 (1942).
15. V. E. Cosslett, "The Electron Microscope". Sigma Press (1947).
16. J. Svedberg, "Colloid Chemistry". Reinhold, New York (1928).
17. A. Wheeler, *in* "Catalysis", ed. by P. H. Emmett, Vol. 2. Reinhold, New York (1955).
18. C. G. Shull, *J. Am. chem. Soc.* **70**, 1405 (1948).
19. E. P. Barrett, L. G. Joyner and P. P. Halenda, *J. Am. chem. Soc.* **73**, 373 (1951).
20. R. Zsigmondy, *Z. anorg. Chem.* **71**, 356 (1911).
21. J. W. McBain, *J. Am. chem. Soc.* **57**, 699 (1935).
22. M. B. Coelingh, *Kolloid. Z.* **87**, 251 (1939).
23. L. H. Cohan, *J. Am. chem. Soc.* **60**, 433 (1938).
24. R. M. Barrer, N. McKenzie and J. S. S. Reay, *J. Colloid Sci.* **11**, 479 (1956).
25. J. H. de Boer, Colston Res. Symp., Bristol (1958).
26. J. H. de Boer, J. J. Steggerda, and P. Zwietering, *Proc. K. ned. Akad. Wet* **B59**, 435 (1956).
27. J. J. Steggerda, Thesis, University of Delft (1955).
28. B. Lambert and A. M. Clark, *Proc. R. Soc.* **A122**, 497 (1929).

29. K. S. Rao, *J. phys. Chem.* **45**, 513 (1941).
30. H. E. Ries, *Adv. Catalysis* **4**, 87 (1952).
31. D. H. Everett, Colston Res. Symp., Bristol (1958).
32. A. V. Kiselev, *Dokl. Akad. Nauk SSSR* **98**, 431 (1954).
33. A. V. Kiselev, Proc. 2nd Int. Cong. Surface Activity, Vol. 2, p. 223. Elsevier, Amsterdam (1957).
34. A. Wheeler, *Adv. Catalysis* **3**, 249 (1951).
35. M. Knudsen, *Annln Phys.* **28**, 75 (1909).
36. M. Knudsen, *Annln Phys.* **35**, 389 (1911).
37. K. Herzfeld and M. Smallwood, *in* "Treatise on Physical Chemistry", ed. by Taylor, Vol. 1, p. 169. Princeton University Press (1931).
38. G. Hagen, *Annln Phys.* **46**, 423 (1939).
39. E. W. Thiele, *Ind. Engng Chem.* **31**, 916 (1939).
40. K. W. Sykes and P. White, *Trans. Faraday Soc.* **52**, 660 (1956).
41. P. B. Weisz, *Adv. Catalysis* **13**, 137 (1962).
42. C. N. Satterfield and T. K. Sherwood, "The Role of Diffusion in Catalysis". Addison-Wesley, Reading, Massachusetts.
43. G. W. Roberts and C. N. Satterfield, *Ind. Engng Chem. Fund. Ed.* **4**, 288 (1965).
44. N. Wakao and J. M. Smith, *Chem. Engng Sci.* **17**, 825 (1962).
45. J. A. Currie, *Brit. J. appl. Phys.* **11**, 318 (1960).
46. C. Wagner, *Z. phys. Chem.* **193**, 1 (1943).
47. N. L. Smith and N. R. Amundson, *Ind. Engng Chem.* **43**, 2156 (1951).
48. P. B. Weisz and J. S. Hicks, *Chem. Engng Sci.* **17**, 265 (1962).

CHAPTER 5

The Role of Lattice Imperfections in Heterogeneous Catalysis

5.1. Introduction

It has been known for some time [1, 2] that the reactivity of a solid passes through a maximum in the region of a crystallographic transition, and that certain treatments such as mechanical straining, which diminish the crystalline perfection, also affect the reactivity. It was, no doubt, the appreciation of such facts which prompted many of the earlier investigators to suggest [3, 4] that the "active centres", conceived by H. S. Taylor, were in some way related to lattice imperfections, and that such centres might be created by irradiation with high-energy projectiles [5]. As a result of the researches conducted during the last decade or so, we are now in possession of several illuminating facts concerning the relationship between lattice imperfections and catalysis, although we are still unable to formulate a satisfactory general theory to account for such a relationship. Nevertheless, this topic invites discussion because, in the first place, by examining the properties of imperfections, we are likely to clarify our thoughts concerning "active centres"—a phrase that has become all too convenient in the literature of catalysis. Secondly, if we are to make any significant progress in our understanding of

the influence of radiation on catalysts, it is imperative that we first comprehend how structural irregularities modify the properties of a solid.

We shall make little headway in our assessment of the importance of
imperfections in the present context unless we define clearly what we mean by
the term imperfection.

5.2. Classification of Lattice Imperfections

Shortly after the discovery of X-ray diffraction, it became apparent [6], as a
result of quantitative measurements of the intensities of reflected rays, that the
architecture of real crystals was far from perfect. In other words, the ideal
crystals first conceived by the X-ray crystallographers, and pictured as being
composed of properly stacked and regularly packed unit cells, seldom, if ever,
exist in nature. Later experiments in the investigation of both mechanical
properties [7–11] and growth [11] of crystals led to the recognition of *dislocations* (or line defects, as they are often called) of two main types, where the
perfection of the lattice is interrupted.

5.2.1. Edge dislocations

Dislocations are of two basic types, edge and screw, although intermediate
varieties of these types frequently exist. The edge type, or Taylor-Orowan,
dislocation is best envisaged using a diagram such as that shown in Fig. 1.
The dislocation line is, effectively, the boundary line which demarcates the

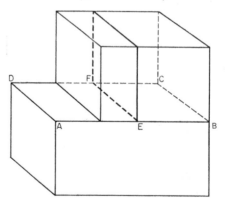

FIG. 1. ABCD is the slip plane; EF is the edge dislocation.

region that has undergone the process of slip from the region that has not
slipped. The edge dislocation in Fig. 1 clearly constitutes such a boundary.
It can also be looked upon as an additional lattice plane that has been inserted
part way into the crystal. (It is the edge of the additional plane which is being
referred to in the term edge dislocation.) Figure 2 reveals that each of the lattice

points (molecules, ions or atoms) on the dislocation line is faced by a gap instead of an adjacent lattice point. Straightaway we can appreciate that, for geometrical reasons, impurity atoms (or molecules) larger than those of the lattice will tend to concentrate at the gaps adjacent to edge dislocations. Conversely, impurity atoms smaller than those of the lattice can be accom-

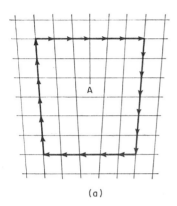
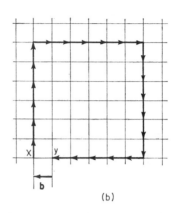

(a)   (b)

FIG. 2. (a) Burgers circuit around an edge dislocation A. (b) The analogous circuit around the corresponding atoms in an ideal lattice. The gap XY is equal to the Burgers vector **b**.

modated in the compressed regions of the lattice on the other side of the line from the gaps. The preferential concentration of impurities in the vicinity of dislocation lines is known as the Cottrell cloud effect [12].

If we trace a complete loop composed of unit lattice translations around an edge dislocation in a real lattice (Fig. 2(a)), and trace an analogous loop around the corresponding atoms in an ideal lattice (Fig. 2(b)), we find that one less lattice translation is required in the real than in the ideal lattice. These loops or circuits are usually termed Burgers circuits, and the vector translation YX required to close the gap is called the Burgers vector **b**. Note that its direction is always perpendicular to the direction of an edge dislocation. If **b** corresponds to one atomic distance (unit lattice vector), the strength of the dislocation is then unity and the dislocation is a "perfect" one; dislocations can be of multiple strength, or of fractional strength (when the dislocations are then known as partials).

The dislocation line at A in Fig. 2(a) is, by definition, a *positive* edge dislocation. If the figure were turned through 180 degrees, that is, if the additional lattice plane had been inserted part way into the crystal *below* the slip plane, the dislocation would be negative. The distinction between positive and negative dislocations becomes important when considering the annihilation of dislocations during heat-treatment.

Associated with an edge dislocation is an elastic strain field and, therefore, an elastic strain energy which decreases with distance from the dislocation. The strain energy E per unit length of dislocation line can be represented [12, 13] by an equation of the form:

$$E \approx \frac{Gb^2}{4\pi} \ln\frac{R}{r_0} \qquad (1)$$

where b is the Burgers vector, G is the shear modulus, R is the limiting radius of the stress field and r_0 is the radius of the core or cylindrical region down the centre of the dislocation axis (where Hooke's law is invalid). The value of r_0 is generally taken to be a few multiples of the Burgers vector. It is the energy of a dislocation—and this energy may amount to several electron volts per atomic plane crossed by the dislocation—that is essentially responsible for the production of well-defined etch pits during chemical reaction at the points of emergence of dislocation lines. The photomicrograph shown in Fig. 3 illustrates how readily, using etching techniques, an estimate can be made of the number of dislocation lines intersecting a given crystal surface. More reliable methods of measuring dislocation content utilize transmission electron

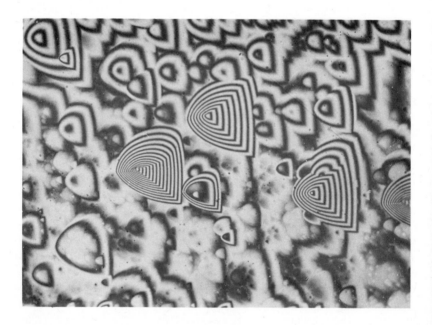

FIG. 3. Typical example of the etch pits produced at the regions of emergence of dislocation lines. This interferogram of the (100) face of a calcite crystal was taken with sodium light after the surface had been exposed to aqueous D-tartaric acid for 30 sec. Dislocation density is about 10^3 cm^{-2}. Top edge is parallel to the [100] direction. ($\times$1000)

microscopy, X-ray microscopy and X-ray diffraction, and "decoration" techniques (see Refs. *14* and *15* for fuller details).

5.2.2. Screw dislocations

This type of dislocation was first described by Burgers [*10*] and is illustrated in Fig. 4. Unlike the edge type, the screw dislocation has its axis parallel to its

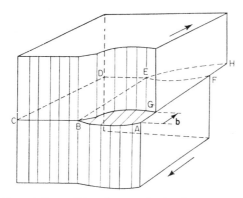

FIG. 4. Screw dislocation, BE, in the slip plane ACDF.

Burgers vector. Slip has occurred over the region ABEF of the slip plane ABCD in the direction AG, and AG is numerically equal to the magnitude of the Burgers vector. The strain pattern of a screw dislocation is quite different from that of an edge, and can be visualized as a spiral surface having a pitch equal to the Burgers vector **b** extending along its axis. The energy of a screw dislocation has been estimated [*15*] to be approximately two-thirds that of an edge dislocation. Just as edge dislocations can be positive or negative, screws can be either right-handed (positive) or left-handed (negative). The arrangement of the atomic planes in a crystal containing a screw is best thought of as being in the form of a helical surface winding through the crystal along the screw axis, rather than a stacking of separate atomic planes.

Dislocations are often referred to as being of mixed or composite type, meaning that they consist of screw and edge components. In Fig. 5, XY is the mixed dislocation out of which have been extracted the pure screw and pure edge components. For topological reasons [*12*], a dislocation must run entirely through a crystal (that is, end at a free surface) or, alternatively, form a complete loop within the body of the crystal.

5.2.3. Point defects

In discussing defects in crystals, it is convenient to distinguish four subdivisions, as suggested by the classification of Rees [*16*]. These divisions are:

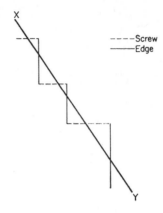

FIG. 5. XY is a mixed (or composite) dislocation.

imperfect crystals, non-stoichiometric compounds, impurity systems, and dual-valency semiconductors.

5.2.3.1. *Imperfect Crystals: Frenkel and Schottky Defects*

We have already noted that ideal crystals, such as that represented by Fig. 6(a), seldom, if ever, exist. Indeed, at any temperature greater than absolute zero we would expect, on thermodynamic grounds, to find some form of lattice disorder: the increased disorder brings about an increased entropy and thence a decrease in free energy:

$$F = U - TS \tag{2}$$

where F, U and S are, respectively, the Helmholtz free energy, the internal energy and the entropy, T being the absolute temperature. The Frenkel [17] and Schottky [18] types of disorder are the simplest and most clearly recognized forms of point defect. A Schottky defect is simply an unoccupied lattice site (Fig. 6(c)). This type of defect can appear in a structure composed of

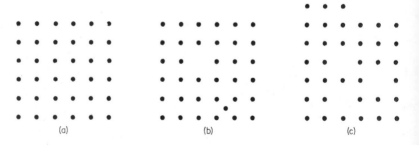

FIG. 6. Two-dimensional representation of (a) an ideal lattice, (b) a lattice containing a Frenkel defect, and (c) a lattice containing three Schottky defects.

identical atoms (such as in metals and in the non-metallic elements), in ionic solids with vacancies either in the cation or anion lattice, in a covalently bonded structure (for example, silicon or germanium), or in a molecular crystal such as anthracene or phthalocyanine. Note that for each Schottky defect that exists in a lattice built up from identical atoms a "detached" surface atom also exists. In an ionic solid, it often happens that the number of Schottky defects in the cation lattice balances out the number in the anion lattice (equal number of cation and anion vacancies, Fig. 7(a)).

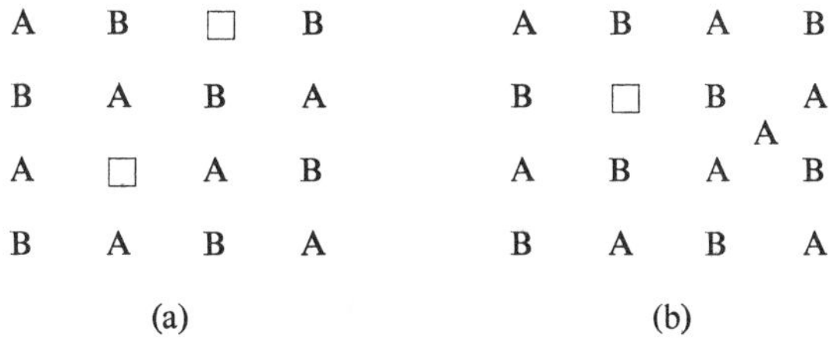

A	B	☐	B		A	B	A	B
B	A	B	A		B	☐	B	A
A	☐	A	B		A	B	A	B
B	A	B	A		B	A	B	A

(a)   (b)

FIG. 7. (a) Vacancies on two types of lattice sites. (b) Vacancy on one site plus an interstitial atom.

A Frenkel defect may be visualized as in Figs. 6(b) and 7(b), where we have, in each case, a combination of a vacancy and a so-called interstitial. An interstitial (or interstitialcy) is the name given to an atom that has been displaced from a normal lattice site into one of the spaces between these sites. From geometrical considerations, it can be concluded that Schottky defects will occur more generally than Frenkel defects. Indeed, for crystals of identical atoms or of two kinds of ion of roughly the same size, the accommodation of an atom or ion in an interstitial site and, therefore, the formation of a Frenkel defect, requires a prohibitive energy.

Point defects may be combined or associated in several distinct ways. Two or more vacancies may come together to form a vacancy pair or cluster. Other names given to clusters of vacancies are vacancy aggregates or condensed vacancies. The aggregates may form either spherical pores, as in some ionic solids, or flat-circular cavities, as in layer structures such as graphite [19] and zinc [15].

In a perfectly annealed solid, at any finite temperature, there must be an equilibrium concentration of both Schottky and Frenkel defects, and the relative concentration of each depends upon the respective energies of formation. Mott and Gurney [20] calculated the number of Frenkel defects in thermal equilibrium at a temperature T in the following way. Let N be the total number of atoms of identical kind in the crystal, and N' the total number of possible interstitial sites. If n atoms have left their "regular" positions in the lattice to produce n Frenkel defects at equilibrium, then, provided that we can assume that (i) the concentration of defects is so small that they are essentially independent of one another, (ii) the volume of the crystal remains constant so that the energy of the defects is independent of temperature, and (iii) the vibrational frequencies of the lattice are uninfluenced by the presence of the Frenkel defects, we may write:

$$P = \frac{N!}{(N - n)!\, n!} \qquad (3)$$

and

$$P' = \frac{N'!}{(N' - n)!\, n!} \qquad (4)$$

where P and P' are, respectively, the number of ways in which the vacancies and interstitial atoms can be arranged. From elementary statistical mechanics, it follows that the entropy associated with n Frenkel defects is given by:

$$S = k \ln P + k \ln P' \qquad (5)$$

so that

$$S = k \left[\ln \frac{N!}{(N - n)!\, n!} + \ln \frac{N'!}{(N' - n)!\, n!} \right] \qquad (6)$$

k being the Boltzmann constant.

Using Stirling's formula:

$$\ln X! \approx X \ln X - X \qquad (7)$$

(where X is any large variable) equation (6) becomes:

$$S = k[N \ln N - (N - n) \ln (N - n) - n \ln n]$$
$$+ k[N' \ln N' - (N' - n) \ln (N' - n) - n \ln n] \qquad (8)$$

The internal energy U associated with a Frenkel defect may be written as:

$$U = n\, W_F \qquad (9)$$

where W_F is the work required to form a single Frenkel defect. At equilibrium, under conditions of constant temperature and volume, the Helmholtz free energy, defined by equation (2), should be a minimum with respect to n. That is,

$$\left(\frac{\partial F}{\partial n} \right)_{T,V} = 0 \qquad (10)$$

so that from equations (2), (8) and (10)

$$W_F = kT \ln \frac{(N - n)\,(N' - n)}{n^2} \qquad (11)$$

or

$$\frac{n^2}{(N' - n)\,(N - n)} = \exp \left(-\frac{W_F}{kT} \right) \qquad (12)$$

Provided that n is very small in comparison with N and N'—a condition met at all except the most elevated of temperatures—then

$$n = \sqrt{(NN')} \exp \left(-\frac{W_F}{2kT} \right) \qquad (13)$$

By making the same assumptions as for the evaluation of n, we can proceed in an analogous manner to evaluate n_S, the number of Schottky defects at equilibrium in a crystal at temperature T. It emerges that

$$n_S = N \exp \left(-\frac{W_S}{kT} \right) \qquad (14)$$

where W_S is the work required to form a Schottky defect, and N, as before, is the total number of atoms. Similar procedures can be adopted [14] to evaluate the equilibrium concentration of vacancy clusters and various other point-defect configurations.

It is useful to note that, when W_S is about 1 eV (a typical value) and T is 1000°K, then equation (14) yields $n_S/N \approx 10^{-5}$.

Both equation (13) and equation (14) reveal that the concentration of point defects in a crystal is going to be at its very highest at a temperature just below the melting-point.

5.2.3.2. *Non-stoichiometric Compounds* [21]

This is unquestionably the most important group of defective solids from the standpoint of the chemist. By definition, a non-stoichiometric compound is a solid in which there is an excess or a deficiency of one component as revealed by the composition (found by chemical analysis). Some non-stoichiometric compounds display only a very small departure from the stoichiometric composition, others exhibit marked departures, but in all of them the number of atoms of one component to the number of atoms of the other component does not correspond exactly to the whole number ratio expressed by the formula. (It must not be thought that perfectly stoichiometric compounds are always the most stable: one of the greatest contributions of X-ray crystallography to chemistry has been to demonstrate that there is nothing unique in the ideal composition [22].)

The simplest type of non-stoichiometry, displayed extensively by alkali metal halides, involves an excess of metal ions due to anion vacancies. Such a non-stoichiometric compound is readily produced by heating sodium chloride in sodium vapour. The lattice incorporates sodium atoms, which then occupy the same regular cation sites as Na^+ ions, and the freed electrons are trapped in the vicinity of the newly generated vacant anion sites (see Fig. 8) which, so long as they are vacant, act as centres of effective positive charge in the crystal. The name given to an electron trapped at an anion vacancy is an F-centre [23, 24].

Na^+	Cl^-	Na^+	Cl^-
Cl^-	Na^+	Cl^-	Na^+
Na^+	$\square e$	Na^+	Cl^-
Cl^-	Na^+	Cl^-	Na^+
Na^+	Cl^-	Na^+	Cl^-

FIG. 8. Schematic model of anion vacancy with trapped electron (the F-centre).

Zn^{2+}	O^{2-}	Zn^{2+}	O^{2-}	Zn^{2+}	O^{2-}
O^{2-}	Zn^{2+}	O^{2-}	Zn^{2+}	O^{2-}	Zn^{2+}
	$(e\ Zn^+)$				
Zn^{2+}	O^{2-}	Zn^{2+}	O^{2-}	Zn^{2+}	O^{2-}
O^{2-}	Zn^{2+}	O^{2-}	Zn^{2+}	O^{2-}	Zn^{2+}

FIG. 9. Excess of zinc in zinc oxide.

Another type of non-stoichiometry also involves an excess of metal ions due, not to anion vacancies, but to interstitial cations. The oxides of zinc and cadmium give rise to this type of non-stoichiometric solid [25]. Figure 9 illustrates the lattice arrangement believed [16] to exist in zinc oxide. It will be noted that an electron is trapped in the vicinity of the ionized interstitial zinc atom. An electron trapped at an F-centre (Fig. 8) or at an interstitial ion may be released (just as an electron trapped by a proton in the hydrogen atom may be released) provided that sufficient energy is absorbed thermally, photo-chemically or during irradiation with more powerful electromagnetic radiation such as X-rays. The electrons freed in the process are known as quasi-free electrons and they contribute to the electronic conductance of the crystal. If the interaction responsible for the trapping of the electrons at the lattice defects is small, thermal energy alone may be sufficient to release electrons. This indicates why non-stoichiometric ionic solids such as those depicted in Figs. 8 and 9 may exhibit an electronic conductance that increases exponentially with increasing temperature; that is, the metal excess compounds display what is called n-type semiconductivity. If the conductance of one of these solids is increased by irradiating with light, it is said to exhibit photoconductance; the absorption of the light quanta is the process now responsible for the release of trapped electrons.

Two other types of non-stoichiometric solid are known where semiconductivity also prevails, but where the mechanism of conductance involves the migration of positive charges rather than that of negative charges (the electrons) as occurs in n-type semiconductivity. The first, and more widely occurring, so-called p-type compound semiconductor is one with a deficiency of metal ions. Nickel oxide, and many other ionic solids including cuprous oxide, ferrous oxide, ferrous sulphide and cupric iodide, fall into this category (Fig. 10(a), (b)). In order to preserve electroneutrality in this type of defective solid, it is necessary for the requisite number (e.g. one per cation vacancy for Cu_2O, two per cation vacancy for NiO) of lattice cations to lose an extra electron. Each "missing" electron is termed a "positive hole" and, like an electron, it can wander through the crystal. If, in Fig. 10(a), one of the Ni^{2+} ions adjacent to one of the $Ni^{2+\oplus}$ (i.e. Ni^{3+}) ions were to pass an electron to the latter, the location of the positive hole would be moved by as much as the location of the electron, but in the opposite direction. Positive holes do move freely under the influence of an electric field, and this is the process which characterizes a p-type semiconductor. Note that the conduction is still of the electronic kind.

Ni^{2+}	O^{2-}	Ni^{2+}	O^{2-}	Ni^{2+}	O^{2-}		Cu^+	O^{2-}	Cu^+	O^{2-}	Cu^+	O^{2-}
O^{2-}	Ni^{2+}	O^{2-}	□	O^{2-}	Ni^{2+}			$Cu^{+\oplus}$		Cu^+		Cu^+
Ni^{2+}	O^{2-}	$Ni^{2+\oplus}$	O^{2-}	$Ni^{2+\oplus}$	O^{2-}		Cu^+	O^{2-}	□	O^{2-}	Cu^+	O^{2-}
O^{2-}	Ni^{2+}	O^{2-}	Ni^{2+}	O^{2-}	Ni^{2+}			Cu^+		Cu^+		

(a)   (b)

FIG. 10. Non-stoichiometric nickel oxide and cuprous oxide, both p-type semiconductors (⊕ denotes quasi-free positive hole).

The second, less widely occurring, type of compound semiconductor (p-type) is also one with a deficiency of metal ions (strictly speaking with a surfeit of electronegative component) due to the presence of interstitial anions. The reason why such defective solids are so rare—uranium dioxide is probably the best-known example—is that, generally, anions are much larger in size than cations, so that the energy and lattice rearrangement required to accommodate interstitial anions tends to be prohibitive. A schematic representation of a uni-univalent non-stoichiometric compound M^+A^- of this type is shown in Fig. 11.

M^+	A^-	M^+	A^-	M^+	A^-
		A^-			
A^-	M^+	A^-	M^{2+}	A^-	M^+
M^+	A^-	M^+	A^-	M^+	A^-
A^-	M^{2+}	A^-	M^+	A^-	M^+
	A^-				
M^+	A^-	M^+	A^-	M^+	A^-
A^-	M^+	A^-	M^+	A^-	M^+

FIG. 11. Non-stoichiometric compound with excess of electronegative component A.

We are not here concerned with the various methods of determining [21] which type of non-stoichiometry dominates in a particular crystal, but two points may be made in this connection. Firstly, so small is the deviation from the ideal whole number ratio expressed by the composition of some compounds that the ability of a substance to act as a semiconductor is often the only criterion of the prevailing non-stoichiometry. Secondly, superimposed upon the concentrations of the various point defects associated with non-stoichiometry is a small, but usually negligible (at most temperatures), concentration of Frenkel and Schottky defects of the kind discussed in Section 5.2.3.1.

5.2.3.3. *Impurity Systems*

Under this heading we may note two very important categories of defective solid: the first involves solid solution of one elemental species in another; the second embraces controlled-valency semiconductors which have been used widely [25] in studies of catalysis.

The point defects which occur in germanium and silicon when small quantities of certain elements from Groups III and V are substitutionally dissolved in them fall into the first category. Take germanium, which crystallizes in the diamond structure, as an example. Small quantities of phosphorus or arsenic can be accommodated in the parent germanium structure with each impurity atom occupying a "regular" lattice site (see Fig. 12). Now there are

Ge	Ge	Ge	Ge
Ge	$\boxed{P^+e}$	Ge	Ge
Ge	Ge	Ge	Ge
Ge	Ge	Ge	Ge

FIG. 12. A donor impurity (substitutional phosphorus) in germanium (*e* denotes a quasi-free electron).

five valence electrons to each phosphorus atom and these must be accounted for: four are shared with adjacent germanium atoms (forming covalent bonds similar to those between adjacent germanium atoms), while the fifth becomes quasi-free. It will be attracted weakly by the phosphorus, but it can wander quite freely through the lattice under the influence of an electric field (effectively, at large separation distances from the phosphorus atom, the energy state of the fifth electron is in the conduction band—see Section 5.2.4). This type of impurity (phosphorus or arsenic in germanium or silicon) is known as a *donor impurity* and it confers n-type semiconductivity on the host material. If group III elements were used in place of group V ones (see Fig. 13) there would be insufficient valence electrons from the impurity material to share with those of the host, so that, for each trivalent impurity atom accommodated, there would be one positive hole formed. Just as the electron would be weakly attracted by the phosphorus in the previous example, here the hole would be weakly attracted by the group III element, e.g. boron. This type of impurity (boron, indium or gallium in germanium or silicon) is known as an *acceptor impurity*, because each impurity atom can "accept" an electron from a nearby host atom in attempting to form a covalent bond, thereby introducing holes into the valence band (see Section 5.2.4).

Ge	Ge	Ge	Ge
Ge	B-*p*	Ge	Ge
Ge	Ge	Ge	Ge
Ge	Ge	Ge	Ge

FIG. 13. An acceptor impurity (substitutional boron) in germanium (*p* stands for a quasi-free positive hole).

Controlled-valency semiconductors are best typified if we consider nickel oxide (Fig. 14). When produced under normal conditions, this oxide, as mentioned in the previous section, exhibits a metal-deficient type of non-stoichiometry, there being a number of cation vacancies, and twice that

(a)

Ni^{2+}	O^{2-}	Ni^{2+}	O^{2-}	Ni^{2+}	O^{2-}
O^{2-}	□	O^{2-}	Ni$^{2+\oplus}$	O^{2-}	Ni^{2+}
Ni$^{2+\oplus}$	O^{2-}	Ni^{2+}	O^{2-}	Ni^{2+}	O^{2-}
O^{2-}	Ni^{2+}	O^{2-}	□	O^{2-}	Ni$^{2+\oplus}$
Ni$^{2+\oplus}$	O^{2-}	Ni^{2+}	O^{2-}	Ni^{2+}	O^{2-}

(b)

Ni^{2+}	O^{2-}	Ni^{2+}	O^{2-}	Ni^{2+}	O^{2-}
O^{2-}	Li$^+$	O^{2-}	Ni^{2+}	O^{2-}	Ni^{2+}
Ni$^{2+\oplus}$	O^{2-}	Ni^{2+}	O^{2-}	Ni$^{2+\oplus}$	O^{2-}
O^{2-}	Ni^{2+}	O^{2-}	□	O^{2-}	Ni^{2+}
Ni^{2+}	O^{2-}	Ni$^{2+\oplus}$	O^{2-}	Ni^{2+}	O^{2-}

(c)

Ni^{2+}	O^{2-}	Ni^{2+}	O^{2-}	Ni^{2+}	O^{2-}
O^{2-}	□	O^{2-}	Ni$^{2+\oplus}$	O^{2-}	Ni^{2+}
Ni$^{2+\oplus}$	O^{2-}	Ni^{2+}	O^{2-}	Ni^{2+}	O^{2-}
O^{2-}	Ni^{2+}	O^{2-}	□	O^{2-}	Ni$^{2+\oplus}$
Cr^{3+}	O^{2-}	Ni^{2+}	O^{2-}	Ni^{2+}	O^{2-}

FIG. 14. (a) p-Type nickel oxide with cation vacancies and positive holes. (b) p-Type nickel oxide doped with lithium ions. (c) p-Type nickel oxide doped with tervalent chromium ions.

number of quasi-free positive holes. If monovalent cations, such as Li$^+$ ions, were introduced into the lattice (process of introducing an "impurity" is now widely known as doping) each Li$^+$ ion could take up a regular Ni^{2+} lattice site, the radius of these two ions being the same (0·78 Å). What happens when the lithium first becomes assimilated by the nickel oxide is that the Li$^+$ ions fill up the vacancies in the Ni^{2+} ion sublattice. During this process, a corresponding number of Ni^{3+} (i.e. Ni$^{2+\oplus}$) ions, which have a smaller radius than

Ni^{2+}, are reduced to the Ni^{2+} oxidation state. The lattice must consequently expand, and the conductance must decrease. But when all the vacant lattice sites have been filled up, further addition of lithium can be achieved only by substitution of Ni^{2+} by Li^+. Under these circumstances, for each Li^+ introduced, a Ni^{2+} must be converted to the Ni^{3+} state, so that the lattice must now contract and the conductance increase. This picture has been verified by the experimental work of Verwey and his co-workers [76, 72], who have observed the inversion in lattice size and conductance. It is interesting to note that the composition of p-type nickel oxide doped with lithium oxide to yield a controlled-valency semiconducting oxide may be represented by the equation

$$\frac{x}{2} Li_2O + (1 - x) NiO + \frac{x}{4} O_2 \rightleftharpoons (Li_xNi_{1-2x}^{2+} Ni_x^{3+})O \tag{15}$$

The introduction on a cation site of an impurity cation of higher charge than the host ion (depicted in Fig. 14(c)) serves to reduce the number of residual Ni^{3+} ions and also reduces the conductance.

5.2.3.4. *Dual-valency Intrinsic Semiconductors* [16]

Certain compounds, chiefly of the ionic type, have some of the properties associated with defective solids, but are nevertheless quite stoichiometric. They have little or no structural defects of the kind described in the previous three subsections. The properties of such compounds stem from the fact that ions of the same element in two different valence states are present in the crystal in stoichiometric proportions.

The higher oxides of iron and cobalt, Fe_3O_4 (magnetite) and Co_3O_4, are examples of this type of defective solid. These oxides crystallize in the spinel structure [22]; in the unit cell there are thirty-two oxygen ions and twenty-four positively charged iron ions, eight of which are Fe^{2+} and sixteen Fe^{3+}. The ideal composition is therefore represented by $Fe^{3+}(Fe^{2+}Fe^{3+})O_4$. It was Verwey and his collaborators who demonstrated [26–29] that, in magnetite, all the divalent ions and half the tervalent ones were distributed statistically over the octahedral sites (that is, cation sites surrounded by six oxide ions at the corners of an octahedron), the remaining tervalent cations being situated in tetrahedral sites. The electronic conductance, together with the magnetic and optical properties of these dual-valency compounds arise from the ease of electron transfer between the di- and tervalent cations. The word intrinsic is included in the description of these compounds because no impurities (donor or acceptor) are required to produce the semiconductivity.

5.2.3.5. *Notation for Description of Lattice Imperfections*

Rees has introduced a comprehensive method of formulating, symbolically, the various types of point defect now known to exist in real crystals. The reader is referred to Rees's monograph [16] for a full treatment of this topic; we shall simply quote a few pertinent examples here.

Cation sites of types a and b are denoted, respectively, by $\square_a^+$ and $\square_b^+$. Anion sites of types c and d by $\square_c^-$ and $\square_d^-$ respectively. Interstitial sites of types e and f are represented by Δ_e and Δ_f. An electron trapped in the field of an excess positive charge in the lattice (that is, a quasi-free electron) is simply written as e (see, for example, Fig. 9); a positive hole trapped in the field of an excess negative charge (a quasi-free positive hole, as in Fig. 13) is denoted by p. An F-centre, mentioned in Section 5.2.3.2, is written as $(e/\square^-)$ signifying that this type of defect consists of an electron trapped by a vacant anion site. A V-centre, which is the name given for a positive hole trapped by a vacant cation site, is written as $(p/\square^+)$. When a second electron is trapped at a vacant anion site, that is, when an F-centre traps an electron, the resulting defect is called an F'-centre and is described either by $(e/e/\square^-)$ or $(e_2/\square^-)$. According to the notation of Rees we should describe the production of a Frenkel defect by the following equation:

$$\Delta + (A^+/\square^+) \rightleftharpoons (A^+/\Delta) + \square^+ \tag{16}$$

5.2.4. Energy band diagrams

In Chapter 2 we discussed how potential energy diagrams could be used to depict the energetics of molecular systems. At this stage it is instructive to consider briefly the type of potential energy diagram used to discuss the electronic and other properties of crystals. It is important to appreciate that, in discussing the various energy changes accompanying processes which occur within, or at the surface of, crystals, we cannot, strictly speaking, talk of energy levels. We must consider energy bands, which may be thought of as having been produced as a result of the coalescence of the energy levels associated with discrete species when they condense to form a crystal. The fundamentals of band theory are discussed fully in other texts [16, 20, 30, 31].

The two bands of interest to us are the so-called valence band and conduction band which, in all crystals, are separated to a greater or lesser extent by what is termed the forbidden energy gap (see Fig. 15). Electron energies are plotted against distance through the crystal. In Fig. 15(a) is illustrated the state of affairs expected at absolute zero: there are no free carriers (electrons in the conduction band or positive holes in the valence band), because there is no thermal energy available to produce an electron–hole pair (contrast Fig. 15(b), which refers to higher temperatures). Free carriers can also be produced by absorption of light quanta; and recombination may occur spontaneously. In Fig. 15, the carriers are shown as though they were localized; in reality, they may move at random throughout the crystal, as implied in Section 5.2.3.2.

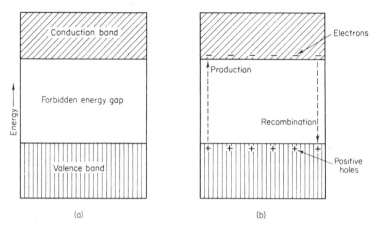

FIG. 15. Energy band diagrams showing state of affairs in an intrinsic semiconductor at (a) low, and (b) high temperatures.

Figure 16 summarizes the relevant energetics of semiconductors which are of the donor, acceptor or partially compensated type. Only the region between the top of the valence band and the bottom of the conduction band is of interest. Figure 16(a) would describe the energetics of, say, n-type germanium or non-stoichiometric zinc oxide, while Fig. 16(b) would describe that of p-type germanium or non-stoichiometric nickel oxide. The origin of the levels (often called impurity levels even when they are not strictly due to impurities, as in zinc oxide) may be appreciated thus. When substitutional phosphorus has been introduced into germanium to generate n-type semiconductivity, the fifth electron lies in a higher energy state than the normal valence electrons of the germanium atoms. Hence, a localized extra level associated with the phosphorus impurity must lie above the top of the filled valence band. But since there is some binding energy for an electron in this state to remain on the phosphorus, the level must lie below the lowest free

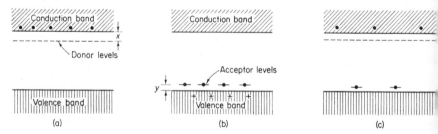

FIG. 16. Energy band diagrams for extrinsic semiconductors, showing the situation for a donor type (a), acceptor type (b), and for a partially compensated type, (c), with both donor and acceptor levels. x and y are the ionization energies.

electron state in the conduction band. An analogous argument would explain why the acceptor levels, arising from boron impurity in germanium, for example, must lie as shown in Fig. 16(b). When both acceptors and donors are present in the solid, there is a "compensation" effect, as illustrated in Fig. 16(c).

Situations can arise when the electron and the positive hole are not completely independent of each other, as they generally are. The word used to describe the state of the electron–hole pair in such a situation is the exciton. The exciton, which is a pre-ionization state in which an excited electron and its hole are trapped by the interaction of their own fields, was first pictured by Frenkel [32]. Excitons are readily produced [33] by irradiating crystals with radiation corresponding to the tail of the fundamental absorption band.

5.2.4.1. *Energetics of Adsorption on Semiconductors: the Boundary-layer Theory and Surface States*

When chemisorption occurs on a semiconductor surface, the resulting change in electrical conductance of the solid yields unambiguous information concerning the type of electronic rearrangement, or the direction of the charge transfer, at the surface. A fall in the conductivity of a p-type semiconductor signifies an electron transfer from the chemisorbed gas to the solid, and such a change is only to be explained [33, 34] by a fall in the surface concentration of positive holes in the full band due to entry of electrons from the adsorbate. A fall in the conductivity of an n-type semiconductor as a result of chemisorption signifies an electron transfer from the conduction band of the semiconductor to the adsorbed species. It will be noticed that, although the direction of charge transfer is opposite in these two processes, each is accompanied by a depletion of carriers in the semiconductor, and both are examples of *depletive chemisorption*. Adsorption on the p-type semiconductor has led to the production of a *cationic layer*, adsorption on the n-type semiconductor to an *anionic layer*. The reverse process, namely *cumulative adsorption*, distinguished by a rise in conductivity as a result of chemisorption on both p-type and n-type semiconductors, signifies that the surface concentration of carriers has increased. Cumulative adsorption clearly leads to anionic chemisorption on a p-type and to cationic chemisorption on an n-type semiconductor.

It is possible to make use of energy band diagrams to illustrate the energetics of depletive and cumulative chemisorptions on semiconductor surfaces. The way in which we shall now do this underscores the essentials of the boundary-layer theory of chemisorption first formulated, independently, by Aigrain and Dugas [35], Hauffe [36] and Weisz [37]. This theory postulates that depletive chemisorption on a semiconductor may be treated as an electronic boundary-layer problem of the type encountered in the theory of metal–semiconductor contacts. Although Engell and Hauffe [*36*] have argued that the boundary-layer theory can be applied equally well to cumulative adsorption, it is in the interpretation of anionic chemisorption on an n-type, and cationic chemisorption on a p-type conductor (both depletive) that the theory invites attention.

Consider depletive chemisorption on n-type and p-type conductors (see Figs. 17 and 18). The situation prior to adsorption is represented [*33*] in Figs. 17(a) and 18(a). Note the positions of the Fermi level, which can be regarded either as the potential energy of the electrons in the solid or as the energy level having a probability of being occupied equal to one half. If α is the electron affinity of the adsorbed atom and ϕ the work function of the semiconductor, the energy of chemisorption of the first atom on the n-type conductor will be $(\alpha-\phi)e$, where e is the electronic charge. As more atoms are adsorbed, a space-charge builds up in the boundary layer, since the donor levels deeper in the semiconductor are called upon to yield their electrons. Consequently, the potential energy of electrons in the solid becomes modified, and, in going from the solid to the adsorbate, electrons have to surmount a potential-energy barrier V. The height of this barrier is increased with each atom adsorbed, so that the Fermi level in the semiconductor is progressively depressed. However, once the potential energy of the electrons in the adsorbate equals that of the electrons in the solid, equilibrium will be established, and no further net adsorption will be possible (see Fig. 17(b)). If, at equilibrium, the barrier has a height V_f, and the number of atoms adsorbed (as anions)

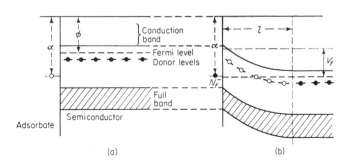

Fig. 17. Diagrammatic representation of anionic chemisorption on an n-type semiconductor; (a) before chemisorption, and (b) after chemisorption [*33*].

is N_f, it follows that N_f electrons have been transferred from the semiconductor, there being a depletion of carriers in a boundary layer of depth l.

For cationic chemisorption on a p-type semiconductor, the energy of chemisorption of the first adatom will be $(\phi-I)e$, where I is the ionization potential of the adsorbate. As chemisorption progresses, the Fermi level of the solid gradually rises and, at equilibrium, equals the potential energy of the electrons in the adsorbate. The potential energy barrier built up during the attainment of equilibrium when N_f atoms are adsorbed is again denoted by V_f, and the surface layer of the semiconductor has been depleted of carriers to a depth l (Fig. 18(b)).

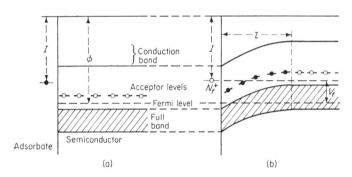

FIG. 18. Diagrammatic representation of cationic chemisorption on a p-type semiconductor; (a) before chemisorption, and (b) after chemisorption [33].

By assuming that the space-charge density is constant in a layer of thickness l, it can be shown (see Ref. 33 for an elegant derivation) that

$$N_f = \left(\frac{K}{2\pi e} \, n_0 \, V_f \right)$$

(17)

where K is the dielectric constant, and n_0 is the concentration of the carriers in the interior of the semiconductor. Using typical values for the parameters in equation (17), it emerges that the boundary-layer theory predicts coverage values at equilibrium, for depletive chemisorptions, of about 1 %. Although there are many systems [34] which seem to conform to equation (17) and in other respects also to the predictions of the boundary-layer theory, it is now generally felt that too much emphasis can be placed on a purely physical model. Such an approach tends to minimize important chemical characteristics associated with semiconductor and metal surfaces.

In some of the energy-band diagrams discussed so far, we have drawn energy levels within the forbidden energy gap, these levels being ascribed to lattice imperfections of the types enumerated in Section 5.2.3. It was Tamm [38] who first pointed out that the discontinuity at the surface of a *perfect* lattice could itself give rise to energy levels in the forbidden gap, and that electrons in these levels (often called Tamm levels) could then move freely on the surface but not in the bulk. Much theoretical discussion [39–42] has centred on this topic, and it is only within the last few years that definite evidence for the existence of surface states has come to light [43]. A schematic

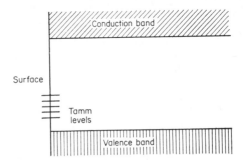

F<small>IG</small>. 19. Illustration of a solid with Tamm levels.

illustration of a solid—not necessarily a semiconductor—with Tamm levels is shown in Fig. 19.

5.3. Specific Examples Illustrating the Part Played by Lattice Imperfections in Catalysed Reactions

Having discussed the most prominent types of lattice imperfection, we may now proceed to consider the evidence for believing that such imperfections play a significant role in heterogeneous catalysis. We shall first examine how, and under what circumstances, dislocations affect catalyst behaviour, and then deal with the corresponding effects of point defects. Often, as we shall see, it proves difficult to separate the influence of these two types of defect. However, we ought to enquire more closely whether we should expect lattice imperfections to function as the "active centres" conceived by H. S. Taylor (see Section 5.1).

There are two main reasons for suspecting that the lattice imperfections present on the surface of a catalyst may turn out to be the active centres for certain reactions on that surface. The first of these is the fact that, at the sites of emergent dislocations and, to a different extent, at surface point defects, the geometrical arrangement of the catalyst atoms is different from that prevailing elsewhere on the surface. It will be shown in Chapter 6 that the atomic separation distances at the surface of a catalyst, together with the bond lengths and general stereochemical features of the reactants molecules and transition complex, can be important factors in deciding the extent of catalytic activity. It is therefore apparent, at least in qualitative terms, why some emergent edge and screw dislocations, possessing the appropriate Burgers vector, may facilitate heterogeneous reaction more readily than other sites on the surface. The second reason for expecting greater catalytic activity at the location of lattice imperfections is an electronic one. The current status of the electronic factor in catalysis is also discussed in Chapter 6; suffice it to say here that modification of the electronic properties of a solid can occur at surface sites associated with dislocations on point defects.

5.3.1. Role of dislocations

Apart from the classical work of Hedvall referred to in Section 5.1, there have been several reports in the past claiming that catalysis by solids, particularly metals, is profoundly influenced by the imperfection content of the solid. Thus, Rienäcker [44], in 1940, found an increase in the rate of decomposition of formic acid and also a simultaneous increase in the activation energy after cold-rolling the nickel catalyst. Seven years earlier, Eckell [45], who used polycrystalline nickel as a catalyst in the hydrogenation of ethylene, had observed that the velocity of the reaction was increased by a factor of from 600 to 1000 when the nickel was plastically deformed by cold-rolling. There is no doubt that the number of dislocations is increased by cold-rolling [12, 46, 47]. There is some likelihood, however, that, in such early work, the surfaces may have been badly contaminated or not cleaned satisfactorily after being rolled.

In recent years, some attempts have been made to investigate anew the influence of dislocations on catalytic activity. Particular attention has been given to surface cleanliness and the use of well-characterized solids. Using the ion-bombardment technique outlined in Section 3.2.2.6, Farnsworth and Woodcock [48] prepared clean nickel and platinum surfaces, and showed that the catalytic activity of these metals in the hydrogenation of ethylene was enhanced by ion-bombardment. It was not unambiguously established, however, that the enhanced activity arose from individual dislocations or from a complex network of dislocations and other defects. Probably one of the most rewarding and interesting of studies has been carried out by Sosnovsky [49], using well-characterized and cleaned single crystals of silver. As Sosnovsky's work provides a striking demonstration of the importance of dislocations, and as it also highlights a ubiquitous phenomenon in heterogeneous catalysis, the compensation effect, it will now be given especial attention.

5.3.1.1. *Sosnovsky's Work on the Influence of Dislocations on the Catalytic Decomposition of Formic Acid* [49]

Single crystals of high-purity silver were grown from the melt in a graphite mould; they were electro-polished, acid-washed and then bombarded by positive argon ions in a low-pressure discharge [50]. The potentials used varied from 14 to 4000 V. After ion bombardment, the specimens were transferred to a reaction vessel where they were annealed *in vacuo* at 250°C for 18 h prior to the commencement of rate measurements [51, 52] on the decomposition of the formic acid vapour. The crystal plates used by Sosnovsky had (111),

(100) and (110) planes, respectively, parallel to their surfaces, and for each of these planes the catalytic activity was measured as a function of the energy of the bombarding ions.

Arrhenius plots of log rate versus reciprocal temperature (from data recorded in the temperature range 150–250°C) were obtained for the three distinct crystallographic faces, a range of applied voltages from 22·5 to 4000 V (maximum ion energies between 15 and 3000 eV) being employed in each case. Table 5.1 summarizes the results obtained, A and E being defined by the well-known Arrhenius equation: rate $= A \exp(-E/RT)$. (The errors estimated by Sosnovsky are also given.)

TABLE 5.1. Effect of bombarding voltage on the catalytic activity of silver crystals of different orientations [49]

Applied potential (V)	Max. ion energy (eV)	(111)		(110)		(100)	
		$\log_{10}A$	E (kcal mole^{-1})	$\log_{10}A$	E (kcal mole^{-1})	$\log_{10}A$	E (kcal mole^{-1})
22·5	13	2·3±0·6	12·2±1·2	7·1±0·2	24·1±0·5	9·2±0·9	29·9±3·0
46	38	6·1±0·5	17·6±0·9	9·1±0·3	26·8±0·7	12·4±0·8	32·5±1·7
86	77	7·3±0·4	20·4±0·8	9·5±1·0	27·1±2·2		
130	118	6·9±0·7	19·9±1·5	12·4±0·6	32·8±1·3		
300	280	6·7±0·9	19·7±1·8			13·2±0·3	34·4±0·6
500	450			13·4±0·4	34·8±0·8		
4000	3000	7·7±1·1	22·5±2·3	11·1±0·3	30·2±0·6	13·0±0·5	35·2±1·1

It is evident from Table 5.1 that the catalytic activity of a silver surface can be considerably altered by bombarding it with positive argon ions. The systematic changes in A and E with changes in bombarding energy are to be noted. In view of the fact that the surface area of the silver crystals remained essentially unaltered by ion bombardment, it may be concluded that the changes in A of several orders of magnitude imply corresponding changes in the number of active sites (if it can be assumed that the reaction mechanism remains the same irrespective of the severity of the bombardment). The active sites (produced by bombardment) could, in principle, be either point defects such as Frenkel and Schottky defects, dissolved argon atoms or combinations of these, or dislocations. The concentration of point defects, at any temperature, can be computed from equations such as (13) and (14) (Section 5.2.3.1), and knowing [53] the work required to form single vacancies, Sosnovsky showed that, after annealing at 250°C, the concentration of point defects should be negligible. We are therefore left with the possibility that the sites of emergence of dislocations may constitute the active sites in the reaction. This is the first piece of evidence quoted by Sosnovsky for accepting the role of dislocations in catalysis. The second piece of evidence comes from observations made by Ogilvie [50], using transmission electron microscopy, on identical samples of silver prepared and pretreated in a manner strictly analogous to that used by Sosnovsky. The work of Ogilvie demonstrated that ion bombard-

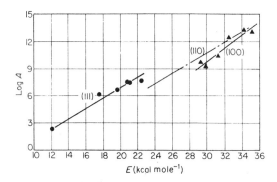

FIG. 20. $Log_{10}A$ versus E for (111), (110) and (100) surfaces.

ment gives rise to a greatly increased number of dislocations. Sosnovsky points out that the number of dislocations introduced into a metal by positive-ion bombardment is at least as large as the number resulting from cold-rolling, and that, in particular, the magnitude of A for the (111) surface increases owing to ion bombardment in approximately the same ratio as the dislocation density. All these facts seem to suggest that the active sites for the catalysed reaction are those where dislocation lines come to the surface. Figure 20 shows how, for all three crystallographic faces, an increase in the pre-exponential term A is always accompanied by an increase in activation energy E. This is an example of the well-known compensation effect (not to be confused with compensation mentioned in Section 5.2.4 in reference to a semiconductor consisting of both donor and acceptor centres). We shall consider the significance of Fig. 20 after first outlining the properties of the (kinetic) compensation effect.

5.3.1.2. The Compensation Effect

It was Constable [54] in the course of his studies of the dehydrogenation of ethanol on copper catalysts, who first reported the existence of a compensation effect. By reducing cupric oxide at a series of temperatures, he prepared a number of catalysts of differing catalytic activity. However, he found that an increase in the activation energy of the reaction frequently did not lead to the expected decrease in the rate constant, because there occurred a simultaneous increase in the pre-exponential term A, which compensated for the change in the exponential term of the Arrhenius equation. The compensation effect has been observed to operate in a variety of systems, and it has received considerable attention in the literature [55–62]. It often happens that, for a series of catalysts and for a given reaction, the relation between A and E takes the form:

$$\log A = a E + b \qquad (18)$$

where a and b are constant. Schwab [58] refers to the compensation effect, which is embodied in equation (18), as the "theta rule". The compensation effect is not restricted to heterogeneous catalysis (see, for example, Ref. 63).

A universally applicable explanation for the occurrence of the compensation effect is still awaited, although it seems possible [61] that it arises because of the effect of temperature on the distribution of active sites on the surface. (It can also arise because of bad experimentation! For a given value of the specific rate constant, any error in E will clearly produce a corresponding error in A such that $\delta \log A \propto \delta E$, which leads, therefore, to equation (18).)

We shall now return to the discussion of the results of Sosnovsky (Fig. 20) and see how the observed compensation effect can be interpreted in terms of the nature of the active sites.

Sosnovsky proposed [49] that the observed increase in E with increase in A, that is (see Section 5.3.1.1) with the increase in the number of active sites, or, putting it differently, with the increase in the density of dislocations at the surface, could be due to an interaction between the dislocations. Since the new dislocations produced by ion bombardment are believed [49, 50] to be confined to the boundaries (tilt boundaries) of small crystallite "blocks" within the single-crystal plates, their local density (surface concentration) is likely to be very high and their interaction correspondingly strong. Hence, the energy of each site could be lowered, resulting in an increased value of E associated with the increase in A.

Let us, following Sosnovsky, assume for simplicity that there are but two types of active site present on any of the silver crystal planes: one with a lower activation energy E_1, typical of the isolated dislocation, the other with a higher activation energy E_2, typical of the new dislocation crowded into boundaries. If c_1 and c_2 are, respectively, the concentrations of the active sites at isolated dislocations and dislocations in a boundary, then the observed rate constant k may be written:

$$k = \gamma \left\{ c_1 \exp \left(\frac{-E_1}{RT} \right) + c_2 \exp \left(\frac{-E_2}{RT} \right) \right\} \qquad (19)$$

where γ is a factor that depends on the crystallographic orientation of the face under consideration. It follows that the observed values of A and E (Table 5.1) depend on the relative magnitude of the two terms on the right-hand side of this equation. If, for a particular orientation, the extreme experimental values of E_1 and E_2 are chosen, we may discuss the significance of the measured A values in terms of c_1 and c_2. For the (111) face, upon which we shall now concentrate, the E values are: $E_1 = 12$ and $E_2 = 22$. Let us now suppose that the concentration c_1 is 10^6 sites cm^{-2} (this is a reasonable figure based on the measured dislocation density at the surface of an undeformed crystal). When we are dealing with an undeformed crystal, the total concentration of sites is c_1 ($c_2=0$). Only under these circumstances will the observed values of A and E be representative of all sites on the surface. Now it can be shown by calculation [49] that, if the new dislocations with activation energy E_2 are introduced in increasing numbers, their effect on the measured values of A and E will remain negligible until c_2 reaches $10^{9 \cdot 3}$ sites cm^{-2} for the (111) face. A further increase in c_2 by a factor of 10 makes the two terms in equation (19) comparable so that, under such conditions, the measured A and E values are composite and larger than before. Upon increasing c_2 still further (e.g. to $10^{11 \cdot 3}$) the second term predominates to such an extent that the measured values of A and E are now representative of the higher energy sites only.

The analysis just given tells us that, although a substantial amount of deformation may be present in the silver surface without having any effect on its catalytic activity, enhanced deformation should produce increases in the values of A and E provided that this deformation gives rise to a large number of sites of activation energies at or closely distributed around E_2. With reference to Table 5.1, we may now express our conclusions quantitatively. At 22·5 V, for the (111) face, $c_2=0$ and we can, as before, write $c_1=10^6$ sites cm^{-2}, each of these sites having an associated activation energy of 12 kcal mole^{-1}. The changes in A by a factor of 10^5 and in E of 10 kcal mole^{-1} after bombardment at 4000 V can now be explained by the plausible supposition that $10^6 \times 10^5 = 10^{11}$ sites cm^{-2} have been introduced, each site having an associated activation energy of about 22 kcal mole^{-1}. It is evident, therefore, that, if the original 10^6 sites cm^{-2} remain on the (111) face after extensive bombardment, their effect is masked by the 10^{11} new dislocations per square cm. From Table 5.1 we may, using the above argument, deduce that, for the (111) face, the value of γ, defined by equation (19), is $10^{2 \cdot 3}/10^6 = 10^{-3 \cdot 7}$ in arbitrary units. For the (110) and (100) faces, respectively, α is equal to $10^{7 \cdot 1}/10^6 = 10^{1 \cdot 1}$ and $10^{9 \cdot 2}/10^6 = 10^{3 \cdot 2}$. Sosnovsky has suggested [49] that the reason for the variations in γ, E_1 and E_2 with crystallographic orientation may be due to an anisotropy of properties associated with a dislocation line intersecting a free surface of different orientations. In the light of the work carried out by Livingston [64] and others [15, 65, 66] on the importance of stereochemical considerations in dislocation etching, this suggestion seems reasonable.

5.3.1.3. *Other Examples Illustrating the Role of Dislocations*

Uhara and his co-workers [67–72] have recently carried out a series of catalytic studies using cold-worked metals. Samples of copper, platinum and nickel have, after cold-working, been annealed at a series of temperatures and then used as catalysts in (gas–solid) systems such as the dehydrogenation of ethanol, and the para–ortho hydrogen conversion, as well as in a variety of other systems (liquid–solid) including the hydrogenation of cinnamic acid, the decomposition of diazonium salts and the electrolytic generation of gaseous hydrogen from solution. Almost without exception, it has been found that a marked decrease in catalytic activity of the metals occurs after annealing at a temperature which, from other evidence [73–76], is known to remove dislocations from the solid. We shall return to a brief discussion of this work when considering the influence of point defects on the catalytic activity of metals. (In Uhara's work, the influence of lattice vacancies and dislocations is observed and discussed concurrently.)

Gwathmey and his collaborators have shown that the progressive morphological changes that occur on the surfaces of single crystals of copper during the catalytic reaction of hydrogen and oxygen are initiated at the sites of emergence of dislocation lines [77–79]. Another example of a catalysed reaction which appears to depend on the presence of dislocations is the dehydrogenation of ethanol on lithium fluoride. It has been found [80] that the rate of this reaction is approximately proportional to the number of dislocation terminations at the surface, this number being determined by the etch-pit method discussed in Section 5.2.1.

It would be wrong to conclude that whenever a catalyst surface has a significant number of points of emergence of dislocations these points will constitute the active sites. There is no theoretical reason for expecting such a state of affairs, if only because, in the words of H. S. Taylor [81], "the amount of surface which is catalytically active is determined by the reaction catalysed". A thorough study by Bagg, Jaeger and Sanders [82] of the influence of arrays of emergent dislocations on evaporated silver surfaces in the catalysis of formic acid decomposition led [82] to the conclusion that points of emergence of dislocations are unimportant. The reasons why the conclusions of Bagg *et al.* [82] seem to be somewhat at variance with those of Sosnovsky [49] are partly attributable to an increased understanding of the mechanism of sputtering [83]. According to Sanders [83], it is no longer legitimate to explain the results of Sosnovsky entirely in terms of dislocations, for there is now doubt as to whether the energies of the bombarding ions utilized by Sosnovsky (see Table 5.1) were sufficient to introduce dislocations into the silver catalyst. This illustrates how important it is that systematic studies be carried out in the future to try to assess the true role of dislocations in catalysis. Great care must be exercised before deciding whether dislocations have genuinely contributed to catalysis. A simple numerical estimate illustrates well how easy it would be to mistake the catalysing effects of a minute amount of surface contamination for an example of emergent dislocations functioning as "active sites". All solids have about 10^{15} lattice sites per cm^2 of surface: hence, if a surface is contaminated only to the extent of having a millionth of a monolayer of impurity at the surface, there will be roughly 10^9 "impurity sites" per cm^2. This value of 10^9 is comparable with the dislocation density at small-angle boundaries, but far in excess of that expected for a solid which, because of its use as a catalyst at relatively high temperatures, has had a large proportion of its dislocations annealed out during pretreatment. The concentration of point defects, on the other hand (see equation (14)) is much greater than the density of dislocation lines and much more comparable with, if not greater than, the concentration of impurities in the catalyst. Moreover, the concentration of point defects must, for thermodynamic reasons, increase with increasing temperature (see Section 5.2.3.1), whereas dislocations are, basically, thermodynamically unstable [84].

5.3.2. Role of point defects

In contrast to the voluminous literature describing the influence of variations in point-defect concentrations on the catalytic activity of semiconductors (see below), relatively few unambiguous examples are known of the role of point defects in the catalytic activity of metals. With semiconductors, it is difficult, if not meaningless, to distinguish between point defects and so-called electronic factors. Hence, in assessing the role of point defects it is necessary, with such compounds, to consider the wide diversity of results obtained using doped and undoped oxides, sulphides, chlorides, etc., and many of the various examples of imperfect crystal classified in Section 5.2.3. With metals, however, it is meaningful to talk of point defects *per se*, although, in the final analysis, it may prove difficult to distinguish between point-defect clusters on the one hand and dislocation loops on the other [15]. The work of Robertson and his associates [85–87], who observed the phenomenon termed "catalytic superactivity", and of Uhara [67–70], mentioned earlier, illustrates well the catalytic importance of point defects in metals.

5.3.2.1. *Influence of Point Defects in Metals*

Upon realizing that many substances exhibit a greatly enhanced catalytic activity after they have been quenched from relatively high temperature, Robertson [87] decided to investigate how the activity of copper and nickel wires, in the decomposition of formic acid, varied with the thermal history of the metal. The probability P of reaction for a molecule of formic acid striking the bare catalyst surface was determined under a variety of conditions. It transpired that normal activity is shown by metal wires which have

not been flashed at a high temperature. However, after flashing at high temperature, both copper and nickel displayed "catalytic superactivity", the
activity of nickel, for example, having been increased 10^5 times. Moreover, on
nickel, the probability P was unity over a large temperature range (see Fig. 21).
The superactivity of nickel is frozen into the wire; it is not removed by

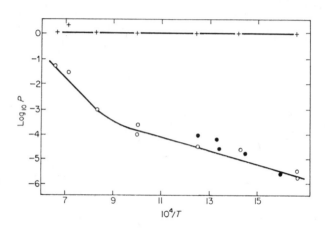

FIG. 21. P for the decomposition of formic acid on nickel: ○, a nickel wire; ●, a nickel
disc; +, a nickel wire with superactivity [85–87].

annealing for a time in a vacuum, nor by cooling to room temperature, nor
by slight oxidation. But superactivity disappears rapidly when formic acid
undergoes decomposition, and the "normal" activity soon prevails.

Bearing in mind that when a metal such as nickel is rapidly cooled from a
temperature of about 1300°C the dynamic disorder of the surface is largely
retained [88] and, more important, a high non-equilibrium concentration
(~ 1 in 10^4) of quenched-in vacancies can be produced, Robertson concluded
that the process of aggregation of point defects (as vacancies) is a significant
one in the production of catalytic superactivity. Although he does not suggest
why, exactly, a vacancy or vacancy cluster should act as an "active site",
Robertson's results can be convincingly explained in terms of the special
activity of vacancies. The rapid disappearance of superactivity when formic
acid is decomposed is explicable if there is an induced surface mobility of the
metal atoms which arises when a catalytic action occurs. In other words,
reaction itself brings about annihilation of the active sites, and time is needed
for more vacancies to diffuse from the bulk when the wire is hot in a vacuum.
(The regeneration of superactivity in a vacuum is a slow process [87].)
Gwathmey's study of surface morphology [77–79] produces ample evidence
for a special mobility of surface atoms when catalysis occurs.

Uhara's results [67–70] corroborate those of Robertson, for they show that a marked decrease in catalytic activity of cold-worked nickel wire occurs when the metal is annealed in the range 200–400°C—a range at which, according to Clarebrough [73–76] vacancies can be annealed out from cold-worked nickel.

5.3.2.2. Influence of Point Defects in Nonmetals

There is no shortage of data relating to adsorption and catalysis involving defective nonmetals: but there is an acute lack of systematic data involving just one reaction (the mechanism of which has been securely established) and several related imperfect nonmetals. Only for the decomposition of nitrous oxide ($2N_2O \rightarrow 2N_2 + O_2$) and the oxidation of carbon monoxide ($2CO + O_2 \rightarrow 2CO_2$) has a real attempt been made to compile comprehensive data which permit us to correlate catalytic activity with point defects. All the work has been done with semiconducting oxides [89], so we shall deal almost exclusively in this section with such nonmetals. It is pertinent to bear in mind the background information contained in Section 5.2.4.1, and the remark that in dealing with the catalytic significance of defects in semiconductors we are opening up the whole question of the "electronic factor" in catalysis by semiconductors [33, 90, 91]—see Section 6.3.

We shall deal first with a selection of results obtained in the study of the decomposition of nitrous oxide on metal-oxide semiconductors. For a very long time [92] there has been strong evidence for believing that the mechanism of this reaction may be written:

$$N_2O + \text{electron (from catalyst)} \rightleftharpoons N_2 + O^{-\text{(asd)}}$$

followed by

$$2O^- \rightarrow O_2 + 2e \text{ (to catalyst)}$$

or by

$$O^-_{ads} + N_2O \rightarrow N_2 + O_2 + e$$

It can be seen that the first step is an electronic one, a fact which prompted Dell, Stone and Tiley [93] to consider the relative merits of various semiconducting oxides as catalysts in this reaction. These workers, upon summarizing prior work by Schwab [94] and others [95], were able to compile an activity series (see Fig. 22). The most remarkable feature of the series is that it divides

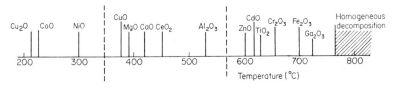

Fig. 22. The relative activity of oxides for the decomposition of N_2O, showing the temperature at which reaction first becomes appreciable [33].

into three distinct groups, the most active group being made up entirely of p-type semiconductors. The insulators (e.g. MgO and Al_2O_3) are of inter-mediate activity, and the third group is composed mostly of n-type semi-conductors. Baker and Jenkins [96], in their interpretation of the high activity of the p-type oxides, and the low activity of the n-type oxides, draw attention to the fact that, in the presence of oxygen, nitrous oxide decomposing over p-type oxides such as NiO leads to an increased conductivity of the solid. The same decomposition when it occurs on n-type oxides such as ZnO leads to a decrease in conductivity. Each of these observations is explicable in terms of the formation of a layer of chemisorbed oxygen during chemisorp-tion. Baker and Jenkins pointed out, in addition, that the results of "doping" experiments (see Section 5.2.3) are understandable if it is assumed that the slow step of the nitrous-oxide decomposition is a desorption process by which oxygen ions (O^-_{ads}) transfer their electrons to the catalyst and escape as gaseous oxygen (i.e. one of the two written above). If this transfer of electrons is indeed the slow step, it follows logically that p-type oxides would be superior to n-type ones as catalysts since the lower lying energy levels of p-type oxides would be in a better position to accept an electron from O^-_{ads} than n-type oxides, which have higher energy levels (compare Figs. 17 and 18).

We shall now turn to the comparison of the catalytic activities of semi-conducting oxides in the oxidation of carbon monoxide. Schwab has demon-strated [97] that the defect character (electronic properties) of metal oxides plays a dominant part in the catalytic action of these solids. Figures 23 and 24, reproduced from a summarizing article by Schwab [97], illustrate well the type of correlations which many have sought [90]. As explained previously in this chapter, Verwey [26, 27] showed that the addition of monovalent cations such as lithium to a nickel-oxide matrix results in an increase in the number of positive holes in the oxide and thereby increases the p-type conductivity. Conversely, the addition of trivalent ions (such as chromium) decreases the number of positive holes and hence decreases the conductivity. In the lower part of Fig. 23, Schwab illustrates the influence of added ions upon the conductivity of nickel oxide by quoting the results of Verwey [26] and of Hauffe [98]. The upper part shows the results obtained for the catalytic oxida-tion of carbon monoxide. Not only does the addition of lithium ions increase the conductivity, but the activation energy for the oxidation is also lowered in the range 300–450°C. The addition of trivalent chromium ions, on the other hand, lowers the conductivity and raises the activation energy for the catalytic reaction. These observations are consistent with the conclusion that the rate-determining step in the oxidation of carbon monoxide is the forma-tion of positive ions on the surface of p-type semiconducting nickel oxide.

The influence of the addition of altervalent ions on the conductivity of zinc oxide and on the activation energy of the oxidation of carbon monoxide

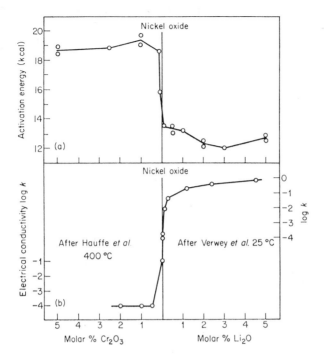

FIG. 23. Activation energy for the catalytic oxidation of carbon monoxide and electrical conductivity for nickel oxide doped with oxides of lithium and chromium (cf. Fig. 14) [97].

over zinc oxide is shown in Fig. 24. This time, addition of lithium ions causes a decrease in the number of free electrons and hence a decrease in the conductivity of the n-type semiconductor, whereas the addition of tervalent gallium ions increases the conductivity by increasing the number of free electrons. It can be seen that the activation energy of the catalytic reaction increases as the concentration of added lithium ions increases, and that the addition of gallium causes a marked decrease in the activation energy. These observations are consistent with the conclusion, reached by Schwab [97], that the catalytic oxidation of carbon monoxide over zinc oxide involves the chemisorption of oxygen as the slow step. As oxygen is an electron acceptor, the greater the supply of the electrons at the surface of the semiconductor the easier will be the formation of the surface oxygen ions and hence the lower the activation energy for the catalysed reaction.

The kinetics of this reaction are in line with the interpretations proposed by Schwab (see Section 8.1.1 for fuller discussion), but it would be erroneous to conclude, purely from the examples we have adduced here, that the correlation between semiconductivity and catalysis has been completely elucidated.

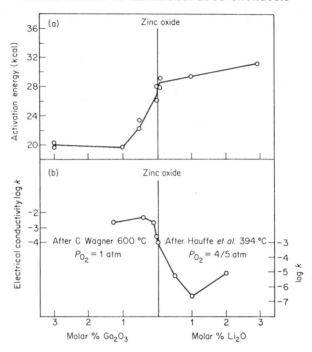

FIG. 24. Activation energy for the catalytic oxidation of carbon monoxide and electrical conductivity for zinc oxide doped with oxides of gallium and lithium [97].

In citing the above examples, an element of tendentiousness has crept into our discussion. Had we, for example, given a full account of Parravano's work [99], the overall picture would have been rather confusing. When Parravano studied the catalytic oxidation of carbon monoxide over doped nickel oxide, he found that the activation energy of the oxidation increased, rather than decreased, as the concentration of added lithium ions was increased. Another result, directly contradicting Schwab's, was the observation that the addition of trivalent chromium ions caused a lowering rather than an elevation of the activation energy (see Fig. 23). The complexity of the picture is enhanced when it is recalled that some careful work, carried out by Roginskii *et al.* [100], has confirmed the essential features of Parravano's work, and meticulous researches, executed by Dry and Stone [101], have confirmed Schwab's results. It emerges that the most important difference between the work of Parravano and that of Schwab is that the former studied the oxidation in the range 180–240°C and the latter discusses the range 250–450°C. Apparently, the mechanisms of reaction are quite sensitive to the temperature of reaction.

A seemingly unambiguous example of the direct relationship between catalytic activity and p-type character is to be found in the work of Krauss [102]. Krauss arrived at the extent of p-type semiconductivity in a series of oxides by measuring (titrimetrically) the amount of excess oxygen in the lattice. The catalytic activity of each oxide he recorded by measuring the percentage nitrous oxide produced in a given time during the oxidation of ammonia over that oxide. His results are summarized in Fig. 25. It can be seen

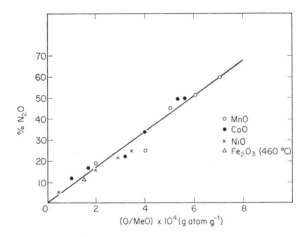

FIG. 25. The catalytic activity of a series of metallic oxides MeO for the catalytic oxidation of ammonia to nitrous oxide, as a function of the excess oxygen content of the oxides [102].

that there is a linear increase in catalytic activity with increased concentration of excess oxygen in the lattice.

More recently, a thorough study has been made by Hart and Ross [103] of the relative efficacy of various oxides as catalysts in the decomposition of hydrogen peroxide vapour. The activity sequence was found to be: $Mn_2O_3 > PbO > Ag_2O > CoO > Cu_2O > Fe_2O_3 > CdO \gg ZnO = MgO > \alpha\text{-}Al_2O_3$. The sequence resembles that which obtains for the decomposition of nitrous oxide (see Fig. 22) in that the p-type oxides are again the most active, but the fact that, here, the n-type oxides turn out to be more active than the insulators, emphasizes the extent of our ignorance of the crucial link between the catalytic activity and the defect structure of a solid.

5.3.2.3. *Current Trends in the Assessment of the Catalytic Activity of Semiconducting Oxides*

There are several reasons why we should not rest content with correlations such as those discussed in the previous subsection and summarized by Figs. 22–25. In the first place, evidence has come to light [104, 105] which strongly suggests that the precise degree of semiconductivity of a metal oxide may differ considerably in the bulk phase from that which prevails at the surface: the correlations so far noted may, therefore, represent nothing more than fortunate concomitances. Secondly, it is obviously inadequate merely to seek a correlation between catalytic activity and such properties as conductance, diffusion, etc. Much remains to be learned about the nature of the defects themselves. For example [106], questions such as the extent of localization of the holes and electrons, and the tendency of individual point defects to interact with one another need to be investigated more thoroughly. (One can only hope that refined magnetic susceptibility, nuclear magnetic resonance and electron spin resonance measurements (see Chapter 3) will help to clarify such issues.) The third difficulty with the traditional approach to the study of semiconducting oxide catalysts is that, as Haber and Stone [107] have stressed, too much attention has been paid to the *number* or *concentration* of positive holes and other point defects in the lattice. Many of the apparent anomalies which arise in catalytic studies involving oxides—including the differences observed by Schwab and by Parravano (see previous Section)—might have been resolved if greater attention had been paid to the *mobility* of defects and less to their concentration (or to statements concerning the height of the Fermi level). Lastly, the question of the extent of *d*-orbital overlap, or exchange-interaction between nearest-neighbour (or next nearest-neighbour) cations in catalytic-oxide semiconductors has been tacitly ignored. Since this constitutes one of the most significant omissions in the assessment of the catalytic activity of metal oxides, the current position will now be summarized.

The phenomenon of nearest-neighbour interaction (for further details consult the text by Goodenough [108]) is exhibited by transition metal oxides such as NiO, V_2O_3, Cr_2O_3, etc. If a diamagnetic ion (such as Mg^{2+}) is added to the lattice of a transition metal oxide, the exchange interaction is disrupted and ultimately, at high magnetic dilution, disappears almost completely. The effect has been strikingly demonstrated [108] for the systems: CoO–MgO; Cr_2O_3–Al_2O_3 and UO_2–ThO_2. Evidence concerning the precise effect of diluting (i.e. doping) NiO with Li_2O is not indisputable [109], but there is no doubt that such doping greatly modifies the exchange interaction between adjacent nickel ions. In the words of Vrieland and Selwood [110], who have very recently attempted to disentangle the complexities of the situation: "similar effects (modifications to exchange interaction) have been observed, or may certainly be expected, in almost all transition metal oxides doped (or 'promoted' in catalytic terminology) with foreign atoms or ions. It appears, therefore, that in virtually all the work tending to relate conductivity to catalysis in transition metal oxides there is at least one (the exchange effect), and possibly other, parameters which change simultaneously with the conductivity whenever a doping agent is added. In the absence of a truly convincing theoretical basis for the relationship, it is impossible to say that catalytic activity is more closely related to the one parameter than to another."

In the light of the situation adumbrated above, Vrieland and Selwood [110] decided to examine, as far as was possible, the separate influence of the semiconductivity and the exchange effect. This involved selecting a system in which semiconductivity might be made to change in a manner which was essentially independent of exchange effects, as measured by the magnetic susceptibility. It also involved finding a system in which exchange effects could be altered significantly without extensive modification of the conductivity. (An attempt to investigate the effects of exchange interaction upon catalytic activity had previously been made by De and Stone [111]: their results were not, however, unequivocal [110].)

Vrieland and Selwood [110] recognized that the rare-earth oxides Eu_2O_3 and Gd_2O_3 are about as nearly identical as two catalytic solids can be, except for the difference in their semiconductivity. Eu_2O_3 differs from Gd_2O_3 in that the former has one less electron in the $4f$-shell, and the consequent possibility of reduction to the $2+$ oxidation state. Owing to the shielding of the outer electrons, exchange effects are minimal in both these oxides. Both are n-type oxides, but the semiconductivity of Eu_2O_3 is greater by a factor of 19 at 800°C than that of Gd_2O_3. Since it transpired [110] that the rate of decomposition of ammonia and the activation energy for this decomposition are almost identical for each of the oxides, the important conclusion can be drawn that there is here no correlation between semiconductivity and catalytic activity.

In their search for a system in which conductivity effects are minimal but exchange effects are pronounced, Vrieland and Selwood [110] found the MgO–MnO system to be most appropriate. When MgO is incorporated into MnO the degree of exchange interaction between Mn^{2+} ions is appreciably diminished [112], but the electrical conductivity remains essentially unaltered throughout the series of solutions from pure MnO to pure MgO. Moreover, there is very little change of lattice spacing as the composition of the solid solution is varied. Again, it transpired [110] that the rate of decomposition of ammonia, and the activation energy, were not perceptibly altered by changes in the composition of the MnO–MgO catalyst system. A further far-reaching conclusion may, therefore, be drawn: that there is no correlation between degree of exchange interaction and catalytic activity. There is little doubt that, in the near future, many experiments will be designed to discover whether the startling results of Vrieland and Selwood are universally valid.

Before terminating this section on current trends in assessing the catalytic activity of semiconductors, mention ought to be made of the increasing tendency to consider adsorption on, and catalysis by, semiconductors in terms of coordination at surface cations. Crystal field theory has been invoked, and the interpretations appear cogent, if occasionally confusing. We shall make use of crystal field theory in Section 5.4.1 below, where the phenomenon of photodesorption is considered.

5.3.3. The views of Wolkenstein [113, 114]

An elegant theory designed to account for observed adsorption and catalytic phenomena on semiconductors has been formulated by Wolkenstein. The theory is firmly grafted on the band theory of solids, and its essence can best be judged by reciting some of the tenets and interpretations which Wolken-stein has himself made [113].

A cardinal feature of his theory is that adsorbed species may be chemi-sorbed to the surface of a semiconductor in essentially three ways. "Weak" chemisorption involves an adsorbed species which remains electrically neutral, and the free electrons or holes of the lattice do not contribute to the chemi-sorption bond. In "strong" chemisorption the chemisorbed particle captures a free electron or a free hole of the crystal lattice—thus representing an electrically charged system—and the free electron or free hole participates in the chemisorption bond. Two "strong" bonds may be distinguished, depend-ing upon whether a free electron or a free hole is involved in the bonding. The "strong" n-bond (not to be mistaken for adsorption on an n-type conductor) or "acceptor bond", is formed when a free electron is captured by an adsorbed particle. Such a bond is denoted by CeL, where the symbol eL denotes a free electron of the lattice. The "strong" p-bond (again, quite unrelated to p-type conductors) or "donor bond" is formed if a hole is captured by the adsorbed species. It is often denoted by CpL where pL is the symbol for the free hole. The various types of chemisorption are believed to differ not only in character and strength but also in reactivity; one form may change into another, under certain conditions of temperature, pressure, or as a result of an external agency (e.g. irradiation, addition of impurities, etc.).

Wolkenstein conceives the electrons and holes to "perform the functions of free valencies capable of breaking the valence bonds in the chemisorbed particles and themselves becoming saturated by these bonds" [113]. More-over, by invoking Fermi statistics [115], he can compute the fraction of the total number of species adsorbed that is held weakly, or strongly (acceptor), or strongly (donor), in terms of the distances from the Fermi level to the conduction band and to the valence band.

The three different ways in which Wolkenstein pictures a sodium atom to be adsorbed on a semiconductor is illustrated by Fig. 26. In "weak" bonding, the valence electron of the sodium atom remains unpaired (a), and, in this sense, the valence of the metal atom may be considered unsaturated. A "strong" n-type or acceptor bond is seen in (b) and a "strong" p-type one in (c).

Wolkenstein points out that, since the Fermi level plays a dominant role in this theory, it is understandable why the catalytic activity and adsorbability of semiconductors possessing widely different Fermi levels may be so different. The Fermi level, since it determines the magnitude and sign of the surface charge for a given coverage of chemisorbed particles, the total number of chemisorbed particles, and the reactivity of these particles, acts as a regulator of the chemisorptive and catalytic properties of the surface.

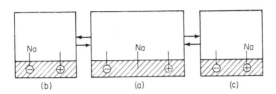

FIG. 26. Three ways in which, according to Wolkenstein [113], a sodium atom may be adsorbed by a semiconductor.

Without going into elaborate detail, we may illustrate the heuristic value of Wolkenstein's theory by citing one example of the way in which it can account for the radical mechanism of some heterogeneous reactions [116]. Any heterogeneous reaction may, for convenience, be regarded as proceeding by a radical mechanism—a statement which must not be taken to imply that non-radical mechanisms are precluded in heterogeneous catalysis. Consider a reaction between two molecules AB and CD, where A, B, C, and D denote separate atoms or atomic groups. If A and B, as well as C and D, are connected by single bonds, then the exchange reaction:

$$AB + CD \rightleftharpoons AC + BD$$

involves the rupture of two bonds and the formation of two others. An example would be the chlorination of ethane, which is catalysed by, e.g., solid $ZnCl_2$

$$C_2H_6 + Cl_2 \rightleftharpoons C_2H_5Cl + HCl$$

Figure 27 shows a possible radical mechanism of the reaction which proceeds by way of the dissociation of the two reacting molecules on a free valence of the surface. This is really an example of a chain mechanism, the chain being sustained by the free valencies of the catalyst. Using this type of scheme, it is a simple matter to construct a "mechanism" for the hydrogenation of ethylene (over MoO_3, say) or for the addition of hydrogen halides to olefins. Taken a stage further, depending upon whether a catalysed reaction proceeds faster the higher or lower is the Fermi level, Wolkenstein proceeds to classify *reactions* into acceptor or donor types. An acceptor reaction proceeds faster the higher the Fermi level: such reactions are accelerated by electrons. A donor reaction, on the other hand, is one which proceeds faster the lower the Fermi level: these reactions are accelerated by positive holes. The dehydrogenation of ethanol is an acceptor reaction, whereas the dehydration of the same substance is a donor reaction.

It is very difficult to find real fault with Wolkenstein's views, because they are so general and logical. However, one of the weaknesses of the theory is that, quantitatively, it contains too many adjustable parameters, and it is

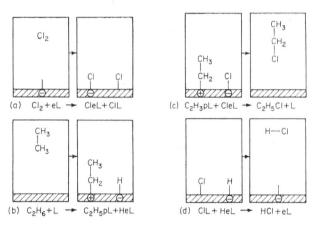

FIG. 27. Wolkenstein's radical mechanism [*113*] for the catalysed reaction: C_2H_6 + + Cl_2 ⇌ C_2H_5Cl + HCl.

not clear in what way the presence of surface states affects determinations of the positions of the Fermi level [*117*]. Another feature which, quite naturally, is lacking is reference to the role of local coordination (of orbitals) between atoms in the catalyst surface and the adsorbed particles.

5.3.4. Lattice imperfections and catalysis of polymerization

Both dislocations and point defects have recently been postulated to play an important role in several polymerization reactions. So far as the part played by dislocations is concerned, the evidence is largely indirect. But the mode of participation of point defects in Ziegler-Natta catalysis (see below) has been convincingly demonstrated. We shall first consider the role of dislocations.

5.3.4.1. *Importance of Dislocations in the Polymerization of Vinyl Monomers* [*118*]

Bamford and his co-workers [*118, 119*] observed that the application of mechanical stress to crystals of acrylic and methacrylic acids exerted a profound influence upon the kinetics of the polymerization of these substances by ultraviolet irradiation. In general, the polymerization reaction was retarded or stopped if the monomer crystals were subjected to small mechanical compressive stresses. For example, a stress equal to about 7 atm applied before irradiation (and maintained) suppressed the photopolymerization of acrylic acid crystals at 4°C almost completely. On removing the stress, the reaction started again after an induction period similar to that normally encountered at the beginning of the polymerization. Both the initiation and the propagation reactions are probably stress dependent.

Although the nature of the dislocations in acrylic and methacrylic acids is, as yet, completely unknown, Bamford et al. [119] have argued cogently in favour of interpreting the stress-dependence of the polymerization in terms of the dislocations which are probably present in these solids. The suggestion is that polymerization occurs predominantly along dislocation lines— just as the CO_2 is liberated preferentially along dislocation lines in the thermal decomposition of calcite [120]—and that, on application of stress, the dislocations are moved away from the growing radicals, so that reaction ceases. It is not yet known whether it is a "pure" dislocation, or one possessing a Cottrell cloud [12] (say of water) that functions as the site of polymerization. A further difficulty with the hypothesis is that it is not easy to visualize why reaction should not be initiated in a dislocation in its new position. Obviously, a systematic study of the role of dislocations in vinyl monomers—their origin, movement, annihilation and slip planes—requires to be performed.

5.3.4.2. *Importance of Point Defects in Stereoregular Addition Polymerization* [121–125]

There is little doubt that one of the most exciting recent developments in heterogeneous catalysis started with the discovery, by Ziegler et al. [126], that ethylene can be polymerized at low pressure under the influence of a mixture (suspended in a hydrocarbon solvent) of transition metal compound and a metal alkyl derived from a strongly electropositive metal. Natta and his collaborators [127, 128] extended the possibilities of the so-called Ziegler catalysts by showing that three kinds (see Fig. 28) of polymeric α-olefins (alk-1-enes), which have different steric arrangements of the carbon backbone chain, can be produced if the Ziegler catalysts are suitably modified. The discovery of Ziegler-Natta catalysts invites us to speculate on the nature of the active centre—as does the discovery of any new catalyst—but, in addition, it focuses our attention on the mechanism for the production of stereospecific polymers. This is altogether a different kind of challenge.

We must first become acquainted with the notion of, and the language for describing, stereoisomerism in polymer molecules [122]. When a mono-substituted ethylene (CH_2—CHR) or 1,1-disubstituted ethylene (CH_2—CRR′) undergoes addition polymerization according to the reaction:

$$n(CH_2-CRR') \rightarrow -CH_2CRR'[CH_2CRR']_{n-2}CH_2CRR'-$$

where R′ may be H, and R′ ≠ R, then every alternate carbon atom in the polymer chain will be an asymmetric centre because, in effect, the two poly-meric substituent groups will be of different chain lengths. These carbon atoms can consequently display two different configurations which we may, for convenience, designate as the D- and L-configurations. Figure 28 shows the three possible stereochemical arrangements of the substituents in the polymer

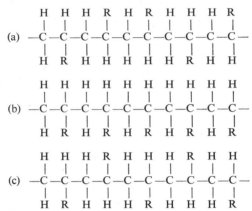

FIG. 28. Three possible stereochemical arrangements of the substituents in a polymer molecule: (a) atactic; (b) isotactic; and (c) syndiotactic.

molecule. The zigzag carbon backbone is projected on to the plane of the paper, there being alternate CH_2 and CHR groups. If the substituent R is below the chain (in the D-configuration say), then we have the following possibilities. (i) The arrangement of the D- and L-configurations is random, in which case we refer to the structure as being atactic. (ii) The arrangement is such that the asymmetric centres are either all D- or all L-; such a structure is isotactic. If (iii) the substituents alternate regularly above and below the planar zigzag backbone, i.e. the asymmetric centres are alternately D- and L-, it is a syndiotactic structure. Ziegler-Natta catalysts are remarkable in that they lead to the production of isotactic polymers. In practice, however, 100% isotacticity is never obtained, although very high degrees of stereoregularity are possible.

Rather than attempt a comprehensive appraisal of the various types of Ziegler-Natta catalysts that have been utilized to produce stereoregular poly-mers of several α-olefins and dienes, we shall, instead, concentrate upon one typical system, the production of isotactic polypropylene, using a catalyst system consisting of α-$TiCl_3$ and $Al(C_2H_5)_3$. Although these constituents are mixed in the liquid phase, it has been established beyond dispute that iso-tactic polymers require, for their production, a solid catalyst derived from these substances [121, 129]. Any theory which seeks to account for the facts of stereoregularity in heterogeneous Ziegler-Natta catalysis must account for the fact that when the different crystalline modifications of $TiCl_3$ (α, β and γ) are used in the catalysis, the highest percentage isotacticity is obtained when well-crystallized α-$TiCl_3$ is present. The theory of Cossee is firmly based on the notion of crystal (point) defects and explains, in a logical manner, the enhanced activity of α-$TiCl_3$. This theory is a particularly elegant one, as it incorporates the essentials of ligand-field theory and lattice imperfections in a way that may well be beneficial for heterogeneous catalysis in general.

5.3.4.3. *Cossee's Model of the Mechanism of Stereoregular Polymerization*
Unlike almost all other theories that have been formulated (see Ref. *123* for fuller details) to account for stereoregular polymerization, Cossee's theory postulates (see below) that the active centre is an essentially octahedrally coordinated ion of a transition element (in the special system considered by us, a titanium ion), with empty or at least partly empty t_{2g} orbitals (i.e. d_{xy}, d_{yz} or d_{zx}, see Fig. 37), carrying in its coordination sphere one alkyl group and having one vacant octahedral position. The concept of π-bonding between olefins and transition elements [*130*] and the crystallographic likelihood of vacancies (see Section 5.2.3) are an integral part of Cossee's interpretation.

Figure 29 illustrates the spatial arrangement that Cossee envisages in a "π-bond" (of the type first mooted by Chatt [*130*]) formed between a transition metal and the simplest olefin, ethylene. The characteristic features of this π-bond are the position of the olefin molecule perpendicular to the free

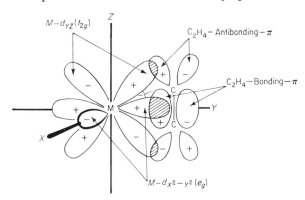

FIG. 29. Schematic picture showing spatial arrangement of the relevant orbitals in a π-bond between a transition metal and C_2H_4 [*121*].

valency of the metal ion, which allows the π-electrons of the olefin to be used in the σ-bond formation with an e_g orbital (i.e. with the $d_{x^2-y^2}$-orbital of the metal), and the possibility of overlap of the d_{xy}-orbital of the metal with the antibonding orbital of the olefin. In a manner of speaking, a kind of double bond is formed between the metal and the olefin, one of the bond components having σ- and the other π-symmetry. The results of both X-ray diffraction and infrared spectroscopy are compatible with this picture of the π-bond, there being little doubt that the C—C distance of the olefin is considerably increased (with consequential decrease in bond strength) in the π-complex [123]. On the basis of such information, it follows logically that the first step

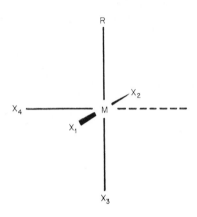

FIG. 30. Configuration supposed to be the active centre in a Ziegler-Natta catalyst. M is a transition metal ion; R is an alkyl group (a growing polymer chain); and X_1 to X_4 are anions [123].

in Ziegler-Natta catalysis probably involves the coordination of a monomer molecule to a transition metal compound via a π-complex.

Following on from the first step comes the propagation step in Ziegler-Natta catalysis. This must involve the interposition of an olefin molecule between a metal atom and an alkyl group, and many authors [123, 131] have pointed out that this requires a transition metal–alkyl bond and the possibility of coordinating the monomer to the transition element. Cossee suggests that the simplest configuration meeting these requirements is shown in Fig. 30, which represents an essentially octahedrally coordinated transition metal ion of which one of the octahedral positions is vacant while another is filled by an alkyl group. Reverting to our original consideration of the α-TiCl$_3$–Al(C$_2$H$_5$)$_3$ system, X_1 to X_4 in Fig. 30 are Cl-ions, and R is C$_2$H$_5$.

In the special case of propene polymerization using the above catalyst, the propagation reaction proceeds according to the scheme at the top of the opposite page. This mechanism entails only small amounts of nuclear displacements, and it offers a satisfactory explanation for the driving force of the reaction, because, on theoretical grounds [132], one may expect the transition metal-to-carbon bond to become more susceptible to radical breaking at the moment the π-bond between metal ion and olefin is formed. Moreover, the mechanism relates catalytic activity to the size of the t_{2g} orbitals of the transition metal, for it is clear that, in order to exhibit the desired catalytic effect, the empty d-orbitals of the metals must be large enough to overlap sufficiently with the antibonding orbitals of the olefin. Therefore, only ions with a comparatively

low effective nuclear charge are expected to be good catalysts. (This is essentially equivalent to Natta's observation [133] that metals with a first ionization potential smaller than 7 eV are particularly suitable in Ziegler-Natta catalysis.)

To understand why the catalysis turns out to be stereoregular, it is necessary to consider, in detail, the situation likely to exist at the surface of α-TiCl₃. Originally, Cossee [121] pictured the active centre to be situated at the basal face of the solid, i.e. at the {000l} planes. He formulated a cogent explanation for the occurrence of isotacticity on this basis. Such an explanation cannot, however, be entirely satisfying, despite its heuristic value, because it has recently been firmly established [124a, b] that catalysis of the polymerization may take place at all but the {000l} planes of the α-TiCl₃. We shall, therefore, discuss the more recent formulations of Cossee and Arlman [123–125].

Arlman [124], working from general principles of inorganic crystal chemistry, calculated that the chances of there being surface chlorine vacancies □

would be very small at basal faces, but quite appreciable at prismatic {10$\bar{1}$0} faces; a fact which, in conjunction with the observed crystallographic preference for catalysis at {10$\bar{1}$0}, supports the original picture for the active centre. It is instructive to reiterate the explanation of how the active centre shown in Fig. 30 is generated from a surface site of the solid TiCl₃. The process of generation consists of the alkylation of the pentacoordinated Ti ions according to the reaction scheme shown below. Arlman and Cossee [125] consider this reaction to represent the essential role of the trialkyl aluminium in these catalyst systems, and Arlman [124] has himself demonstrated the feasibility of such reactions by experiments using radioactively labelled

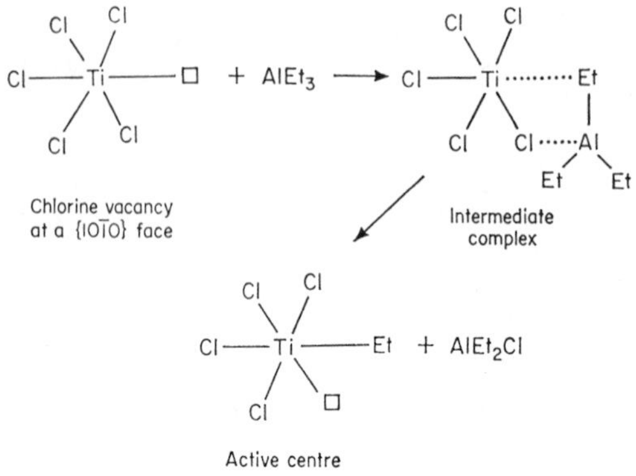

$$
\begin{array}{ccc}
\text{Cl} \diagdown \diagup \text{Cl} & & \text{Cl} \diagdown \diagup \text{Cl} \\
\text{Cl——Ti———□} + \text{AlEt}_3 & \longrightarrow & \text{Cl——Ti} \cdots\cdots \text{Et} \\
\diagup \diagdown & & \diagup \diagdown \quad | \\
\text{Cl} \quad \text{Cl} & & \text{Cl} \quad \text{Cl} \cdots\cdots \text{Al} \\
& & \diagup \diagdown \\
& & \text{Et} \quad \text{Et}
\end{array}
$$

Chlorine vacancy   Intermediate
at a {10$\bar{1}$0} face   complex

$$
\begin{array}{c}
\text{Cl} \diagdown \diagup \text{Cl} \\
\text{Cl——Ti———Et} + \text{AlEt}_2\text{Cl} \\
\diagup \diagdown \\
\text{Cl} \quad \square
\end{array}
$$

Active centre

molecules. It is to be noted that the crystallographic location where the alkylation takes place will be that where chlorine vacancies are already present.

Cossee and Arlman [123–125] have argued that, of the five chlorine atoms situated around a surface titanium ion (at {10$\bar{1}$0} faces), one is always loosely bound. It is highly probable that this loosely bound chlorine is the atom removed by the alkylation process. Moreover, as the remaining four chlorine atoms are firmly bound, only the sites of the loosely bound chlorine atom and of the vacancy are available for the alkyl group and the monomer participating in the propagation step. These two sites are not, however, equivalent, a fact which is portrayed in the models constituting Fig. 31 (a) and (b). When the alkyl group is in one of these two alternative sites, only four of its twelve surrounding sites are occupied, whereas in the other site seven of the surrounding sites are occupied. If the plane of the crystal surface is drawn through ions 1, 2 and 3, the alkyl group in Fig. 31 (a) lies outside the plane, but in Fig. 31 (b) it lies in the plane.

Thus, according to the views of Cossee and Arlman, the essential feature of the active centre is that it consists of a square base having three chlorine atoms and the alkyl group at its corners and the titanium atom in the middle (refer to Fig. 30 for clarity). The square is itself anchored by the fourth chlorine to the inside of the crystal and it has a chlorine vacancy at its outside. The plane of the square base always includes an angle of about 55 degrees with the {000l} planes (see Ref. 124a) and is usually in an oblique position towards the other crystal planes. Consequently, two of the chlorine atoms of the

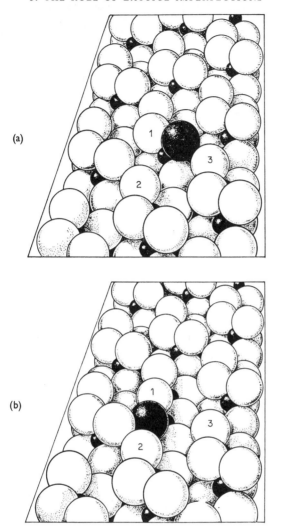

(a)

(b)

FIG. 31. The alkyl group (black sphere) has four of its twelve surrounding sites occupied in the configuration shown in (a), whereas seven of the sites are occupied in (b) [125].

square base lie more or less blocked, whereas one is, to a certain extent, exposed.

We now return to the stereochemistry of the propagation reaction which proceeds at the active centre according to the scheme drawn on p. 283. Arlman and Cossee [125] postulate that the propagation reaction can proceed only provided that the C=C double bond is parallel to the bond adjoining the titanium atom and the alkyl group. This gives four alternatives for the position of the propene molecule. Two of the alternatives may be immediately discounted for steric reasons—they would require space, which is not available, for the bulky methyl group over the blocked chlorine atoms.

Upon examining the reactive position of a propene molecule on an active centre (Fig. 32), it may be seen that on the square base around an exposed titanium ion the plane perpendicular to the square and passing through the titanium–carbon bond is not a plane of symmetry (the exposed and the corresponding blocked chlorine atoms have different surroundings). Hence, for the reacting propene molecule, the two remaining orientations are not equivalent. In one of the orientations the methyl group would tend to cover a blocked chlorine atom. This will, however, be practically impossible because of the considerable steric hindrance. In the other orientation the methyl group of the propene will protrude out of the crystal: this will therefore be the preferred one (see Fig. 32).

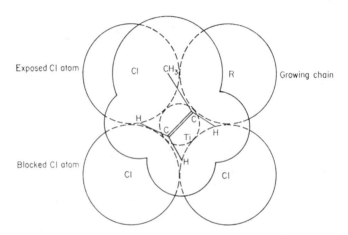

Exposed Cl atom

Blocked Cl atom

Growing chain

F IG. 32. The reactive position of a propene molecule on an active centre. The propene is projected perpendicularly on to the TiCl₃R plane [*125*].

The requirement necessary, but not sufficient, for isotactic polymerization —the fixation of a monomer at an active centre exclusively in one orientation—is fulfilled by the mechanism proposed by Arlman and Cossee. The second, and sufficient, requirement is that the process must consist of identical steps, and this occurs [*125*] through the fact that one of the two non-equivalent sites shown in Fig. 31 is a favoured position. After the incorporation of the monomer, the alkyl group and the chlorine vacancy have changed sites (see alkylation scheme), but, after one propagation step, the alkyl group will move back to the favoured position so that the process may be continued with a step identical to the previous one. (The fact that [128] the rate of polymeriza-tion of the monomer is directly proportional to the partial pressure of the monomer indicates that, on the average, the active sites are unoccupied. There is consequently ample time for the alkyl group to move to the favoured octahedral position between two successive growth steps.)

The great merit of the Cossee-Arlman interpretation of isotactic poly-merization is that it unifies several approaches in a convincing and logical manner. The principles of quantum chemistry as well as those of crystal chemistry are utilized.

5.4. The Influence of Radiation on Catalytic Activity

There have been numerous recent publications on the effects of various types of corpuscular and electromagnetic radiation upon the catalytic activity of metals, semiconductors and insulators. There is a growing interest in this branch of research because it is felt that radiation catalysis [135–137] will contribute—largely through focusing attention on the interplay between chemical reactivity and crystalline imperfections—to a better understanding of heterogeneous catalysis in general. Although the number of experimental studies conducted to date is already very large, it is not yet possible to estab-lish an all-embracing theory of the influence of radiation upon catalysis.

This section is intended briefly to review some experimental results of, and tentative theoretical assessments made in, radiation catalysis. We shall begin by considering some recent work dealing with the consequences of ordinary electromagnetic radiation.

5.4.1. Photocatalysis and photodesorption with visible or ultraviolet radiation

Many solids that yield an enhanced number of electron–hole pairs at their surfaces when illuminated by radiation in the visible or ultraviolet regions of the spectrum are known to display photocatalysis. Thus, zinc oxide, under these conditions, catalyses reactions such as the formation of carbon monoxide [138] or hydrogen peroxide [139]. It is also known that some solids that function as good catalysts in the absence of electromagnetic radiation are more liable to desorb certain species from their surfaces when exposed to light of a particular wavelength. Nickel oxide, studied recently by Haber and Stone [140] and by Jongepier and Schuit [141] is in this category. It is because the interpretation of chemisorption on, and subsequent photodesorption from, NiO surfaces seem to be explicable in terms of crystal field theory that we now examine this topic (see Section 5.3.2.3).

The crystal-field theory, like its more sophisticated progeny, the ligand-field theory, sets out to explain the origins and consequences of the splitting (spectroscopically) of inner orbitals of ions by their surroundings in chemical compounds [*142*]. The crystal-field theory, using as it does the point-charge approximation and the assumption that all bonds are effectively totally electrovalent, is known to offer no more than a semi-quantitative interpretation of various chemical phenomena. Nevertheless, Dowden and Wells [*143*], by utilizing the theory and incorporating some further approximations, showed that it provided a valuable interpretation of chemisorptive and catalytic phenomena (see also Chapter 6). Following the lead given by Dowden and Wells, Haber and Stone [*140*] set out to estimate the crystal-field stabilization (CFS) afforded by the chemisorption of oxygen on the various crystallographic faces of a nickel-oxide crystal.

Consider, in turn, the situation that prevails during the chemisorption of oxygen ions (as O^{2-}) on the three principal families of planes {100}, {110} and {111}. A {100} face is shown in Fig. 33(a). The middle one of the three diagrams in Fig. 33(a) represents a {100} face as it appears at the instant of cleavage. In the surface, a Ni^{2+} ion has five nearest O^{2-} neighbours disposed in the shape of a square pyramid (shown in the diagram as heavy lines). Haber and Stone [*140*] point out that, after cleavage, the surface relaxes in such a way that the Ni^{2+} ion enters the pyramid, giving a state more akin to that depicted on the left of Fig. 33(a). (For simplicity, the process of relaxation has been illustrated so as to imply that the Ni^{2+} ions alone move: the O^{2-} will, in reality, also move.) The chemisorption of a single O^{2-} ion (shown shaded) can lead to the completion of the octahedron about the Ni^{2+} ion (Fig. 33(a), right) and a reversion of the cation to the position which it occupied in the cleavage surface. Haber and Stone point out [*140*] that the change in CFS accompanying chemisorption is going to be approximated by the change from square-pyramidal stabilization (which, for transition elements with a d^8-configuration is given by Dowden and Wells [*143*] as $-10\ Dq$) to octahedral stabilization ($-12\ Dq$ for d^8).† The gain in CFS energy is thus approximately $2\ Dq$ per atom adsorbed on a {100} type plane. The value of Dq for an octahedral complex around a central Ni^{2+} ion is 910 cm⁻¹, which corresponds to 2·6 kcal mole⁻¹ [*140*]. Figure 33(b) and (c) illustrate the analogous configurational changes for {110} and {111} planes. On the {110} planes the relaxed position approximates to tetrahedral symmetry for the Ni^{2+} ion, and adsorption of two O^{2-} ions now leads to full octahedral stabilization. The gain in CFS energy for this change is $8·4\ Dq$. (It is interesting to observe that Ni^{2+} ions situated at the top edge of a monatomic step on a {100} face (see Fig. 35) would behave, so far as CFS on adsorption goes, in exactly the same way as a Ni^{2+} ion situated in a {110} plane.) On the {111} planes, a Ni^{2+} ion will relax towards the centre of the triangle of O^{2-} lying immediately

† It is to be noted that Ni^{2+} has a d^8-configuration. The magnitude Dq of the splitting is usually obtained from optical data [*144*]. Figure 34 illustrates the type of orbital splitting for an octahedral arrangement of ions.

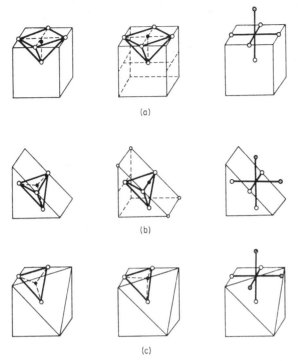

(a)

(b)

(c)

FIG. 33. Changes in nickel ion co-ordination during the chemisorption of oxygen on nickel oxide: (a) {100} plane; (b) {110} plane; (c) {111} plane. Diagrams in the middle column refer to a bare surface at the instant of cleavage. Left column, bare surface, relaxed position; right column, after adsorption of oxygen: ●, nickel ion; ○, lattice oxygen ions; ⊘, adsorbed oxygen ions [140].

beneath it (trigonal symmetry). Adsorption of three O^{2-} ions now provides octahedral symmetry, and the gain in CFS energy is 1·1 Dq. (Again, it is instructive to note that a Ni^{2+} ion situated at the corner of a kink in a mona-tomic step on a {100} face would experience essentially the same type of CFS, following adsorption, as a Ni^{2+} ion situated in a {111} plane.) The CFS energies accompanying the various changes considered are summarized in Table 5.2, from which it can be seen that the {110} planes (or sites at monato-mic steps on {100} planes) provide the opportunity for the greatest stabiliza-tion (8·4 Dq or approximately 20 kcal mole^{-1}) during the chemisorption of oxygen. One of the instructive deductions we can make from all this is that, since adsorption on {110} planes is particularly strong, when the *overall* coverage of NiO with adsorbed oxygen is low, we may expect the {110} planes to be preferentially covered at the expense of the {100} and {111} planes. We have here a theoretical foundation for expecting crystallographic in-equality.

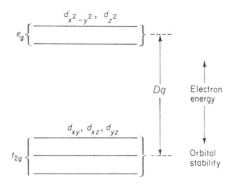

FIG. 34. Nature of orbital splitting for an octahedral arrangement around a central ion.

Haber and Stone [*140*] report that illumination of a nearly stoichiometric nickel oxide carrying adsorbed oxygen gives rise to oxygen desorption; and the region of the visible spectrum reported to be the most active is the wavelength range 650–900 mμ. By using an Orgel diagram [*145*], which refers to the various energy levels associated with octahedral and tetrahedral fields, Haber and Stone suggest a mechanism for the oxygen photodesorption in terms of a transition from the octahedrally coordinated nickel in its ground state $^3A_{2g}$ to the excited state $^3T_{1g}$. This transition has its spectral absorption centred around 650 mμ. They argue that the excited state $^3T_{1g}$, which corresponds to an activated complex, is unstable with respect to the ground state of a tetrahedrally coordinated Ni^{2+} ion. In other words, a surface Ni^{2+} ion which had previously attained octahedral coordination as the result of the

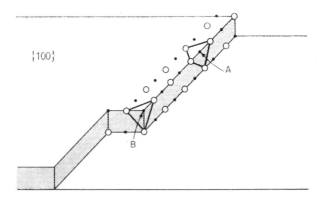

FIG. 35. Diagrammatic representation of the roughly tetrahedral disposition of O^{2-} ions around a Ni^{2+} ion (A) in a monatomic step on a {100} type face, and the roughly trigonal disposition around the cation (B) at the kink site.

TABLE 5.2. Changes in crystal-field stabilization energy (CFSE) during the chemisorption of oxygen on nickel oxide [140]

Plane	Initial field	CFSE	Atoms adsorbed	Final field	CFSE	Gain in CFSE	Gain per atom adsorbed
{100}	Square pyramid	$-10\,Dq$	1	Octahedral	$-12\,Dq$	$2\,Dq$	$2\,Dq$
{110}	Tetrahedral	$-3{\cdot}6\,Dq$	2	Octahedral	$-12\,Dq$	$8{\cdot}4\,Dq$	$4{\cdot}2\,Dq$
{111}	Trigonal	$-10{\cdot}9\,Dq$	3	Octahedral	$-12\,Dq$	$1{\cdot}1\,Dq$	$0{\cdot}4\,Dq$

$Dq = 2{\cdot}6$ kcal mole^{-1}.

chemisorption of two oxygen atoms can readily lose the two atoms, as an O_2 molecule, and return to the tetrahedral coordination when irradiated with the wavelength corresponding to $^3A_{2g} \rightarrow {}^3T_{1g}$ transitions.

This model is most plausible, and is indeed likely to act as an inspiration for interpretation of future work. At present, however, it is not without its difficulties. Jongepier and Schuit [141] have repeated the work of Haber and Stone and they report satisfactory agreement with the previous investigation so far as experimental details are concerned. Their interpretation is, however, at variance with that of Haber and Stone: the most serious criticism being that no allowance has been made for charge transfer and valence changes of the nickel ion upon desorption of (neutral) oxygen. Moreover, according to Jongepier and Schuit, it is not possible to have a continuous transition from O_h to T_d symmetry during which the $^3T_{1g}$ wave function conserves its symmetry; they conclude that a more comprehensive theoretical model, based on a molecular orbital treatment, is required before the full details of photodesorption and, presumably, photocatalysis can be interpreted.

5.4.2. General comments concerning the influence of high-energy radiation on catalytic processes

It has been emphasized by Coekelbergs et al. [135] that, in studying the influence of radiation upon catalytic processes, it is convenient to envisage two distinct situations. The first considers preliminary irradiation of the catalyst in the absence of reactants, and the second the catalyst and the reactants when they are simultaneously irradiated. These distinctions are useful, irrespective of whether the radiation is effected using α-, β- or γ-rays, high energy electrons and X-rays, proton and deuteron beams, or thermal and fast neutrons.

In the first situation, the preliminary irradiation activates the catalyst by causing radiation damage (and consequential electronic modification) to the lattice. This damage is often described as being of quasi-permanent character. As a result of such damage, new catalytic properties may be imparted to the irradiated solid; but such disturbance does not cause any modification of the thermodynamic laws applicable to the reactant gases, even though reaction rates, mechanisms and products may be altered profoundly.

When, however, the catalyst and reactants are simultaneously irradiated, imperfections (both structural and "electronic") of a more transient nature assume considerable significance. Three types of phenomenon may be distinguished, depending upon the characteristics of the heterogeneous system under consideration [135, 146].

(a) The radiation damage wrought in the solid serves merely to create a new catalyst, the properties of which are determined by the nature and stationary concentration of the imperfections with transient character. In effect, here we are concerned essentially with the question of catalyst activation.

(b) A significant fraction of the energy absorbed by the solid may be transferred to the gaseous species by three possible mechanisms: (i) by electronic excitation, (ii) through the agency of so-called thermal spikes (thermally disturbed tracks in the lattice), and (iii) by selective photon absorption. All this is equivalent to saying that the reaction is radiation induced. And in much the same way as it is possible, via "hot" atoms, for irradiation to stimulate reactions which are thermodynamically unfavourable in homogeneous phases, so also in heterogeneous systems can irradiation give rise to thermodynamically "unexpected" products.

(c) The radiation-induced reactions may be initiated exclusively in the gas phase. Here, the solid may influence only the evolution of excited species (ions or radicals), which were first produced in the gas phase by the radiation. For example, the adsorption of excited species on the solid surface may lead to the formation of new activated complexes.

Summarizing, the effect of radiation can usually be assigned either to *catalyst activation* on the one hand or to *energy transfer* on the other.

5.4.3. Specific examples of radiation catalysis

It has been something of a surprise for radiation chemists to observe that, very often, a given catalyst may be altered in much the same way, and often to the same extent, when subjected to different kinds of pre-irradiation. Take, for example, the way in which a silica-gel catalyst for hydrogen–deuterium exchange is altered by pre-irradiation. Kohn and Taylor [147] found that the rate of exchange, which is very slow at room temperature over an ordinary silica-gel catalyst, is increased approximately a thousandfold when the catalyst is pre-irradiated either with neutrons or with γ-rays. Moreover, both types of irradiation succeed in increasing the catalytic activity of the solid at

—195°C by a factor of about 10^9—the largest change in catalytic activity, as a result of radiation, so far recorded.

But it must not be thought that irradiation always increases the catalytic activity of a solid. There are well-documented instances in which irradiation is shown to decrease the activity [148–150]. Increased activity following irradiation seems to be associated with an increased number of lattice imperfections—Frenkel and Schottky defects, cascades of such defects, a foreign atom produced by a nuclear reaction, excitons, ion–hole pairs (dissociated), thermal spikes, etc.—some or all of which may enhance the catalytic activity. Often, however, the activity is not significantly increased [151] even when such defects are produced, and the reason for this may be simply that the defects tend to be annealed out of the lattice at the temperatures at which the catalysis is being studied. Diminution in activity following irradiation is almost invariably the result of destruction of active centres by the radiation itself. Thus, Schwab et al. [148, 149] found that the catalytic activity of both copper and nickel in the hydrogenation of ethylene was virtually eliminated by prior irradiation with α-, β- or γ-radiation. Since the catalytic activity could be restored by hydrogen treatment, it was concluded that the active centres in the catalytic reactions consisted of surface metal atoms on to which hydrogen was adsorbed. Schwab and Konrad [150], in studying the influence of neutron bombardment on the catalytic activity of alumina surfaces in the conversion of para-hydrogen to ortho-hydrogen, and of ortho-deuterium to para-deuterium, found that here also preliminary irradiation diminished the catalytic activity. They concluded that the active sites on the alumina surface were those regions at which hydrogen had been adsorbed (yielding paramagnetic centres). (It is interesting to observe that Schwab and Konrad, in commenting on the increasing rate of conversion with decreasing temperature on un-irradiated surfaces, point out that endothermic adsorption, such as that described in Chapter 2, could be partly responsible for the negative temperature coefficient of the reaction.)

More seems to be known about the phenomenological aspects of radiation catalysis of the insulating and semiconducting oxides than those of metals [137, 152, 153], this being a consequence of the higher annealing temperatures of the oxides. Graham [153] has, however, succeeded in comparing the influence of β-irradiation, X-rays, deuteron-irradiation, and neutron bombardment upon the catalytic activity of a series of metal hydrogenation catalysts. Active centres appear to be both created and destroyed by the irradiation: the net effect is a small, but significant, improvement in the catalytic activity.

References

1. J. A. Hedvall and E. Gustafsson, *Z. anorg. allg. Chem.* **170,** 71 (1928).
2. J. A. Hedvall, E. Garping, N. Lindekrantz and L. Nelson, *Z. anorg. allg. Chem.* **197,** 399 (1931).
3. G. M. Schwab and L. Rudolph, *Z. phys. Chem.* **B12,** 427 (1937).
4. K. Morikawa, *J. ind. Chem. Soc. Japan* **41,** 694 (1938).
5. P. Günther, *Ergebn. tech. Roentgenk.* **4,** 100 (1934); *Chem. Abstr.* **30,** 4093 (1936).
6. C. G. Darwin, *Phil. Mag.* (6) **27,** 315, 675 (1914).
7. G. I. Taylor, *Proc. R. Soc.* **A145,** 362, 388 (1934).
8. E. Orowan, *Z. Phys.* **89,** 605 (1934).
9. M. Polanyi, *Z. Phys.* **89,** 660 (1934).
10. J. M. Burgers, *Proc. K. ned. Akad. Wet.* **42,** 293 (1939).
11. F. C. Frank, *Disc. Faraday Soc.* **5,** 48 (1949).
12. A. H. Cottrell, "Dislocations and Plastic Flow in Crystals". Clarendon Press, Oxford (1953).
13. P. Haasen and A. Seegar, *in* "Halbleiterprobleme" IV, ed. by W. Schottky, p. 68. Vieweg, Braunschweig (1958).
14. R. G. Rhodes, "Imperfections and Active Centres in Semiconductors", Ch. 3. Pergamon Press, Oxford (1964).
15. S. Amelinckx, "The Direct Observation of Dislocations". Academic Press, New York (1964).
16. A. L. G. Rees, "Chemistry of the Defect Solid State", Ch. 1. Methuen, London (1954).
17. J. Frenkel, *Z. Phys.* **35,** 652 (1926).
18. C. Wagner and W. Schottky, *Z. phys. Chem.* **B11,** 163 (1930).
19. S. Amelinckx and P. Delavignette, *in* "Direct Observation of Imperfections in Crystals", ed. by J. B. Newkirk and J. H. Wernick, p. 295. Wiley, New York (1962).
20. N. F. Mott and R. W. Gurney, "Electronic Processes in Ionic Crystals". Clarendon Press, Oxford (1948).
21. L. Mandelcorn (ed.), "Non-Stoichiometric Compounds". Wiley, New York (1964).
22. A. F. Wells, "Structural Inorganic Chemistry". Clarendon Press, Oxford (1950).
23. J. H. de Boer, *Recl Trav. chim. Pays Bas Belg.* **56,** 301 (1937).
24. R. W. Pohl, *Proc. phys. Soc.* **49,** 3 (1937).
25. T. J. Gray, *in* "Chemistry of Solid State", ed. by W. E. Garner, Ch. 5. Butterworths, London (1955).
26. E. J. W. Verwey, "Semiconducting Materials", p. 151. Butterworths, London (1951).
27. E. J. W. Verwey, P. W. Haayman and F. C. Romeijn, *Chem. Wkbl.* **44,** 705 (1948).
28. E. J. W. Verwey and E. L. Heilmann, *J. chem. Phys.* **15,** 174 (1947).
29. J. H. de Boer and E. J. W. Verwey, *Proc. phys. Soc.* **49,** 59 (1937).
30. J. Callaway, "Energy Band Theory". Academic Press, New York (1964).
31. N. B. Hannay, *in* "Semiconductors", Ch. 1. Reinhold, New York (1959).
32. J. Frenkel, *Phys. Rev.* **37,** 17 (1931).

33. F. S. Stone, *in* "Chemistry of the Solid State", ed. by W. E. Garner, Ch. 15. Butterworths, London (1955).
34. F. S. Stone, *in* "Chemisorption", ed. by W. E. Garner, p. 179. Butterworths, London (1957).
35. P. Aigrain and C. Dugas, *Z. Elektrochem.* **56,** 363 (1952).
36. K. Hauffe and H. J. Engell, *Z. Elektrochem.* **56,** 366 (1952).
37. P. B. Weisz, *J. chem. Phys.* **20,** 1483 (1952).
38. I. Tamm, *Phys. Z. Sowjet* **1,** 733 (1932).
39. R. H. Fowler, *Proc. R. Soc.* **A141,** 56 (1933).
40. E. T. Goodwin, *Proc. Camb. Phil. Soc.* **35,** 205 (1939).
41. J. Koutecky, *Phys. Rev.* **108,** 13 (1957).
42. E. Aerts, "Solid State Physics in Electronics and Telecommunications", p. 628. Academic Press, New York (1960).
43. W. Dekeyser, *in* "Reactivity of Solids", ed. by J. H. de Boer, p. 376. Elsevier, Amsterdam (1961).
44. G. Rienäcker, *Z. Electrochem.* **46,** 369 (1940).
45. J. Eckell, *Z. Electrochem.* **39,** 433 (1933).
46. J. J. Gilman, *in "Progress in Ceramic Science"*, Vol. 1, p. 146. Pergamon Press, Oxford (1961).
47. L. E. Cratty and A. V. Granato, *J. chem. Phys.* **26,** 96 (1957).
48. H. E. Farnsworth and R. F. Woodcock, *Adv. Catalysis* **9,** 123 (1957).
49. H. M. C. Sosnovsky, *J. Phys. Chem. Solids* **10,** 304 (1959).
50. G. J. Ogilvie, *J. Phys. Chem. Solids* **10,** 222 (1959).
51. H. M. C. Sosnovsky, G. J. Ogilvie and E. Gillam, *Nature, Lond.* **182,** 523 (1958).
52. H. M. C. Sosnovsky, *J. chem. Phys.* **23,** 1486 (1955).
53. F. Seitz, *Adv. Phys.* **1,** 43 (1952).
54. F. H. Constable, *Proc. R. Soc.* **A108,** 355 (1923).
55. T. Kwan, *Adv. Catalysis* **6,** 67 (1954).
56. A. A. Balandin, *Z. phys. Chem.* **B19,** 451 (1932).
57. G. M. Schwab, *Trans. Faraday Soc.* **42,** 689 (1946).
58. G. M. Schwab, *Adv. Catalysis* **2,** 251 (1950).
59. E. Cremer, *Z. Phys. Chem.* **A144,** 231 (1929).
60. E. Cremer and E. Marschall, *Mh. Chem.* **82,** 840 (1951).
61. E. Cremer, *Adv. Catalysis* **7,** 75 (1955).
62. D. D. Eley and D. R. Rossington, *in* "Chemisorption", ed. by W. E. Garner, p. 137. Butterworths, London (1957).
63. J. M. Thomas and E. E. G. Hughes, *Carbon* **1,** 209 (1964).
64. J. D. Livingston, *in* "Direct Observation of Imperfections in Crystals" ed. by J. B. Newkirk and J. H. Wernick, p. 115. Wiley, New York.
65. J. M. Thomas, G. D. Renshaw and C. Roscoe, *Nature, Lond.* **203,** 72 (1964).
66. R. E. Keith and J. J. Gilman, *Acta metal.* **8,** 1 (1960).
67. I. Uhara, S. Yanagimoto, K. Tani and G. Adachi, *Nature, Lond.* **192,** 867 (1961).
68. I. Uhara, S. Kishimato, T. Hikino, Y. Kageyama, H. Hamada and Y. Numata, *J. phys. Chem.* **67,** 996 (1963).
69. I. Uhara, T. Hikino, Y. Numata, H. Hamada and Y. Kageyama, *J. phys. Chem.* **66,** 1374 (1962).
70. I. Uhara, S. Yanagimoto, K. Tani, G. Adachi and S. Teratani, *J. phys. Chem.* **66,** 2691 (1962).

71. S. Kishimoto, *J. phys. Chem.* **66**, 2694 (1962).
72. S. Kishimoto, *J. phys. Chem.* **67**, 1161 (1963).
73. L. M. Clarebrough, M. E. Hargreaves and G. W. West, *Proc. R. Soc.* **A232**, 252 (1955).
74. L. M. Clarebrough, M. E. Hargreaves and G. W. West, *Phil. Mag.* **1**, 528 (1956).
75. W. Boas, "Defects in Crystalline Solids", p. 212. The Physical Society, London (1955).
76. L. M. Clarebrough, M. E. Hargreaves, M. H. Loretto and G. W. West, *Acta metal.* **8**, 797 (1960).
77. H. Leidheiser, Jr., and A. T. Gwathmey, *J. Am. Chem. Soc.* **70**, 1200 (1948).
78. A. T. Gwathmey and R. E. Cunningham, *Adv. Catalysis* **10**, 57 (1958).
79. R. Y. Meelheim, R. E. Cunningham, K. R. Lawless, S. Azim, R. H. Kean and A. T. Gwathmey, Proc. 2nd Int. Congr. Catalysis, Paris, Vol. 1, p. 2005. Editions-Technip (1961).
80. J. F. Hall and H. F. Rase, *Nature, Lond.* **199**, 585 (1963).
81. H. S. Taylor, *Proc. R. Soc.* **A108**, 105 (1925).
82. J. Bagg, H. Jaeger and J. V. Sanders, *J. Catalysis* **2**, 449 (1963).
83. J. V. Sanders, Private communication to J.M.T. (1965).
84. R. A. Swalin, "Thermodynamics of Solids". Wiley, New York (1964).
85. D. J. Fabian and A. J. B. Robertson, *Proc. R. Soc.* **A237**, 1 (1956).
86. A. J. B. Robertson and D. Crocker, *Trans. Faraday Soc.* **54**, 931 (1954).
87. M. J. Duell and A. J. B. Robertson, *Trans. Faraday Soc.* **57**, 1416 (1954).
88. E. W. Müller, *J. appl. Phys.* **26**, 732 (1955).
89. R. L. Burwell, Jr. and J. A. Peri, *A. Rev. phys. Chem.* **15**, 131 (1964).
90. P. H. Emmett, "New Approaches to the Study of Catalysis", p. 58. The Pennsylvania State University, University Park (1962).
91. W. E. Garner, T. J. Gray and F. S. Stone, *Proc. R. Soc.* **A197**, 294 (1949).
92. C. Wagner and K. Hauffe, *J. chem. Phys.* **18**, 69 (1950).
93. R. M. Dell, F. S. Stone and P. F. Tiley, *Trans. Faraday Soc.* **49**, 201 (1953).
94. G. M. Schwab, *Z. phys. Chem.* **25**, 411, 418 (1934).
95. G. Schmid and N. Keller, *Naturwissenschaften* **37**, 43 (1950).
96. M. McD. Baker and G. I. Jenkins, *Adv. Catalysis* **7**, 47 (1955).
97. G. M. Schwab, "Semiconductor Surface Physics", p. 283. University of Pennsylvania Press, Philadelphia (1957).
98. K. Hauffe, R. G. Long, and H. J. Engell, *Z. phys. Chem.* **201**, 721 (1952).
99. G. Parravano, *J. Am. chem. Soc.* **75**, 1352, 1448 (1953).
100. N. P. Keier, S. Z. Roginskii and P. S. Sazonova, *C. r. Acad. Sci., U.R.S.S.* **106**, 859 (1956).
101. M. E. Dry and F. S. Stone, *Disc. Faraday Soc.* **28**, 192 (1959).
102. W. Krauss, *Z. Elektrochem.* **53**, 320 (1948).
103. A. B. Hart and R. A. Ross, *J. Catalysis* **2**, 251 (1963).
104. A. Bielanski, J. Deren, J. Haber and J. Sloczynski, *Trans. Faraday Soc.* **58**, 166 (1962).
105. H. B. Charman, R. M. Dell and S. S. Teale, *Trans. Faraday Soc.* **59**, 453 (1963).
106. M. O'Keeffe and F. S. Stone, *Proc. R. Soc.* **A267**, 501 (1962).
107. J. Haber and F. S. Stone, *Trans. Faraday Soc.* **59**, 192 (1963).
108. J. B. Goodenough, "Magnetism and the Chemical Bond". Interscience, New York (1963).

109. N. Perakis, J. Wucher and G. Parravano, *C. r. hebd. Séanc. Acad. Sci., Paris* **248,** 2306 (1959).
110. E. G. Vrieland and P. W. Selwood, *J. Catalysis* **3,** 539 (1964).
111. K. S. De and F. S. Stone, *Nature, Lond.* **194,** 570 (1962).
112. H. Bizette, *Ann. Phys. Paris* **1,** 233 (1946).
113. Th. Wolkenstein, *Adv. Catalysis* **12,** 189 (1962).
114. F. F. Vol'kenshtein, "The Electron Theory of Catalysis on Semiconductors", transl. by E. J. H. Birch. Pergamon Press, Oxford (1963).
115. Th. Wolkenstein, *J. phys. Chem. U.S.S.R.* **32,** 2383 (1958).
116. See p. 215 of Ref. *113.*
117. Ya. T. Eidus and B. K. Nefedov, *Russ. chem. Revs* **9,** 445 (1963).
118. C. H. Bamford, *J. appl. Chem.* **13,** 525 (1963).
119. C. H. Bamford, G. C. Eastmond and J. C. Ward, *Proc. R. Soc.* **A271,** 357 (1963).
120. G. D. Renshaw and J. M. Thomas, *Nature, Lond.* **209,** 1196 (1966).
121. P. Cossee, *Trans. Faraday Soc.* **58,** 1226 (1962).
122. C. E. H. Bawn and A. Ledwith, *Q. Rev.* **16,** 361 (1962).
123. P. Cossee, *J. Catalysis* **3,** 80 (1964).
124. E. J. Arlman, *J. Catalysis* (a) **3,** 89 (1964); (b) **5,** 178 (1966).
125. E. J. Arlman and P. Cossee *J. Catalysis* **3,** 99 (1964).
126. K. Ziegler, E. Holzkamp, H. Breil and H. Martin, *Angew. Chem.* **67,** 541 (1955).
127. G. Natta, *Angew. Chem.* **68,** 393 (1956).
128. G. Natta and I. Pasquon, *Adv. Catalysis* **11,** 1 (1959).
129. G. Natta, *J. Polymer Sci.* **34,** 21 (1959). A. Simon, L. Kollar and L. Böröcz, *Mh. Chem.* **95,** 842 (1964).
130. J. Chatt and L. A. Duncanson, *J. chem. Soc.* 2939 (1953).
131. W. L. Carrick, *J. Am. chem. Soc.* **80,** 6455 (1958).
132. P. Cossee, *Tetrahedron Lett.* **17,** 12 (1960).
133. G. Natta, *Angew. Chem.* **68,** 393 (1956).
134. G. Natta and I. Pasquon, Proc. 2nd Int. Congr. Catalysis, Paris, p. 1373 (1960).
135. R. Coekelbergs, A. Crucq and A. Frennet, *Adv. Catalysis* **13,** 55 (1962).
136. E. H. Taylor, *J. chem. Educ.* **36,** 396 (1959).
137. V. G. Baru, *Russ. chem. Revs* 590 (1963).
138. W. Doerffler and K. Hauffe, *J. Catalysis* **3,** 156, 171 (1964).
139. T. R. Rubin, J. G. Calvert, G. T. Rankin and W. MacNevin, *J. Am. chem. Soc.* **75,** 2850 (1953).
140. J. Haber and F. S. Stone, *Trans. Faraday Soc.* **59,** 192 (1963).
141. R. Jongepier and G. C. A. Schuit, *J. Catalysis* **3,** 464 (1964).
142. F. A. Cotton and G. Wilkinson, "Advanced Inorganic Chemistry, A Comprehensive Text". Interscience, New York (1962).
143. D. A. Dowden and D. Wells, "Actes du Deuxieme Congress de Catalyse", p. 1499. Editions-Technip, Paris (1961).
144. F. Basolo and R. G. Pearson, "Mechanisms of Inorganic Reactions". Wiley, New York (1958).
145. L. E. Orgel, "An Introduction to Transition Metal Chemistry, Ligand Field Theory". Wiley, New York (1960).
146. E. H. Taylor, *Nucleonics* **20,** 53 (1962).
147. H. W. Kohn and E. H. Taylor, *J. phys. Chem.* **63,** 500 (1959).

148. G. M. Schwab, R. Sizmann and N. Todo, *Z. Naturf.* **16a,** 985 (1961).
149. J. Ritz, G. M. Schwab and R. Sizmann, *Z. phys. Chem.* NF **39,** 45 (1963).
150. G. M. Schwab and A. Konrad, *J. Catalysis* **3,** 274 (1964).
151. T. I. Barry and R. Roberts, *Nature, Lond.* **184,** 1061 (1959).
152. C. C. Roberts, A. Spilners and R. Smoluchowski, *Bull. Am. phys. Soc.* **3,** 116 (1958).
153. D. Graham, *J. phys. Chem.* **66,** 510 (1962).

CHAPTER 6

Geometric, Electronic and
Related Factors in Heterogeneous Catalysis

6.1. Introduction

In the last chapter considerable prominence was given to the view that lattice defects of various types may be highly significant in determining the catalytic activity of a solid. However, the reader who has had past acquaintance, however slender, with heterogeneous catalysis is by now probably impatient to ask what has become of such notions as the geometric and electronic factors in interpreting or predicting catalytic activity. After all, the idea of the importance of the geometry of atoms in a catalyst surface, along with the electronic configuration of these atoms, has been repeatedly discussed for several decades. Our objective in this chapter is, therefore, to summarize current views on this aspect of the subject. We shall see that *inter alia* the promise of fifteen years ago, when it appeared likely that heterogeneous catalysis would develop in a framework logically determined by the electronic theory of solids, has not been fulfilled. We shall indicate, whenever possible, principles and interpretations that seem to have general validity. However, as will become increasingly obvious by the end of this chapter, valid generalizations are few; also, when we talk of the mechanism of catalysis it usually means that we are restricting our attention to very few, often one, systems. For this reason Chapter 8 is devoted to a discussion of the mechanisms of certain specific catalytic reactions.

6.1.1. What is catalytic activity?

Consider, for simplicity, a reaction $G_2(g) \rightleftharpoons P_2(g)$ which is catalysed by a solid S. Let us suppose that a Langmuir-Hinshelwood mechanism (see

Section 1.2) prevails where (see Sections 2.4.1.1–2.4.1.3) the reactant G_2 forms a surface complex very rapidly, and the surface complex itself decomposes slowly to yield surface products which, in turn, are rapidly liberated to regenerate vacant surface sites:

$$G_2(g) + \begin{array}{cc} & \\ | & | \\ -S\!-\!S- \end{array} \underset{k_2}{\overset{k_1}{\rightleftharpoons}} \begin{array}{cc} G & G \\ | & | \\ -S\!-\!S- \end{array} \qquad \text{(rapid)}$$

$$\begin{array}{cc} G & G \\ | & | \\ -S\!-\!S- \\ | & | \\ P & P \end{array} \underset{k_3}{\rightarrow} \begin{array}{cc} P & P \\ | & | \\ -S\!-\!S- \end{array} \qquad \text{(rate determining)}$$

$$\begin{array}{cc} P & P \\ | & | \\ -S\!-\!S- \end{array} \rightleftharpoons P_2(g) + \begin{array}{cc} & \\ | & | \\ -S\!-\!S- \end{array} \qquad \text{(rapid)}$$

Numerous reactions are known to conform to this scheme [1]. If N_S is the total number of sites available on the surface of the catalyst, p the pressure of the reactant, and θ the fraction of the available sites covered by surface complex, then:

$$k_1 p N_S (1-\theta) = (k_2+k_3) N_S\theta \qquad (1)$$

since, under steady-state conditions, the concentration of surface complex must be constant. Rearranging equation (1) we have:

$$\theta = \frac{k_1 p}{k_2 + k_3 + k_1 p}$$

so that the rate of production of $P_2(g)$, $k_3 N_S\theta$, may be written:

$$k_3 N_S\theta = \frac{k_1 k_3 N_S p}{k_2 + k_3 + k_1 p} \qquad (2)$$

from which it follows that enhanced reaction rates (i.e. increased catalytic activity) may be brought about by increasing N_S. Clearly, an increased surface area of catalyst would lead to a commensurately faster rate. But it also follows that increasing the number of active sites per unit area—roughly known as the activity—enhances the rate of reaction. It is not only the number of sites per unit area that must be increased. In order to produce a good catalyst such sites must have associated with them large values of k_3 and optimal values of the ratio of k_1/k_2, so as to make reversible chemisorption feasible, but not predominant.

Since we have already seen (see Section 2.4.1.3) that a catalyst lowers the activation energy of a reaction, it may, at first, seem odd that we do not automatically accept the magnitude of the value of the activation energy of a heterogeneously catalysed reaction as a quantitative measure of the catalytic activity. The prime reason for not adopting this criterion is the widespread occurrence of compensation effects (see Section 5.3.1.2). A given reaction may have a lower activation energy over a catalyst A than over a catalyst B. The rate of reaction over B may, however, be greater than over A, because the pre-exponential factor with catalyst B may be correspondingly larger.

One widely used index of catalytic activity is the temperature required for the reaction under consideration to attain an arbitrary degree (or rate) of conversion. This approach can be misleading. Relative activities of a series of catalysts will vary with the degree of conversion chosen unless all the catalysts exhibit the same activation energy, a situation which is not very likely to prevail. There is something to be said, therefore, for using as an index of activity the relative efficiencies of different catalysts at the same temperature, although, ideally, one ought to relate activity to energies of activation and pre-exponential factors. But fixing a certain temperature may turn out to be just as arbitrary as, and perhaps more misleading than, fixing a certain degree of conversion—quite different orders of catalytic efficiency may be obtained if a different standard temperature is fixed. Schuit [2] has summarized the advantages of preferring the criterion of temperature at fixed conversion rather than conversion (or rate) at fixed temperature. He points out that (i) the temperatures of equal conversion are measured quantities, whereas the conversions at equal temperatures are partly extrapolations; (ii) from the Langmuir-Hinshelwood mechanism it follows that energies and entropies of activation are not constant over a relatively wide temperature range, but depend on the conditions of the experiment. It is worth elaborating this point, for it serves to distinguish apparent and true activation energies. If a log rate versus reciprocal temperature plot is constructed over a temperature range where θ is about unity throughout, a true activation energy is obtained from the slope of the plot. Such conditions are more likely to prevail (i.e. $\theta \rightarrow 1$) at low temperatures. If, however, θ is fractional and small—a state of affairs likely to obtain at high temperatures—θ will itself be temperature dependent, increasing temperature at constant pressure decreasing θ. Hence

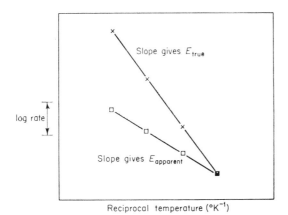

FIG. 1. Relation between true and apparent energies of activation; □, observed rates; ×, rates obtained after correcting for increased coverage (θ) with decreasing temperature.

302   INTRODUCTION TO HETEROGENEOUS CATALYSIS

the log rate versus reciprocal temperature plot under these conditions embraces two terms and the true activation energy is obtained only after adding the appropriate heat of adsorption to the apparent activation energy (see Fig. 1). The nub of Shuit's second remark is therefore seen to be a warning against extrapolation, since both energies and entropies of activation will tend to be considerably greater at lower rather than at higher temperature.

In the ensuing sections of this chapter we shall, in our efforts to comprehend what factors determine the magnitude of k_3 and the ratio k_1/k_2, make use of both the rate at a fixed temperature and the temperature at a fixed conversion as criteria of catalytic activity.

6.2. The Geometric Factor

It is unfortunate that, in the field of catalysis, the term geometric factor has become synonymous with lattice spacing of the atoms at a catalyst or other adsorbent surface. Gwathmey and Cunningham [3] have, very properly, pointed out that lattice spacing is only one aspect of the geometric factor and is by no means a true measure of the surface microtopography of a catalyst. Since, however, the term has become hollowed by convention, we shall in this section—unless otherwise stated—regard the geometric factor as being concerned simply with lattice spacing.

6.2.1. Theoretical basis for the operation of a geometric factor

Over thirty years ago Eyring [4] and Horiuti [5] calculated the magnitude of the activation energies of adsorption for the systems:

$$
\begin{array}{cc}
\text{H—H} & \text{H \ H} \\
 & | \ \ | \\
\text{—C—C—} \ \rightarrow & \text{—C—C—} \\
| \ \ | & | \ \ | \\
\text{(a)} &
\end{array}
\qquad
\begin{array}{cc}
\text{H—H} & \text{H \ H} \\
 & | \ \ | \\
\text{—Ni—Ni—} \ \rightarrow & \text{—Ni—Ni—} \\
| \ \ | & | \ \ | \\
\text{(b)} &
\end{array}
$$

and concluded that the lattice spacing (separation distance of the carbon or nickel atoms) played a dominant role in determining the magnitude of the activation energy. In system (a), for example, there was a minimum of about 14 kcal mole^{-1} at a separation distance of 3·6 Å. At low separation distances the energy is high, because repulsion forces (between adsorbate species) retard adsorption. At large separations it is again high, because the H$_2$ molecule has effectively to be completely severed before adsorption can occur. One can readily conceive of numerous systems involving chemisorption where separation distance is likely to play an important part; for example, the adsorption of oxygen and carbon monoxide on oxide surfaces (see Chapter 8), and of hydrocarbons on metals.

But lattice spacing may affect the magnitude of the heat of adsorption as well as the energy of activation. Let us consider, as did Twigg and Rideal [6], the following intermediate which, for argument's sake, may be regarded as a typical surface complex in a heterogeneously catalysed reaction.

$$CH_2 - CH_2$$

$$M\text{———}M$$

(For present purposes it is supererogatory whether ethylene adsorption over most metal adsorbents is of this form.) If the metal is nickel, and we take the Ni—C bond length to be the same as that in Ni(CO)$_4$, viz. 1·82 Å, then, with the C—C bond length taken to be the normal value for paraffins, 1·54 Å, the Ni—C—C valence angle can be readily computed for various values of the Ni—Ni distance. From the representation of the principal lattice planes of nickel depicted in Fig. 2, we see that two values of Ni—Ni are of particular interest, 2·48 and 3·51 Å. The Ni—C—C angle is very close to 105 degrees on the 2·48 Å spacing and to 123 degrees on the 3·51 Å spacing. The former angle is not much different from the strain-free tetrahedral angle of 109° 28' and such strain as is present may be ameliorated by twisting the surface complex so that the Ni—Ni and C—C axes are slightly inclined to each other. The complex would, however, be strained to a much greater extent if the Ni—Ni distance were 3·51 Å. Moreover, the strain would be intensified by twisting. The heat of adsorption on the 2·48 Å spacing may consequently be expected to be somewhat larger than that on the 3·51 Å spacing. Since a low heat of

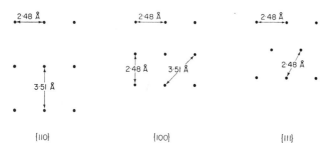

FIG. 2. Representation of the spacings in three lattice planes in a crystal of nickel.

adsorption is conducive to high reactivity, 3·51 Å spacings might be expected to be more active, catalytically, than the 2·48 Å spacing in reactions involving adsorbed ethylene.

6.2.2. Experimental evidence favouring the operation of a geometric factor

After Beeck, Smith and Wheeler [7] succeeded in producing orientated evaporated metal films (see Section 3.2.2.2), they soon discovered striking differences in catalytic activity between non-orientated and orientated films. For example, films of nickel which preferentially exposed {110} type planes were found to be five times as active in ethylene hydrogenation as non-orientated films which, in all probability, were composed of approximately equal amounts of {100}, {110} and {111} type planes. Since {110} planes in

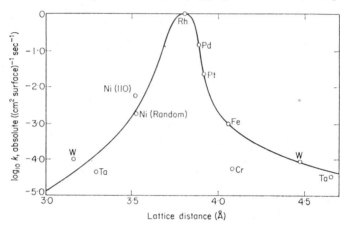

FIG. 3. Activity at 0°C of a series of thin metallic films (per unit area) as a function of the distance between particular pairs of atoms in the lattice (after Beeck [8]).

crystallites of nickel contain the optimal 3·51 Å spacing discussed in the previous section, the results of Beeck, Smith and Wheeler did much to strengthen the view that lattice spacing was an important key to the understanding of catalytic activity.

Upon exploring further the role of the lattice spacing, Beeck and his associates [8] discovered that when the logarithm of the activity of the various metal films used for catalysing the hydrogenation of ethylene was plotted against the lattice spacing, a relatively smooth curve (Fig. 3) was obtained. (Note that rate of reaction at fixed temperature is used as the criterion of activity here. Note also that, for some of the body-centred metals, e.g. Fe, Ta, W, the lattice distance is taken as a distance across a face of the {110} type planes of the unit cell, whereas for face-centred cubic metals, e.g. Rh, Pd and Pt the a_0 distance of the unit cell is taken as the lattice spacing.) The most favourable spacing is clearly 3·75 Å, that of rhodium. It can be seen from Fig. 3 that, per unit area, the activity of rhodium exceeds that of nickel by a factor of approximately 10^3. (We shall shortly see that when the same rate measurements are plotted against the per cent d-character (see Section 6.3.1), the points for all the metals, with the exception of tungsten, which is obtrusively nonconformist, fall on a smooth curve, and rhodium has again the highest catalytic activity.)

Earlier evidence for the operation of a geometric factor had been formulated by Balandin [9–12] in 1929. When we talk of Balandin's ideas, in particular his multiplet theory of catalysis, we recognize that by geometric factor something more than just lattice spacing is involved. There is also the question of arrangement or disposition of atoms in the catalyst surface as well as crystallographic symmetry. Similar concepts were mooted in the 1930s by other Russian workers, notably Pisarshevskii and Dankov [13]. The essence of Balandin's theory is that the activity of a catalyst depends to a large degree on the presence in the lattice of correctly spaced groups (or multiplets) of atoms to accommodate the various reactant molecules and their decomposition or reaction to form various products. Balandin first paid especial attention to the reactions of six-membered rings at metal surfaces, e.g. the hydrogenation of benzene and the dehydrogenation of cyclohexane. He proposed that, in these reactions, the benzene skeleton lay flat on the surface, adsorbed to six metal-to-carbon bonds. For such an arrangement to be possible the crystal face would have to display octahedral symmetry in addition to possessing prescribed values of the interatomic distances of the surface metal atoms. Only the {111} type planes of face-centred cubic and the habit faces of hexagonal close-packed crystals exhibit this symmetry. Balandin therefore argued that the metals active in the benzene ⇌ cyclohexane system would possess both the required lattice spacing and the required crystal symmetry. He computed the distance between the hydrogens attached to the carbon atoms of benzene and the metal atoms from which additional hydrogen atoms could be transferred to the benzene to form cyclohexane, and he plotted this distance against the atomic radii of a number of metals (see Fig. 4). The metals which displayed good catalytic activity were noted by Balandin to be within a certain square, marked on Fig. 4. He pointed out that metals such as calcium, thorium or iron were not good catalysts for the benzene ⇌ cyclohexane reaction, because the atoms of these metals lacked the requisite spacing to accommodate the adsorbed benzene skeleton. On the other hand, metals such as platinum, palladium and nickel were good catalysts, because they did possess the requisite spacings. Further support for Balandin's concept of the geometric factor came from Emmett's work, in which it was shown [14] that powdered catalysts of metals, such as iron, which crystallize with the body-centred cubic structure, were inactive in benzene hydrogenation. More recent work [15, 16] on iron has shown, however, that, under certain conditions, e.g. as an evaporated film [16], iron can be a very active catalyst for the above-mentioned hydrogenation–dehydrogenation. Such experiments have tended to cast doubt on the validity of Balandin's original theory.

Before we proceed to attempt a summary of the current status of the geometric factor in catalysis mention ought to be made here of another useful contribution, also based on the importance of lattice geometry, that of

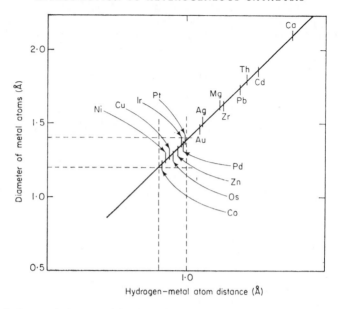

F I G. 4. Geometrical nature of the dimension involved in adsorbing benzene on to a range of metal surfaces. The abscissa represents the numerical value of the distance from the nearest metal atom to the hydrogen bonded to the closest carbon atom of the adsorbed benzene. The prime hydrogenating catalysts are situated within the square drawn by Balandin [9].

Kobozev. The Kobozev theory of catalysis, which is now receiving the attention of carbon chemists (see Section 8.7), envisages a catalyst as a solid consisting of a well-defined crystalline phase and a so-called amorphous phase. (In the language of present-day solid-state theory, these amorphous phases may be visualized as the grain boundaries mentioned in Chapter 5 when dealing with Sosnovsky's work.) According to Kobozev [17], some of the atoms of the amorphous phase of the catalyst constitute an "ensemble", which acts as the vehicle of catalytic activity; the crystalline phase serves merely as a kind of catalyst support. The ensemble is, in a way, regarded as the smallest group of catalytically active atoms. This notion, as we shall see in Chapter 8, is useful in interpreting the dependence which apparently exists between degree of catalytic activity of a metal catalyst in the gasification of carbon and the geometric size of the catalyst particle. Catalyst poisoning can be readily interpreted in terms of the Kobozev theory [18].

6.2.3. Current status of geometric factor

There is no doubt that lattice spacing and surface geometry play an important part in some heterogeneously catalysed reactions, and the role of crystal symmetry, rightly emphasized by Balandin, ought not to be ignored. To doubt the validity of this statement would be to question a wide range of reliable experimental observations. How, otherwise, would we interpret the fact that the kinetic parameters for the decomposition of formic acid on silver crystals vary significantly amongst the {100}, {110} and {111} planes [19] (see Section 5.3.1.1); or the report that the stereoregular polymerization of propene occurs only on the edges of {000l} planes of $\alpha TiCl_3$ (i.e. probably on prismatic or pyramidal planes of the type {11$\bar{2}$l} or {10$\bar{1}$l}) [20] (see Section 5.3.4.2)? Moreover, we are compelled to accept the basic notion of the so-called geometric factor by the powerful arguments, incorporating crystal-field stabilization, enunciated by Haber and Stone [21] in their discussion of oxygen adsorption on nickel oxide (see Section 5.4.1). Indeed, we are now fully aware that the differences in chemisorptive capacity and catalytic activity between two crystal faces of one solid may be significantly larger than the differences between the corresponding crystal faces of two different solids. Nevertheless, it is necessary to display a healthy degree of scepticism concerning the widespread use of the geometric factor, and it is important to be familiar with some of the more obvious deficiencies.

Less than a decade ago it was fashionable to criticize the traditional approach to the geometric factor in catalysis by metals solely on the score that the electronic approach—involving d-character of the metallic bond (see Section 6.3.1 below)—was so much more percipient. It was confidently felt that the results achieved using the geometric approach could also be obtained, and extended, using instead the electronic concept. This trend was not surprising once it is realized that electronic factors must, in turn, affect the separation distances and other properties of atoms in a solid—the work of Haber and Stone [21], described in Section 5.4.1, illustrates this point rather well. From the Pauling theory of metals, mentioned in Section 6.3.1, there emerges a direct relationship between the d-character and the single-bond radius of atoms in a metal crystal. Although this criticism still retains much of its tenability, we must, in the light of recent experiments, draw attention to some other, more serious, defects.

Low-energy electron diffraction (Section 3.3.8) on the one hand and optical and electron microscopy (Section 3.3.9) on the other are the two main techniques which have led to grave doubts being expressed concerning the validity of the geometric factor. In the first place it has been shown that, for nickel at least, when certain gases are adsorbed on the {110} planes a drastic rearrangement of the atoms of nickel takes place. It is also known that the surface atoms of some metals, such as copper, undergo very considerable rearrangements during the actual course of catalysis. Some metals, notably silver, when heated at temperatures just below their melting-point (as in outgassing), develop, as a result of surface migration, crystal planes which may not be present initially [22]. Furthermore, accurate measurements of the lattice spacing between the outer and next layer of nickel atoms in a "clean" nickel surface reveals that the interatomic distance is increased about 5% over that in the bulk. All these facts demonstrate that, if there does appear to be (as in Fig. 3) a correlation between the catalytic activity of a series of metals and their lattice spacing (measured, by X-ray diffraction, using bulk specimens), this *may* be fortuitous, for the *actual* lattice spacing *of the surface atoms active in catalysis* may be significantly different from the lattice spacing of the bulk specimen.

6.3. The Electronic Factor

The idea that the collective electronic properties of metals, alloys or semiconductors are in some way intimately related to the catalytic activity of these solids forms the basis of the so-called electronic factor in catalysis. Reduced to its simplest form, it may be said that the electronic approach seeks a relationship between the catalytic activity and the electronic structure of the *bulk* solid. To appreciate why, in the first place, it was felt that such a relationship may exist, it is necessary to recall some relevant properties of metals and semiconductors.

6.3.1. Theoretical and experimental basis for the operation of an electronic factor

6.3.1.1. *Metals and Alloys*

According to the band theory [23, 24], the electrons in metals retain much of the character that they possess in the isolated atoms, although the valency electrons are thought to move quite freely through the assembly of positively charged nuclei and their associated (closed) electron shells. It is still legitimate to talk of s-, p- and d-electrons in the solid, but, whereas in the isolated atoms each energy state is discrete and single-valued, in the metal crystal the energy of each state has a band of permitted values. Furthermore, the number of electrons per atom in a band of the metal crystal may differ from the number in the corresponding shells of the isolated atoms. For example, the electronic configuration of the $3d$- and $4s$-orbitals of the isolated iron atom is $3d^6 4s^2$, whereas, in the metallic state, the band structure is, on average, $3d^{7.8} 4s^{0.2}$. We note that there is an incomplete $3d$-band and, in the terminology of the band theory, we talk of 0·2 "holes" in the d-band. A feature of the transition metals in general is that their crystals possess incomplete d-bands.

According to the Pauling theory [25, 26], metal crystals may be pictured as being held together by essentially covalent bonds between adjacent atoms. From the bulk physical properties of metals (e.g. their cohesive energies or melting-points) Pauling concluded that there are three types of d-orbitals associated with each atom in the solid state: *bonding* d-orbitals, which can participate in dsp-type hybrid bonds; *metallic* d-orbitals, involved in electrical conduction; and *atomic d*-orbitals, which are non-bonding and into which electrons may be placed. In Pauling's theory the transition metals have some vacant atomic *d*-orbitals; and their cohesive energy is attributed to the formation of *dsp*-orbitals. The so-called percentage of *d*-character [25] δ represents the extent to which *d*-electrons participate in *dsp*-orbitals: the higher the value of δ, the fewer the number of atomic *d*-orbitals there are available at each atom. In other words, δ is a measure of the unavailability of electrons in atomic *d*-orbitals. Some typical values of the percentage *d*-character are tabulated below

TABLE 6.1. Percentage *d*-character (δ) in the metallic bond of some elements

Cr	Mn	Fe	Co	Ni	Cu	Ta	W	Rh
39	40·1	39·7	39·5	40	36	39	43	50

It is to be noted that, whereas on the band theory transition metals are said to have vacant or incomplete *d*-bands, on the Pauling theory these metals are said to possess vacant atomic *d*-orbitals.

In 1949, Boudart [27] drew attention to the fact that the variation of the catalytic activity of a number of transition metals for the hydrogenation of ethylene could be represented in the form of plots of the percentage *d*-character against the logarithm of the rate constant for the reaction. Beeck [8] also observed such dependence (see Fig. 5). As mentioned previously (see Section 6.2.2), most of the points fall on a smooth curve, and rhodium, as in Fig. 3, has the highest activity (see also p. 387).

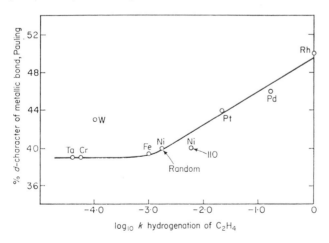

FIG. 5. Variation of the percentage *d*-character of a range of metals against the logarithm of the activity of the metals per unit area for the reaction: $C_2H_4 + H_2 \rightleftharpoons C_2H_6$ (after Beeck [8]).

The importance of the electronic factor in the catalytic activity of certain alloy systems had been observed earlier by Schwab [28], who reported that the activation energy for the decomposition of formic acid over a copper catalyst increased when minute quantities of such metals as Cd, In, Tl, Pb and Hg were added to the copper. Schwab argued that the higher the electron concentration in the catalyst the greater would be the activation energy for the decomposition of the formic acid, and he demonstrated that the above-mentioned alloy components increased the electron concentration in the catalysts. Shortly after the publication of Schwab's work came Kemball's investigation [29] of the rate of exchange of deuterium with ammonia over a series of metal surfaces. Here again it transpired that there appeared to be a rough correlation between the percentage d-character and the activation energy of the reaction.

The trend away from the "geometric approach" and towards an electronic one went a stage further when it was realized that, according to Pauling's theory, the d-character controls the interatomic spacing. Thus, R_1, the single bond radius of the atoms in a metal crystal, δ, and z, the number of electrons in the neutral atom outside the inert gas shell, are related as follows:

$$R_1 = 1 \cdot 825 - 0 \cdot 043z - (1 \cdot 600 - 0 \cdot 100z)\delta \qquad (3)$$

It is evident from this equation that, since the constants of the terms containing z are small, the value of R_1 is largely determined by δ. But it must be remembered that δ controls only the atomic radius or lattice spacing: it does not decide the precise arrangement of the active centres at a given surface. If this fact is borne in mind it becomes easier to appreciate why approaches to the problem of catalytic activity based on such concepts as d-character can, at best, explain only the general level of activity of *polycrystalline* catalysts and not why the differences in activity between different faces on one metal with a particular d-character may be much greater than the differences in activity between two metals with significantly different d-characters [3].

Despite this obvious limitation, the electronic approach was pursued with considerable enthusiasm for about a decade after the notion was first mooted. And there was much hope that the specificity of the chemical elements in general, and of the metals in particular, in their chemisorption of simple gases such as hydrogen and oxygen would be interpretable in terms of the theories associated with the names of Mott and Jones and Pauling. For example, it was at one time confidently believed [30, 31] that the pronounced tendency for transition metals, unlike "sp" metals, to participate in the chemisorption of molecular hydrogen (see Fig. 6), was interpretable in terms of vacant d-orbitals or holes in the d-band. We shall see, in Section 6.3.2 that such hopes were fanciful.

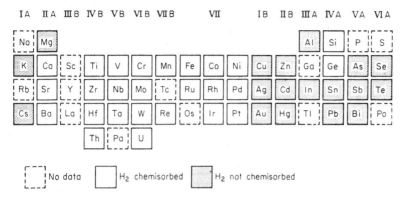

FIG. 6. Specificity of the elements in hydrogen chemisorption. The coding is based on the observation of perceptible adsorption from the molecular gas at 300°K and ~10⁻¹ torr (after Ehrlich [31]).

6.3.1.2. Semiconductors

A full account of what is meant by the electronic factor in adsorption and catalysis on semiconductors has been given in the previous chapter. We have seen, in Section 5.2.4.1, how various electronic rearrangements occur following adsorption on p-type and n-type semiconductors; and how the catalytic activity of semiconducting oxides may be interpreted in terms of the properties of the point defects existent in such solids (Section 5.3.2) or in terms of the Wolkenstein (or electron) theory of catalysis (Section 5.3.3). We have also summarized some current trends in the assessment of the catalytic activity of semiconductors (Section 5.3.2.3). It will be recalled that, for semiconductors, the electronic factor and the role of lattice defects in catalysis are more or less synonymous.

That the electrons of the d-shell of transition-metal ions may determine the catalytic activity of transition-metal oxide semiconductors first became clearly apparent following the study by Dowden and co-workers [32] of the ease of hydrogen–deuterium exchange ($H_2 + D_2 \rightleftharpoons 2HD$) at 90°K over the oxides of the first series of transition elements. In traversing the series from TiO_2 to CuO, two peaks of activity are found; the first occurs at Cr_2O_3 and the second is centred around Co_3O_4 and NiO (see Fig. 7). It is to be noted that low activity is encountered with MnO and Fe_2O_3. Broadly speaking, the same type of activity pattern has been shown to hold true for propane dehydrogenation [33] and cyclohexane disproportionation [34]. The ions of highest activity are those of the $3d^3(Cr^{3+})$, $3d^6(Co^{3+})$, $3d^7(Co^{2+})$ and $3d^8(Ni^{2+})$ configurations, whereas the ions of lowest activity are associated with $3d^5(Mn^{2+}$ and $Fe^{3+})$, $3d^0(Ti^{4+})$ and $3d^{10}(Zn^{2+})$ configurations. Dowden and Wells [33] showed that, if the simple concepts of electron transfer are augmented by the application of crystal field theory (see Section 5.4.1), it is possible to interpret the activity patterns in a logical manner. According to Dowden and Wells [33], any rate-controlling step which involves the adsorption of a species so polarized as to restore near octahedral symmetry about a surface cation (in a transition metal oxide) contributes to higher energies of

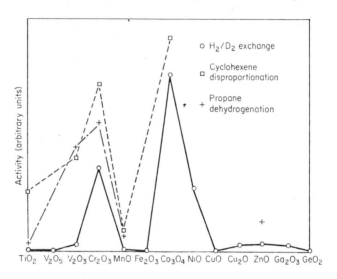

FIG. 7. The catalytic activity of the oxides of the first long period (after Dowden [35]).

activation and lower activities at the cation electron-configurations d^0, d^5 and d^{10}. (Concerning the comments made in Section 6.1.1 about the difficulty of deciding upon a reliable criterion of catalytic activity, it is of interest to note that Steiner et al. [34], in their analysis of activity patterns in the disproportionation of cyclohexane, used both temperature at a fixed rate and rate at a fixed temperature as indexes of activity.)

6.3.2. Current status of the electronic factor

So far as metals and alloys are concerned, the results of the last twenty years have firmly indicated that attempts to correlate bulk electronic and surface properties have been largely unsuccessful. For elemental semiconductors, too, the situation is hardly more encouraging; but for compound semiconductors, especially transition metal oxides, some semi-quantitative theories appear to offer a measure of hope.

Consider the question of the specificity of the elements in their chemisorption of hydrogen (Fig. 6). If the electronic theory of solids is to be of any value at all in interpreting heterogeneous catalysis, it should first be capable of accommodating the observed facts concerning chemisorption—the precursor of catalysis. It may at first sight appear [35, 36] that the selectivity apparent in Fig. 7 can be accounted for if we postulate that covalent bonds are formed between hydrogen atoms and the solid, provided that the latter has holes in the d-band. This conclusion does not appear at all convincing once it is realized that Fig. 6 merely demonstrates the specificity of the elements for the chemisorption of *molecular* hydrogen as atoms (see Section 2.2.1.1). If we work on the approximate basis that the free energy of adsorption is the same as the heat of adsorption (i.e. ignoring the entropy term $T\Delta S$ in equation (22) of Chapter 2 so that $\Delta G = \Delta H$), it becomes evident that chemisorption of molecular hydrogen as individual adatoms can only occur if the binding energy of an adatom exceeds one-half of the dissociation energy of the H_2 molecule [31]. If, therefore, we are to expect a correlation between the d-band structure and aptitude for hydrogen chemisorption, we should employ the binding energies of hydrogen *atoms* to the elements. Figure 8 demonstrates that there is no particularly definite dependence of selectivity upon the number of holes in the d-bands, because quite strong bonds can be formed between hydrogen atoms and elements that have no d-character.

Returning to the question of catalytic activity of metals and alloys, one of the most serious deficiencies of the electronic factor is that it can embrace only the general level of activity of polycrystalline catalysts. The work of Cox, Lawless and Gwathmey [37] on carbon-monoxide decomposition over

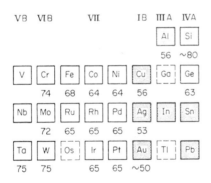

FIG. 8. Binding energies in kcal mole^{-1} of hydrogen adatoms to the elements. The coding is the same as in Fig. 6 (Ehrlich [31]).

single-crystal alloys of copper and nickel proves this point unequivocally. They studied the catalytic activity of different crystallographic faces of alloy crystals ranging in composition from 0% copper (pure metal) to 84·3% copper. As for other reactions [38], a decrease in the percentage of d-character of the nickel was accompanied by a decrease in activity for most of the crystal faces. However, certain faces such as the {100} planes in pure nickel and alloy crystals remained largely inactive at all compositions of the alloy. It is apparent, therefore, that the electronic interpretation alone is inadequate, because different crystal faces of an alloy with a given percentage d-character have different activities.

It is obvious that, in future assessments of the mechanisms of catalysis, we must pay much less attention to the general electronic properties of the solid and more to the properties of individual atoms or complexes at the surface. Already we are beginning to appreciate the wisdom of such an approach in the work of Rooney et al. [39–41] on π-bonded intermediates. By concentrating, in general, on the nature of the bonds around the carbon atoms in adsorbed species from hydrocarbons or other organic molecules, and by postulating, in particular, that intermediates can be π-bonded to a metal catalyst in a manner similar to the bonding in compounds of the ferrocene type, Rooney has gone a long way towards relating the general chemistry of transition metals to their catalytic activity in a wide range of heterogeneously catalysed reactions.

As a consequence of the recent emphasis on the role of two-dimensional "compounds" in heterogeneous catalysis, evidence has steadily accumulated for believing that the original suggestion of Sabatier [42]—that heterogeneous catalysis functions through an intermediate compound—may, in some measure, prove to be valid for a number of reactions on metal and oxide surfaces. The work of the Dutch school [43–46] on the catalysed decomposi-

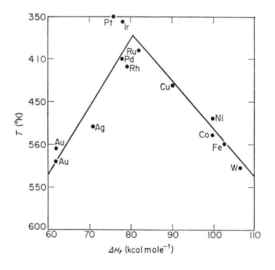

Fig. 9. Activity of various metals for the decomposition of formic acid as a function of the heat of formation of their formates (Fahrenfort et al. [44]).

tion of formic acid constitutes the most cogent evidence for accepting the dominant role of such factors as the ease of formation and the stability of surface intermediates. Sachtler [46] has emphasized that this trend (or reversion) is motivated by attempts to relate regularities observed in catalysis to regularities in the general field of chemistry rather than directly to basic "electronic" principles. To illustrate the success of the present application of the intermediate compound theory we shall recall here the salient features of Dutch work.

The Dutch school took as their point of departure the fact that infrared spectroscopy proves incontrovertibly that a formate is formed when formic acid is adsorbed on a catalyst surface (see Section 3.3.1), and a central feature of the interpretation of the catalysis is that decomposition proceeds via the formate intermediate on all metals (see also Section 3.2.8.4). Figure 9 shows the activity (taken as the temperature T at which a certain prefixed reaction rate is attained) of various metals plotted against the heat liberated when one gram molecule of formic acid is adsorbed on their surface. (The latter is really the enthalpy of formation of the metal formate.) The mechanism of the decomposition involves three stages:

First

$$HCOOH(g) \rightarrow HCOO(ads) + H(ads);$$

then the formate fragment decomposes:

$$HCOO(ads) \rightarrow CO_2(g) + H(ads);$$

and finally the hydrogen adatoms desorb:

$$2H(ads) \rightarrow H_2(g)$$

Schematically we may write the following sequence [45]

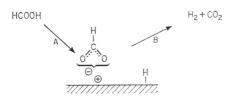

When the kinetics of the decomposition are first-order, step A, the formation of formate, is rate-determining. When, on other metals, zero-order kinetics are obeyed, the rate determining step is B, the decomposition of the surface formate.

A metal such as gold is a poor catalyst (large value of T for a fixed decomposition rate), because very little formic acid is adsorbed. With metals such as tungsten, the formic acid is rapidly adsorbed but the formate decomposes slowly. On both metals the catalysis is low. However, under steady-state conditions the gold is sparsely covered, whereas the tungsten is well covered.

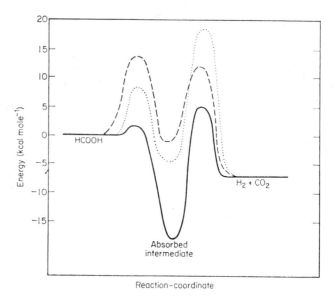

FIG. 10. Possible potential energy curves for the decomposition of formic acid on nickel (full line), copper (dotted line) and gold (dashed line) (after Mars *et al.* [45]).

The metals Ru, Pt and Ir are good catalysts, since chemisorption and decomposition (stages A and B above) on these metals are quite rapid and have approximately equal specific rates. These are the reasons for the occurrence of the volcano-shaped curve in Fig. 9. Figure 10, taken from the work of Mars *et al.* [45], illustrates the energy pattern for a number of typical catalysts.

Although the approach pioneered by the Dutch workers has been criticized [48], it has much promise [47] and the principles that obtain for the formic-acid decomposition appear to apply to a variety of systems. A full test of the theory awaits the compilation of trustworthy data on the kinetics of the thermal decomposition of numerous "bulk" compounds [49] analogous to those produced as surface compounds during catalysis; but there is little doubt that there is much to be gained by envisaging intermediate compounds in catalysis (see Section 8.7).

References

1. K. J. Laidler, "Catalysis", Vol. I, Ch. 4. Reinhold, New York (1954).
2. G. C. A. Schuit, Proc. 2nd Int. Cong. Catalysis, p. 916. Editions Technip, Paris (1961).
3. A. T. Gwathmey and R. E. Cunningham, *Adv. Catalysis* 10, 57 (1958).
4. A. Sherman and H. Eyring, *J. Am. chem. Soc.* **54**, 2661 (1932).
5. G. Okamoto, J. Horiuti and K. Hirota, *Sci. Papers Inst. phys. chem. Res. Tokyo* **29**, 223 (1936).
6. G. H. Twigg and E. K. Rideal, *Trans. Faraday Soc.* **36**, 533 (1940).
7. O. Beeck, A. E. Smith and A. Wheeler, *Proc. R. Soc.* A177, 62 (1940).
8. O. Beeck, *Disc. Faraday Soc.* **8**, 118 (1950).
9. A. A. Balandin, *Z. phys. Chem.* **132**, 289 (1929).
10. A. A. Balandin, *Adv. Catalysis* 10, 96 (1958).
11. A. A. Balandin, *Russ. chem. Rev.* **11**, 589 (1962).
12. A. A. Balandin, A. Bielanski and others, "Catalysis and Chemical Kinetics", Ch. 1. Academic Press, New York and Wydawnictura Nankowo-Techniczne, Warszawa (1964).
13. L. V. Pisarshevskii, Report at the Meeting on Physical Chemistry 1928, p. 135, cited by P. D. Dankov in *Progr. phys. Sci. U.S.S.R.* **14**, 63 (1934).
14. P. H. Emmett and N. Skau, *J. Am. chem. Soc.* **65**, 1029 (1943).
15. P. H. Emmett, "New Approaches to the Study of Catalysis", Ch. 3 (Priestley Lecture). The Pennsylvania State University, University Park (1962).
16. O. Beeck and A. W. Ritchie, *Disc. Faraday Soc.* **8**, 159 (1950).
17. N. I. Kobozev, *Acta physiochim. URSS* **21**, 294 (1946).
18. J. G. Tulpin, G. S. John and E. Field, *Adv. Catalysis* **5**, 217 (1953).
19. H. M. C. Sosnovsky, *J. Phys. Chem. Solids* **10**, 304 (1959).
20. E. J. Arlman and P. Cossee, *J. Catalysis* **3**, 99 (1964).
21. J. Haber and F. S. Stone, *Trans. Faraday Soc.* **59**, 192 (1963).
22. J. G. Allpress and J. V. Sanders, *Phil. Mag.* **9**, 645 (1964).
23. N. F. Mott and H. Jones, "The Theory of the Properties of Metals and Alloys". Oxford University Press, London (1936).
24. F. Seitz, "Modern Theory of Solids". McGraw-Hill, New York (1940).
25. L. Pauling, *Proc. R. Soc.* A196, 343 (1949).
26. W. Hume-Rothery, *A. Rep.* **46**, 42 (1949).
27. M. Boudart, *J. Am. chem. Soc.* **72**, 1040 (1950).
28. G. M. Schwab, *Disc. Faraday Soc.* **8**, 166 (1950).
29. C. Kemball, *Proc. R. Soc.* A124, 413 (1952).
30. B. M. W. Trapnell, "Chemisorption". Butterworths, London (1955).
31. G. Ehrlich, General Electric Report No. 64-RL-3771M, Schenectady (1964).
32. D. A. Dowden, N. McKenzie and B. M. W. Trapnell, *Proc. R. Soc.* 237A, 245 (1956).
33. D. A. Dowden and D. Wells, Actes 2ᵉ Congr. Int. Catalysis, p. 1499, Paris (1961).
34. G. M. Dixon, D. Nichols and H. Steiner, Proc. 3rd Int. Congr. Catalysis, Amsterdam (1964).
35. D. A. Dowden, *Endeavour* **24**, 69 (1965).
36. M. H. Dilke, E. B. Maxted and D. D. Eley, *Nature, Lond.* **161**, 804 (1948).
37. E. Cox, K. R. Lawless and A. T. Gwathmey, Actes 2ᵉ Congr. Int. Catalysis, p. 1605, Paris (1961).

38. M. McD. Baker and G. I. Jenkins, *Adv. Catalysis* **7**, 47 (1955).
39. J. J. Rooney, F. G. Gault and C. Kemball, *Proc. chem. Soc.* 407 (1960).
40. J. J. Rooney, *J. Catalysis* **2**, 53 (1963).
41. J. J. Rooney and G. Webb, *J. Catalysis* **3**, 488 (1964).
42. P. Sabatier, "La Catalyse en Chemie Organique". Librairie Polytechnique, Paris (1913).
43. J. Fahrenfort, L. L. van Reijen and W. M. H. Sachtler, *Z. Electrochem.* **64**, 216 (1960).
44. J. Fahrenfort, L. L. van Reijen and W. M. H. Sachtler, "The Mechanism of Heterogeneous Catalysis", ed. by J. H. de Boer *et al.*, p. 23. Elsevier, Amsterdam (1960).
45. P. Mars, J. J. F. Scholten and P. Zweitering, *Adv. Catalysis* **14**, 35 (1963).
46. W. M. H. Sachtler, *Discovery* **26**, 16 (1965).
47. G. C. A. Schuit, L. L. van Reijen and W. M. H. Sachtler, Actes 2e Congr. Int. Catal., p. 893, Paris (1961).
48. G. C. Bond, "Catalysis by Metals", p. 478. Academic Press, London (1962).
49. A. K. Galwey, *J. chem. Soc.* 4235 (1965).

CHAPTER 7

The Dynamics of Selective and Polyfunctional Catalysis

7.1. Catalyst Selectivity

The unceasing quest for the catalyst which performs a particular function is a manifestation of the more general but ubiquitous problem of the nature and reactivity of the chemical bond. Whether or not two molecules react to give a product is really determined by the ability of the reactants to undergo electronic and structural rearrangements. The fundamental question of chemical reactivity has to be solved in terms of the mutual forces between molecules. In the presence of a third component, the catalyst, the complexity of the problem is exacerbated. All that we can hope for is that chemical reactivity can be correlated with characteristic yet accessible parameters describing the electronic and geometrical properties of the catalyst. The subject of catalysis has not matured sufficiently to predict the best catalyst to expedite the conversion of reactants into products in any given chemical reaction. However, as is evident from previous chapters, some concepts concerning the role of the catalyst do allow of certain classifications which enable a reasonable choice to be made in a limited variety of cases which conform to an established pattern. Despite the advantage gained by empirical and theoretical observations, it is rare that a catalyst can be chosen so that it is entirely specific or unique in its behaviour. For this reason, it is important to consider

† In this Chapter, velocity constants for chemical reaction are referred to on the basis of (i) unit volume and (ii) unit surface area. The reader is advised to consult the list of symbols.

those factors which operate in such a way as to be selective in their function, thus providing a means by which a catalytic reaction can be controlled to give a high yield of some preferred product.

There are numerous heterogeneous catalytic reactions which involve either (i) successive kinetic steps leading ultimately to an end product, but in which an intermediate product may be isolated and identified, or (ii) alternative, but simultaneous, reaction paths yielding two or more products. The former kinetic scheme may be represented

$$A \xrightarrow{k_{v1}} B \xrightarrow{k_{v2}} C \tag{1}$$

and exemplified by the dehydrogenation of cyclohexane to cyclohex-1-ene and further dehydrogenation to cyclohex-2-ene and, ultimately, benzene; such reactions are catalysed by transition metals and metal oxides. Thus, Herington and Rideal [1, 2] demonstrated that cyclohexene is produced in measurable quantities as an intermediate product when cyclohexane is passed over a chromia–alumina catalyst to yield benzene. However, if a supported platinum catalyst is used, considerably less cyclohexene is produced and there is a corresponding increase in the amount of benzene.

When two simultaneous reaction paths are involved, the kinetic scheme may be written

$$A \begin{array}{c} \xrightarrow{k_{v1}} B \\ \xrightarrow{k_{v2}} C \end{array} \tag{2}$$

A classical example of this type of reaction is the dehydration of ethanol by a copper catalyst at about 300°C. The principal product is acetaldehyde, but ethylene is also evolved in smaller quantities. If an alumina catalyst is used at the same temperature, ethylene is the preferred product rather than acetaldehyde.

These examples show that catalysts are rarely specific in their ability to favour a single desired product to the exclusion of either intermediate or secondary products. A convenient method of estimating the selectivity of a catalyst is to compare the chemical rate constants of the individual steps. Thus if B is the desired product, the ratio k_{v1}/k_{v2}, termed the kinetic selectivity, is a measure of the extent to which B is produced at the expense of C.

It may be necessary sometimes to convert only one component in a mixture of components into a required product. For example, it is possible to dehydrogenate six-membered cycloparaffins in the presence of five-membered cycloparaffins without affecting the latter. In this case, it is desirable to select a catalyst which favours the reaction

$$A \xrightarrow{k_{v1}} B \tag{3}$$

when it might be possible for the reaction

$$X \xrightarrow{k_{v2}} Y \tag{4}$$

to occur simultaneously. Once again the ratio k_{v1}/k_{v2}, sometimes designated σ, defines the kinetic selectivity.

In the case of two parallel competing reactions, such as (3) and (4), high selectivity demands the preferential adsorption of A. Both A and X are, nevertheless, adsorbed simultaneously and are competing for the surface. Application of the principles embodying Langmuir's adsorption isotherm (see Section 2.3.1) show that, for the case of two species being simultaneously adsorbed, the surface coverages of A and X are given by:

$$\theta_A = \frac{b_A p_A}{1 + b_A p_A + b_X p_X} \tag{5}$$

$$\theta_X = \frac{b_X p_X}{1 + b_A p_A + b_X p_X} \tag{6}$$

respectively, where the coefficient b is similar to the constant of the Langmuir equation and p_A and p_X are the partial pressures of A and X. The ratio of the coverages is thus:

$$\frac{\theta_A}{\theta_X} = \frac{b_A p_A}{b_X p_X} \tag{7}$$

Now b has the properties of an equilibrium constant since, at adsorption equilibrium, the rates of adsorption and desorption from the surface are equal. Hence we may write:

$$b_A = \frac{(k_a)_A}{(k_d)_A} = \exp\left(\frac{-(\Delta G_a)_A}{RT}\right) \tag{8}$$

where k_a and k_d are the kinetic constants for adsorption and desorption, and $(\Delta G_a)_A$ is the Gibbs free energy change associated with such an equilibrium for the species A. A similar expression can be written for the species X so that

$$\frac{b_A}{b_X} = \exp\left(\frac{\delta(-\Delta G_a)}{RT}\right) \tag{9}$$

where $\delta(-\Delta G_a)$ is the difference between the free energy changes associated with the species A and X. From equation (7), therefore, the relative rates of formation of products B and Y depend on $\delta(-\Delta G_a)$ and so

$$\frac{dp_B}{dp_Y} = \frac{k_{v1}}{k_{v2}} \frac{\theta_A}{\theta_X} = \frac{k_{v1}}{k_{v2}} \frac{p_A}{p_X} \exp\left(\frac{\delta(-\Delta G_a)}{RT}\right) \tag{10}$$

This effect has been termed the thermodynamic factor in selectivity [3]. It is also important for reactions of type (1) where it is enhanced by (i) the ability of the intermediate B to desorb rapidly before it reacts further, and (ii) the reluctance of the catalyst to readsorb B after it has vacated the surface. Reactions of type (2) are unaffected by adsorption–desorption equilibrium effects.

Finally, we consider the situation where the catalyst may perform a dual function in one single operation, and the researches of Weisz [4] are relevant. Thus, for example, a composite catalyst can effect the hydrogenative cracking of hydrocarbons in which a high molecular weight hydrocarbon is converted via cracking and hydrogenation steps to a low molecular weight hydrocarbon. Another example is the dehydrogenation of cyclohexane to benzene and its simultaneous conversion to methylcyclopentane by the use of platinum metal supported on a silica–alumina catalyst. Both the metal and the acidic type catalysts have their separate roles to play in this latter conversion. Unless some consideration is given to catalyst pellet size and to the proportions of each catalyst component, the selectivity may be such that the formation of benzene is preferred to the formation of methylcyclopentane.

7.1.1. An empirical and a kinetic method of estimating selectivity in consecutive reactions

An empirical method of determining the most-favoured conditions for the production of a desired product B from a reactant A and at the expense of an undesired product C was formulated by Weber [5] and discussed by Waterman [6]. If y is the fraction of A transformed into the desired product B and x is the fraction converted into undesired product C, then a single empirical equation,

$$y = \frac{x(1 - x)}{a + bx} \qquad (11)$$

in which a and b are constants, describes adequately the levels of conversion in a series of selective catalytic reactions in which only one reaction parameter is varied. A right-angle isosceles triangle in which y is the vertical axis and x the horizontal axis is a convenient way of representing the experimental points (x, y). Suppose the reaction parameter which is varied is the contact

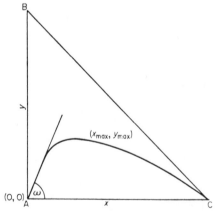

FIG. 1. Empirical method of estimating selectivity.

time, then for every value of the contact time there will be a corresponding conversion y and x related by equation (5). Figure 1 shows a typical curve which is described mathematically by the empirical equation (11) and which indicates the conversion of A into B and C.

van der Borg [7] showed that a kinetic scheme describes the data equally well as an empirical equation containing two constants. In place of the consecutive reaction given by equation (1) the scheme

$$
\begin{array}{ccc}
 & \text{B} & \\
k_{v1}\nearrow & & \searrow k_{v2} \\
\text{A} & \xrightarrow{\hspace{1cm}} & \text{C} \\
 & k_{v3} &
\end{array}
\tag{12}
$$

allows a second selectivity factor k_{v1}/k_{v3} to be introduced. If each reaction is first order, then, in terms of the conversion of A into B and C, the rate equations are

$$\frac{-d(1 - x - y)}{dt} = (k_{v1} + k_{v3})(1 - x - y) \tag{13}$$

$$\frac{dy}{dt} = k_{v1}(1 - x - y) - k_{v2}y \tag{14}$$

$$\frac{dx}{dt} = k_{v3}(1 - x - y) + k_{v2}y \tag{15}$$

With the initial conditions,

$$y = 0, \; x = 0 \tag{16}$$

equation (13) immediately gives

$$1 - x - y = \exp\{-(k_{v1} + k_{v3})t\} \tag{17}$$

and substitution of this into equation (14) gives

$$\frac{dy}{dt} + k_{v2}y = k_{v1} \exp\{-(k_{v1} + k_{v3})t\} \tag{18}$$

which is a simple linear first-order differential equation. With the boundary condition (16), the solution is

$$y = \frac{k_{v1}}{k_{v2} - (k_{v1} + k_{v3})} [\exp\{-(k_{v1} + k_{v3})t\} - \exp\{-k_{v2}t\}] \tag{19}$$

From equations (17) and (19) there results

$$(1 - x - y)^{k_{v2}/(k_{v1}+k_{v3})} = 1 - x - \left(\frac{k_{v2} - k_{v3}}{k_{v1}}\right)y \tag{20}$$

This latter equation relates y and x by the two selectivities k_{v1}/k_{v2} and k_{v1}/k_{v3} which we will denote σ_2 and σ_3 respectively. The values of σ_2 and σ_3 may be evaluated by plotting the experimental data in a right-angle isosceles triangle as in the empirical method of Weber [5]. A curve such as illustrated in Fig. 1 is thus obtained. Dividing equation (14) by equation (15) the maximum values of y and x are

$$\frac{dy}{dx} = 0 = \frac{k_{v1}(1 - x_{max} - y_{max}) - k_{v2}y_{max}}{k_{v3}(1 - x_{max} - y_{max}) + k_{v2}y_{max}} \tag{21}$$

and so

$$\frac{y_{max}}{1 - x_{max} - y_{max}} = \frac{k_{v1}}{k_{v2}} = \sigma_2 \tag{22}$$

The slope of the curve at the origin (0, 0) is, from Fig. 1 and equations (14) and (15),

$$\frac{dy}{dx} = \frac{k_{v1}}{k_{v3}} = \sigma_3 = \tan \omega \tag{23}$$

van der Borg [7] compared the kinetic and empirical approaches and showed that similar values of y are obtained from various values of x whichever method is used.

7.1.2. Kinetics of the selective formation of the intermediate product in consecutive reactions

Consider a heterogeneous catalytic reaction in which the reactant A is first adsorbed at the catalyst surface and subsequently reacts to form the final product C. The intermediate product B can be detected only if it is desorbed from the surface, and, if C is formed via the intermediate product, then B has to be readsorbed. Nevertheless, B may be formed at the surface and react to form C before it is desorbed. In this latter case, the kinetic equations will not contain a term for the gaseous concentration of B and the direct route A→C, described by a single kinetic constant, is apparently followed.

de Boer and van der Borg [8, 9] proposed a kinetic scheme for pseudo-first-order catalytic reactions involving two reactants A and R. They assumed a Langmuir-Hinshelwood mechanism (see Chapter 1) and supposed that the concentration of R (a second reactant such as hydrogen) does not vary much during the course of the reaction. This condition may be fulfilled if R is strongly adsorbed at the surface so that the reaction is zero order with respect to this component. Alternatively, a slow diffusion of the reactants to the surface may cause pseudo-first-order (overall) kinetics. Taking into account the adsorption and desorption of A, B and C as well as the surface reactions, the scheme may be written

$$A \underset{(k_d)_A}{\overset{(k_a)_A}{\rightleftharpoons}} A(ads) \tag{24}$$

$$A(ads) + R \xrightarrow{(k_s)_A} B\,(ads) \tag{25}$$

$$B(ads) \underset{(k_d)_B}{\overset{(k_d)_B}{\rightleftharpoons}} B \tag{26}$$

$$B(ads) + R \xrightarrow{(k_s)_B} C(ads) \tag{27}$$

$$C(ads) \underset{(k_a)_C}{\overset{(k_a)_C}{\rightleftharpoons}} C \tag{28}$$

de Boer and van der Borg [8, 9] showed that under certain conditions the kinetics can be adequately described by either of the simple schemes outlined by the routes (1) and (12). The four cases examined are (i) slow surface chemical reactions, (ii) slow adsorption and desorption, (iii) fast adsorption and desorption of the intermediate product and (iv) the rates of the surface reactions comparable with the rate of adsorption and desorption of the intermediate product.

7.1.2.1. Case (i). Slow Surface Chemical Reaction

In this case all steps, except those which involve surface chemical reaction, are fast. Adsorption–desorption equilibrium is therefore established for each of the species A, B and C. The net rate of adsorption of A is the difference between the rates of adsorption r_α and desorption r_δ,

$$(r_a)_A = (r_\alpha)_A - (r_\delta)_A = (k_a)_A c_A \left(1 - \sum_j \theta_j\right) - (k_d)_A \theta_A \qquad (29)$$

where $(1 - \sum_i \theta_i)$ represents the free surface available for adsorption. If equilibrium is established, the net rate of adsorption may be set equal to zero, so

$$K_A = \frac{(k_a)_A}{(k_d)_A}\left(1 - \sum_j \theta_j\right) = \frac{\theta_A}{c_A} \qquad (30)$$

where c_A is the surface concentration of species A. Similarly, for B and C,

$$K_B = \frac{(k_a)_B\left(1 - \sum_j \theta_j\right)}{(k_d)_B} = \frac{\theta_B}{c_B} \qquad (31)$$

$$K_C = \frac{(k_a)_C\left(1 - \sum_j \theta_j\right)}{(k_d)_C} = \frac{\theta_C}{c_C} \qquad (32)$$

The rate of the surface reactions of the adsorbed species A and B with R will be slow compared to the rate at which adsorption–desorption equilbria are established. For reaction of adsorbed A and B with R we may write

$$-\frac{dc_A}{dt} = (k_s)_A \theta_A \qquad (33)$$

and

$$\frac{dc_B}{dt} = (k_s)\theta_A - (k_s)_B \theta_R \qquad (34)$$

The rate of formation of gaseous product C is, by the same token,

$$\frac{dc_C}{dt} = (k_s)_B \theta_B \qquad (35)$$

Making use of the equilibria defined by equations (30), (31) and (32), we have

$$\frac{dc_A}{dt} = (k_s)_A K_A c_A = k_{v1} c_A \qquad (36)$$

and
$$\frac{dc_B}{dt} = (k_s)_A K_A c_A - (k_s)_B K_B c_B = k_{v1} c_A - k_{v2} c_B \tag{37}$$

$$\frac{dc_C}{dt} = (k_s)_B K_B c_B = k_{v2} c_B \tag{38}$$

In the above equations, only two rate constants k_{v1} and k_{v2} are employed, so the overall kinetics may be described by the simple reaction scheme (1). The kinetic selectivity factor in this case is

$$\frac{k_{v1}}{k_{v2}} = \frac{(k_a)_A (k_d)_B (k_s)_A}{(k_a)_A (k_a)_B (k_s)_B} \tag{39}$$

Provided that the surface reaction of R with adsorbed A is much faster than with adsorbed B, so that $(k_s)_A/(k_s)_B$ is large, the formation of gaseous product B will be preferred to that of C. Even if this ratio is unity, i.e. the rates of reaction of R with adsorbed A and B are comparable, a high selectivity for product B is obtained, provided the surface is capable of adsorbing A much more strongly than B. This implies $(k_{dB}) \gg (k_d)_A$, while at the same time $(k_a)_A \gg (k_A)_B$.

7.1.2.2. *Case (ii). Slow Adsorption and Desorption of Reactant*

In this case both the adsorption and desorption steps are regarded as measurably slow. The surface concentration of A will reach a steady value which we may find by supposing that, in the steady state, $d\theta_A/dt$ is zero. Hence,

$$\frac{dc_A}{dt} = \frac{1}{K_A} \frac{d\theta_A}{dt} = (k_a)_A (1 - \sum_j \theta_j)_A c_A - (k_d)_A \theta_A - (k_s)_A \theta_A = 0 \tag{40}$$

so that
$$\theta_A = \frac{(k_a)_A (1 - \sum_j \theta_j)}{(k_s)_A + (k_d)_A} c_A \tag{41}$$

Similarly, the fraction of surface occupied by B will be, in the steady state,

$$\frac{dc_B}{dt} = \frac{1}{K_B} \frac{d\theta_B}{dt} = (k_s)_A \theta_A - (k_d)_B \theta_B + (k_a)_B (1 - \sum_j \theta_j) c_B - (k_s)_B \theta_B = 0 \tag{42}$$

and so
$$\theta_B = \frac{(k_s)_A}{(k_s)_B + (k_d)_B} \theta_A + \frac{(k_a)_B (1 - \sum_j \theta_j)}{(k_s)_B + (k_d)_B} c_B \tag{43}$$

The rate of formation of gaseous B is

$$\frac{dc_B}{dt} = (k_d)_B \theta_B - (k_a)_B (1 - \sum_j \theta_j) c_B \tag{44}$$

Substituting the expressions for θ_A and θ_B, equation (44) becomes

$$\frac{dc_B}{dt} = \frac{(k_d)_B (k_s)_A (k_a)_A (1 - \sum_j \theta_j)}{\{(k_s)_A + (k_d)_A\}\{(k_s)_B + (k_d)_B\}} c_A + \frac{(k_a)_B (k_s)_B (1 - \sum_j \theta_j)}{(k_s)_B + (k_d)_B} c_B = k_{v1} c_A - k_{v2} c_B \tag{45}$$

By similar reasoning,

$$\frac{dc_C}{dt} = \frac{(k_a)_A(k_s)_A(k_s)_B(1-\sum_j\theta_j)}{\{(k_s)_A+(k_a)_A\}\{(k_s)_B+(k_a)_B\}}c_A + \frac{(k_a)_B(k_s)_B(1-\sum_j\theta_j)}{(k_s)_B+(k_a)_B}c_B = k_{v3}c_A + k_{v2}c_B \tag{46}$$

In this case the rates of formation of gaseous B and C are described in terms of three overall rate constants and is equivalent to the reaction scheme (12). If B is the desired product then a high selectivity is obtained if both k_2 and k_3 are small in comparison with k_1. In terms of the rate constants for individual steps, the two selectivity factors are

$$\frac{k_{v1}}{k_{v2}} = \frac{(k_a)_A}{\{(k_s)_A + (k_a)_A\}} \frac{(k_d)_B}{(k_a)_B} \frac{(k_s)_A}{(k_s)_B} \tag{47}$$

and

$$\frac{k_{v1}}{k_{v3}} = \frac{(k_d)_B}{(k_s)_B} \tag{48}$$

As for case (i), provided $(k_s)_A/(k_s)_B$ is large, or, if this ratio is unity and A is adsorbed more readily than B, then k_{v1}/k_{v2} is also large and the selectivity is such that a high proportion of B is produced. From equation (48) it is obvious that B must desorb easily in order to maintain a high gaseous concentration of this intermediate product. Since the rate constant $(k_a)_B$ is related to the heat of adsorption Q_B of B (see Section 2.3.1) by

$$(k_a)_B = A_d \exp\left(\frac{-(E_d)_B}{RT}\right) = A_d \exp\left(\frac{-(E_a)_B + Q_B}{RT}\right) \tag{49}$$

then if B is the desired product then the heat of adsorption of B at the catalyst should be low.

7.1.2.3. Case (iii). Fast Adsorption and Desorption of the Intermediate Product

In this case the extent of surface occupied by A will be determined by its rate of adsorption and desorption, as well as by its rate of consumption in the surface chemical reaction step. Thus θ_A will be given by equation (41). Since adsorption–desorption equilibrium is established rapidly for B, then its surface coverage will be given by equation (31). Utilizing the steady-state method, we obtain the rate equations

$$\frac{dc_A}{dt} = (k_s)_A\theta_A = \frac{(k_s)_A(k_a)_A\left(1 - \sum_j\theta_j\right)}{(k_s)_A + (k_a)_A}c_A = k_{v1}c_A \tag{50}$$

$$\frac{dc_B}{dt} = (k_s)_A\theta_A - (k_s)_B\theta_B = k_{v1}c_A - (k_s)_BK_Bc_B = k_{v1}c_A - k_{v2}c_B \tag{51}$$

$$\frac{dc_B}{dt} = (k_s)_B\theta_B = k_{v2}c_B \tag{52}$$

It is apparent that in this case also the set of equations is reduced to that of reaction (1) in which only two kinetic constants are defined. The selectivity k_{v1}/k_{v2} is governed by equation (47) and the same criteria apply for obtaining a high proportion of intermediate product.

7.1.2.4. *Case (iv). Rates of Surface Reactions Comparable with Rates of Adsorption and Desorption of Intermediate Product*

We assume that adsorption–desorption equilibrium is established rapidly for A. Hence equation (30) is applicable for the amount of surface occupied by A. In the steady state the surface occupied by B will be given by equation (43). The rate of formation of B is then

$$\frac{dc_B}{dt} = (k_d)_B \theta_B - (k_a)_B (1 - \sum_j \theta_j) c_B$$

$$= \frac{(k_d)_B (k_s)_A K_A}{(k_s)_B + (k_d)_B} c_A - \frac{(k_a)_B (k_s)_B (1 - \sum_j \theta_j)}{(k_s)_B + (k_d)_B} c_B$$

$$= k_{v1} c_A - k_{v2} c_B \qquad (53)$$

and of C is

$$\frac{dc_C}{dt} = (k_s)_B \theta_B = \frac{(k_s)_B (k_s)_A K_A}{(k_s)_B + (k_d)_B} c_A + \frac{(k_a)_B (k_s)_B (1 - \sum_j \theta_j)}{(k_s)_B + (k_d)_B} c_B$$

$$= k_{v3} c_A + k_{v2} c_B \qquad (54)$$

As in case (ii), three overall rate constants are sufficient to characterize the kinetics which reduce to the simple reaction scheme (12). For a high yield of B, k_{v1} should be large in comparison with both k_{v2} and k_{v3}. The selectivity k_{v1}/k_{v2} is given by equation (39) in this case, while k_{v1}/k_{v3} is obtained from equation (48). If B is the desired product, then, for high selectivity, k_{v1}/k_{v3} must be large and the heat of adsorption of B at the catalyst surface must be small.

7.1.3. Localized and mobile adsorption of the intermediate product: effect on selectivity

7.1.3.1. *Localized Adsorption*

Applying the absolute rate theory to the localized adsorption of both B and R at the surface, de Boer and van der Borg [9] obtained expressions for the desorption of B and the surface interaction of adsorbed B and R. For simplicity, we assume that no rotational degrees of freedom need to be taken into account and that the frequency of vibration in a plane at right-angles to the surface is low. Following the results of Chapter 2 (Section 2.4.1.4) the rate of desorption of B is

$$(r_d)_B = c_B \frac{kT}{h} \frac{f_\pm}{f_B} \exp\left(\frac{-(E_d)_B}{kT}\right) \qquad (55)$$

where c_B is the surface concentration of B, f_+ is the partition function of the activated complex, f_B is the partition function of adsorbed B and $(E_d)_B$ the activation energy of desorption of B. Similarly, the rate of reaction of adsorbed B with adsorbed R is

$$(r_s)_B = \frac{c_B c_R}{c_S} z \frac{kT}{h} \frac{(f_+)_{BR}}{f_{BR}} \exp\left(\frac{-(E_s)_B}{kT}\right) \qquad (56)$$

where c_S is the total surface concentration of active sites, z the number of nearest neighbour sites, $(f_+)_{BR}$ is the partition function of the activated complex formed between B, R and the surface, f_{BR} is the partition function of adsorbed pairs of B and R and $(E_s)_B$ the activation energy for the surface reaction. We assume f_+ and $(f_+)_{BR}$ are approximately equal. Because the adsorption is localized, $f_B = f_{BR} = 1$. If, with these simplifications, equation (55) is divided by equation (56), there results

$$\frac{(r_d)_B}{(r_s)_B} = \frac{c_S}{c_R} \frac{1}{z} \exp\left(\frac{-[(E_d)_B - (E_s)_B]}{kT}\right) \qquad (57)$$

7.1.3.2. Mobile Adsorption

For mobile adsorption, the rate $(r_d)_B$ of desorption of B is the same as that given by equation (55). However, for the rate $(r_s)_B$ of the surface reaction in this case

$$(r_s)_B = c_B c_R \frac{kT}{h} \frac{f_{+m}}{f_B f_R} \exp\left(\frac{-(E_s)_B}{kT}\right) \qquad (58)$$

where f_{+m} is the partition function of the mobile activated complex. Dividing equation (49) by equation (52) and assuming that $f_+ \approx f_{+m}$, we have

$$\frac{(r_d)_B}{(r_s)_B} = \frac{f_R}{c_R} \exp\left(\frac{-[(E_d)_B - (E_s)]_B}{kT}\right) \qquad (59)$$

It is evident from equations (57) and (59) that the difference between localized and mobile adsorption is determined by the pre-exponential factors. The complete partition function of R may be written in terms of the translational, rotational and vibrational partition functions

$$f_R = \frac{2\pi M_R kT}{h^2} (f_R)_r (f_R)_{vi} \qquad (60)$$

Neglecting the rotational and vibrational contributions to the partition function, f_R is of the order of 10^{-6} at 300°K. The total concentration of active sites c_S on an active catalyst is usually of the order of $10^{15} cm^{-2}$ while z, the number of nearest neighbour sites, is, depending on the surface geometry, 3, 4 or 6. Therefore, the pre-exponential factor for mobile adsorption is at least 30 times the value for localized adsorption. If the activation energies of desorption and for the surface reaction are equal, then the ratio $(r_d)_B/(r_s)_B$ (which is the ratio of rate of desorption of B to the rate of the surface reaction of B with R) will also be 30 times higher if mobile adsorption occurs than if the adsorption were localized. This means that the selective formation of product B is enhanced by mobile adsorption. Furthermore, an increase in temperature promotes mobile adsorption and hence increases the selectivity in favour of B. If, in addition, $(E_d)_B < (E_s)_B$, then an increase in temperature will have an even more pronounced effect on selectivity. On the other hand, when the activation energy of desorption is considerably less than the activation energy for the surface reaction, the catalyst is not such an active one in spite of favourable selectivity.

7.1.4. Effect of pore size on catalyst selectivity

The above considerations do not take into account the possibility that, in the case of a microporous catalyst, diffusion of reactants into and products out of the pores may impede the reaction rate of one reaction step to a greater extent than that of another. If this does occur, then clearly the pore structure will have a direct influence on selectivity.

7.1.4.1. *Isothermal Conditions*

Wheeler [10] considered the problem of selectivity in porous catalysts for three distinct kinetic schemes.

Consider two simultaneous reactions

$$A \xrightarrow{k_{v1}} B + C \tag{61}$$

$$X \xrightarrow{k_{v2}} Y + Z \tag{62}$$

occurring on the same catalyst. Suppose the desired product is B. If the rates (measured as moles per unit volume per unit time) of each reaction are expressed in differential form and divided, then

$$\frac{\dfrac{dc_A}{dt}}{\dfrac{dc_X}{dt}} = \frac{k_{v1}c_A}{k_{v2}c_X} \tag{63}$$

where c_A and c_X are the molar concentrations of A and X per unit volume. If the initial concentrations of A and X are $c_A{}^0$ and $c_X{}^0$ respectively, then integration gives

$$\frac{c_A}{c_A{}^0} = \left(\frac{c_X}{c_X{}^0}\right)^{k_{v1}/k_{v2}} \tag{64}$$

which, if expressed in terms of the fraction y of reactants reacted, becomes

$$y_A = 1 - [1 - y_x]^{k_{v1}/k_{v2}} \tag{65}$$

Equation (59) tells us that, for a given fraction of X converted, the fraction of A reacted is a function only of the ratio k_{v1}/k_{v2}. The kinetic selectivity is therefore governed by the relative rate of the two reactions.

On a porous catalyst, the rate of one of the reactions may be impeded by the rate of diffusion of reactant into the pore structure, so that the progress of the other reaction is favoured. Consider the single pore model of Wheeler; then, as shown in Chapter 4, if both reactions are first order, the reaction rate depends in each case on the quantity $h_1 \tanh h_1$ where h_1 is the parameter $L(2k_1/rD)^{1/2}$, L is the pore length, r its radius, D the diffusion coefficient and k_1 the first-order rate constant per unit surface area. The selectivity in a porous catalyst is thus

$$\frac{h_{11} \tanh h_{11}}{h_{12} \tanh h_{12}} = \sqrt{\left(\frac{k_{11}}{D_A}\frac{D_X}{k_{12}}\right)} \frac{\tanh h_{11}}{\tanh h_{12}} \tag{66}$$

where h_{11} and h_{12} are the Thiele moduli for the first and second reaction steps and k_{11} and k_{12} are the first-order velocity constants per unit area for each corresponding step. For large values of h_1, $\tanh h_1$ is equal to h_1 and the selectivity observed would be the same as on a non-porous catalyst, i.e. simply equal to k_{11}/k_{12}. However, for a catalyst with small pores, $\tanh h_1$ is approximately unity, so that the selectivity now depends on the square root of the ratio of the rate constants. The corollary is that, for simultaneous reactions, maximum selectivity is displayed by catalysts with large pores.

If the reactant yields two different products, selectivity is unaffected provided that the order of the two reactions is the same. Suppose that in reaction scheme (2) both reactions show first-order kinetics

$$A \underset{k_{12}}{\overset{k_{11}}{\lessgtr}} \begin{matrix} B \\ C \end{matrix} \tag{2}$$

then we can say that the rate at which A is consumed is balanced by the rate of diffusion into the pellet. Hence, the material balance equation for component A is

$$\pi r^2 D \frac{\partial^2 c_A}{\partial x^2} = 2\pi r(k_{11} + k_{12})c_A \tag{67}$$

Similarly, the rates at which the products B and C are formed will, in the steady state, be balanced by their rates of diffusion out of the pore. For the products, the diffusion equations are therefore

$$\pi r^2 D \frac{\partial^2 c_B}{\partial x^2} = -2\pi r k_{11} c_A \tag{68}$$

and

$$\pi r^2 D \frac{\partial^2 c_C}{\partial x^2} = -2\pi r k_{12} c_A \tag{69}$$

where the diffusion coefficients for each component are considered numerically equal. With the usual boundary conditions for equation (67), we can now compute the rate of disappearance of A by a single integration. If the boundary condition for equation (68) is $\partial c_B/\partial x = 0$ at $x = L$, then

$$\pi r^2 D \left(\frac{\partial c_B}{\partial x}\right)_{x=0} = 2\pi r k_{11} c_A{}^0 L \frac{\tanh h'_1}{h'_1} \tag{70}$$

where

$$h'_1 = L \left(\frac{2(k_{11} + k_{12})}{rD}\right)^{1/2} \tag{71}$$

The rate of formation of C will be given by a similar expression with k_{12} replacing k_{11} in equation (70). Thus, the relative rate of formation of B with respect to C at the pore mouth is simply k_{11}/k_{12}. Hence, the selectivity is independent of pore size when both reactions are of the same order. However, if the formation of C proceeds by a reaction of zero order, we replace equations (67)–(69) by the set of equations

$$\pi r^2 D \frac{\partial^2 c_A}{\partial x^2} = 2\pi r(k_{11} c_A + k_{02}) \tag{72}$$

$$\pi r^2 D \frac{\partial^2 c_B}{\partial x^2} = -2\pi r k_{11} c_A \tag{73}$$

$$\pi r^2 D \frac{\partial^2 c_C}{\partial x^2} = -2\pi r k_{02} \tag{74}$$

where k_{02} refers to the zero order rate constant for the second reaction step. With the boundary conditions $x=0$, $c_A = c_A{}^0$; and $x=L$, $\partial c_B/\partial x = \partial c_C/\partial x = 0$, the relative rate of formation of B with respect to C at the pore mouth may be found by solving the above equations (72–74). Thus we obtain

$$\frac{R_B{}^0}{R_C{}^0} = \frac{\pi r^2 D \left(\frac{\partial c_B}{\partial x}\right)_{x=0}}{\pi r^2 D \left(\frac{\partial c_C}{\partial x}\right)_{x=0}} = c_A{}^0 \frac{k_{11}}{k_{02}} \frac{\tanh h_{11}}{h_{11}} - 1 \tag{75}$$

where h_{11} is the Thiele modulus for the first-order reaction and is defined by $L(2k_{11}/rD)^{1/2}$. If the pores are sufficiently small, the relative rate of formation is equal to $\{(c_A{}^0 k_{11}/k_{02}h_{11}) - 1\}$. This may be compared with the relative rate if the reaction were to proceed in very large pores $\{(c_A{}^0 k_{11}/k_{02}) - 1\}$. It is evident that the rate of formation of B with respect to C is impeded when the reaction occurs in a catalyst with small pores, and provided that $h_{11} > 1$. The zero-order reaction is therefore enhanced. A similar conclusion may be drawn when one of the reactions is second order and the other first order: the reaction of the lowest kinetic order is favoured. Such an effect is apparent when one considers that the rate of a second-order reaction will fall considerably in comparison with a first-order reaction when there is only a low concentration of reactant available in the interior of the pellet.

We now consider consecutive reactions (scheme (1), p. 320). Again B is considered to be the desired product, while C is wasteful. The diffusion equation for B is, for first-order kinetics,

$$\pi r^2 D \frac{\partial^2 c_B}{\partial x^2} = 2\pi r (k_{12} c_B - k_{11} c_A) \tag{76}$$

while that for reactant A is identical with the case for a reaction in which only A is present, and which is given by equation (103), Chapter 4,

$$\pi r^2 D \frac{\partial^2 c_A}{\partial x^2} = 2\pi r k_{11} c_A.$$

Solving these two simultaneous linear differential equations with the usual boundary conditions,

$$\frac{c_B}{c_B{}^0} = 1 + \frac{c_A{}^0}{c_B{}^0} \left\{ \frac{k_{11}}{(k_{11} - k_{12})} \right\} \frac{\cosh\left\{ L\left(\frac{2k_{12}}{rD}\right)^{1/2} - x\left(\frac{2k_{12}}{rD}\right)^{1/2} \right\}}{\cosh\left\{ L\left(\frac{2k_{12}}{rD}\right)^{1/2} \right\}}$$

$$- \frac{c_A{}^0}{c_B{}^0} \left\{ \frac{k_{11}}{k_{11} - k_{12}} \right\} \frac{\cosh\left\{ L\left(\frac{2k_{11}}{rD}\right)^{1/2} - x\left(\frac{2k_{11}}{rD}\right)^{1/2} \right\}}{\cosh\left\{ L\left(\frac{2k_{11}}{rD}\right)^{1/2} \right\}} \tag{77}$$

where $c_A{}^0$ and $c_B{}^0$ are the concentrations of A and B at the pore mouth. The solution for c_A is equation (106) of Chapter 4. The rate of formation of B with respect to A at the pore mouth is then

$$\frac{R_B{}^0}{R_A{}^0} = \frac{\pi r^2 D \left(\frac{\partial c_B}{\partial x}\right)_{x=0}}{\pi r^2 D \left(\frac{\partial c_A}{\partial x}\right)_{x=0}} = \frac{k_{11}}{k_{11} - k_{12}} - \left(\frac{c_B{}^0}{c_A{}^0} + \frac{k_{11}}{(k_{11} - k_{12})}\right) \left(\frac{k_{12}}{k_{11}}\right)^{1/2} \frac{\tanh\left\{ h_{11}\left(\frac{k_{11}}{k_{12}}\right)^{1/2} \right\}}{\tanh h_{11}} \tag{78}$$

For a fast reaction in small pores h_{11} is very large and then

$$\frac{R_B{}^0}{R_A{}^0} = -\frac{\partial c_B{}^0}{\partial c_A{}^0} = \frac{\left(\frac{k_{11}}{k_{12}}\right)^{1/2}}{1 + \left(\frac{k_{11}}{k_{12}}\right)^{1/2}} - \left(\frac{k_{12}}{k_{11}}\right)^{1/2} \frac{c_B{}^0}{c_A{}^0} \tag{79}$$

Integrating this equation to obtain the fraction α_B of B converted in terms of the fraction α_A of A converted,

$$\alpha_B = \frac{k_{11}}{k_{11} - k_{12}} \, (1 - \alpha_A) \left\{ (1 - \alpha_A)^{\sqrt{(k_{12}/k_{11})} - 1} - 1 \right\} \qquad (80)$$

For slow reactions in very large pores $h_{11} \ll 1$ so, under these conditions,

$$\frac{R_B^0}{R_A^0} = - \frac{\partial c_B^0}{\partial c_A^0} = 1 - \frac{k_{12} c_B^0}{k_{11} c_A^0} \qquad (81)$$

Integration of equation (81) gives

$$\alpha_B = \frac{k_{11}}{k_{11} - k_{12}} \, (1 - \alpha_A) \left\{ (1 - \alpha_A)^{(k_{12}/k_{11}) - 1} - 1 \right\} \qquad (82)$$

Comparison of equations (80) and (82) indicates that, for small pores, the selectivity is less than for large pores, since the relative rate in this case depends on the square root of the ratio of rate constants. When $k_{11} > k_{12}$, for a given fraction of A converted the conversion to B is impeded by using a catalyst with small pores. In a practical case, if the catalyst activity and yield of B are both greater on small pellets than on large pellets, then the pore size must be influencing selectivity. This corresponds to $0.3 < h_{11} < 3.0$ in equation (78). If the activity depends on pellet size, while selectivity is independent of pellet size, the selectivity is decreased considerably by the pore structure, for in this case equation (80) shows that when h_{11} is large the selectivity is constant, but much smaller than the selectivity on a non-porous catalyst. Wheeler [10] suggests using fluidized beds in the latter case to improve yields, since a drastic reduction in pellet size is necessary. The operation of a fixed bed reactor containing small particles causes large pressure drops across the reactor and this is obviated if a fluidized bed is used.

7.1.4.2. Non-isothermal Conditions

The influence of the simultaneous transfer of mass and heat in the selectivity of first-order concurrent heterogeneous catalytic reactions has recently been investigated by Ostergaard [11]. As shown by equation (72), for two concurrent first-order reactions

$$A \underset{k_{12}}{\overset{k_{11}}{<}} \begin{array}{c} B \\ \\ C \end{array}$$

the selectivity is not affected by any limitations due to mass transfer. However, if heat is transferred between the interior and exterior of catalyst pellets (made possible by temperature gradients resulting from the slow diffusion of components into the pellet catalysing a very exothermic reaction), the selectivity may be substantially altered. The model which Ostergaard adopted to investigate this problem is similar to the flat plate model of Thiele and the single-pore model of Wheeler discussed in Chapter 4. The steady-state material balance equations required to be solved are similar to those described previously, but the velocity constants are now functions of temperature and a pore of unit cross-sectional area is assumed. For component A, diffusion along a single pore gives the equation

$$\pi r^2 D \frac{\partial^2 c_A}{\partial x^2} - 2\pi r \left\{ k_{11} \exp\left(-\frac{E_1}{RT}\right) + k_{12} \exp\left(-\frac{E_2}{RT}\right) \right\} c_A = 0 \quad (83)$$

where k_{11} and k_{12} are now temperature-independent constants, E_1 and E_2 are the activation energies of the respective reactions and r is the pore radius.

For component B the corresponding equation is

$$\pi r^2 D \frac{\partial^2 c_B}{\partial x^2} + 2\pi r \left\{ k_{11} \exp\left(-\frac{E_1}{RT}\right) \right\} c_B = 0 \quad (84)$$

A similar equation may be written for component C. The diffusivities of components A, B and C are assumed to be equal. The steady-state equation expressing the conservation of heat within the pore is

$$\pi r^2 \kappa \frac{\partial^2 T}{\partial x^2} - 2\pi r \left\{ \Delta H_1 k_{11} \exp\left(-\frac{E_1}{RT}\right) + \Delta H_2 k_{12} \exp\left(-\frac{E_2}{RT}\right) \right\} c_A = 0 \quad (85)$$

where ΔH_1 and ΔH_2 are the respective enthalpies of reaction and κ is the thermal conductivity. The sets of boundary conditions for the problem are

$$\frac{\partial c_A}{\partial x} = \frac{\partial c_B}{\partial x} = \frac{\partial c_C}{\partial x} = \frac{\partial T}{\partial x} = 0 \text{ at } x = L \quad (86)$$

and

$$c_A = c_A{}^0, \; c_B = c_B{}^0, \; c_C = c_C{}^0, \; T = t^0 \text{ at } x = 0 \quad (87)$$

where $2L$ is the length of the open-ended pore. This is a two point boundary condition problem and, because of the nonlinearity of the equations, cannot be solved analytically. If, however, $E_1 = E_2$, it follows from the above equations that

$$\left(\frac{\partial c_A}{\partial x}\right)_{x=0} = \frac{\left(1 + \dfrac{k_{12}}{k_{11}}\right)}{\left(1 + \dfrac{\Delta H_2 k_{12}}{\Delta H_1 k_{11}}\right)} \frac{\kappa}{D \Delta H_1} \left(\frac{\partial T}{\partial x}\right)_{x=0} \quad (88)$$

and

$$\left(\frac{\partial c_B}{\partial x}\right)_{x=0} = \frac{1}{\left(1 + \dfrac{\Delta H_2 k_{12}}{\Delta H_1 k_{11}}\right)} \frac{c_B{}^0 \kappa}{c_A{}^0 D \Delta H_1} \left(\frac{\partial T}{\partial x}\right)_{x=0} \quad (89)$$

The selectivity, determined by the ratio of the reaction rates at the pore entrance, is therefore

$$\frac{c_B{}^0 \left(\dfrac{\partial c_C}{\partial x}\right)_{x=0}}{c_C{}^0 \left(\dfrac{\partial c_B}{\partial x}\right)_{x=0}} = -\frac{c_B{}^0 \left(\dfrac{\partial c_A}{\partial x}\right)_{x=0}}{c_A{}^0 \left(\dfrac{\partial c_B}{\partial x}\right)_{x=0}} - 1 = \left(1 + \frac{k_{12}}{k_{11}}\right) - 1 = \frac{k_{12}}{k_{11}} \quad (90)$$

and this is identical to the selectivity which would have been observed had there not been any resistance to mass and heat transfer.

For the case $\Delta H_1 = \Delta H_2$, but $E_1 \neq E_2$, the selectivity is determined by simultaneous mass and heat transfer effects. The equations for this particular case were solved numerically [11] for a pore of unit cross-section. The selectivity and the effectiveness factor η (defined as the ratio of the actual rate to that which would occur if there were no effects due to mass and heat transfer) were computed as a function of the Thiele modulus for various parameters β and E_2/E_1. The Thiele modulus is here defined by

$$h_T = L \left(\frac{k_{11}}{D}\right)^{1/2} \left\{ \exp\left(-\frac{E_1}{RT}\right) + \frac{k_{12}}{k_{11}} \exp\left(-\frac{E_2}{RT}\right)^{1/2} \right\} \qquad (91)$$

and the parameter β as

$$\beta = -\frac{\Delta H_1 \, D c_A^0}{\kappa T^0} \qquad (92)$$

When $E_1 < E_2$ it is apparent from Fig. 2 that the selectivity increases from unity towards an upper limit as h_T increases. However, when $E_1 > E_2$ a

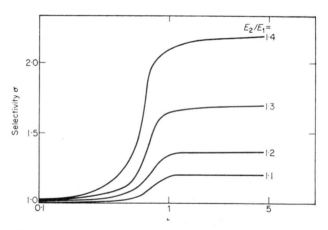

FIG. 2. Kinetic selectivity as a function of the Thiele modulus for various values of E_2/E_1 as parameter.

decrease in selectivity results. Figure 3 shows that a higher value of β also corresponds to a more rapid increase in selectivity and to a higher numerical value for the selectivity at high values of the modulus h_T. The curves in Figs. 4 and 5 show the change in effectiveness factor η characteristic of exothermic reactions. The effectiveness factors are nearly unity at low values of h_T, pass through maxima as h_T increases and then decrease with further increase in the modulus h_T.

The conclusions drawn from the above analysis are: (i) if the activation energy of the desired reaction is lower than that of the reaction leading to the wasteful product, the best selectivity is obtained for low values of the Thiele modulus; (ii) if the activation energy of the desired reaction is higher than that of the competing reaction, the best selectivity is obtained for high values of the Thiele modulus. In the latter case, however, the selectivity approaches an upper limit asymptotically and it is not worth while increasing the Thiele

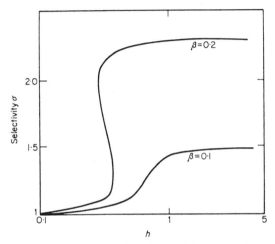

FIG. 3. Kinetic selectivity as a function of the Thiele modulus for various values of β as parameter.

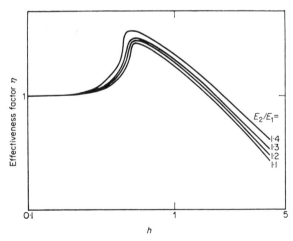

FIG. 4. Effectiveness factor as a function of the Thiele modulus for various values of E_2/E_1 as parameter.

modulus beyond a value where there would be a significant decrease in the efficiency of conversion. The optimum value of the Thiele modulus will be close to that for which the effectiveness factor is a maximum.

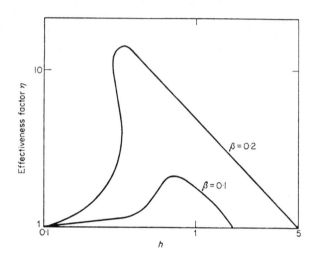

FIG. 5. Effectiveness factor as a function of the Thiele modulus for various values of β as parameter.

7.2. Polyfunctional Catalysis

In the kinetic scheme discussed in Section 7.1.2, the intermediate gaseous product B can be formed only if adsorbed B is desorbed from the surface in a detectable amount. If its further reaction at the surface to form adsorbed C occurs more readily than desorption, then no gaseous B is produced, although adsorbed B exists as an intermediate in the reaction. If the conversion of A to B requires different catalyst sites from the conversion of B into C then, unless there is surface migration of adsorbed B from one type of site to another, gaseous B must be formed, even if only in traces so small as to be analytically undetectable. The presence of two or more chemically distinct catalyst sites constitutes a system which is capable of performing more than one function. The conversion of A to B might be effected by a catalyst X, while the further conversion of B into C might not take place at a measurable rate unless catalyst Y is present. Hence a compaction of catalysts X and Y into a single composite catalyst will provide a bifunctional catalyst capable of converting A into C via an intermediate B according to scheme (I),

$$A \overset{X}{\longrightarrow} B \overset{Y}{\longrightarrow} C \tag{I}$$

Because B must be desorbed from X and readsorbed at Y before C can be formed, it is a necessary condition that B should have at least a transitory existence in the gas phase.

Polyfunctional catalysis makes possible, in one single operation, a wide variety of syntheses which would otherwise require separate chemical processes. For example, Heinemann et al. [12] showed that a bifunctional catalyst is capable of producing butane from heptane in one operation. Weisz and Swegler [13] showed that a composite catalyst compacted from a silica–alumina cracking catalyst and a platinum catalyst supported on high-area carbon will convert hexadecane into a mixture of octanes. The mechanism postulated may be described by the reactions

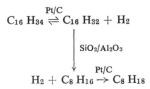

The supported platinum dehydrogenates the hexadecane to hexadecene, which is then cracked by the silica–alumina to octene: in the presence of hydrogen the supported platinum will convert octene to octane. This course of events illustrates one of the uses of a polyfunctional catalyst.

Another important use of a polyfunctional catalyst is to intercept the course a reaction would take if only a monofunctional catalyst were used. Thus, the dehydrogenation of cyclohexane to benzene using a platinum catalyst may be diverted by means of a silica–alumina catalyst to yield methylcyclopentane [14, 15]. The diversion depends upon the silica–alumina catalyst intercepting the intermediate product cyclohexene and isomerizing it to methylclyclopentene, which is in turn hydrogenated to methylcyclopentane by the platinum catalyst already present. Whether or not methylcyclopentane is produced in greater amounts than benzene depends on the selectivity of the composite catalyst. As will be shown later, pellet size, temperature and pressure have a profound influence on the selectivity of a polyfunctional catalyst.

7.2.1. The non-trivial polystep reaction

Weisz [4] defines a non-trivial polystep reaction as one in which the conversion of reactants is greater than the product of conversions attainable in separate successive steps proceeding via an intermediate gaseous component. This is in contrast to a trivial polystep reaction in which the conversion of reactants in one single step is the same as the conversion obtained by carrying out the reaction in two separate successive steps.

Clearly, if a high conversion is attainable for each step in reaction scheme (I) (p. 338), using two separate reactors, one containing catalyst X and the other catalyst Y, then physical combination of the catalysts in a single reactor may be expected to produce a similar overall conversion of A into C. However, if the conversion of A into B is limited by chemical equilibrium according to scheme (II),

$$A \underset{k'_{v1}}{\overset{k_{v1}}{\rightleftharpoons}} B \overset{k_{v2}}{\longrightarrow} C \tag{II}$$

the conversion of A into C in a single operation may, under appropriate conditions, be considerably greater than the limited conversion using two separate operations.

Consider two separate reaction zones, one of which contains catalyst X and allows the equilibrium reaction

$$A \underset{k'_{v1}}{\overset{k_{v1}}{\rightleftharpoons}} B$$

to occur, and the other contains catalyst Y, so converting the B produced in the first zone into C

$$B \overset{k_{v2}}{\longrightarrow} C$$

For the first reaction zone the rate of conversion of A into B is written

$$\frac{dx}{dt} = k_{v1}(1 - x) - k'_{v1}x \tag{93}$$

where x represents the fractional conversion of A. With the initial condition $x=0$, the conversion of A into B in the first stage is given by integrating equation (93). Thus,

$$x = \frac{1}{1 + \dfrac{k'_{v1}}{k_{v1}}} \{1 - \exp(-k_{v1}t)\} \tag{94}$$

In the second reaction zone the product B which emerges from the first zone is converted into C by catalyst Y, so we may write

$$\frac{dy}{dt} = k_{v2} \left[\frac{1}{1 + \dfrac{k'_{v1}}{k_{v1}}} \{1 - \exp(-k_{v1}t)\} - y \right] \tag{95}$$

where y represents the fractional conversion of B. Multiplying through by the integrating factor $\int \exp(k_{v2}t)dt$ and utilizing the second initial condition $y=0$, we obtain for the solution

$$y = \frac{k_{v2}}{1 + \dfrac{k'_{v1}}{k_{v1}}} \left[\frac{1}{k_{v2}} - \frac{1}{(k_{v2}-k_{v1})} \exp(-k_{v1}t) - \frac{k_1}{k_{v2}(k_{v2}-k_{v1})} \exp(-k_{v2}t) \right] \tag{96}$$

The maximum possible value of y will be obtained after an infinite time has elapsed and so

$$y_\infty = \frac{1}{1 + \dfrac{k'_{v1}}{k_{v1}}} = \frac{1}{1 + \dfrac{1}{K}} \tag{97}$$

where K is the equilibrium constant for the equilibrium $A \rightleftharpoons B$. However large K is, the fraction converted is always less than unity. If K is small, then the fraction of A converted to C may be considerably less than unity, even after an infinite time.

If catalysts X and Y are intimately mixed in equal proportions with respect to surface area and packed into a single reacting zone, the reaction scheme (II) occurs in a continuous manner. Under these conditions, two simultaneous kinetic equations must be solved.

$$\frac{dx}{dt} = k_{v1}(1 - x) - k'_{v1}(x - y) \tag{98}$$

$$\frac{dy}{dt} = k_{v2}(x - y) \tag{99}$$

With the initial conditions $x=0$, $y=0$ the solution becomes

$$y = \frac{m_2}{m_1 - m_2} \exp(m_1 t) + \frac{m_1}{m_1 - m_2} \exp(m_2 t) + 1 \tag{100}$$

where m_1 and m_2 are the roots of

$$m^2 + (k_{v1} + k'_{v1} + k_{v2}) m + k_{v1} k_{v2} = 0 \tag{101}$$

The maximum possible value of y in this case will be unity.† For a non-trivial polystep reaction it is therefore possible to obtain a greater overall conversion by using a single-stage operation. The above analysis is only strictly correct provided that the diffusion of B through the gas phase from site X to site Y is fast in comparison with the actual surface reactions. When the surface reactions are comparable in speed to the diffusion of B an additional set of equations must be written to allow for the transport of gaseous B. This is discussed in Section 7.2.2.

7.2.2. Mass transport of the intermediate product in non-trivial polystep reactions

Further consideration of the non-trivial polystep reaction (78) reveals that, unless sites X and Y are geometrically indistinguishable, the overall reaction rate is limited by the rate at which B, desorbed from X, is transported through the gas phase and along catalyst pores to be adsorbed at Y. Even in the case of a single reaction zone containing a composite bifunctional catalyst, the overall rate of a polystep reaction, such as scheme (II), may be limited by mass transport of B between X and Y, although the catalyst particles are intimately mixed. In general, the particles of X and Y, even for a compacted mixed catalyst, will be separated from one another by a distance many times greater than molecular dimensions, so that a diffusion process for transport of B

---

† This would be untrue if the intermediate product B were not being continually drained away by the formation of C in an irreversible step.

through the gas phase must be invoked. The problem has been considered in detail by Weisz [4] and by Gunn and Thomas [16]. Three cases are considered.

7.2.2.1. *Slow Mass Transport through Pores*

If the catalyst particles X and Y exist as separate small pellets in close contact with one another and are themselves porous, then a different situation arises. If both X and Y particles are porous, the problem becomes one of the diffusion of reactants and products through particles rather than transport from the gas phase to the surface. The diffusion equations solved are analogous to those deduced in Section 4.5 when the case of an equilibrium reaction occurring in a spherical porous catalyst was examined. Considering the reaction scheme (II), in this case a second kinetic step takes place at a physically separate surface. The diffusion equations are deduced, as in Section 4.5.3, by considering the conservation of mass within an infinitesimally thin spherical shell, radii a and $(a+\delta a)$, of a catalyst particle. We suppose that the pellets X and Y have radii R_x and R_y respectively. Assuming all rate processes are fast, except that of surface chemical reaction and diffusion in pores, three diffusion equations are written. First A diffuses along pores and reacts within the catalyst X, so for one diffusion equation we obtain

$$D_x \left\{ \frac{\partial^2 c_A}{\partial a^2} + \frac{2}{a} \frac{\partial c_A}{\partial a} \right\} - k_{v1} c_A + k'_{v1} c_B = 0 \tag{102}$$

where D_x is the effective diffusivity of reactant in X catalyst particles. The k'_vs are expressed as chemical rate constants per unit bulk volume of catalyst pellet and, for first-order reactions, have the dimensions of reciprocal time. A similar equation may be written for the diffusion of B concomitant with surface reaction within X,

$$D_x \left\{ \frac{\partial^2 c_B}{\partial a^2} + \frac{2}{a} \frac{\partial c_B}{\partial a} \right\} + k_{v1} c_A - k'_{v1} c_B = 0 \tag{103}$$

Within the porous structure of Y, B is transformed into C, so the third diffusion equation is

$$D_y \left\{ \frac{\partial^2 c_B}{\partial a^2} + \frac{2}{a} \frac{\partial c_B}{\partial a} \right\} - k_{v2} c_B = 0 \tag{104}$$

The boundary conditions are,

$$a = R_x, \ c_A = c_A{}^0, \ c_A \text{ finite at } a = 0 \tag{105}$$

$$a = R_y, \ c_B = c_B{}^0, \ c_B \text{ finite at } a = 0 \tag{106}$$

$$a = 0, \frac{\partial c_A}{\partial a} = 0 \tag{107}$$

The solution of equations (102), (103) and (104) gives the concentration of components as a function of the particle radii. Useful information can be obtained from this set of equations by considering the rate of reaction within the separate particles of X and Y packed in a tubular reactor. The rate of diffusion of each component into an individual catalyst particle is a measure of the rate of reaction of that component, so that the flux of molecules passing through a packed section of a reactor will be determined by (i) the rate of diffusion at the particle periphery and (ii) the fraction of catalyst of a particular type contained in unit reaction volume. If the flux per unit volume of component A past particles of type X is written $N_A(X)$ and ξ is the fraction of reactor volume occupied by catalyst X,

$$N_A(X) = \xi\,(1 - \psi_b)\,\frac{4\pi R_x^2}{\dfrac{4\pi R_x^3}{3}}\,D_x\left(\frac{\partial c_A}{\partial a}\right)_{a=R_x} \tag{108}$$

where ψ_b is the porosity of the catalyst bed contained in the reactor. For component B reacting in catalyst X,

$$N_B(X) = \xi\,(1 - \psi_b)\,\frac{3D_x}{R_x}\left(\frac{\partial c_B}{\partial a}\right)_{a=R_x} \tag{109}$$

For component B reacting in catalyst Y particles

$$N_B(Y) = (1 - \xi)\,(1 - \psi_b)\,\frac{3D_y}{R_y}\left(\frac{\partial c_B}{\partial a}\right)_{a=R_y} \tag{110}$$

since the fraction of catalyst particles per unit reactor volume which are of type Y is $(1 - \xi)(1 - \psi_b)$. A material balance for component A over an element of length δz of the reactor gives

$$u_l\left(\frac{\partial c_A{}^0}{\partial z}\right)\delta z = -N_A(X)\,\delta z \tag{111}$$

where u_l is the superficial velocity of reactant through the reactor.

Similar equations are written for components B and C,

$$u_l\left(\frac{\partial c_A{}^0}{\partial z}\right)\delta z = N_B(X)\,\delta z \tag{112}$$

$$u_l\left(\frac{\partial c_C{}^0}{\partial z}\right)\delta z = N_B(Y)\,\delta z \tag{113}$$

The equations (111), (112) and (113) are simultaneous linear differential equations which may be solved in closed form when $R_x = R_y$ and $D_x = D_y$ to give the concentration profiles of components A, B and C as functions of the length of reactor. The functional form of these profiles is complex and the original paper of Gunn and Thomas [16] should be consulted for details. Each equation describing a concentration profile contains the parameters k_v, R and ξ, the reaction velocity constants, the radii of the particles and the fraction of catalytic material of type X respectively. Here it suffices to say that solution of equations (102–113) inclusive gives a formal answer to the problem. The important point which emerges from such an analysis is that there exists an optimum value of ξ which will give rise to the best yield of desired product C. That there should be an optimum value is obvious, for if there were no X-type particles present, then B, and hence C, would not be produced, whereas if there were no Y-type particles in the reactor, no C could be formed. The optimum value of ξ which will produce the best yield of C is determined by the condition

$$\frac{\partial c_C^0}{\partial \xi} = 0 \qquad (114)$$

Similar equations may be written for reaction scheme (I) and also for the scheme

$$
\begin{array}{c}
X \\
A \underset{k'_{v1}}{\overset{k_{v1}}{\rightleftharpoons}} B \overset{k_{v2}}{\underset{k_{v3}}{<}} \begin{array}{c} C \\ D \end{array} \\
X \quad Y
\end{array}
\qquad (III)
$$

These cases were also investigated by Gunn and Thomas [16]. For each of these reaction schemes an optimum value of ξ exists which will produce the best yield of desired product. However, the optimum value of ξ varies and depends on the kinetics of the particular system. A numerical solution was obtained for each scheme by arbitrarily assuming the radii of spherical catalyst particles to be 0·25 cm, the volumetric flowrate 1cm^{3}sec^{-1}, and the effective diffusivities 0·02 cm^{2}sec^{-1}. The kinetic constants chosen for the schemes were $k_{v1}=k_{v2}=k_{v3}=1$ sec^{-1} and $k'_{v1}=10$ sec^{-1}. Although other numerical values will give different concentration profiles and catalyst composition optima, depending on the relative rates of reaction and the diffusivity within the particles, the general trend of results may be noted.

The concentration profiles along the reactor length for the optimum catalyst composition are compared in Figs. 6, 7 and 8, while the optimum catalyst composition obtained for each of the reaction schemes (I), (II) and (III) are shown in Fig. 9. It is apparent from Figs. 6, 7 and 8 that when B gives rise to a product C through catalyst X and to D through catalyst Y, as in scheme (III), the increase in concentration of the intermediate B is delayed in comparison to scheme (II) in which only one end product C is formed: the maximum concentration of B is not reached until a larger fraction of the length of the reactor has been traversed. This is because the optimum catalyst formulation required to produce the maximum output (moles per unit reactor volume per unit time) of D in scheme (III) contains a smaller

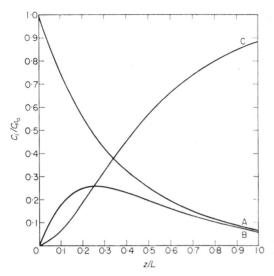

FIG. 6. Reaction scheme (I). Concentration profiles obtained using the optimum catalyst formulation.

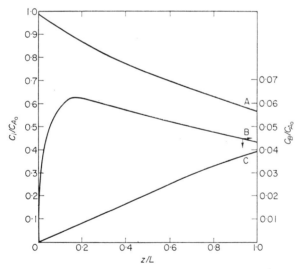

FIG. 7. Reaction scheme (II). Concentration profiles obtained using the optimum catalyst formulation.

fraction of catalyst X than the formulation required to maximize the output of C in scheme (II). However, because two end products are formed in scheme (III), then the rate at which B disappears, once it has attained its maximum value, is greater than in scheme (II) where only one end product

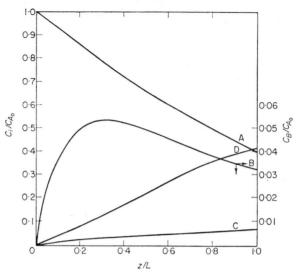

FIG. 8. Reaction scheme (III). Concentration profiles obtained using the optimum catalyst formulation.

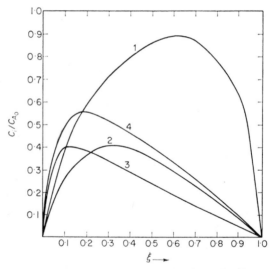

FIG. 9. Optimum catalyst compositions: (i) curve 1, scheme (I); (ii) curve 2, scheme (II); (iii) curve 3, scheme (III); (iv) curve 4, composite catalyst in scheme (III).

is formed. Figure 8 also shows that the total amount of C produced via catalyst X in scheme (III) amounts only to a small fraction of the gases at the reactor exit: if D is the desired product, then the optimum catalyst formulation also serves to suppress, more effectively than any other catalyst composition, the formation of the undesired product C. Figure 9 shows the effect of a change in catalyst composition upon the exit concentration of desired products for the three reaction schemes. Curve 1, referring to reaction (I), is fairly symmetrical and indicates that the optimum catalyst composition should contain equal parts of X and Y. This arises purely from the choice of reaction velocity constants. The influence of the reverse reaction in scheme (II) is immediately apparent when curve 2 is inspected. The optimum composition has been displaced so that a smaller proportion of X is required. This trend is continued in curve 3, which shows the effect of catalyst composition upon the concentration of D for scheme (III).

It is thus apparent that an optimum catalyst composition exists which, for a given reaction scheme, will produce the best yield of desired product. Some further consideration, however, shows that the restriction of constant catalyst composition along the length of the reactor may be undesirable. Intuitively, if the reactant entering the reactor consists only of pure A, there is no point in the presence of catalyst Y at the reactor inlet, since there is no intermediate product to convert. Similarly, at the reactor exit the presence of catalyst X serves no useful purpose, since there is no possibility of further conversion. This therefore suggests that the catalyst composition should change from pure X at the reactor inlet to pure Y at the reactor outlet. This line of thought is

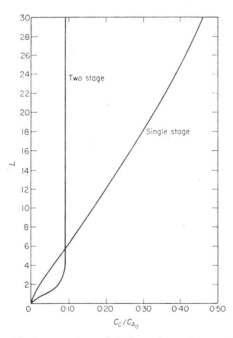

FIG. 10. A comparison of single- and two-stage reactors.

confirmed by Fig. 10, which compares the reactor volume requirements calculated for scheme (II), using a mixture of catalysts X and Y in a single reactor, with the reactor volume requirements for the same reaction scheme in a two-stage reactor (A being converted to B by a reactor containing catalyst X followed by conversion of B into C in a second stage). The most obvious point which Fig. 10 illustrates is the limitation of the maximum possible conversion in the two-stage reactor. This is restricted by the equilibrium $A \rightleftharpoons B$. However, at low conversions the two-stage reactor is superior, since a smaller reactor volume is required to achieve a desired outlet concentration. This confirms the earlier intuitive reasoning, since the catalyst composition profile given by the two-stage reactor is superior to the single-stage reactor at low conversions. The problem of finding the optimum catalyst profile for a given reaction scheme is analogous to the problem of finding the optimum temperature profile for some classes of consecutive and competing reactions. Although the problem of a catalyst composition profile has, as yet, not been considered, Bilous and Amundson [17, 18] have discussed the analogous problem concerning temperature profiles in tubular reactors.

To calculate the fraction of the surface of catalyst X available for reaction, the diffusion-limited rate of reaction is divided by the uninhibited chemical reaction rate. Hence,

$$f_x = \frac{4\pi R_x^2 \left(\dfrac{\partial c_A}{\partial a}\right)_{a=R_x}}{\dfrac{4\pi R_x^3}{3}(k_{v1}c_A^0 - k'_{v1}c_B^0)\,\xi} \tag{115}$$

For negligible resistance to the diffusion of A or B into X the condition is that $f_x = 1$ (but see Section 7.2.3). Similarly, the fraction of the surface of Y available for reaction is

$$f_y = \frac{4\pi R_y^2 \left(\dfrac{\partial c_B}{\partial a}\right)_{a=R_y}}{\dfrac{4\pi R_y^3}{3}(k_{v2}c_B^0 - k_{v1}c_A^0)\,\xi} \tag{116}$$

For negligible resistance to the diffusion of B into catalyst Y, the condition is that $f_y \neq 1$.

7.2.2.2. Slow Mass Transport through Interparticle Space

When small particles of X and Y are compressed together under force to produce composite pellets containing both X and Y, evenly dispersed, a different situation arises. Catalyst pellets prepared in this way will contain a honeycomb of pores consisting of the small voids between the ultimate particles of X and Y. If it is supposed that transport of species through such voids between X and Y is slow and comparable with the rates of surface chemical reaction, then the problem becomes one of diffusion through inter-particle space. If the composite pellet is a spherical pellet of radius R and contains a fraction ζ of the X component, then, for scheme (II), the diffusion equation for A becomes,

$$D_e \left\{ \frac{\partial^2 c_A}{\partial a^2} + \frac{2}{a} \frac{\partial c_A}{\partial a} \right\} - \zeta k_{v1} c_A + \zeta k'_{v1} c_B = 0 \qquad (117)$$

where D_e is the effective diffusivity in the composite catalyst. The diffusion equation for B becomes

$$D_e \left\{ \frac{\partial^2 c_B}{\partial a^2} + \frac{2}{a} \frac{\partial c_B}{\partial a} \right\} + \zeta k_{v1} c_A - \zeta k'_{v1} c_B - (1 - \zeta) k_{v2} c_B = 0 \quad (118)$$

Equation (118) differs from equation (103) in the previous case in so far as B is transformed into A and C within the same catalyst pellet. The boundary conditions are

$$a = R, \; c_A = c_A{}^0, \; c_B = c_B{}^0 \qquad (119)$$

$$a = 0, \; c_A \text{ finite}, \; c_B \text{ finite} \qquad (120)$$

The concentration of C may be determined by a simple material balance. The solution to the above equations will give the concentrations of species A, B and C as a function of the distance of penetration into the pellet interior and the reader is referred to the original paper [16] for full details. The rates of diffusion of each component into the composite catalyst particle is a measure of the rate of reaction of that component. The flux per unit reactor volume for component A is given by

$$N_A = \frac{3(1 - \psi_b)}{R} D_e \left(\frac{\partial c_A}{\partial a} \right)_{a=R} \qquad (121)$$

with similar equations for B and C. From these equations, together with material balances effected over an infinitesimal section of a packed tubular reactor, concentration profiles for the four components as a function of reactor length may be obtained by proceeding in exactly the same way as for the previous case. Figure 11 shows the concentration profiles (full lines) obtained for the case of an isothermal tubular reactor packed with a compo-site catalyst and compares these with the profiles (broken lines) obtained for the case of a reactor packed with a mixture of discrete particles of catalysts X and Y. The optimum catalyst composition for the case of the composite catalyst is shown in Fig. 9 (curve 4). The same numerical values for the reaction velocity constants, particle radius and diffusivities were assumed in order to calculate the illustrated curves for this case as for the previous cases discussed.

It is seen that the optimum catalyst composition for reaction scheme (II), using a composite catalyst, contains a smaller proportion of X than for the case of a mixed catalyst consisting of separate discrete particles. The inter-mediate product B does not have to be transported through interparticle space when X and Y are compounded, so that, although B may be formed from A at the same rate in both cases, an opportunity is afforded to B to decompose more rapidly to the product C when Y is contained within the same porous particle as X.

Referring to Fig. 11, in which concentration profiles (corresponding to the optimum ratio of X to Y) as a function of reactor length are compared for both types of catalyst, the rate of disappearance of B, after it has reached its maximum concentration, is considerably accelerated if a composite cata-lyst replaces a mixed catalyst. The rate of disappearance of A is similarly enhanced, as also is the rate of formation of C. Furthermore, for the numerical values chosen for this calculation, the exit concentration of C is increased by a factor of about 1·5 for the same size reactor. This is due to the fact that the optimum amount of X which is used decreases from 30%, for the mixed cata-

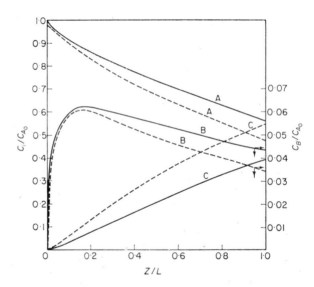

FIG. 11. Comparison of a composite and a mixed catalyst.

lyst, to 20% for the composite catalyst. It is therefore to be expected that the output of product is greater when a given reactor is packed with a com-posite catalyst than when packed with a mixture of catalysts. Alternatively, for the same conversion to product C, one could use a reactor containing a composite catalyst with the optimum ratio of X and Y almost two-thirds the size of a reactor containing the optimum ratio of mixed catalysts. This illustrates the important effect which interparticle diffusion has on the ultimate formation of desired product.

7.2.2.3. *Slow Mass Transport through the Gas Phase*

We next examine the case of a reaction scheme such as (II) in which the rates of diffusion of components to the surfaces X and Y are comparable to the rates of surface reaction, all other rate processes being considered fast. Suppose that the resistance to gaseous diffusion is confined to a thin gaseous film, thickness z, surrounding the catalyst particle. The model considered here is one in which the species A, B and C diffuse to and from the non-porous catalyst particles X and Y prior to the respective surface chemical reactions. For catalysts in the form of flat plates, if we consider an element δx of the gaseous film surrounding the particle, then the steady state diffusion equation to be solved for each component is,

$$D \frac{\partial^2 c_i}{\partial x^2} = 0 \tag{122}$$

where c_i represents the concentration of any one of the components. The boundary conditions are $c_i = c_i^0$ at the surface where $x = 0$, and $c_i = c_i^\delta$ at $x = \delta$ in the bulk gas. The solution to equation (122) is easily obtained for each of the components. At the surface of catalyst X the solution gives

$$c_A = \frac{(c_A^\delta)_x - (c_A^0)_x}{\delta_x} x + (c_A^0)_x \tag{123}$$

and

$$c_B = \frac{(c_B^\delta)_x - (c_B^0)_x}{\delta_x} x + (c_B^0)_x \tag{124}$$

At the surface of catalyst Y,

$$c_B = \frac{(c_B^\delta)_y - (c_B^0)_y}{\delta_y} x + (c_B^0)_y \tag{125}$$

and

$$c_C = \frac{(c_C^\delta)_y - (c_C^0)_y}{\delta_y} x + (c_C^0)_y \tag{126}$$

Equations (123) to (126) express the concentration of the species A, B and C in terms of the film thickness and concentrations at both the fluid side and solid side of the boundary layer. In the steady state the diffusive flux of gas to and from the surface will be balanced by the rates of the respective surface reactions,

$$D \left(\frac{\partial c_A}{\partial x} \right)_x = k_{v1}(c_A^0)_x - k'_{v1}(c_B^0)_x = D \left(\frac{\partial c_B}{\partial x} \right)_x \tag{127}$$

$$D \left(\frac{\partial c_B}{\partial x}\right)_y = k_{v2}(c_B{}^0)_y = - D \left(\frac{\partial c_C}{\partial x}\right)_y \qquad (128)$$

From equations (127) and (128) in conjunction with equations (123) to (126) the diffusive flux may be expressed in terms of concentration in the bulk gas. If ζ is the fraction of the total catalyst surface area which X offers to the reactant, then the diffusive flux of A (which in the steady state equals the reaction rate) is

$$- D \left(\frac{\partial c_A}{\partial x}\right)_x = \frac{D \zeta}{\delta_x \left(k_{v1} + k'_{v1} + \dfrac{D}{\delta_x}\right)} \{k_{v1}(c_A{}^\delta)_x - k'_{v1} (c_B{}^\delta)_x\} \qquad (129)$$

The effectiveness factor, or fraction of surface available for reaction, is the ratio of the inhibited rate to the uninhibited rate. So, for catalyst X,

$$f_x = \frac{D \left(\dfrac{\partial c_A}{\partial x}\right)_x}{k_{v1}(c_A{}^\delta)_x - k'_{v1}(c_B{}^\delta)_x} = \frac{D \zeta}{\delta_x \left(k_{v1} + k'_{v1} + \dfrac{D}{\delta_x}\right)} \qquad (130)$$

At catalyst Y,

$$f_y = \frac{D \left(\dfrac{\partial c_B}{\partial x}\right)_y}{k_{v2}(c_B{}^\delta)_y} = \frac{D(1 - \zeta)}{\delta_y \left(k_{v2} + \dfrac{D}{\delta_y}\right)} \qquad (131)$$

The effectiveness factors can therefore be calculated provided that an estimate can be made for the magnitude of the film thickness from the hydrodynamic nature of the flow and the packing arrangement of catalyst.

The above equations may also be used to obtain a concentration profile for each of the components in a tubular reactor packed with an intimate mixture of catalysts X and Y. If u_l is the superficial flowrate, then, assuming piston-type flow, a material balance across an element of length δz in the reactor gives for each component the steady-state equations

$$u_l \frac{\partial c_A}{\partial z} = D \left(\frac{\partial c_A}{\partial x}\right)_x \qquad (132)$$

$$u_l \frac{\partial c_B}{\partial z} = D \left(\frac{\partial c_B}{\partial x}\right)_{net} = D \left\{\left(\frac{\partial c_B}{\partial x}\right)_x - \left(\frac{\partial c_B}{\partial x}\right)_y\right\} \qquad (133)$$

$$u_l \frac{\partial c_C}{\partial x} = D \left(\frac{\partial c_C}{\partial x}\right)_y \qquad (134)$$

Equations (132–134) with the boundary conditions $c_A = (c_A)_i$ and $c_B = c_C = 0$

at $\delta=0$ give a formal solution to the problem in so far as it gives the concentration profile of each species as a function of the tube length. There exists an optimum value of ζ which will give, for fixed values of the chemical rate constants, the best conversion of A into C. This can easily be calculated by utilizing the condition

$$\left(\frac{\partial c_C}{\partial \zeta}\right) = 0 \tag{135}$$

For any given value of ζ, the maximum conversion of A into C depends on the resistance to mass transfer from the gas phase to the solid catalysts. The high levels of conversion to be expected purely from the chemical kinetics, as discussed in Section 7.2.1, can be obtained only if this resistance to mass transfer is negligible. Suitable criteria for negligible resistance to mass transfer in the gas phase are obtained from equations (130) and (131): these are that $f_x \sim 1$ and $f_y \sim 1$. Slow mass transport through the gas phase will only be important for non-porous catalysts in a static or batch reactor when the boundary layer between gas and solid is not disturbed by turbulent flow conditions.

7.2.3. Simplified criteria for negligible resistance to mass transfer

Weisz [4] also considered the problem of mass transport in the pores of bifunctional catalysts, but imposed an arbitrary restriction that the steady-state flux of B across the boundaries of X and Y are equal and identical with the overall reaction rate. By making this assumption, the condition that there is no resistance to diffusion in the Y catalyst becomes

$$1 > \frac{R_y{}^2}{D_y} k_{v2} = \frac{R_y{}^2}{D_y} \frac{r_o}{c_B} \tag{136}$$

where r_o is the measured overall reaction rate. For negligible resistance to diffusion in the X catalyst,

$$1 > \frac{R_x{}^2}{D_x} (k_{v1} + k'_{v1}) \tag{137}$$

Now we can write, from Weisz's supposition,

$$r_o = \frac{dc_B}{dt} = k_{v1}c_A - k'_{v1}c_B \tag{138}$$

Defining

$$K = \frac{k_{v1}}{k'_{v1}} = \frac{c_B{}^*}{c_A} \tag{139}$$

where $c_B{}^*$ is the concentration of B in equilibrium with a concentration c_A of gaseous A, the criterion (137) becomes

$$1 > \frac{R_x{}^2}{D_x} \frac{(1 + K)r_o}{(c_B{}^* - c_B)} \tag{140}$$

If $k'_{v1} \gg k_{v1}$ and $c_B{}^* \gg c_B$, which is true for a non-trivial system, a genera order of magnitude criterion would be

$$1 > \frac{R^2 r_0}{D c_B{}^*} \tag{141}$$

where it is assumed that the radii of the X and Y particles have the same value ($= R$). In terms of partial pressures, the inequality (141) becomes

$$p^*{}_B > RT \frac{R^2}{D} r_0 \tag{142}$$

where R is the gas constant per mole. Figure 12 shows the equilibrium partial pressure of the intermediate B as a function of the maximum particle size for assumed values of diffusivity and reaction rate. For particle sizes of 1μ the lowest partial pressure of the intermediate which will allow a non-trivial polystep reaction to proceed is as low as 10^{-7} atm.

For the isomerization of n-hexane, which proceeds (under dehydrogenative conditions with a mixed silica–alumina platinum catalyst) via the intermediate n-hexene, the rate of conversion is severely limited by the rate of diffusion into the porous catalyst structure. Since the thermodynamics of the hexane–hexene equilibrium indicates a maximum attainable concentration of hexene ($\sim 2 \cdot 4 \times$ $\times 10^{-8}$ mole cm^{-3}) corresponding to only $0 \cdot 6 \%$ of the hexane concentration, it is possible for a non-trivial reaction to occur. The reaction rate on particles of 6×10^{-2} cm radius is about $0 \cdot 5 \times 10^{-6}$ moles sec^{-1} cm^{-3}. Assuming an effective diffusivity of 2×10^{-3} cm^2 sec^{-1}, the numerical value of the right-hand side of the inequality (141) is $37 \cdot 5$. Now the fraction of surface available for reaction is approximately equal to the reciprocal of the dimensionless quantity $R^2 r_0/c^* D$ (which Weisz designates Φ—see Section 9.2.4) for large values of this parameter. Hence, only 3% ($\approx 1/37 \cdot 5$) of the maximum possible catalytic reactivity is realized because of in-pore diffusion limiting the reaction rate.

Although Weisz imposed arbitrary restrictions upon the set of simultaneous differential equations (102–104), the general order of magnitude criterion (142) is a useful one. This was demonstrated by an experiment in which n-heptane was converted into isomeric heptanes using a compacted silica–alumina platinum catalyst. The component particle sizes were varied from 1000μ to 5μ and it was shown that for, the smaller particles, the conversion approached that for a silica–alumina catalyst directly impregnated with platinum. For a 90% approach to equilibrium, the general criterion shows that the particle size must not exceed 40μ. If independent estimates of the chemical rate constants can be made under conditions where there is no resistance to diffusion, then it is probable that the values of f_x and f_y, calculated from

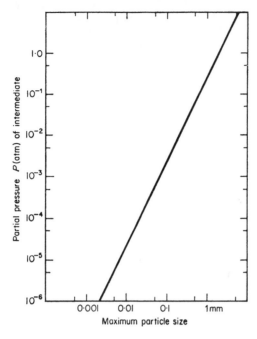

FIG. 12. Partial pressure of n-hexene as a function of maximum particle size of catalyst (mixed Pt/silica–alumina) for the dehydrogenation of n-hexane.

equations (115) and (116), may give a better indication of the appropriate particle sizes.

Mass transport through interparticle space will be important when the pores formed by the voids between particles of X and particles of Y are smaller than any original pores which may have existed in these particles. Thus, a compacted catalyst formed from two comparatively non-porous catalysts would come into this category.

7.2.4. Selectivity of polyfunctional catalysts

If the intermediate B in a polystep reaction can be transformed into two products C and D by the action of two catalysts X and Y according to scheme (III), then it is clear that Y has intruded into the scheme in which the mono-functional catalyst X converts A into C. By introducing catalyst Y, the intermediate B is intercepted and an additional product D is formed. Scheme (III) is now a polystep reaction in which two end products result from the use of two catalysts. Whether or not it is a non-trivial polystep reaction will depend on the equilibrium constant $K(=k_{v1}/k'_{v1})$. If K is large, the reaction is trivial and the conversion to C and D is just as high for two separate catalytic reaction zones as it is if the catalysts were mixed: if K is sufficiently small, the reaction is non-trivial and a mixed catalyst is superior to separate zones. As shown by Fig. 8, a compounded catalyst, containing both X and Y in the same pellet, is even more efficient. In either case, a larger amount of D can be produced, at the expense of C, by increasing the activity of the Y catalyst in relation to the X catalyst. We now consider the consequences of such selectivity for a polyfunctional catalyst.

Neglecting mass transfer effects, the formal kinetics of the above reaction scheme may be described by the two simultaneous differential equations

$$- \frac{dc_A}{dt} = k_{v1}c_A - k'_{v1} c_B \tag{143}$$

$$- \frac{dc_B}{dt} = (k'_{v1} + k_{v2} + k_{v3})c_B - k_{v1} c_A \tag{144}$$

With the boundary conditions $c_B = 0$, $dc_B/dt = k_{v1}c_A{}^0$ at $t = 0$, the concentration of gaseous B at any time t is given by solving (143) and (144) for c_B. The solution is

$$\frac{c_B}{c_A{}^0} = \frac{k_{v1}}{k'_v} \{1 - \exp(-k'_v t)\} \tag{145}$$

where

$$k'_v = (k'_{v1} + k_{v2} + k_{v3}) \tag{146}$$

c_A can now be found by substitution into (144) and using the initial condition $c_A = c_A{}^0$. The result is

$$\frac{c_A}{c_A{}^0} = \exp(-k_v t) - \frac{k_{v1}k_{v2}}{k'_v(k'_v - k_v)} \{\exp(-k_v t) - \exp(-k'_v t)\} \tag{147}$$

where

$$k_v = \frac{k_{v1}(k_{v2} + k_{v3})}{k'_v} = \frac{k_{v1}(k_{v2} + k_{v3})}{k'_{v1} + k_{v2} + k_{v3}} \tag{148}$$

The above solution traces the course of events for a trivial reaction. For a non-trivial polystep reaction, in which at least $k_{v1} \ll k'_{v1}$, all the last two terms in equation (147) are approximately zero so the conversion becomes

$$1 - \frac{c_A}{c_A{}^0} = 1 - \exp(-k_v t) \tag{149}$$

This latter result could also be deduced quite simply by putting $dc_B/dt = 0$ and finding the stationary concentration of B and thence solving the single differential equation (143). The amounts of c_C and c_D may be found from the equation

$$\frac{\dfrac{dc_C}{dt}}{\dfrac{dc_D}{dt}} = \frac{k_{v2}}{k_{v3}} \qquad (150)$$

and the stoichiometric condition

$$c_A{}^0 - c_A - c_B = c_C + c_D \qquad (151)$$

At any time t the concentration of c_D is therefore,

$$\frac{c_D}{c_A{}^0} = \frac{k_{v3}}{k_{v2} + k_{v3}} \left[\left(1 - \frac{k_{v1}k_{v2}}{k'_v(k'_v - k_v)} \right) \left(1 - \exp - k_v t \right) + \right.$$

$$\left. + \frac{k_{v1}}{k'_v} \left(1 - \frac{k_{v2}}{(k'_v - k_v)} \right) \left(1 - \exp - k'_v t \right) \right] \qquad (152)$$

while that of c_C is k_{v2}/k_{v3} times that for c_D as given by equation (150). For a non-trivial reaction, since $k_{v1} \ll k'_{v1}$, then we simply have

$$\frac{c_D}{c_A{}^0} = \frac{k_{v3}}{(k_{v2} + k_{v3})} \left(1 - \exp - k_v t \right) \qquad (153)$$

Weisz [19] considered the consequences of varying the ratio k_{v2}/k_{v3} which determines the selectivity. Figure 13 traces the effect of gradually increasing the activity of the Y catalyst producing D in a non-trivial reaction in which the gas phase concentration of the intermediate is negligible. It is obvious from this diagram that D can be produced in amounts increasing with the activity of the Y catalyst and at the same time causing a more gradual decrease in the amount of C produced by the X catalyst. However, if catalyst Y is only available at a certain level of activity, then, as the activity of catalyst X is increased, the amount of D passes through a maximum, as shown in Fig. 14. This is because catalyst X is responsible for generating the necessary intermediate which is a precursor to D as well as being responsible for the competing step in which C is formed. The amount of D produced at the optimum activity of X depends on the available level of activity of catalyst Y.

An example of the way in which selectivity operates with polyfunctional catalysts is provided by the work of Hindin et al. [20] and of Weisz [4, 19, 21]. Methylcyclopentane can be dehydrogenated at atmospheric pressure and at 500°C to yield methylcyclopentene and methylcyclopentadiene. If, under the same physical conditions, a silica–alumina catalyst is used in addition to a supported platinum catalyst, the methylcyclopentene is intercepted and converted to cyclohexene by the acid-type catalyst and further converted to

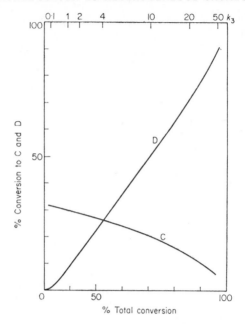

FIG. 13. Variation of conversion with increasing k_{v3} in the absence of mass transfer effects. $k_{v1}=k_{v2}=4$ sec^{-1}; $k'_{v1}=100$ sec; residence time 10 sec.

benzene under dehydrogenative conditions by the metal catalyst. Thus, if we implicitly assume the participation of H_2 in these reactions the scheme is illustrated as follows:

where X is the supported platinum catalyst and Y is the silica–alumina catalyst. The yield of cyclohexene and hence benzene will depend on the reactivity of Y in relation to X. Because X is involved in the formation of the intermediate methylcyclopentene as well as being responsible for the competitive step in which methylcyclopentadiene is formed, the yield of cyclohexene and hence benzene will pass through a maximum as the activity of X is varied. Provided that all steps can be regarded as pseudo-first-order re-

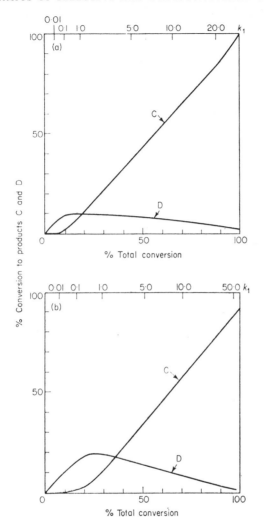

FIG. 14. Variation of conversion with increasing k_{v1} in the absence of mass transfer effects. (i) $k_{v3}=1$ sec^{-1}; $k_{v1}/k'_{v1}=0.01$; $k_{v1}/k_{v2}=1$; residence time 10 sec. (ii) $k_{v3}=2$ sec^{-1}; $k_{v1}/k'_{v1}=0.01$; $k_{v1}/k_{v2}=1$; residence time 10 sec.

actions, equations similar to those discussed will give the concentrations of cyclohexene and benzene for a particular contact time and set of catalytic reactivities.

The above conversion, when carried out under dehydrogenative conditions, represents a trivial polyfunctional catalytic reaction. However, if hydrogenative conditions are employed (12 atm, 350°C) the equilibrium reaction lies well to the left and the equilibrium partial pressure of the cyclopentene is about 10^{-3} atm. If X is the only catalyst present, the equilibrium conversion to methylcyclopentadiene is restricted to about 0·1 %, but about 10 % of a mixture of paraffins, ranging from methane to hexane, is produced by the platinum-catalysed reaction of H_2 with the diene. By introducing the acidic type catalyst Y, the latter hydrogenolysis reaction (in which paraffins are produced) is intercepted and cyclohexene, cyclohexane and benzene are formed. The cyclohexane arises because of the platinum-catalysed hydrogenation of hexene. Under hydrogenative conditions the reaction is non-trivial and consequently larger conversions than in the case of the trivial reaction may be expected. To achieve non-trivial conditions low temperatures and high pressures are employed, yet under these conditions there will be a high partial pressure of hydrogen present and so hydrogenolysis is inevitable to some extent. When the particle size of both catalyst components is reduced to about 5μ the conversion is no longer inhibited by the slow mass transport of the intermediate and the conversions established by equations such as (149–152) are satisfied. The yield of benzene is then substantially increased at the expense of the unwanted paraffins.

These effects were demonstrated by Weisz [4], who used a microcatalytic reactor in which the two catalyst components were intimately mixed. The reactivity of the platinum component was gradually reduced by progressive poisoning with H_2S. As the reactivity of the platinum was decreased, so the amount of paraffins produced decreased with a concomitant increase in the quantity of benzene and hexane. Extensive poisoning diminished the yield of all products, since the generating step responsible for the formation of intermediate cyclopentene is eventually retarded to such an extent that conversion is arrested. Figure 13 illustrates the results obtained. Figure 14 compares the results obtained for similar experiments when the component particle size is varied from 500μ to 5μ. In the case of the catalyst of particle size 500μ, as the platinum activity is varied so the yield of benzene and hexane first increases and then very gradually diminishes. For a particle size of 5μ the yield of benzene and hexane is greater at the maximum than for the larger particles, thus illustrating the advantage of eliminating slow mass transfer effects and so enhancing the transfer of intermediate from catalyst X to catalyst Y.

When a monofunctional catalyst generates a small amount of side product C which is in equilibrium with the main product B,

$$A \xrightarrow{\text{X}} B \overset{\text{X}}{\rightleftharpoons} C \qquad\qquad \text{(IV)}$$

it may be conceivable that C could be converted into some other desired product D by operating on the side product with a second catalyst Y. Although the equilibrium partial pressure of C may be very low, because of an equilibrium favouring B almost entirely, it is nevertheless possible to take advantage of its formation by coupling the catalytic system in which X produces B (and hence C) from A, with a catalytic system in which Y is capable of converting C into D,

$$A \xrightarrow{k_{v1}} B \underset{k'_{v2}}{\overset{k_{v2}}{\rightleftharpoons}} C \xrightarrow{k_{v3}} D \qquad \qquad \text{(V)}$$
$$\phantom{A \xrightarrow{k_{v1}}} X X \phantom{\xrightarrow{k_{v3}}} Y$$

Provided that the rate constant k_{v3} is sufficiently high, C will be consumed to form D, and catalyst Y, by removing C, causes more of B to be converted into C so as to maintain the equilibrium. If equilibrium is established rapidly between B and C, then the above scheme reduces to a consecutive reaction,

$$A \xrightarrow{k_{v1}} B \xrightarrow{K_2 k_{v3}} D$$
$$\phantom{A \xrightarrow{k_{v1}}} X \phantom{\xrightarrow{K_2 k_{v3}}} Y$$

where the product of K_2 (the equilibrium constant k_{v2}/k'_{v2}) and k_{v3} forms a composite rate constant. In this case the amount of D produced increases exponentially with increase in the ratio $K_2 k_{v3}/k_{v1}$ for a given residence time.

Clark [22] and Keulemans and Schuit [23] discussed the possibility that some catalysts, which were originally supposed to be monofunctional, may, in fact, behave as bifunctional catalysts having two different types of sites which are distinguished physically and geometrically and which therefore offer two alternative types of adsorption sites for reactants. Clark [22] found that either Cr_2O_3 or MoO_2 on a silica–alumina catalyst enhances the acidic properties of the support if the catalyst combination is treated with oxygen. If treated with hydrogen, however, the catalyst assumed hydrogenation–dehydrogenation properties. He considered anionic vacancies as possible hydrogenation–dehydrogenation sites, but did not propose suggestions for the nature of the acid sites. Kloosterziel [24] proposed a model for the exchange of deuterium with ethylene on a deuterated alumina catalyst. Since increase in the deuterium content of the catalyst decreases the reaction rate, and since it is obvious that deuterium on the surface is necessary to obtain any exchange at all, it is probable that the catalyst has two functions. It was suggested that one function is an anionic vacancy in the oxygen lattice and the other a surface proton or hydroxyl group.

Weisz and Prater [21] examined the cracking of cumene in a differential reactor using a silica–alumina catalyst and also a silica–alumina catalyst impregnated with platinum. The introduction of platinum into the catalyst hardly altered the reaction rate, but the distribution of products was affected. In the presence of the silica–alumina catalyst the products were benzene and propylene, but when platinum was present methylstyrene and hydrogen were also produced. If the silica–alumina and platinum catalysts were present as separate pellets, the product distribution was such that benzene and propylene on the one hand, and methylstyrene and hydrogen on the other hand, were produced in accordance with the proportion of each catalyst component. When the silica–alumina and platinum were compacted into single small-diameter pellets, the composition of products indicated that there was an increase in concentration of methylstyrene and hydrogen over the amount which could be produced by the platinum catalyst alone or the mixture of separate components of silica–alumina and platinum. It was suggested that an intermediate surface compound was responsible for such behaviour and that its presence in the gas phase was undetectable because of some equilibrium which is not in its favour. This proposition, advanced by Weisz and Prater [21], was equivalent to assuming the coupling of two catalytic systems such as in scheme (V). The coupling was supposed to occur via an intermediate compound C (unspecified) which is present in minute amounts in the gas phase. The mechanism propounded was

in which X is the silica–alumina component and Y the platinum component.

7.2.5. Thermodynamics of polystep reactions

If we consider the change in free energy that occurs when a reactant forms, via an equilibrium reaction, an intermediate compound which subsequently decomposes into products, two useful conclusions may be drawn.

First, we examine the system

$$A \rightleftharpoons B \rightarrow C \qquad (II)$$

bearing in mind that a decrease in Gibbs free energy means that ΔG is negative. Two possibilities exist for the standard free-energy change from A to B: (i) $\Delta G^0(A \rightarrow B)$ is positive, or (ii) $\Delta G^0(A \rightarrow B)$ is negative. If $\Delta G^0(A \rightarrow B)$ is negative, then the equilibrium concentration of the intermediate B will be comparatively high and the conditions are appropriate for a trivial polystep reaction. If, however, $\Delta G^0(A \rightarrow B)$ is positive, the equilibrium concentration of B will be low and the requirements for a non-trivial polystep reaction are met. By substituting the value of $\Delta G^0(A \rightarrow B)$

$$\Delta G^0(A \rightarrow B) = -RT \ln \frac{p_B{}^*}{p_A} \qquad (154)$$

into equation (142) Section 7.2.3, a thermodynamic condition for the occurrence of a non-trivial polystep reaction may be written. For example, in the case of Weisz's physical model for a catalyst composed of a mixture of X and Y components,

$$\varDelta G^0(A \to B) < RT \ln \frac{Dp_A}{R^2 r_0 RT} \tag{155}$$

Equation (157) therefore predicts how positive $\varDelta G^0(A \to B)$ must be before a non-trivial catalytic polystep reaction can occur.
For the selective reaction

$$A \rightleftharpoons B \overset{X}{\underset{Y}{<}} \begin{matrix} C \\ \\ D \end{matrix} \tag{III}$$

the free-energy change from A to C, $\varDelta G(A \to C)$, and the free-energy change from A to D, $\varDelta G(A \to D)$, must both be negative. However, the difference in free energy between C and D may be such that $\varDelta G(C \to D)$ is positive. In this case, C is more stable than D. Nevertheless D may be made at the expense of C by intercepting the intermediate product with catalyst Y, as previously considered. The corollary to this theorem is that, in a non-trivial catalytic reaction, D is not likely to have arisen by any other course other than directly from the intermediate product.

References

1. E. F. G. Herington and E. K. Rideal, *Proc. R. Soc.* **A190**, 289 (1947).
2. E. F. G. Herington and E. K. Rideal, *Proc. R. Soc.* **A190**, 309 (1947).
3. G. C. Bond, D. A. Dowden and N. Mackenzie, *Trans. Faraday Soc.* **54**, 1537 (1958).
4. P. B. Weisz, Proc. 2nd Int. Cong. Catalysis, Vol. 1, p. 937. Editions Technip, Paris (1961).
5. A. B. R. Weber, Thesis, University of Delft (1957).
6. H. I. Waterman, *Analytica chim. Acta* **18**, 395 (1958).
7. R. J. A. M. van der Borg, *Proc. K. ned. Akad. Wet.* **B62**, 299 (1959).
8. J. H. de Boer, and R. J. A. M. van der Borg, Proc. 2nd Int. Cong. Catalysis, Vol. 1, p. 919. Editions Technip, Paris (1961).
9. J. H. de Boer and R. J. A. M. van der Borg, *Proc. K. ned. Akad. Wet.* **B62**, 308 (1959).
10. A. Wheeler, *Adv. Catalysis* **3**, 249 (1951).
11. K. Ostergaard, Proc. 3rd Int. Cong. Catalysis, Paper II, p. 12, North Holland Publishing Company, Amsterdam (1964).
12. H. Heinemann, G. A. Mills, J. B. Hattman and F. W. Kirsch, *Ind. Engng Chem.* **45**, 130 (1953).
13. P. B. Weisz and E. W. Swegler, *Science* **126**, 31 (1957).
14. H. S. Bloch and C. L. Thomas, *J. Am. Chem. Soc.* **66**, 1589 (1944).
15. B. S. Greensfelder and H. H. Voge, *Ind. Engng Chem.* **37**, 983 (1945).
16. D. J. Gunn and W. J. Thomas, *Chem. Engng Sci.* **20**, 89 (1965).
17. O. Bilous and N. R. Amundson, *Chem. Engng Sci.* **5**, 81 (1956).

18. O. Bilous and N. R. Amundson, *Chem. Engng Sci.* **5,** 115 (1956).
19. P. B. Weisz, *Adv. Catalysis* **13,** 137 (1962).
20. S. G. Hindin, S. W. Weller and G. A. Mills, *J. phys. Chem.* **62,** 244 (1958).
21. P. B. Weisz and C. D. Prater, *Adv. Catalysis* **9,** 583 (1957).
22. A. Clark, *Ind. Engng Chem.* **45,** 1476 (1953).
23. A. I. M. Keulemans and G. C. A. Schuit, *in* "The Mechanism of Heterogeneous Catalysis", ed. by J. H. de Boer, p. 159. Elsevier, Amsterdam (1960).
24. H. Kloosterziel, *in* "Chemisorption", ed. by W. E. Garner, p. 76. Butterworths, London (1957).

CHAPTER 8

The Mechanism of Some Typical Heterogeneous Catalytic Reactions†

8.1. Catalytic Oxidation

The oxidation of carbon monoxide using a copper oxide catalyst [1] and the oxidation of ethylene using a silver catalyst [2] are classical examples of heterogeneous oxidation reactions. A careful and assiduous study of surface reactions involving carbon monoxide has led to a better understanding of the role which the catalyst plays. More recently, the catalytic oxidation of a variety of hydrocarbons using metal oxide catalysts has added to information concerning the mechanism of heterogeneous oxidations [3]. Many heterogeneous catalytic oxidation reactions form the basis of important industrial processes. The catalytic oxidation of toluene, xylene and naphthalene using metal oxide catalysts [4] are now well established as convenient methods for producing, on a large scale, phthalic and maleic anhydrides. The catalytic

† In this chapter velocity constants are represented quite generally by k, and any subscript refers to the number of the reaction.

oxidation of ammonia using a platinum catalyst yields nitric oxide and is therefore used in the manufacture of nitric acid [5, 6]. Similarly, the oxidation of sulphur dioxide by means of either a platinum [7] or vanadium pentoxide [8] catalyst is commercially important. Whereas all of these reactions have been studied to a considerable extent, only a selection of these are discussed in this section, sufficient to illustrate those fundamental principles which are important in catalysis.

Most oxidation catalysts are metal oxides in which the metal is capable of existing in more than one valency state. The exceptions appear to be silver, which oxidizes ethylene to ethylene oxide [2], and platinum, a good oxidation catalyst for ammonia [5, 6]. However, there is some evidence [9, 10] that these latter two reactions proceed through a chemisorbed layer of oxygen at the metal surface. Stone [11] has studied the decomposition of nitrous oxide over a series of metal oxide catalysts. The decomposition involves distinct chemisorption and desorption steps which are also involved in oxidation reactions, It was suggested (see p. 269) that nitrous oxide is rapidly adsorbed by accepting a quasi-free electron from the surface

$$e + N_2O \rightarrow N_2 + O^-(ads)$$

and gaseous oxygen formed by positive holes within the oxide accepting electrons

$$O^-(ads) \rightarrow \tfrac{1}{2} O_2 + e$$

Stone [11] was able to classify catalysts active in decomposing nitrous oxide into p-type and n-type electrical semiconductors and also insulators. As assessed from the temperature at which decomposition first occurred, the most active catalysts are p-type semiconductors, whereas n-type semiconductors are least effective, requiring temperatures near the region where homogeneous gas phase decomposition occurs. Insulators have an activity intermediate between those of p- and n-type semiconductors. The addition to a p-type semiconductor of a cation of lower valency than the cations of the host lattice enhances the catalytic activity as predicted by the electronic theory of catalysis outlined in Sections 5.3.2.2 and 6.3. In contrast, adding a cation of a higher valency impairs the catalytic activity. The relationship between N_2O decomposition and oxidation reactions has been discussed by Stone [11] and is helpful in assessing the role which an oxidation catalyst plays.

Thus, an active oxidation catalyst will have the ability to provide the molecule to be oxidized with oxygen via the catalyst surface. This augurs well for a lower activation energy than the homogeneous oxidation, since the surface contains oxygen in a more labile form. Although most oxidation catalysts are p-type semiconductors, it may be that silver (which oxidizes ethylene) and platinum (which oxidizes ammonia and sulphur dioxide) occupy a peculiar position because, in spite of interaction probably occurring through a surface oxide layer, the electronic configurations of ammonia, sulphur dioxide and ethylene are such that the presence of chemisorbed oxygen ions is not a prerequisite to oxidation. Covalent attachment of these molecules to the surface seems to be preferred to adsorption as ions.

Some typical heterogeneous oxidation reactions will now be discussed and some of these problems considered in more detail.

8.1.1. Oxidation of carbon monoxide

It has long been known that hopcalite catalysts, containing admixtures of the oxides of manganese and copper, are active in oxidizing carbon monoxide at temperatures as low as 20°C. The early investigations of the mechanism of this oxidation forms one of the classical examples of researches in the field of heterogeneous catalysis. Stone [1] has pointed out that although an irreversible chemisorption in which a surface carbonate ion is produced explains the facts in the temperature range 100–200°C; other processes have to be invoked to explain the catalytic oxidation of CO at 20°C. The fact that CO_2 could be removed from the catalyst after a low-temperature oxidation was not consistent with an interpretation of experimental results in terms of a simple Langmuir-Hinshelwood mechanism (see Section 1.2), since CO_2 does not inhibit the reaction. Furthermore, although adsorbed CO_2 seems to be a stable state during a low-temperature catalytic oxidation, such a stable surface state cannot be reached simply by admitting CO_2 gas to the outgassed metal oxide catalyst.

In Section 3.2.8.3 it was outlined how thermochemical data could be utilized to deduce the structure of intermediates at a catalyst surface. This approach has proved to be extremely valuable in the study of the oxidation of carbon monoxide on transition metal oxides, and we shall now consider in greater detail the thermochemistry of this system.

It is instructive to discuss first the adsorption of CO at metal oxides and subsequently the oxidation of CO at both high and low temperatures, since it will emerge that there is a strong correlation between catalytic activity and the stability of the adsorbed species.

8.1.1.1. *Adsorption of CO*

The earliest experiments on the adsorption of CO on the oxides of zinc, chromium and manganese established that the gas could be adsorbed in two distinct ways. At room temperature CO was adsorbed reversibly on ZnO and the gas could be recovered without change on heating the oxide to about 100°C. The heat of adsorption, as measured by calorimetric experiments, was between 10 and 20 kcal mole^{-1}, so that chemisorption rather than physical adsorption was occurring [12]. On the oxides Mn_2O_3 and Mn_2O_3–Cr_2O_3, CO was irreversibly adsorbed in the sense that CO_2 was desorbed on heating. The heat of adsorption was over 30 kcal mole^{-1} and a considerable amount of oxygen could be adsorbed after CO adsorption in spite of the fact that little was adsorbed prior to CO adsorption [13]. This unsaturation towards oxygen was carefully measured and it was found that the quantity of oxygen taken up amounted to about one-half of the amount of presorbed CO. In conjunction with the calorimetric data, summarized in Table 8.1, it was possible to postulate a mechanism of adsorption and oxidation involving the CO_3^{2-} surface complex. Table 8.1 shows that the same adsorbed state is probably

TABLE 8.1. Heats of adsorption on $Mn_2O_3.Cr_2O_3$ [13]

Heat of adsorption of CO, Q_{CO}	46 kcal mole^{-1}
Heat of adsorption of O_2 after presorbing CO, Q_{O_2}	78 kcal mole^{-1}
$Q_{CO} + \frac{1}{2}Q_{O_2}$	85 kcal mole^{-1}
Heat of adsorption of a mixture of 1 mole of CO and $\frac{1}{2}$ mole of O_2, $Q_{(CO + \frac{1}{2}O_2)}$	85 kcal mole^{-1}
Heat of adsorption of CO_2, Q_{CO_2}	20 kcal mole^{-1}

reached when oxygen is admitted subsequent to CO adsorption and also when a mixture of $2CO:1O_2$ is adsorbed at an oxidized surface. Furthermore, if the heat of adsorption of CO_2 is subtracted from the quantity $(Q_{CO}+\frac{1}{2}Q_{O_2})$, which represents the sum of (i) the heat of adsorption of CO and (ii) the heat of adsorption of oxygen on a surface at which CO had been presorbed, then the resulting value 65 kcal mole^{-1} is in close agreement with the heat of combustion of CO(g). Noting that the heat of adsorption of CO_2 on Mn_2O_3 corresponds very closely with the heat of dissociation of manganous carbonate, it was concluded that the adsorbed state common to these processes is a surface carbonate ion. It was supposed that the CO interacted with an oxygen ion of the metal oxide lattice, producing the surface carbonate ion, and that unsaturation towards oxygen after CO adsorption occurred as a result of an anion vacancy produced by the CO adsorption. The process is represented by Garner [12] as

$$
\begin{array}{ccc}
M^{2+}\ O^{2-}\ M^{2+} & M^{2+}\ CO_3^{2-}\ M^{2+} & M^{2+}\ CO_3^{2-}\ M^{2+} \\
\xrightarrow{\;\;CO\;\;} & \xrightarrow{\;\;\frac{1}{2}O_2\;\;} & \\
O^{2-}\ M^{2+}\ O^{2-} & O^{2-}\ M^{2+}\ \ 2e & O^{2-}\ M^{2+}\ \ O^{2-}
\end{array}
$$

For p-type oxides (e.g. Cu_2O), such an irreversible adsorption of CO will, in accordance with the theory outlined in Chapter 5 (see Sections 5.2.4.1 and 5.3.2.2), cause a decrease in the number of positive holes, since those electrons released by formation of the surface complex will tend to associate themselves with $Cu^{2+}-Cu^+$ positive hole units. The adsorption will therefore result in a decrease in semiconductivity and is depletive, since the electrons released will gradually form a potential barrier to further adsorption. On the other hand, for n-type oxides the electrons released by irreversible CO adsorption will enter the conduction band of the solid and result in an increase in conductivity. Since there are vacant levels present in the conduction band which enable the incorporation of more electrons, the adsorption will be cumulative.

More elaborate calorimetric experiments by Stone and co-workers [14–16] showed that, with cuprous oxide, pre-adsorption of oxygen considerably increased the heat of adsorption of CO. Similarly, pre-adsorption of CO enhanced the heat of adsorption of oxygen. The surface complex thus produced was stable in the presence of excess oxygen, but CO_2 was desorbed if excess CO was present. Experiments in which CO_2 was adsorbed provided confirmatory evidence that the surface complex has the formula CO_3. Only when there was some pre-adsorbed oxygen present at the surface was CO_2 adsorbed, and this was adsorbed in the greatest amounts when CO_2 and O_2 were premixed in the ratio 2:1 before adsorption. The following equations† can thus be written for the reaction between presorbed O_2 and CO_2:

$$\tfrac{1}{2}O_2(g) = O(ads) + 28 \text{ kcal} \tag{1}$$

$$CO_2(g) + O(ads) = CO_3(ads) + 21 \text{ kcal} \tag{2}$$

Using the accepted data for the heat of combustion of CO

$$CO(g) + \tfrac{1}{2}O_2(g) = CO_2(g) + 67 \text{ kcal} \tag{3}$$

it therefore follows, from Hess's law of heat conservation, that 60 kcal is liberated when CO is adsorbed on a surface containing presorbed oxygen,

$$(3) - (1) + (2); \qquad CO(g) + 2O(ads) = CO_3(ads) + 60 \text{ kcal} \tag{4}$$

This may be compared with the experimental value of 49 kcal. From the calorimetric data for the adsorption of CO on a baked-out surface

$$CO(g) = CO(ads) + 20 \text{ kcal} \tag{5}$$

one therefore predicts that the heat of adsorption of CO on a surface containing presorbed oxygen is 96 kcal

$$\{2 \times (1)\} + (4) - (5); \qquad O_2(g) + CO(ads) = CO_3(ads) + 96 \text{ kcal} \tag{6}$$

in comparison with the 100 kcal mole^{-1} obtained by calorimetry. These reactions were also studied at the surfaces of nickel oxide and cobalt oxide and similar results found. Oxygen was also admitted after CO adsorption had taken place and, as before, surface unsaturation towards oxygen was evident.

After CO_2 adsorption, however, the surface would not adsorb any further quantities of oxygen. The results are therefore in accord with the view that CO_3 complex formation occurs by CO interacting with two surface oxygen atoms. A vacant surface site, which will display unsaturation towards oxygen, therefore remains unoccupied left according to this mechanism. CO_3 complex formation also occurs by CO_2 being adsorbed at a surface oxygen site, but in this case no vacancy remains and the surface will not be unsaturated with respect to oxygen. The essential difference between this mechanism and that

† In this discussion it will emerge that there remains some doubt concerning the participation of *lattice oxygen ions* during chemisorption and oxidation: Labile *adsorbed oxygen* may play a dominant role. Where such a possibility is considered we write the surface complex as CO_3(ads). It should be emphasized that this complex is also negatively charged but it is to be distinguished from the carbonate ion.

of the theory proposed by Garner [12] is that no lattice oxygen ions have been invoked. The mechanism is represented

$$
\begin{array}{ccccccccc}
O & O & O & & CO & & CO_3 & & O \\
| & | & | & \xrightarrow{CO} & | & & | & \xrightarrow{\frac{1}{2}O_2} & | \\
M & M & M & & M & & M & & M
\end{array}
\qquad
\begin{array}{ccc}
\frac{1}{2}O_2 & CO_3 & O & O \\
\end{array}
$$

O   O   O   CO   CO₃   O   ½O₂   CO₃   O   O
| | | → | | → | | | |
M   M   M   M   M   M   M   M   M   M

Although there seems to be substantial evidence in favour of the formation of a CO_3 surface complex, there remains some doubt as to the exact role of lattice oxide ions. Winter [17] showed that when outgassed nickel oxide, prepared as an oxide film on a metal base, was exposed at 540°C to oxygen containing ^{18}O, complete exchange of all the surface oxide ions of the oxide occurred. CO was then admitted to the oxide, now containing $^{18}O^{2-}$ ions, at 200°C, and no measurable exchange of oxygen was observed. Similarly, CO_2 did not exchange its oxygen with the oxide ions of the solid. When a mixture of $2CO:1O_2$ was in contact with the solid at 50°C the mixture was converted to CO_2, but only a small fraction of ^{18}O appeared in the gas phase. It seems, therefore, that for NiO at least, oxide ions of the lattice play no part in CO oxidation, as was assumed by Stone in the scheme outlined above.

Winter [17, 18] also studied the exchange of oxygen with outgassed cuprous oxide and in this case found that both CO and CO_2 readily exchanged oxygen with the whole oxide surface at room temperature, and there was some exchange at temperatures as low as -78°C. Since there would not have been any adsorbed oxygen at the surface of the outgassed oxide, it must be assumed that both CO and CO_2 are adsorbed on cuprous oxide in such a way that it is in close association with the oxygen ions of the crystal lattice. It was assumed, therefore, that the adsorbed species is the carbonate ion, as originally proposed by Garner. However, Stone postulates an alternative explanation in view of the difficulty of accounting for the energetics of dissociation of the carbonate ion to give desorbed CO, which is the stable gas-phase species remaining after the oxygen-exchange experiments. Cuprous oxide is a unique

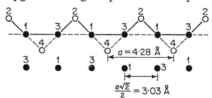

FIG. 1. CO oxidation. Cross-section through a (001) plane of cuprous oxide. ○, oxygen atoms; ●, copper atoms.

oxide, since it has a body-centred anion lattice with cations arranged in tetrahedral coordination around them. Figure 1 shows, diagrammatically, the arrangement of copper and oxygen ions in the (001) and (111) planes. Half of the oxygen anions present at the surface protrude from the geometric plane containing the cuprous ions and half are buried beneath this plane. Stone suggested that at a temperature of 200°C, at which Winter outgassed his oxide, some of the oxygen ions in the "buried" position are converted to the "protruding" position and are thus favourably disposed for exchange with gas-phase atoms. Furthermore, there is hardly a distinction between such protruding oxygen anions and adsorbed oxygen. In the (011) plane, as depicted in Fig. 2, oxygen

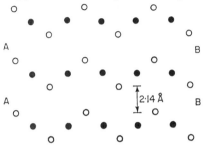

FIG. 2. CO oxidation. Plan view of a (011) face of cuprous oxide. ○, oxygen atoms; ●, copper atoms.

ions are in suitable positions along AB for exchange with reversibly adsorbed CO or CO_2. Cuprous oxide therefore has such a geometric structure that one can reasonably expect oxygen to be exchanged with comparative ease at each of the three main geometrical planes. On the other hand, there are no favourably displaced oxygen ions in the NiO lattice and nickel ions in the flat (111) plane are somewhat inaccessible. Thus, it is understandable in these terms why there is a much lower activity in oxygen exchange with CO and CO_2 when NiO is used as opposed to Cu_2O. Further support for this argument comes from the work of Teichner [19–21], who found that CO and CO_2 exchanged oxygen with high specific surface area NiO which had not been sintered. If NiO is prepared in this way, it is to be expected that there will be an adequate number of oxygen ions suitably displaced for exchange with gaseous oxygen. Also, there is enhanced chemisorption of CO, and the labile oxide ions may again be responsible for this increase in activity.

The structure for the CO_3 complex proposed by Eischens and Pliskin [22], using the technique of infrared spectroscopy (see Chapter 3), is of a form similar to a bicarbonate ion with binding to the metal atom through an oxygen atom, as shown below

$$O\diagdown_{\!\!C}\diagup^{O}$$
$$|$$
$$O$$
$$|$$
$$Ni$$

This structure is difficult to envisage in view of the previous evidence purporting to a carbonate type complex in which CO_2 is bonded through an adsorbed oxygen ion. More recent evidence concerning the nature of the surface complex present under oxidation conditions is provided by the work of Blyholder [23]. Nickel was evaporated into an oil film and this was subsequently oxidized prior to admitting CO. The infrared spectra showed that the nickel film was oxidized at the surface only and that an absorption band at 1330 cm^{-1}, additional to the ones detected by Eischens and Pliskin, appeared. By comparing the observed absorption bands with those shown by monodentate and bidentate carbonates, it was concluded that the spectra was more in accord with a surface carbonate complex than the structure proposed earlier.

8.1.1.2. Mechanism of Oxidation of CO

At temperatures of about $25°C$, a mechanism involving the CO_3 complex explains the observed facts:

$$CO(g) = CO(ads) \tag{7}$$
$$O_2(g) = 2O(ads) \tag{8}$$
$$CO(ads) + 2O(ads) = CO_3(ads) \tag{9}$$
$$CO_3(ads) + CO(g) = 2CO_2(g) \tag{10}$$

If the catalyst is evacuated after oxidizing CO, then the steady state concentration of $CO_3(ads)$ remains unaltered, but if the catalyst is strongly outgassed carbon dioxide and oxygen are desorbed, as would be expected. A novel feature of this mechanism is that the product of reaction can actually enhance formation of the intermediate surface complex. The complex CO_3 may be attained by gaseous CO_2 reacting with adsorbed oxygen, as shown by equation (2). However, there exists some evidence [18] to show that CO_2 may not be desorbed spontaneously and that it inhibits the reaction rate. Winter [18] studied the oxidation of CO over Cu_2O and confirmed the presence of CO_2 in the gas desorbed from the catalyst after an experiment. The catalyst was active in oxygen exchange with CO and CO_2, but only 10% of heavy oxygen from the labelled oxide appeared in CO_2 during CO oxidation. Stone [1] attributes the results of Winter's oxygen-exchange experiments to the fact that freshly adsorbed oxygen, rather than lattice oxygen, is playing the major role during the catalysis.

Calorimetric measurements have, in fact, provided additional evidence in favour of the scheme represented by equations (7) to (10). When admitting CO to a cobaltous oxide surface at which oxygen had been presorbed, the heat of adsorption gradually fell for the first few incremental doses, but no CO_2 was desorbed. Another few incremental additions, however, caused the heat of adsorption to fall drastically by as much as $20 \text{ kcal mole}^{-1}$ with the concomitant formation of gaseous CO_2. From the two thermochemical equations (4) and (5), written here again for convenience,

$$CO(g) + 2O(ads) = CO_3(ads) + 60 \text{ kcal} \tag{4}$$
$$CO(ads) = CO(g) - 20 \text{ kcal} \tag{5}$$

the heat of formation of the CO_3 complex is $40 \text{ kcal mole}^{-1}$,

$$(4) + (5); \qquad CO(ads) + 2O(ads) = CO_3(ads) + 40 \text{ kcal} \tag{9a}$$

Now the standard heat of formation of CO(g) is 27 kcal mole^{-1},

$$C + \tfrac{1}{2}O_2(g) = CO(g) + 27 \text{ kcal} \tag{11}$$

and, rewriting equation (1) for the adsorption of oxygen

$$\tfrac{1}{2}O_2(g) = O(ads) + 28 \text{ kcal} \tag{1}$$

the standard heat of formation of the CO_3 complex may be computed from Hess's law,

$$\{2 \times (1)\} + (9a) + (11) - (5); \quad C + \tfrac{3}{2}O_2(g) = CO_3(ads) + 143 \text{ kcal} \tag{12}$$

Recalling that the standard heat of formation of $CO_2(g)$ is 94 kcal mole^{-1},

$$C + O_2(g) = CO_2(g) + 94 \text{ kcal} \tag{13}$$

the heat of reaction when gaseous CO is admitted to a surface containing the CO_3 complex must be 18 kcal,

$$2 \times (13) - (12) - (11); \quad CO_3(ads) + CO(g) = 2CO_2(g) + 18 \text{ kcal} \tag{10a}$$

From these thermochemical considerations, it appears that the decrease in heat of adsorption to be expected when reaction (10) takes over from reaction (9) is $40 - 18 = 22$ kcal, in close agreement with the observed value of 20 kcal.

For nickel oxide the heat of formation of the surface CO_3 complex was greater than for either CoO or Cu_2O. Because of this increase in the heat of formation, reaction (10), in which CO_2 is formed by the interaction of the complex with gaseous CO, is now endothermic and is susceptible to the effects of poisons. Such a poisoning effect has been noted by Winter [18]. The decreased activity of NiO is thus explained in such terms. Some activity is possible, however, by the reactants utilizing an alternative reaction path involving an Eley-Rideal mechanism (see Chapter 1). Thus Teichner and co-workers [19, 20] have explained their results using high area by supposing that NiO in terms of a Rideal-type mechanism in which the CO_3 complex is decomposed by gaseous CO and so affording a possible alternative route for the reaction. Further detailed kinetic and calorimetric studies by Teichner and co-workers [24], using high area NiO prepared at a relatively low temperature showed that the product CO_2 inhibits the reaction rate. When incremental doses of CO were admitted to NiO, in which oxygen had been presorbed, the calorimetric heat of adsorption at half coverage was 70 kcal/mole. Since NiO is a p-type semiconductor and the adsorption of oxygen is cumulative, an increase in the number of positive holes results, thus increasing the conductivity of the oxide. When oxygen is adsorbed, the heat evolved at half coverage is 10 kcal/mole and the first step in the adsorption sequence is written

$$Ni^{2+} + \tfrac{1}{2}O_2(g) = O^-(ads) + Ni^{3+} + 10 \text{ kcal} \tag{14}$$

The subsequent step in which CO interacts with the adsorbed oxygen can either be written as a single process in which CO_2 is spontaneously desorbed or as two distinct steps

$$Ni^{3+} + O^-(ads) + CO(g) = CO_2(ads) + Ni^{2+} + 70 \text{ kcal} \quad (15)$$

$$CO_2(ads) = CO_2(g) - 15 \text{ kcal} \quad (16)$$

By addition of equations (14), (15) and (16) the heat of combustion of gaseous CO appears to be 65 kcal mole^{-1}, in good agreement with the accepted value of 67 kcal mole^{-1}. Thus Teichner considered CO_2 to be adsorbed at the surface of the oxide. Supporting evidence for reaction (15) is indicated by a change in colour of the nickel oxide from black to yellow (indicating a change in the defect structure of the solid) and the fact that no CO_2 was detected in the gas phase.

Teichner and co-workers [24] also examined calorimetrically two other reaction sequences. CO was adsorbed on freshly prepared NiO and then exposed to gaseous oxygen. The initial heat of adsorption was 134 kcal mole^{-1}, at half coverage was 115 kcal mole^{-1}, and gradually decreased thereafter. The reaction proposed to account for these observations may be written:

$$CO(ads) + O_2(g) = CO_3(ads) \quad (17)$$

Next, gaseous CO was admitted to the catalyst at which the surface complex had been preformed. The initial heat of adsorption was 65 kcal mole^{-1} falling to 40 kcal mole^{-1}. At the same time the conductivity of the sample decreased and the solid turned from black to yellow. Gaseous CO_2 was also detected. By writing down the thermochemical equations, it is possible to demonstrate that at the most active part of the surface neither gaseous nor adsorbed CO_2 could be produced. On the part of the surface where the heat of reaction is approximately 100 kcal mole^{-1}, the thermochemical data are accounted for by adsorbed CO_2 being formed at the surface by the reaction

$$CO_3(ads) + CO = 2CO_2(ads) + 100 \text{ kcal} \quad (18)$$

The conductivity would also fall, since the number of positive hole units Ni^{3+}–Ni^{2+} decreases, but no gaseous CO_2 would be released. Teichner and co-workers [24] believe that adsorbed CO_2 is produced by that part of the surface at which the heat of reaction is 100 kcal mole^{-1} or less: it is also possible to demonstrate that gaseous CO_2 is formed by that part of the surface which reacts with a heat of approximately 40 kcal mole^{-1}. Thus the only combination of thermochemical equations which satisfies the formation of both adsorbed and gaseous CO_2 are

$$CO(ads) + O_2 = CO_3(ads) + 100 \text{ kcal} \quad (19)$$

$$CO_3(ads) + CO = 2CO_2(ads) + 65 \text{ kcal} \quad (20)$$

and

$$CO_2(ads) + O_2 = CO_3(ads) + 40 \text{ kcal} \quad (21)$$

$$CO_3(ads) + CO = 2CO_2(g) + 65 \text{ kcal} \quad (22)$$

The average heats of adsorption of CO(g) and CO_2(g) on the NiO powder used were 29 kcal mole^{-1} and 31 kcal mole^{-1}, respectively,

$$CO(g) = CO(ads) + 29 \text{ kcal} \tag{23}$$

$$CO_2(g) = CO_2(ads) + 31 \text{ kcal} \tag{24}$$

It is therefore easy to see that both sets of equations conform to the accepted value for the heat of combustion of CO, viz. 67 kcal mole^{-1}:

$$\tfrac{1}{2}\{(19) + (20) + (23)\} - (24); \quad CO(g) + \tfrac{1}{2}O_2(g) = CO_2(g) + 66 \text{ kcal} \tag{25}$$

$$\tfrac{1}{2}\{(21) + (22) + (23)\}; \quad CO(g) + \tfrac{1}{2}O_2(g) = CO_2(g) + 67 \text{ kcal} \tag{26}$$

It would be interesting to monitor, mass spectrometrically, the gaseous CO_2 formed and to note at what stage in the reaction it is released from the surface. Kinetic measurements made by Teichner during the reaction established that the slow stage in the reaction is the desorption of gaseous CO_2, and hence the electronic structure of the solid has no influence on the catalytic activity. This latter deduction was confirmed experimentally. However, it does not accord with the observations of Dry and Stone [25] and of Roginskii and co-workers [26], in which it was shown that the apparent activation energy (see Section 6.1.1) varied with the addition of foreign ions. At temperatures below 350°C the addition of lithium ions increased the activation energy, while at temperatures above 300°C the addition of lithium ions decreased the activation energy. With regard to such a reversal of behaviour at high and low temperatures, Stone concludes that it is likely that one or other of the species participating in the reaction is adsorbed more strongly than by the pure oxide and that the electronic boundary layer theory (see p. 257) is not the correct premise on which to base an argument.

To summarize, it may be said that there is some measure of agreement in that CO_2 poisons the reaction at low temperature and that such a prediction cannot be made on the basis of the electronic theory of catalysis. Of greater significance is the fact that a compensation effect (see Section 5.3.1.2) operates and this poses the following dilemma. It seems to be a paradox that a change in the lattice which considerably increases the positive hole concentration, and hence the activation energy, also decreases the Arrhenius frequency factor. If the reaction between adsorbed CO and lattice oxygen is rate-determining, the apparent activation energy will contain a contribution from the heat of adsorption, and if this varies with alteration in the Fermi level, so also will the activation energy.

8.1.2. Hydrocarbon oxidation

The catalytic synthesis of oxygenated derivatives of hydrocarbons provides a convenient method of controlling hydrocarbon oxidation. Margolis [3, 27] has discussed catalytic oxidations involving hydrocarbons, and one of the features to emerge from such a discussion is the possibility that these reactions involve the interaction of surface radicals with oxygen. When a molecule is adsorbed by a catalyst a charge transfer reaction may take place and, in terms of the electronic theory of catalysis, the resulting structure of the adsorbate depends on the relative electronic energy levels of the catalyst and substrate. This general question was discussed in Chapter 5 and it was pointed out that, in some cases, the molecule may be chemisorbed weakly and in such a way that the resulting surface species has no net charge. For such a case Wolkenstein [28] postulates an uncharged surface-free radical which is attached by covalent bonding to the catalyst surface. In other cases involving strong chemisorption the adsorbed species may be regarded as effectively charged and so may be thought of as ionic radicals. Lyubarskii [29] has formulated the mechanism of many catalytic reactions involving hydrocarbons in terms of such surface radicals which are supposed to play an analogous role to the chain-carrying centres in homogeneous gas-phase oxidations. On the other hand, it may be argued that such a picture merely provides an alternative representation to the more orthodox electronic theory of catalysis. To illustrate the real proximity of the two approaches, the principal features of some catalytic hydrocarbon oxidations will be briefly discussed.

8.1.2.1. Ethylene Oxidation

The catalytic oxidation of ethylene using a silver catalyst affords an example in which oxygen adds directly to the unsaturated hydrocarbon. Margolis [30] has demonstrated that although ethylene is sparsely adsorbed at about 200°C on a clean silver surface, it is rapidly adsorbed on a silver surface at which oxygen has been presorbed. Further evidence that oxygen, presorbed at a cobaltous oxide surface, enhances ethylene adsorption is provided by the calorimetric investigations of Stone [1, 31]. Successive increments of ethylene admitted to an outgassed and an oxygenated cobaltous oxide surface indicated that the heat of sorption of ethylene decreased from 80 kcal mole^{-1} to 13 kcal mole^{-1} as the catalyst surface was progressively covered with oxygen. Three significant stages of partial oxidation were recognized by observing heats of adsorption. These stages are represented by the formation of (a) ethylene oxide, (b) acetaldehyde, and (c) formaldehyde. Corresponding to the formation of these substances in the adsorbed state, the heats of formation are 15 kcal mole^{-1}, 40 kcal mole^{-1} and 100 kcal mole^{-1} respectively. Thus, in the initial stages of interaction of ethylene with presorbed oxygen, one molecule of ethylene is probably reacting with two atoms of adsorbed oxygen to give formaldehyde while, in the later stages, one molecule of ethylene reacts with a single adsorbed oxygen atom to yield ethylene oxide and acetaldehyde. These results are largely in accord with the earlier deductions of Twigg [32, 33], who investigated the kinetics of ethylene oxidation over a silver catalyst and deduced that ethylene interacted with two adsorbed oxygen atoms to produce formaldehyde

$$H_2C = CH_2$$

$$O \qquad O \longrightarrow CH_2O + CH_2O \qquad (27)$$

$$Ag \quad Ag \quad Ag \quad Ag$$

and with one adsorbed oxygen atom to produce ethylene oxide,

$$H_2C = CH_2$$

$$O \longrightarrow H_2C - CH_2 \qquad (28)$$

$$Ag \quad Ag \qquad O$$

Carbon dioxide and water may then be formed by formaldehyde being dissociatively adsorbed,

$$CH_2O \rightarrow CO(ads) + 2H(ads) \qquad (29)$$

$$CO(ads) + O(ads) \rightarrow CO_2 \qquad (30)$$

$$2H(ads) + O(ads) \rightarrow H_2O \qquad (31)$$

The formation of acetaldehyde was accounted for by the isomerization of ethylene oxide, and the acetaldehyde thus formed is further oxidized to carbon dioxide and water. Twigg also investigated the oxidation of ethylene oxide and concluded that ethylene oxide was adsorbed (in an unspecified form) at the surface during reaction, since the amount of CO_2 and H_2O produced by oxidation was not equivalent to the quantity of C_2H_4O consumed. Furthermore, ethylene oxide decomposed when passed over a silver catalyst, giving ethylene and adsorbed oxygen, thus accounting for ethylene in the gaseous oxidation products.

In contrast to Twigg, who concluded that ethylene oxide was produced in a separate reaction, Orzechowski and MacCormack [34] proposed that a molecule of ethylene interacted only with a single adsorbed oxygen atom producing adsorbed ethylene oxide and an adsorbed molecule isomeric with ethylene oxide. This isomer, suggested as being acetaldehyde, was capable of decomposing rapidly to yield CO_2 and H_2O. Kurilenko and co-workers [35] proposed a similar mechanism to that of Orzechowski and MacCormack, but suggested that the identity of the adsorbed molecule isomeric with ethylene oxide is vinyl alcohol.

Lyubarskii [29] has measured the electrical conductivity of silver films deposited on glass filaments. His work showed that chemisorption of oxygen involves transfer of electrons from silver to the adsorbate, thus causing a decrease in conductivity. The possibility that charged hydrocarbon radicals exist as surface intermediates during hydrocarbon oxidation must therefore be considered. Margolis and Roginskii [36], using labelled carbon (see Section 3.3.7.1) and oxygen, determined the rate of formation of all the individual species and their mixtures. The rate of formation of ethylene oxide is increased substantially in the presence of formaldehyde and acetaldehyde, while the rate of formation of CO_2 is decreased. Consequently, acetaldehyde is not the principal precursor to the formation of CO_2 from either ethylene or ethylene oxide, and it was suggested that decomposition of an adsorbed ionic intermediate such as $(C_2H_4O_2)$ is the major factor causing CO_2 formation. A similar scheme was suggested to account for ethylene oxidation on a V_2O_5 catalyst. Dixon and Longfield [4] have suggested that because the work of Twigg [32, 33] revealed that the conversion of ethylene was dependent on the mass flowrate of reactant, it is possible that mass transfer may, under some conditions, be a factor determining the observed kinetics and therefore may obscure deductions made concerning reaction mechanism. Further information is also necessary before a satisfactory explanation can be offered for the marked improvement in yield which chlorinated hydrocarbons effect [37].

8.1.2.2. Oxidation of Olefins

In contrast to the oxidation of ethylene, the oxidation of propylene over a cuprous oxide catalyst yields acrolein and direct addition of oxygen does not seem to occur. In the presence of vanadium pentoxide, acids together with CO and CO_2 are produced. Other olefins, when oxidized by catalysts such as vanadium pentoxide, yield aldehydes, acids, CO and CO_2 and would appear to follow a similar course to the oxidation of propylene. Evidence that the reaction proceeds by rupture of the double bond is afforded by the presence, amongst the products, of small amounts of acetaldehyde and formaldehyde. Using a radiotracer technique, it was shown [38] that CO_2 could be formed directly from propylene and also via the subsequent oxidation of acrolein,

$$
\begin{array}{ccc}
 & C_3H_4O & \\
{\scriptstyle k_1} \nearrow & & \searrow {\scriptstyle k_2} \\
C_3H_6 & \xrightarrow[\;\;k_3\;\;]{} & CO_2
\end{array}
\tag{32}
$$

Other studies [39, 40], using flow techniques, have described the rate of formation of acrolein as first order with respect to the partial pressure of oxygen, independent of the partial pressure of propylene and retarded by products such as aldehydes. The rate of formation of CO_2, on the other hand, was shown to be first order with respect to oxygen and retarded by both acrolein and propylene. Activation energies for the formation of acrolein and CO_2, corrected for the inhibition effects produced by products, are approximately 15 kcal mole^{-1} and 25 kcal mole^{-1} respectively. An important point which should be emphasized is that it is quite probable that mass-transport effects through the gas phase and in the catalyst pore structure affect the kinetics. Furthermore, the temperature at which the rate of formation of acrolein becomes comparable to the rates of diffusion of reactant and products may differ from the temperature at which the rate of formation of CO_2 is comparable to rates of diffusion. This would have a profound effect on the distribution of products at a given temperature, so it would be important either to assess the magnitude of the effects or alternatively to investigate the reaction kinetics under such conditions that mass transport is negligible.

As discussed previously in Chapters 5 and 6, the electronic properties and the reactivity of a catalyst can be modified by incorporation of specific impurities into the host crystal lattice. Margolis and co-workers [40] varied the work function of cuprous oxide by adding to it ferric, chromic, lithium, chloride or sulphate ions, and measured the consequent change in chemical rates of formation of acrolein and CO_2. In terms of the activation energy E_i necessary to effect a reaction step i, the heat of adsorption Q of the species whose concentration c determines the kinetics, and the change in work function $\Delta\phi$ consequent upon addition of foreign ions to the catalyst crystal lattice, the rate of formation for each of the reaction products may be expressed

$$r_i = k_i c \exp - \frac{(E_i + Q + \Delta\phi)}{RT} \tag{33}$$

Since both k_i and E_i vary with change in work function, then the selectivity for acrolein formation defined as

$$\sigma = \frac{r_1}{r_1 + r_2} \tag{34}$$

is not a simple function of $\Delta\phi$. However, since a compensation effect (Section 5.3.1.2) operates for this reaction, any changes in the activation energy are compensated for by changes in the pre-exponential factor. Figure 3 shows

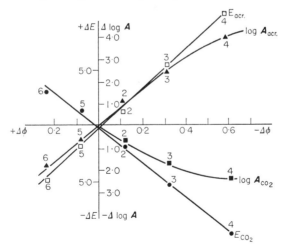

FIG. 3. Oxidation of propylene. Variation of activation energies and pre-exponential factor with work function for (a) the formation of acrolein and (b) carbon dioxide [40].

how the experimental activation energy and the pre-exponential factor for acrolein and CO_2 formation are affected by addition of impurities to a copper catalyst. For the formation of acrolein both the activation energy and the pre-exponential factor increase with decrease in the work function, and conversely the activation energy and pre-exponential factor for CO_2 forma-

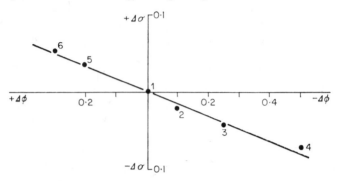

FIG. 4. Oxidation of propylene. Variation in selectivity with work function of the catalyst [40].

tion decrease with decrease in work function. The resulting change in selectivity with change in work function is almost linear as shown in Fig. 4. It is evident that the addition of impurities which cause an increase in work function enhance selectivity, while those which decrease the work function impair selectivity.

8.1.2.3. *Oxidation of Aromatic Hydrocarbons*

The oxidations of benzene to maleic anhydride and of naphthalene and *o*-xylene to phthalic anhydride are catalysed by various types of vanadium pentoxide catalysts. These reactions are important industrial processes and have been studied principally with a view to obtaining kinetic data for the ultimate design of suitable catalytic reactors. Some mechanistic studies have also contributed to an understanding of their reaction mechanisms and the part played by the catalyst surface. It has been suggested that oxidation of benzene to maleic anhydride proceeds through various intermediates such as phenol, hydroquinone and quinone. Dixon and Longfield [4] suggested that only about 10% of the benzene is oxidized via such a route, most of it being oxidized by simultaneous reaction paths to maleic anhydride and to carbon monoxide and carbon dioxide. The oxidation of naphthalene has also been reported as proceeding via simultaneous reaction paths involving phthalic anhydride.

There has been some controversy concerning the role which the catalyst takes during the oxidation of aromatic hydrocarbons. Under normal conditions of oxidation it has been reported that the catalyst is reduced from V_2O_5 to V_2O_4 and V_2O_3 [41]. On the other hand, the rate of oxygen exchange has been compared with the rate of catalytic oxidation of various hydrocarbons and it appears that exchange takes place at temperatures higher than the temperature at which catalytic oxidation proceeds [42]. Rates of oxygen exchange on V_2O_5 and V_2O_4 have been studied and compared with the activity of the catalyst for xylene oxidation [43], but no clear relationship was established. Some evidence which suggests that hydrocarbons act as electron donors to oxidation catalysts such as V_2O_5, NiO, Cr_2O_3 and Cu_2O is provided by electrical conductivity and work function measurements [3]. Conductivity and thermoelectric power measurements have been used to follow changes occurring in the catalyst during the oxidation of xylene [44]. At room temperature it was shown that although the electronic energy levels within the solid were unaffected, the chemisorbed xylene acts as an electron donor to the solid. At temperatures at which oxidation occurs, both the conductivity and the thermoelectric power increased to a marked extent, indicating that the Fermi level of the quasi-free electrons in the catalyst had been reduced to a considerable level below that of the conduction band. Thus, both the surface and interior of the catalyst are affected by the reaction. It is supposed that as an adsorbed o-xylene molecule is oxidized an oxygen-deficient region containing defects is formed. These defects are mobile at temperatures at which the reaction proceeds and migrate throughout the reduced oxide. Chemisorption of oxygen at such a defect would then regenerate the pentoxide. The adsorption of oxygen is believed to be the rate-determining step in the oxidation, since kinetic studies [44] showed that the reaction is zero order with respect to o-xylene and half order with respect to oxygen. Depending on the reaction conditions and average pore size of the catalyst, mass-transfer effects can seriously affect the observed kinetics, so that care should be exercised in interpreting kinetic data for this reaction.

The overall picture thus obtained from such experiments together with kinetic and product distribution data is that xylene oxidation proceeds through the following sequence of events: (1) xylene is chemisorbed, donating an electron to the catalyst, which is an n-type semiconductor containing anion vacancies, (2) the adsorbed xylene reacts with oxygen ions of the lattice, (3) the products of oxidation are desorbed, (4) oxygen from the gas phase replenishes the lattice oxygen lost during interaction in step (2). As indicated, this last step

$$\tfrac{1}{2}O_2 + e \to O^-(\text{ads}) \tag{35}$$

is the one which probably determines the observed kinetics.

Kinetic information obtained from the catalytic oxidation of naphthalene using various vanadia catalysts indicate that the most salient feature of the reaction is the formation of phthalic anhydride, 1,4-naphthoquinone and maleic anhydride in simultaneous reaction steps. Several authors have studied the reaction by observing the initial rate of consumption of naphthalene as a function of both the partial pressures of naphthalene and oxygen [45, 46] and also by measuring the change of naphthalene concentration as a function of contact time [47, 48]. Mars and Van Krevelin [45] interpreted their results in terms of the catalyst being reduced and reoxidized by some unspecified mechanism. They first supposed that naphthalene reduced the oxidized catalyst according to the equation

$$- \frac{dp_N}{dt} = k_1 p_N \theta \qquad (36)$$

where θ is the fraction of surface occupied by oxygen and p_N is the partial pressure of naphthalene at a time corresponding to t. The rate at which the surface was reoxidized was assumed to be proportional to some power n of the oxygen pressure and the fraction of surface not occupied by oxygen

$$- \frac{dp_{O_2}}{dt} = k_2 p_{O_2}^n (1 - \theta) \qquad (37)$$

If v molecules of O_2 are required for the oxidation of one molecule of naphthalene, then, in the steady state,

$$v \, k_1 p_N \theta = k_2 p_{O_2}^n (1 - \theta) \qquad (38)$$

so that

$$- \frac{dp_N}{dt} = \frac{1}{\dfrac{1}{k_1 p_N} + \dfrac{v}{k_2 p_{O_2}^n}} \qquad (39)$$

In most of the experiments excess oxygen is used, so that $p_{O_2}^n$ is constant. This equation therefore shows that for small values of p_N the rate is first order with respect to the partial pressure of naphthalene, whereas for large values of p_N the rate is zero order. These conclusions are in agreement with the measured reaction kinetics. Experiments also show that when p_N is constant and p_{O_2} is varied the reaction is first order with respect to oxygen, so that the exponent n is unity. Pinchbeck [47] concluded from experiments by D'Allesandro and co-workers [48] that mass transport in pores does not influence the measured kinetics when the oxidation occurs on particles smaller than about 0·4 mm diameter.

It was observed [48] that in the initial reaction about 40% of the naphthalene was converted to naphthoquinone, 40% to phthalic anhydride and 20% to maleic anhydride, CO and CO_2. The distribution of products did not change over the temperature range 340–475°C. Calderbank [46] showed that the ratio of the initial rate of formation of naphthaquinone to the initial rate of formation of phthalic anhydride is about 1·75, both reactions having an activation energy of about 26 kcal mole^{-1}. Dixon and Longfield [4] have suggested that if the initial point of attack in the naphthalene molecule is an α-carbon atom, then naphthaquinone is likely to result. On the other hand, if the β-carbon atom suffers attack, phthalic anhydride is formed, while complete combustion occurs if a bridge carbon atom is attacked.

A similar picture has been established for the catalytic oxidation of other aromatics such as toluene. Mars [49] has shown that similar kinetic schemes explain both the catalytic oxidation of toluene and naphthalene. Inherent in such mechanisms is the reduction and reoxidation of the catalyst surface by a similar process to that described for the oxidation of xylene.

8.1.2.4. *Oxidation of Saturated Hydrocarbons*

The conversion of methane and its higher homologues to formaldehyde by direct oxidation using metal or metal oxide catalysts would be economically attractive if sufficiently high yields of formaldehyde were obtained. It has been claimed [50] that formaldehyde may be isolated in significant yields when methane is oxidized over supported metal catalysts. More recently, the oxidation of a number of saturated hydrocarbons were investigated using a copper-oxide catalyst [51]. The results imply that the rate-determining process depends primarily on the hydrocarbon structure which is supporting evidence in favour of a surface radical mechanism of oxidation originally proposed by Lyubarskii [29]. The mechanism follows the ideas of Wolkenstein [28], who proposed that weak chemisorption could be regarded as a sharing of quasi-free electrons between adsorbate and adsorbent (see Section 5.3.3). Thus, the scheme proposed by Lyubarskii for the oxidation of saturated hydrocarbons involves the interaction of chemisorbed hydrocarbon radicals with either gaseous oxygen or chemisorbed oxygen. Intermediate products such as acetaldehydes and acids are accounted for in terms of the decomposition of complex surface radicals or ions which have incorporated oxygen atoms from neighbouring sites. There is some evidence [30] that metal oxides chemisorb saturated hydrocarbons as radicals. Thomas [52] observed that when ethane is adsorbed on stoichiometric nickel oxide the colour changes from green to grey. Such a colour change may be associated with a modification of the defect structure of the solid. Later work [53] in which ethane was permitted to interact with oxygen presorbed at nickel oxide suggested that adsorbed hydrocarbon radicals are present during oxidation and that CO_2 is the main product of oxidation. It therefore seems likely that one possible mechanism for the catalytic oxidation of saturated hydrocarbons is the interaction of adsorbed oxygen with adsorbed hydrocarbon radicals and the subsequent breakdown of the intermediate surface complex into simpler molecules. The nature of the final product obtained seems to depend on the metal oxide used.

8.2. Catalytic Hydrogenation

One of the earliest industrial applications of the hydrogenation of olefinic and acetylenic hydrocarbons originated from the studies of Sabatier, who found that some of the transition metals were active in hydrogenating ethylene and acetylene. Since then, the techniques which have been used for elucidating the mechanism of catalytic hydrogenation reactions include (i) the use of mass spectrometry to identify the intermediate components when deuterium is used to trace the course of the reaction, (ii) infrared spectroscopic examination of the adsorbed state, and (iii) conventional chemisorption and kinetic studies. One of the difficulties which arises in interpreting the results of such studies is the simultaneous occurrence of exchange and polymerization reactions, thus complicating the kinetics and the distribution of products.

8.2.1. Hydrogenation of olefins

8.2.1.1. *Olefin Chemisorption*

Although it is recognized that the results of chemisorption experiments should be regarded with some caution before they are applied to chemical reactions, it is nevertheless important to bear in mind some of the salient features of such measurements. These will at least afford a guide when deciding what surface species are likely to be involved during a reaction.

The chemisorption of ethylene at room temperature has been studied at the surfaces of evaporated metal films of nickel, palladium and tungsten. The first gaseous increments are completely adsorbed, but eventually, after several doses have been admitted, ethane and subsequently ethylene appear in the gas phase. The interpretation of these results is that ethylene is adsorbed dissociatively

$$C_2H_4 \rightarrow C_2H_2(ads) + 2H(ads) \qquad (40)$$

and adsorbed hydrogen then reacts either with gaseous ethylene

$$C_2H_4 + 2H(ads) \rightarrow C_2H_6 \qquad (41)$$

or with the adsorbed complex

$$C_2H_2(ads) + 4H(ads) \rightarrow C_2H_6 \qquad (42)$$

Both of these latter reactions (41) and (42) create fresh sites for the adsorption of ethylene. By following magnetization changes when C_2H_4 is adsorbed at supported nickel it has been shown that, on average, the nickel gains slightly more than two electrons per molecule of adsorbed ethylene [54, 55]. This was explained in terms of an associative two-point attachment to the metal surface M, but with a small fraction existing in a dissociated form requiring four or more sites. The fraction which dissociates on adsorption increases with rise in temperature until, at 100°C, each molecule requires six sites for adsorption,

$$
\begin{array}{ccccccc}
C_2H_4 & \longrightarrow & HC & - & CH & + & H & + & H \\
& & \diagdown\diagup & & \diagdown\diagup & & | & & | \\
& & M\ \ M & & M\ \ M & & M & & M
\end{array} \tag{43}
$$

Infrared spectroscopic studies [56] reveal that the intensity of C—H bands on admitting ethylene to a fresh nickel surface is less than when ethylene is admitted to a surface presorbed with hydrogen. When hydrogen is admitted to a nickel surface at which ethylene has been presorbed the intensity of the C—H bands increases. These results infer a dissociative adsorption with the surface hydrocarbon radicals containing a low ratio of hydrogen to carbon atoms and which ratio increases with the admission of fresh hydrogen. The positions of the bands in the spectra suggest that the adsorbed complex is mainly paraffinic in nature, thus supporting an associative form of adsorption. Polymeric structures have also been detected [56–58], and surface processes such as

$$
\begin{array}{c}
2\ HC - CH \longrightarrow HC - C - C - CH + H_2 \\
\diagdown\diagup\ \ \diagdown\diagup\ \ \ \diagdown\diagup\ \diagdown\diagup\ \diagdown\diagup\ \diagdown\diagup \\
M\ M\ M\ M\ \ \ \ M\ \ M\ M\ M\ M\ M\ M\ M
\end{array} \tag{44}
$$

are envisaged. These polymeric structures may well be the precursors of the higher paraffins which are detected amongst the products of hydrogenation. The variation in rates of hydrogenation with the order of addition of reactants to the catalyst is explained by assuming that, for the case of tantulum, tungsten and nickel, the polymeric complexes reduce the active surface area, so that, if ethylene is admitted first, the rate of hydrogenation is less than if hydrogen were first introduced to the surface. Conversely, if hydrogen is admitted first, there is no opportunity for these complexes to form and a more rapid reaction occurs.

The way in which the adsorbed hydrocarbon packs on the metal surface may be expected to influence the course a hydrogenation reaction takes. Thus, associatively adsorbed ethylene, with its ethane-like structure, would be expected to occupy two metal atom sites, but there would be an appreciable overlap on the (100) and (111) faces of face-centred cubic metals: no overlap would occur on the (110) faces [56]. Hence, on the (100) and (111) faces, where mutual interference is substantial, there will exist gaps in the ethylene layer and on which hydrogen may be adsorbed. In substituted ethylenes overlap will be much greater, with the result that the number of adsorbed molecules in a layer will be smaller. Such a surface could accommodate a larger number of adsorbed interstitial hydrogen atoms at sites where there would be insufficient space to adsorb ethylene. The existence of these sites has been demonstrated by examining the extent to which methylethylenes are capable of inhibiting the ordo-para hydrogen conversion (59).

8.2.1.2. Kinetics and Mechanism of Olefin Hydrogenation

When the relevant information on the kinetics of olefin hydrogenation is examined it is apparent that a variety of rate expressions has been proposed and that the quoted activation energies for the same reactions differ markedly. There seems to be some ground for general agreement, however, since the kinetic order with respect to hydrogen lies between a half and unity, whereas the order with respect to the partial pressure of olefin is usually zero or a negative fractional value. Discrepancies in kinetic orders probably arise because of variable experimental conditions. Differing activation energies are observed because of the operation of a compensation effect, which is not revealed unless the frequency factor has also been measured, and also because of the activation energies varying with the heat of adsorption which is itself a function of coverage. Whether an Eley-Rideal mechanism or a Langmuir-Hinshelwood mechanism (see Chapter 1) operates during the hydrogenation of olefins is not absolutely clear, although, for nickel catalysts, the observed zero-order kinetics with respect to olefin concentration is in favour of a Rideal-Eley mechanism [60, 61]. On the other hand, a negative reaction order is observed with Group VIII metals [62]. A more plausible interpretation is that, because of steric hindrance, ethylene molecules cannot completely cover a nickel surface, so that there are sites at which hydrogen may be weakly adsorbed: because of the larger radii of the Group VIII metals, ethylene may be more tightly packed, and so hydrogen could only adsorb in competition with ethylene, therefore giving rise to fractional reaction orders. Jenkins and Rideal [63] proposed that both hydrogen and ethylene are chemisorbed, but that the ethylene may be adsorbed in an inactive state.

Successive substitution of methyl groups into ethylene effects a lowering of the activation energy for the hydrogenation reaction, and this may be due to two concurrent factors: (i) the heat of adsorption of the olefin decreases with increasing substitution of methyl groups due to stabilization of the π-bond by hyperconjugation; (ii) the number of sites on which hydrogen may adsorb progressively increases with the number of methyl groups substituted, since increased steric effects will cause a larger proportion of the surface area to be exposed. Thus it is quite likely that the true activation energy is influenced by both of these factors. It is difficult, however, to extricate the separate contributions and assess their relative importance. A general phenomenon observed during olefin hydrogenation is the manner in which the activation energy depends on temperature. Above temperatures of about 100°C the activation energy decreases continuously until it is zero at about 150°C and thereafter becomes negative. Rates of reaction therefore increase to a maximum at a particular temperature T_{max} and then gradually decline [63–65]. T_{max} also decreases with decrease in the partial pressure of olefin, and zur Strassen [66] suggested that the heat of adsorption of the olefin was contributing to the observed activation energy. Beeck [67, 68] showed that the rate of hydrogenation of ethylene over evaporated metal films increases as the heat of chemisorption of hydrogen or ethylene decreases and, further, that the rate depends in some way on the pre-exponential factor, for which there was some correlation with the size of the unit metal crystal. Rhodium is the most active metal and it is significant that this metal has the most metallic d-bonding character. Beeck demonstrated that there was a strong correlation between the percentage d-character of metals and their activity for the hydrogenation of ethylene. Because the amount of d-character is closely associated with the number of nearest atom neighbours in a metallic lattice, it is not surprising that there is also a correlation (see p. 309) between activity and lattic dimensions of the unit metal cell. Some studies of ethylene hydrogenation using Cu/Ni, Pd/Cu and Pt/Cu alloys show that changes in the frequency factor rather than the activation energy are responsible for changes in activity; the operation of an electronic factor in enhancing reactivity is not clearly revealed.

Bond [69] has discussed the mechanism of olefin hydrogenations in some detail. The approach which was adopted was to set down logically the possible elementary steps in the process and then to show that the many experimental results conform, to a greater or lesser extent, to such a reaction path. We will briefly review the mechanism as set out by Bond. Since much of the later experimental evidence derives from experiments using deuterium, the distribution of products may best be illustrated by considering the deuteration of an olefin. First it is assumed that the olefin adsorbs associatively and the deuterium in a dissociated form,

$$
\begin{array}{ccc}
CH_2 = CH_2 \rightleftharpoons CH_2 & - & CH_2 \\
| & & | \\
M & & M
\end{array}
\tag{45}
$$

$$
\begin{array}{c}
\tfrac{1}{2}D_2 \rightleftharpoons D \\
| \\
M
\end{array}
\tag{46}
$$

Next, the adsorbed deuterium may add either successively or simultaneously to the associatively adsorbed olefin

$$
\begin{array}{cccc}
CH_2 - CH_2 + D \rightarrow H_2C & - & CH_2D + 2M \\
| \quad\quad | \quad\quad | & & | \\
M \quad\quad M \quad\quad M & & M
\end{array}
\tag{47}
$$

or

$$
\begin{array}{ccc}
CH_2 - CH_2 + 2D \rightarrow DCH_2 & - & CH_2D + 4M \\
| \quad\quad | \quad\quad | & & \\
M \quad\quad M \quad\quad M & &
\end{array}
\tag{48}
$$

According to Twigg [70] the half-hydrogenated state in (47) is obtained via an Eley-Rideal mechanism in which adsorbed olefin reacts with physically adsorbed deuterium. The fate of the adsorbed species in (47) or (48) will depend on whether (i) a surface reaction such as

$$
\begin{array}{ccc}
H_2C - CH_2D + 2M \rightarrow & CH_2 - CD + H \\
| & | \quad | \quad | \\
M & M \quad M \quad M
\end{array}
\tag{49}
$$

occurs, in which case the desorption of the olefin adsorbed in the associative form will account for the appearance of deuterated olefins in the reaction products, (ii) the adsorbed radical in (47) disproportionates

$$
\begin{array}{ccc}
2H_2C - CH_2D \rightarrow & H_2C - CH_2 + H_2DC - CDH_2 \\
| & | \quad | \\
M & M \quad M
\end{array}
\tag{50}
$$

$$
\begin{array}{c}
\rightarrow H_2C - CD + H_3C - CDH_2 \\
| \quad | \\
M \quad M
\end{array}
\tag{51}
$$

(iii) there is interaction between adsorbed deuterium and the adsorbed species from (47),

$$
\begin{array}{cc}
CH_2 - CH_2D + D \rightarrow & CH_2D - CH_2D + 2M \\
| & | \\
M & M
\end{array}
\tag{52}
$$

or (iv) a surface interaction involving exchange occurs,

$$
\begin{array}{cccccc}
CH_2 - CH_2D + CH_2 - CH_2 \rightarrow & H_2C - CHD + CH_2 - CH_3 \\
| \quad\quad | \quad\quad | \quad\quad | & | \quad\quad | \\
M \quad\quad M \quad\quad M \quad\quad M & M \quad\quad M
\end{array}
\tag{53}
$$

$$
\begin{array}{c}
\rightarrow H_2C - CH_2 + CH_2 - CH_2D \\
| \quad | \quad | \\
M \quad M \quad M
\end{array}
\tag{54}
$$

During the very early stages of the catalytic reaction between ethylene and deuterium the product which appears in the greatest proportion is C_2H_6, and this suggests that step (53) is a rapid process, an adsorbed hydrogen and an ethyl radical producing ethane in the gas phase.

With higher olefins more complex situations are likely to arise. Thus, for propylene, either or both of the steps

$$
\begin{array}{ccc}
CH_3 - CH - CH_2 + D \rightarrow & CH_3 - CHD - CH_2 + 2M \\
| \quad\quad | \quad\quad | & | \\
M \quad\quad M \quad\quad M & M
\end{array}
\tag{55}
$$

$$
\begin{array}{c}
\rightarrow CH_3 - CH - CH_2D + 2M \\
| \\
M
\end{array}
\tag{56}
$$

are possible. Whereas the 1-adsorbed 2-deuteropropyl radical has a one-in-two chance of losing a deuterium atom, the 1-deutero 2-adsorbed propyl radical has only a one-in-six chance of losing a deuterium atom and hence the distribution of products is likely to be affected accordingly. For the n-butanes there exist other possible modes of reaction. In the first place double bond migration can take place, and further, *cis–trans* isomerization can occur because of the possibility of free rotation about the central carbon–carbon bond in the half-hydrogenated state. The simultaneous study of the exchange reaction, double bond migration and *cis–trans* isomerization would afford valuable information concerning reaction mechanism.

The experimental evidence which qualitatively supports some of the above reaction steps is provided by kinetic and exchange studies. Thus, in a kinetic investigation of the hydrogenation of ethylene, Twigg [70] proposed that ethylene is reversibly adsorbed in an associative form and that this interacts with physically adsorbed deuterium by an Eley-Rideal mechanism to give the half-hydrogenated state, as described by reaction (47). The hydrogenated state subsequently reacts with adsorbed deuterium according to equation (52). Twigg's reaction scheme is written

$$
\begin{array}{c}
 \overset{k_1}{} \\
C_2H_4 \underset{k_{-1}}{\rightleftharpoons} H_2C - CH_2 \\
 | \quad\quad | \\
 M \quad\quad M
\end{array}
\tag{57}
$$

$$
\begin{array}{c}
\overset{k_2}{}\overset{k_3}{} \\
H_2C - CH_2 + D_2 \underset{k_{-2}}{\rightleftharpoons} CH_2 - CH_2D + D \rightarrow C_2H_5D \\
| \quad\quad | \quad\ | \quad\quad | \quad\quad\quad\quad | \\
M \quad\quad M \quad M \quad\ M \quad\quad\quad\quad M
\end{array}
\tag{58}
$$

and is seen to incorporate many of the steps discussed by Bond. By assuming that the steps k_{-2} and k_3 are slow in comparison with the other steps, a conventional stationary state kinetic treatment gives

$$
r_H = \frac{k_2 k_3}{k_{-2} + k_3} \theta_1 p_{H_2}
\tag{59}
$$

$$
r_E = \frac{k_{-2} k_2}{k_{-2} + k_3} \theta_1 p_{H_2}
\tag{60}
$$

where r_H and r_E are the rate of hydrogenation and exchange respectively and θ_1 is the fraction of surface covered by associatively adsorbed ethylene. If θ_1 is constant, the observed kinetics follow. This scheme also satisfies the experimentally observed fact that the rates of hydrogenation and exchange differ. This led Twigg to suppose that the products, although formed from the same intermediates, ensue from different transition states. Thus, hydrogenation proceeds via k_1, k_2 and k_3, while exchange proceeds through k_1, k_2 and k_{-2}. To account for the formation of H_2 and C_2H_6, Twigg supposed that a fast reaction such as (49) occurred. Ashmore [71] points out that this would remove the need for the step k_{-2} in Twigg's scheme. That reaction (49) does occur is shown by the infrared studies of Eischens and Pliskin [56], who found that the dissociated and associated form of chemisorbed ethylene could co-exist on a reduced surface containing chemisorbed hydrogen atoms.

Further interesting points in connection with the mechanism of ethylene hydrogenation have recently been raised by Halsey [72], who applied his own earlier work [73] on catalysis on non-uniform surfaces to this particular problem. Halsey points out that if chemical reaction succeeds the adsorption of reactants, then the normal conditions pertaining to equilibrium between reactants and surface species are disturbed and the data can only be expressed by a non-equilibrium isotherm. For example, Beeck's original interpretation [74] of his experimental results was based on the assumption that equilibrium between gaseous and adsorbed hydrogen was sustained, as was the equilibrium between gaseous and adsorbed ethylene, even though adsorbed hydrogen was subsequently removed by gaseous ethylene in a chemical reaction forming ethane,

$$H_2 + M \underset{k_2}{\overset{k_1}{\rightleftharpoons}} H_2(ads)$$

$$C_2H_4 + M \underset{k_4}{\overset{k_3}{\rightleftharpoons}} C_2H_4(ads)$$

$$C_2H_4 + H_2(ads) \overset{k_5}{\rightarrow} C_2H_6$$

To illustrate Beeck's assumptions more explicitly we write down the conditions for the occurrence of steady-state equilibria. For hydrogen, the condition is that the rate of adsorption of hydrogen at the free surface (given by $1-\theta_{H_2}-\theta_{Et}$) is balanced by the rate of desorption, so

$$\frac{d\theta_{H_2}}{dt} = k_1(1 - \theta_{H_2} - \theta_{Et})p_{H_2} - k_2\theta_{H_2} = 0 \qquad (61)$$

Similarly, for ethylene

$$\frac{d\theta_{Et}}{dt} = k_3(1 - \theta_{H_2} - \theta_{Et})p_{Et} - k_4\,\theta_{Et} = 0 \qquad (62)$$

Therefore

$$\frac{\theta_{H_2}}{\theta_{Et}} = \frac{k_1k_4p_{H_2}}{k_2k_3p_{Et}} \qquad (63)$$

If, during catalysis, most of the surface is occupied by adsorbed ethylene, the nθ_{Et} is approximately unity. If an Eley-Rideal mechanism is applicable, the rate of hydrogenation is then

$$r = k_5p_{Et}\theta_{H_2} = \frac{k_1k_4k_5}{k_2k_3}\,p_{H_2} \qquad (64)$$

which result is in accordance with the observed reaction orders with respect to hydrogen (first order) and ethylene (zero order). However, the fact that it is very much easier to remove adsorbed species from the surface by chemical reaction rather than by evacuation suggests that the steady state is maintained by adsorption followed by reaction rather than equilibrium between gaseous and adsorbed components,

$$H_2 + M \xrightarrow{k_1} H_2(ads)$$

$$H_2(ads) + C_2H_4 \xrightarrow{k_2} C_2H_6$$

$$C_2H_4 + M \xrightarrow{k_3} C_2H_4(ads)$$

$$C_2H_4(ads) + H_2 \xrightarrow{k_4} C_2H_6$$

Halsey [73] discussed such a non-equilibrium adsorption in relation to the catalytic hydrogenation of ethylene. The rate at which the surface is covered with hydrogen and ethylene is now determined by the respective rates of adsorption and the rates of removal of the components by surface reaction. If the surface reaction in each case follows an Eley-Rideal mechanism, then the stationary surface coverage of hydrogen is given by

$$\frac{d\theta_{H_2}}{dt} = k_1(1 - \theta_{H_2} - \theta_{Et})p_{H_2} - k_2\theta_{H_2}p_{Et} = 0 \qquad (65)$$

and that for ethylene by

$$\frac{d\theta_{Et}}{dt} = k_3(1 - \theta_{H_2} - \theta_{Et})p_{Et} - k_4\theta_{Et}p_{H_2} = 0 \qquad (66)$$

Thus the non-equilibrium isotherm

$$\frac{\theta_{H_2}}{\theta_{Et}} = \frac{k_1k_4}{k_2k_3}\left(\frac{p_{H_2}}{p_{Et}}\right)^2 \qquad (67)$$

is obtained. Hence

$$\frac{\theta_{H_2}}{\theta_{H_2} + \theta_{Et}} = \frac{1}{1 + \dfrac{k_2k_3}{k_1k_4}\left(\dfrac{p_{Et}}{p_{H_2}}\right)^2} \qquad (68)$$

If, during catalysis, the surface is almost saturated with hydrogen and ethylene, then it follows from equation (68) that

$$\theta_{H_2} = \frac{1}{1 + \dfrac{k_2k_3}{k_1k_4}\left(\dfrac{p_{Et}}{p_{H_2}}\right)^2} \qquad (69)$$

On the other hand, if the surface coverage with respect to hydrogen is small and the greater fraction of the surface is covered by ethylene, then θ_{Et} is almost unity and from equation (67) one obtains

$$\theta_{H_2} = \frac{k_1k_4}{k_2k_3}\frac{p_{H_2}^2}{p_{Et}^2} \qquad (70)$$

These non-equilibrium isotherms are quite distinct from the normal Langmuir adsorption isotherm given by equation (63).

The non-equilibrium adsorption mechanism expressed by equations (65) and (66) is, in effect, a quantitative description of the production of ethane either by gaseous or van der Waals adsorbed ethylene reacting with chemisorbed hydrogen gas or by van der Waals adsorbed or gaseous hydrogen reacting with chemisorbed ethylene. The respective contributions from the two independent surface reactions depend on the relative rates of adsorption of hydrogen and ethylene on the available free surface.

Now the rate of adsorption is, from the kinetic theory of gases (see Section 2.3.1),

$$r_a = \frac{\sigma p}{(2\pi m k T)^{1/2}} (1 - \theta) \exp \left\{ - \frac{E_a}{RT} \right\} \tag{71}$$

where σ is the condensation coefficient, $(1-\theta)$ is the extent of free surface available for adsorption, E_a the activation energy for adsorption, m the mass of a single molecule of the adsorbate and k the Boltzmann constant. If the activation energies for both of the two adsorption processes are small and the area of surface available for adsorption remains sensibly constant, then the ratio of the rates of adsorption of hydrogen and ethylene is equal to the square root of the inverse ratio of their molecular weights, i.e. $\sqrt{14}$. Under these conditions the rate of adsorption of hydrogen is faster than that for ethylene by this factor and the rate of formation of ethane is given by $k p_{Et} \theta_{H_2}$. On substituting equation (70) the rate is

$$r_H = k p_{Et} \theta_{H_2} = k' \frac{p_{H_2}^2}{p_{Et}} \tag{72}$$

which, however, does not conform to the experimentally observed kinetics.

Halsey [73] suggests that the non-equilibrium mechanism invoked by equations (65) and (66) may be modified to take into account the possibility of an energetically non-uniform surface. This possibility, discussed at some length in Section 2.4.1.5, is now considered in the hydrogenation of ethylene. It is assumed that the two rate-limiting steps are (a) the interaction of chemisorbed hydrogen with gaseous ethylene (rate-constant k_2 in equation (65)) and (b) reaction between chemisorbed ethylene and gaseous hydrogen (rate-constant k_4 in equation (66)). According to Halsey's analysis outlined previously, the reaction rate will be proportional to $[\theta_2/(1-\theta_4)]^{n/2}$. The term θ_2 may be identified with the fraction of surface occupied by hydrogen and $(1-\theta_4)$ with the fraction of surface remaining unoccupied by hydrogen (i.e. the fraction of surface covered by ethylene θ_{Et}). If an Eley-Rideal mechanism is operating, the reaction rate may therefore be written

$$r = k p_{Et} \left(\frac{\theta_{H_2}}{\theta_{Et}} \right)^{n/2} \tag{73}$$

Substituting the value of θ_{H_2}/θ_{Et} as given by the non-equilibrium isotherm (67), the rate of reaction is

$$k_4 \left(\frac{k_1 k'}{k_2 k_3}\right)^{1/2} = p_{Et} \left(\frac{p_{H_2}^2}{p_{Et}^2}\right)^{1/2} = k' p_{H_2} \tag{74}$$

when $n=1$, and this conforms to the observed kinetics.

We therefore see that, by taking into consideration the very real likelihood of non-equilibrium adsorption processes between the reactant gases and the catalyst surface, it is possible to interpret, still in terms of an Eley-Rideal mechanism, the observed kinetics of a typical heterogeneous catalytic reaction.

The reaction of ethylene and of 1-butane with deuterium has been studied over nickel wires [75, 76]. The distribution of deuterated ethylenes and ethanes as a function of conversion are shown in Fig. 5(a) and (b) for the reaction

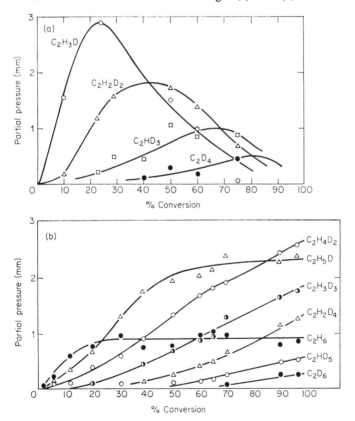

FIG. 5. Reaction of ethylene with deuterium: (a) formation of deuterated ethylenes; (b) formation of deuterated ethanes [75, 76].

of ethylene with deuterium. It is seen that ethane-d_0 and ethane-d_1 are the major products formed during the initial stages of the reaction. The relative rates of formation of ethanes-d_0 to -d_2 decrease with increasing conversion, but by comparison the rates of formation of ethanes-d_3 to -d_6 increase. These results were interpreted as being indicative of the rapidity of reaction (53). Each deuterium atom present in an ethyl radical is rapidly transferred to an adsorbed ethylene molecule which has a good chance of desorbing: the deuterium atom is thus removed from the adsorbed ethyl radical and, if these disproportionate by a reaction such as (50), ethane-d_0 would be expected to be the chief product in the early stages of reaction. The ethylene-d_1, which is desorbed from reactions such as (49) or (51), may then readsorb and reappear either as ethylene-d_2 or a deuteroethane. The observed deuterium content of the ethanes is such that adsorption-desorption equilibrium is maintained between gaseous and adsorbed ethylene. Taylor [77] has reviewed the mechanisms proposed for olefin hydrogenations and it appears that olefin desorption cannot be a rate-determining step in exchange and double-bond migration, since there is a kinetic isotope effect and also *cis–trans* isomerization is faster than either the exchange reaction or the rate of double-bond migration. Thus, the addition of a hydrogen or deuterium atom by a reaction such as (47) or (48) is probably the slowest step in the exchange and double-bond migration processes. Since no isotope effect is found for the isomerization reaction, hydrogen transfer according to (53) is probably the slowest step in this process.

Bond [78] has claimed that a semi-quantitative treatment relating to the distribution of deuterium amongst the observed products is more successful than an approach assuming a binomial distribution of deuterium atoms. Thus it was observed that the yields of deuterated paraffins decrease logarithmically with increasing deuterium content:

$$\frac{\text{Yield } C_nH_{2n+1-x}D_{x+1}}{\text{Yield } C_nH_{2n+2-x}D_x} = y \tag{75}$$

If the fate of adsorbed alkyl radicals is to disproportionate according to (50) and (51), then an alkyl radical has about the same chance of desorbing in the form $C_nH_{2n+2-x}D_x$ as it has of remaining on the surface and suffering further exchange, ultimately appearing as $C_nH_{2n+1-x}D_{x+1}$. Hence y is about 0·5 if this assumption is true. Departures from this value are attributed to (i) unequal occurrence of the exchange reactions, (ii) addition of deuterium by mechanisms other than (50), (iii) zero point energy effects. The contribution of direct addition of deuterium according to (52) is probably inversely related to y. Kemball [61], Horiuti [79] and Keii [80] have also produced general schemes to account for the distribution of deuterium amongst products. Kemball considered the probability of hydrogen and deuterium exchange amongst adsorbed ethylene molecules and ethyl radicals. To each process was assigned a parameter which, when expressed as a simple probability function, describes the probability of such a process occurring. In this way the distribution of deuterium amongst the products of ethylene deuteration was calculated.

8.2.1.3. *Hydrogenation of Alicyclics*

The only product observed when cyclopropane is hydrogenated over metal catalysts is propane. Orders of reaction depend on temperature and hydrogen pressure [81]. The order with respect to cyclopropane increases from 0·2 to 1·0 with increase in temperature and constant hydrogen pressure. With increasing hydrogen pressure, rates increase to a maximum and then either remain constant or decline. These results are interpreted by a Langmuir-Hinshelwood mechanism, the slow step being interaction of adsorbed cyclopropane with strongly adsorbed hydrogen atoms. Methylcyclopropane, when hydrogenated, could break either in the 1,2 position to give n-butane or at the 2,3 position to give isobutane

$$
\begin{array}{c}
CH_2\diagdown \\
| \quad CH - CH_3 \\
CH_2\diagup
\end{array}
\quad
\begin{array}{c}
\xrightarrow{H_2} \\
\\
\searrow \\
H_2
\end{array}
\quad
\begin{array}{cc}
CH_2\diagdown & \\
| \quad CH_2 - CH_3 & \qquad (76) \\
CH_2 & \\
\\
CH_3\diagdown & \\
CH - CH_3 & \qquad (77) \\
CH_3\diagup &
\end{array}
$$

Bond and Newham [81] state that the ring opens predominantly by breaking of the bond opposite to the carbon atom with the greatest number of substituent groups. Steric hindrance at carbon atom 1 in the adsorbed molecule induces attack by hydrogen of carbon atoms 2 or 3. The electrophobic alkyl group also tends to move electrons towards C atoms 2 and 3, and hence this effect also favours breaking of the bond between C atoms 2 and 3.

The reaction between cyclopropane and deuterium is similar in many respects to the exchange reaction between propane and deuterium [82–84]. The predominant feature of the former reaction is, in fact, exchange of hydrogen for deuterium and propane -d_8 is often the major product. For cyclopropane hydrogenation the initiating step is probably

$$
\begin{array}{ccc}
C_3H_6 + D \rightarrow CH_2CH_2CH_2D + M \\
| \quad\quad | \quad\quad\quad | \\
M \quad\quad M \quad\quad M
\end{array}
\qquad (78)
$$

where the cyclopropane may be chemisorbed by forming a π-bond between the delocalized electrons of the ring and the unfilled d-orbitals of the metal. This step is probably succeeded by steps such as (50) written for olefin hydrogenation. Some cyclopropane exchange also occurs and this possibly means that cyclopropane may also chemisorb dissociatively thus

$$
\begin{array}{c}
\text{CH}_2 \\
\diagup \quad \diagdown \\
\text{C}_3\text{H}_6 + 2\text{M} \rightarrow \text{CH} - \text{CH} + \text{H} \qquad\qquad (79) \\
\mid \qquad\quad \mid \\
\text{M} \qquad\quad \text{M}
\end{array}
$$

$$
\begin{array}{c}
\Bigg\downarrow \text{D(ads)} \\
\text{C}_3\text{H}_5\text{D}
\end{array} \qquad\qquad (80)
$$

although there is no kinetic evidence to support the dissociative adsorption of cyclopropane.

8.2.2. Hydrogenation of acetylenes

8.2.2.1. Chemisorption of Acetylenes

Not very much evidence exists concerning the chemisorption of acetylenes at metal surfaces. Infrared studies [56] have revealed that adsorbed ethyl radicals are present at silica-supported nickel catalysts and an associatively adsorbed form of acetylene has been reported [57] on alumina-supported palladium. At room temperature the heat of adsorption is independent of coverage until the molecules are sufficiently close to cause adsorbate–adsorbate interaction: this therefore implies an immobile adsorbed state. One important fact observed is that acetylene is much more strongly adsorbed than ethylene. If the latter is added to acetylene it is the acetylene which adsorbs and reacts to the exclusion of the ethylene. This fact determines the fate of species adsorbed at the surface during acetylene hydrogenation, since any olefins formed are immediately desorbed to appear as gaseous products in the early stages of reaction. Thus the process of ethylene formation is a selective one and has been made use of in the commercial synthesis of ethylene and also the selective removal of acetylene from gases by hydrogenation.

Geometrical considerations of the associatively adsorbed state show that acetylenes are adsorbed with greater strain on platinum and palladium than on nickel. Further, it has been argued that acetylene is not to be expected to adsorb on the (111) plane of face-centred cubic metals, since the interatomic distances are too close. This latter hypothesis has been used to explain the inactivity of the metals Os and Ru, which have a close-packed hexagonal structure. Possible arrangements for packing adsorbed acetylene on (100) and (110) planes of face-centred cubic metals have been proposed by Bond [69], and a staggered arrangement of molecules seems the most preferable, as this avoids excessive overlap.

The associative form of adsorbed acetylene is an ethylene-like structure and therefore has residual unsaturation. Depending on whether the π-bond or a carbon–metal bond is broken when hydrogen attacks adsorbed acetylene, either an adsorbed unsaturated vinyl radical is formed or a free radical,

$$
\begin{array}{c}
\xrightarrow[\text{bond rupture}]{\text{Carbon–metal}} \quad
\begin{array}{c} HC = CH_2 \\ | \\ M \end{array} \tag{81}
\end{array}
$$

$$
\begin{array}{c} HC = CH \\ | \quad\; | \\ M \quad M \end{array}
\qquad
\xrightarrow[\pi\text{-Bond rupture}]{}
\qquad
\begin{array}{c} H\dot{C} - CH_2 \\ | \quad\; | \\ M \quad M \end{array} \tag{82}
$$

The adsorbed free radical may then initiate a surface polymerization process in which the adsorbed acetylene is the monomer unit,

$$
\begin{array}{c}
H_2C - \dot{C}H + HC = CH \rightarrow H_2C - CH - CH - \dot{C}H \\
\,| \quad\;\; | \quad\;\; | \quad\;\; | \qquad\;\; | \quad\;\; | \quad\;\; | \quad\;\; | \\
M \quad M \quad M \quad M \qquad M \quad M \quad M \quad M
\end{array} \tag{83}
$$

Repetition of this process leads to surface polymer units containing an integral number of pairs of carbon atoms and such hydrocarbons do appear amongst the products of hydrogenation. It is more difficult to explain the appearance of branched chain polymers, but a s-butyl free radical could be formed as follows

$$
\begin{array}{c}
H_2C - \dot{C}H + HC = CH_2 \rightarrow H_2C - CH - CH = CH_2 + M \\
\,| \quad\;\; | \quad\;\; | \qquad\qquad | \qquad | \\
M \quad M \quad M \qquad\qquad M \qquad M
\end{array} \tag{84}
$$

$$
\begin{array}{c}
\downarrow \\
\begin{array}{c}
H_2C - CH - CH - CH_2 \\
\,| \qquad\;\; | \qquad\qquad | \\
M \qquad M \qquad\qquad M
\end{array}
\xrightarrow{\;C_2H_2(ads)\;}
\begin{array}{c}
\qquad\qquad CH \\
\qquad\quad /\;\;|\;\;\backslash \\
H_2C - CH\, CH_2\, CH - \dot{C}H \\
\,| \qquad | \quad\;\; | \quad\;\; | \quad | \\
M \qquad M \quad M \quad M \quad M
\end{array}
\end{array} \tag{85}
$$

The latter process (equation (85)) involves an intramolecular rearrangement of electrons in which a π-bond is broken and a new carbon–metal bond formed, and so only steric and geometric effects may be expected to hinder the formation of the branched chain polymer.

8.2.2.2. Selectivity and Kinetics of Acetylene Hydrogenation

The preferential formation of ethylene when acetylene is hydrogenated is a unique feature of this reaction. Ethane is not formed until the later stages of reaction. The selectivity pertaining to the hydrogenation of acetylene has been explained [85] by means of the following kinetic scheme.

$$
\begin{array}{ccccc}
HC = CH & \xrightarrow{+H} & H_2C = CH & \xrightarrow[k_1]{+H} & C_2H_4 \\
\,| \quad\;\; | & & | & & \uparrow \\
M \quad M & & M & & \\
& \Big\Uparrow & \Big\backslash\;{+H} \atop k_2 & \Big\uparrow\, k_4 & \\
& & & & \\
& & H_2C - \dot{C}H & \xrightarrow[\;+H\;]{} H_2C - CH_2 & \xrightarrow[k_5]{+H_2} C_2H_6 \\
& & \,| \quad\;\; | & | \qquad | & \\
& & M \quad M & M \quad M &
\end{array} \tag{86}
$$

Selectivity arises because hydrogen can add to either a normal adsorbed vinyl radical or to the corresponding free radical form. In the former case, either gaseous or adsorbed ethylene is formed, depending on whether a metal–carbon bond or the π-bond is broken. If, for the steady state, θ_v is the fraction of vinyl radicals adsorbed in the normal form and $(1-\theta_v)$ in the free radical form, then a simple steady-state kinetic analysis gives

$$r_{C_2H_4} = \theta_H \left\{ k_1\theta_v + \frac{k_4B}{k_4 + k_5p_{H_2}} \right\} \qquad (87)$$

$$r_{C_2H_6} = \frac{k_5p_{H_2}\theta_H B}{k_4 + k_5p_{H_2}} \qquad (88)$$

where

$$B = k_2\theta_v + k_3(1 - \theta_v) \qquad (89)$$

The kinetic selectivity (see Chapter 7) σ is given by the ratio of the rates of ethylene and ethane formation. This ratio may also be expressed as the factor $S_k(=\sigma/(1+\sigma))$. From the above equations we therefore obtain

$$1 - S_k = \frac{r_{C_2H_6}}{r_{C_2H_6} + r_{C_2H_4}} = \frac{k_5p_{H_2}B}{k_4(k_1\theta_v + B) + k_5p_{H_2}(k_1\theta_v + B)} \qquad (90)$$

Kinetic results [86] for the dependence of selectivity on hydrogen pressure indicate that the equation

$$1 - S_k = kp_{H_2}^n \qquad (91)$$

is obeyed for some systems. The exponent n in this equation varies with both the hydrogen/acetylene ratio and the temperature and it may be regarded as an approximation to equation (90). Observed activation energies for acetylene hydrogenation are greater than those for ethylene hydrogenation, and so the increasing selectivity with rise in temperature is in general accord with this.

The exchange reaction of deuterium with acetylene proceeds more slowly than the corresponding reaction with ethylene. Over nickel–kieselguhr, *cis* and *trans* dideuteroethylene are formed in equal amounts at room temperature [87]. Over a nickel–pumice catalyst, analysis of the distribution of deuterium amongst the products of reaction indicate that ethylene-d₂ is the most abundant product [88]. The deuterium number, which is defined as the average number of deuterium atoms contained by deuterated species, was slightly greater than two, and this means that a net acetylene exchange reaction was occurring. However, its importance is much less than during olefin hydrogenation. Deuterium numbers less than two, indicating a net hydrogen exchange reaction, were only found at high hydrogen to acetylene ratios.

If substituted deutero-acetylenes are associatively adsorbed, then *cis* isomers of deutero ethylene would be expected to appear as products in the gas phase. However, a significant fraction of *trans* isomers are actually formed. For this to occur it is necessary to disrupt the normal planar configuration of the adsorbed vinyl radical, and this takes place when the free-radical form is produced. If this is represented stereochemically as

$$
\begin{array}{cc}
D & H \\
\diagdown & \vdots \\
\overset{|}{\underset{|}{C}}\!-\!\overset{|}{\underset{|}{C}} & \\
\diagup & \diagdown D \\
M & M
\end{array}
\tag{92}
$$

then addition of a hydrogen atom would result in equal parts of *cis* and *trans* dideuteroethylene. On increase in temperature the yield of *trans* isomers increases by the same amount as the increase in yield of polymers suggesting that the same intermediate is involved. This leads one to suppose that the free radical form of the adsorbed vinyl radical is a precursor to the formation of equal amounts of *cis* and *trans* isomers and polymers.

8.2.3. Hydrogenation of aromatics

8.2.3.1. *Adsorption of Benzene and Homologues*

It appears that, on adsorption, benzene loses its resonance energy. The heat of adsorption of benzene is fairly low ($\sim$30 kcal mole^{-1} on nickel powder and 12 kcal mole^{-1} on copper powder at low coverage) [89] and it would be expected to be much higher than this if some or all of the resonance energy were retained. Of some relevance is the difference between the heats of adsorption of thiophen (33 kcal mole^{-1}) and ethyl sulphide (65 kcal mole^{-1}) on platinum [90]. Similarly, the conversion of benzene to a cyclic diene is endothermic (5·6 kcal mole^{-1}) [91] and this compares with the activation energy for the hydrogenation of benzene $\sim$7·4 kcal mole^{-1}. Selwood [92] has shown that when benzene is chemisorbed, at 150°C, on supported nickel the relative magnetization of the catalyst decreases to an extent which is accounted for by benzene being attached to the surface by six bonds. Other important facts which are relevant to an interpretation are (i) when benzene is chemisorbed hydrogen is liberated and appears in the gas phase [93], and (ii) cyclohexene is the primary product in the dehydrogenation of benzene [94]. The evidence therefore seems to favour adsorption processes such as

$$
2M + C_6H_6 \longrightarrow \underset{M\quad\ M}{\bighexagon}
\tag{93}
$$

and

$$
6M + C_6H_6 \longrightarrow M\!-\!\underset{M\quad\ M}{\overset{M\quad\ M}{\bighexagon}}\!-\!M
\tag{94}
$$

It would seem, because cyclohexene is the major product during benzene dehydrogenation, that the first mode of adsorption exists under hydrogenation conditions. Benzene could, however, be chemisorbed in the form of a distorted hexane like molecule, as in (94), and the interatomic metal distances are conveniently disposed for such a structure on the {111} planes of face-centred cubic metals and the {110} planes of body-centred cubic metals (95). The presence of benzene on a surface does not appear to inhibit para-hydrogen conversion and so it must be supposed that packing of the adsorbed molecules leaves vacant surface sites at which there is no overlap of molecular orbitals. From the published work it can only be concluded that benzene is chemisorbed strongly and probably exists at the surface either as phenyl radicals or as the diadsorbed state as in (93). The structure represented in (94) (suggested also by Balandin—see p. 305) may exist in the absence of hydrogen.

8.2.3.2. *Kinetics and Mechanism of Hydrogenation of Benzene and Exchange with Deuterium*

Generally, it is found that the rate of hydrogenation is proportional to the hydrogen partial pressure and independent of the benzene partial pressure [*96, 97*]. The exponent of the hydrogen pressure tends to increase with temperature until it reaches a limiting value of unity [*98*]. These results therefore imply that benzene is strongly adsorbed, while hydrogen is either weakly adsorbed or possibly reacts from a van der Waals layer of physically adsorbed hydrogen.

The exchange reaction proceeds faster than the hydrogenation reaction, in spite of the latter having the lower activation energy [*99, 100*]. The orders of reaction are similar to those found for hydrogenation, but the orders with respect to hydrogen tend to be somewhat lower. The para-hydrogen conversion and hydrogen–deuterium exchange reaction occur readily during benzene hydrogenation, the rates being some 100 times the rate of the exchange reaction: thus, possibly because of the way in which strongly adsorbed benzene packs on the crystal faces, hydrogen has access to active areas of the catalyst surface during hydrogenation. Because the exchange reaction is inhibited by one of the products, cyclohexane, whereas the hydrogenation reaction is not affected in this way, it has been suggested that the exchange reaction proceeds via a different mechanism to that of hydrogenation [*101*]. The deuterium number appears to be between one and two, so that, on average, more than one deuterium atom is introduced into a benzene molecule during each act of exchange. An analysis of the distribution of deuterium amongst the products of exchange shows that CH_3D is produced in least quantity while the most abundant product is $C_6H_6D_6$. This has been interpreted by supposing that two mechanisms operate simultaneously,

$$D_2 + 2M \rightarrow \underset{\underset{M}{|}}{2D} \tag{95}$$

$$C_6H_6 + 2M \rightarrow \;\; \text{[benzene–M ring with =]} \;\; + \underset{\underset{M}{|}}{H} \tag{96}$$

$$\text{[benzene ring with M and M]} + D \rightarrow C_6H_5D + 2M \tag{97}$$

$$\text{[ring with M, M]} + 2M \rightarrow \text{[ring with M, M]} + \underset{\underset{M}{|}}{H} \tag{98}$$

$$\downarrow D_{(ads)}$$

$$\text{[ring with M, D]} + 2M \tag{99}$$

the cyclohexane -d_6 being produced by addition of deuterium atoms via benzene and deuterated benzene chemisorbed at a dual site.

8.3. Fischer-Tropsch Synthesis

Although the synthesis of high molecular weight paraffins and alcohols, accomplished by passing a mixture of carbon monoxide and hydrogen over a metal catalyst, has achieved industrial prominence, the mechanism of reaction and the role the catalyst plays have not yet been established unequivocally (see Section 3.3.7.1). Nevertheless, there are some interesting features peculiar to this synthesis which will be considered in this section.

Metals which are active in the Fischer-Tropsch reaction include iron, cobalt, nickel and ruthenium. Of these only ruthenium, used at high pressure to produce solid paraffins of high molecular weight, is used in the pure state, requiring neither supporting nor promoting. Iron, which yields olefins and alcohols, is usually promoted by means of the addition of either of the irreducible metal oxides ThO_2, MgO, Al_2O_3, MnO, Cr_2O_3, U_3O_8, Ce_2O_3, CaO, SiO_2 or V_2O_5, which act as structural promoters enhancing the stability of the metal phase. In addition, an electronic promoter is added, usually an alkali metal such as sodium or potassium, which effects an alteration in the electronic energy levels of the metal catalyst. Iron catalysts for the Fischer-Tropsch synthesis do not usually require supporting. Cobalt and nickel are invariably supported, often with kieselguhr, and small amounts of promoters are also added. Both of the latter catalysts yield mainly straight-chain hydrocarbons of intermediate molecular weight.

It appears that bulk carbide formation does not occur under synthesis conditions, since (i) the carbides can be hydrogenated to methane and ethane at temperatures of $250°C$ (neither of these paraffins is found as an intermediate product), and (ii) the formation of higher hydrocarbons by the reduction of carbides has a positive free-energy change. In spite of these facts, iron carbides (cementite, Fe_3C, and Hägg carbide, Fe_2C) are good Fischer-Tropsch catalysts and are better than reduced iron catalysts. It is also apparent from X-ray and electron diffraction studies that, as far as cobalt catalysts are concerned, surface carbide is not a true intermediate [102, 103] nor is it formed at the surface of active iron catalysts under synthesis conditions [104]. There is a weight of evidence against surface carbide formation, but, on the other hand, its existence has not been disproved.

Thermodynamic considerations reveal that reactions of the types

$$(2n + 1) H_2 + nCO \rightarrow C_nH_{2n+2} + nH_2O \tag{100}$$

$$2nH_2 + nCO \rightarrow C_nH_{2n} + nH_2O \tag{101}$$

$$2nH_2 + nCO \rightarrow C_nH_{2n+1} OH + (n - 1) H_2O \tag{102}$$

have negative free-energy changes at moderate temperatures. For stoichiometric equations in which CO_2 is written as the product,

$$(n + 1) H_2 + 2nCO \rightarrow C_nH_{2n+2} + nCO_2 \tag{103}$$

$$nH_2 + 2nCO \rightarrow C_nH_{2n} + nCO_2 \tag{104}$$

$$(n + 1) H_2 + (2n - 1) CO \rightarrow C_nH_{2n+1} OH + (n - 1) CO_2 \tag{105}$$

the free-energy change is even more negative, indicating that products formed by elimination of CO_2 are more thermodynamically favourable than those formed by elimination of water.

8.3.1. Chemisorption of reactants

In the temperature range 150–300°C, where the Fischer-Tropsch reaction generally occurs, hydrogen is strongly chemisorbed (for more details see Section 2.2) on metal catalysts in a dissociated form. The hydrogen atoms are possibly located in the interstices between surface metal atoms or at the intersections of lattice planes [105]. Carbon monoxide is chemisorbed on transition metals in two forms, as is evidenced by the infrared studies of Eischens and Pliskin [56]. One structure is linear, while the other is a bridge structure

$$
\begin{array}{ccc}
O & & O \\
\| & & \| \\
C & & C \\
\| & & / \ \backslash \\
M & & M \quad M \\
\text{(a)} & & \text{(b)}
\end{array}
\tag{106}
$$

For platinum metal supported on silica both forms may co-exist [56, 106] but on supported iron only the linear form (a) has been detected [56]. When hydrogen is admitted to a presorbed layer of carbon monoxide at 50°C some hydrogen is taken up, even though no additional CO could have been accommodated [107, 108]: however, no change in the infrared spectrum is observed [56]. When CO is admitted to hydrogen layers, calorimetric measurements [109, 110] indicate that complex formation occurs, although no inference concerning its structure has been made.

8.3.2. Mechanism of Fischer-Tropsch synthesis

Most of the deductions concerning the reaction mechanism have been based on product-distribution analyses. Available kinetic data generally refer only to integral type reactors in which conversions are too large to allow a kinetic analysis of such a complex reaction. Much information has been derived from experiments in which additives, such as saturated and unsaturated paraffins and alcohols, have been introduced to the feed stream and their subsequent fate determined. The added gas may emerge unchanged, reduced, oxidized or incorporated in the products of synthesis. Incorporation of the additive in a product molecule implies that the additive has initiated the growth of hydrocarbon chains or that it has become built into the molecule by some type of condensation or polymerization process. By suitably labelling molecules (e.g. with radioactive species), it is possible, in principle, to decide from what point in the molecule chain growth originates.

Primary alcohols are efficient chain initiators [111]. At temperatures of the order of 200°C, 1-propanol is incorporated in the products of synthesis to the extent of about 30%. From an analysis of the C_4 hydrocarbons in the product, it is apparent that the incorporation has occurred chiefly in the straight-chain hydrocarbons and only to a very much smaller extent in branched isomers. Thus chain growth occurs predominantly at the end of a hydrocarbon chain. Primary alcohols decompose over Fischer-Tropsch catalysts to give products similar to those formed in synthesis, and this is probably due to the surface species resulting from alcohol decomposition undergoing polymerization [112]. Chain growth could occur by the polymerization of methylene radicals formed by reduction of surface carbide, but this hypothesis fails to account for the presence of products containing oxygen [112]. It is more likely that surface species such as CH(OH), arising from the hydrogenation of adsorbed carbon monoxide, are responsible for chain growth. The initiating step may thus be represented

$$\begin{array}{c} O \\ \parallel \\ C \\ \parallel \\ M \end{array} \xrightarrow{\ 2H\ } \begin{array}{c} H \quad OH \\ \diagdown \diagup \\ C \\ \parallel \\ M \end{array} \qquad\qquad (107)$$

This would then be followed by the elimination of water between two surface radicals and subsequent hydrogenation [112–114].

$$
\underset{M}{\overset{H \quad OH}{\underset{\|}{\overset{\diagdown \diagup}{C}}}} + \underset{M}{\overset{H \quad OH}{\underset{\|}{\overset{\diagdown \diagup}{C}}}} \xrightarrow{-H_2O} \underset{M \quad M}{\overset{H \qquad OH}{\underset{\| \; \|}{\overset{\diagdown \qquad \diagup}{C-C}}}} \xrightarrow{2H} \underset{M}{\overset{}{\underset{\|}{H_3C-C-OH}}} \qquad (108)
$$

Chain growth may proceed either by further elimination of water between the above surface species followed by hydrogenation

$$
\underset{M}{\overset{}{\underset{\|}{H_3C-C-OH}}} + \underset{M}{\overset{}{\underset{\|}{H-C-OH}}} \xrightarrow[+2H]{-H_2O} \underset{M}{\overset{}{\underset{\|}{H_3C-CH_2-C-OH}}} \qquad (109)
$$

or alternatively via the surface intermediate $H_3CCH(OH)$,

$$
\underset{M}{\overset{}{\underset{|}{H_3C-CH-OH}}} + \underset{M}{\overset{}{\underset{\|}{H-C-OH}}} \xrightarrow[+2H]{-H_2O} \underset{M}{\overset{CH_3}{\underset{|}{\overset{|}{CH_3-C-OH}}}} \qquad (110)
$$

which explains the formation of branched isomers. Termination may proceed by desorption

$$
\underset{M}{\overset{}{\underset{\|}{R - C - OH}}} \rightarrow RCHO + M \qquad (111)
$$

and alcohols formed by hydrogenation of the aldehyde. Olefins and paraffins may be formed by the simultaneous elimination of water and addition of hydrogen to surface radicals,

$$
\underset{M}{\overset{}{\underset{|}{R - CH_2 - CH - OH}}} \xrightarrow[+2H]{-H_2O} \underset{\underset{+H \downarrow}{M}}{\overset{}{\underset{|}{R - CH_2 - CH_2}}} \xrightarrow{-H} R - CH_2 = CH_2 \qquad (112)
$$
$$
R - CH_2 - CH_3
$$

Herington [115] postulated that the carbon chain grows one atom at a time until the chain is terminated. For a growing chain of n carbon atoms a quantity α_n was defined as the probability that the chain will grow and the quantity $1-\alpha_n$ the probability that the chain will terminate. Now the number of moles, ϕ_n, of product containing n carbon atoms that are present in the products of synthesis is a measure of the probability of termination, while the sum of the number of moles of product containing more than n carbon atoms is a measure of the probability α_n that the chain will grow. The ratio β_n of these probabilities is therefore written

$$
\beta_n = \frac{1 - \alpha_n}{\alpha_n} = \frac{\phi_n}{\sum\limits_{n+1}^{\infty} \phi_i} \qquad (113)
$$

From an analysis of the distribution of products during a synthesis reaction, Herington found that β is substantially independent of the number of carbon atoms in the product. In this case the probability of chain growth declines logarithmically with increasing carbon number, i.e.

$$\alpha = \frac{\phi_{n+1}}{\phi_n} \tag{114}$$

and this is precisely what is found in practice. A semilogarithmic plot of the molar yield of hydrocarbons containing n carbon atoms against the number of carbon atoms almost invariably gives a straight line, and this is only another way of expressing the above equation (114). Weller and Friedel [*116, 117*] assumed that there was an *a priori* probability of addition to a terminal carbon atom in the chain and that this was different from the probability of addition to a penultimate carbon atom. Assigning an adsorbed species a probability a of adding another carbon atom at the end of the chain and a probability b of adding one at the penultimate carbon atom, the ratio b/a is a measure of the extent of branching, f (say). The fraction f was assumed to be independent of chain length, and growth at both ends of the chain was envisaged. Proceeding in this way, the probability of forming the normal C_4 chain from the C_3 group is $2a$ and that for the iso-C_4 chain is b. Hence the fractions of n-C_4 and iso-C_4 in the C_4 fraction of the product are $2a/(a+1)$ and $b/(a+1)$ respectively. The probability of forming n-C_5 from n-C_4 is also $2a/(a+1)$, whereas iso-C_5 can be formed from iso-C_4 with a probability $b/(a+1)$ and from n-C_4 with a chance $2ab/(a+1)$. A list of these fractions may thus be compiled for hydrocarbons containing any number of carbon atoms. For cobalt catalysts the predicted distribution agrees very well with observed data [*118*] if $a=0.961$ and $b=0.039$. For iron catalysts there is less satisfactory agreement between theory and experiment [*119*], particularly for the C_4 fraction.

The case for addition occurring at both ends of the carbon chain is questionable on the grounds that it fails to predict the observed distribution of C_4 isomers. Anderson *et al.* [*114*] therefore considered a scheme in which addition was only allowed at an end or a penultimate carbon atom of the longest chain, but did not allow addition on a penultimate carbon atom already attached to three carbon atoms. The growth pattern may be represented as follows:

$$\begin{array}{l} C\,C \rightarrow C\,C\,C \rightarrow C\,C\,C\,C \rightarrow C\,C\,C\,C\,C \\ \qquad\qquad\quad \lfloor\!\!\!-\!\!\!-\!\!\!\longrightarrow C\,C\,C\,C \\ \qquad\qquad\qquad\qquad\qquad C \\ \qquad\qquad\longrightarrow C\,C\,C \rightarrow C\,C\,C\,C \\ \qquad\qquad\qquad\quad C \qquad\qquad C \end{array} \tag{115}$$

Assigning rate parameters α and β for the rate of addition to an end carbon atom and a penultimate carbon atom respectively, and letting γ be the rate of desorption of the growing chains from the surface, two reaction sequences are considered,

$$(1)\ C\,C\,C\,C \rightarrow C\,C\,C\,C\,C \rightarrow C\,C\,C\,C\,C\,C$$
$$\ \ \ \ \ \ \ \ \ \ \ \ \ \ \ \ \longrightarrow C\,C\,C\,C\,C \tag{116}$$
$$C$$

$$(2)\ C\,C\,C\,C \rightarrow C\,C\,C\,C \rightarrow C\,C\,C\,C\,C \tag{115}$$
$$CC$$

In the steady state the rate of formation of any growing intermediate must equal its rate of growth and desorption; hence we have

$$c_4\alpha = c_5(\alpha + \beta + \gamma) \tag{118}$$

for sequence (1), while for sequence (2)

$$c_4\beta = c_{5'} (\alpha + \gamma) \tag{119}$$

where c_4, c_5 and $c_{5'}$ represent the surface concentrations of the C_4, C_5 and iso-C_5 chains respectively. Now the ratio of the concentration c_5/c_4 and $c_{5'}/c_4$ will be equal to the ratio of the respective rates of formation R_5/R_4 and $R_{5'}/R_4$ and since these may be identified with the parameters a and b we have

$$a = \frac{R_5}{R_4} = \frac{c_5}{c_4} = \frac{\alpha}{(\alpha + \beta + \gamma)} \tag{120}$$

$$b = \frac{R_{5'}}{R_4} = \frac{c_{5'}}{c_4} = \frac{\beta}{(\alpha + \gamma)} \tag{121}$$

For addition to an end carbon atom $R_{n+1}/R_n = a$ and for a penultimate carbon atom $R'_{n+1}/R_n = b = af$. Thus the carbon number and isomer distribution for all hydrocarbons (given up to C_8 in Table 8.2) may be computed since the ratio of the yield of the fractions containing n carbon atoms to the yield of the fraction containing two carbon atoms is

$$\frac{\phi_n}{\phi_2} = 2a^{n-2} F_n \tag{122}$$

where F_n is the appropriate function in parenthesis in Table 8.2. Taking logarithms of equation (122),

$$\log \frac{\phi_n}{F_n} = n\log a + \log \frac{2\phi_2}{a^2} \tag{123}$$

and as Fig. 6 shows, this equation is obeyed by the data obtained for synthesis using both cobalt and iron catalysts with values of $a = 0\cdot836$, $f = 0\cdot035$ for

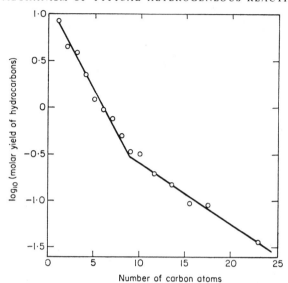

FIG. 6. Fischer-Tropsch synthesis. Logarithm of molar yield distribution as a function of the number of carbon atoms [114].

TABLE 8. 2. Isomer and carbon number distribution in terms of the parameters a and f [114].

Carbon chain		Relative isomer Composition	Relative carbon No. distribution (moles)
C_2		1	1
C_3		1	$2a$
C_4	$n\text{-}C_4$	1	$2a^2(1+f)$
	2-Methyl-C_3	f	
C_5	$n\text{-}C_5$	1	$2a^3(1+2f)$
	2-Methyl-C_4	$2f$	
C_6	$n\text{-}C_6$	1	$2a^4(1+3f+f^2)$
	2-Methyl-C_5	$2f$	
	3-Methyl-C_5	f	
	2,3-Dimethyl-C_4	f^2	
C_7	$n\text{-}C_7$	1	$2a^5(1+4f+3f^2)$
	2-Methyl-C_6	$2f$	
	3-Methyl-C_6	$2f$	
	2,3-Dimethyl-C_5	$2f^2$	
	2,4-Dimethyl-C_5	f^2	
C_8	$n\text{-}C_8$	1	$2a^6(1+5f+6f^2+f^3)$
	2-Methyl-C_7	$2f$	
	3-Methyl-C_7	$2f$	
	4-Methyl-C_7	f	
	2,3-Dimethyl-C_6	$2f^2$	
	2,4-Dimethyl-C_6	$2f^2$	
	3,4-Dimethyl-C_6	f^2	
	2,5-Dimethyl-C_6	f^2	
	2,3,4-Trimethyl-C_5	f^3	

cobalt and $a=0.6$, $f=0.115$ for iron [*112*]. Equation (123) predicts the isomer and carbon number distribution for iron only up to 10 carbon atoms, but thereafter there is a break in the curve and different values of a and f are then required to secure agreement. The implication is that chains containing more than 10 carbon atoms grow more rapidly than shorter chains.

8.4. Synthesis and Decomposition of Ammonia

The synthesis of ammonia is a classic example of the application of catalysis to an important industrial process. A discussion of either the synthesis or the decomposition of ammonia over metal catalysts affords an example in which the technique of isotopic labelling has done much to assist in the interpretation of some conflicting experimental results and to provide a basic pattern upon which further studies may now be confidently developed.

The pioneering work of Haber [*120*], Nernst [*121, 122*] and of Larson and Dodge [*123, 124*] established that the synthesis of ammonia from its elements is exothermic and occurs with a decrease in volume. By Le Chatelier's principle, therefore, it follows that a low temperature and a high pressure are best for the attainment of equilibrium conditions. However, considerations of the output from an ammonia reactor (moles per unit volume per unit time) make it imperative that the temperature be sufficiently high to enable ammonia to be produced at an acceptable rate. The optimum temperature will, of course, also depend on the choice of catalyst in so far as its activity will affect the reaction kinetics, but for a triply promoted iron catalyst the optimum temperature is about 450°C. The efficiency of reaction, defined as the ratio of the %NH_3 formed to the %NH_3 present at equilibrium, decreases slightly with increase in pressure, and it was found that it is best to utilize pressures between 200 and 1000 atm. Most industrial plants operate within such temperature and pressure limits. Since the reaction is fast, space velocities, defined as the volumetric flowrate of a particular component per unit reactor volume, are of the order of 20,000 h^{-1}. Methods of determining optimum reaction conditions for a reaction of known kinetics occurring in a tubular reactor under flow conditions will be discussed in Section 9.7, and it suffices to mention that this reaction has been the subject of investigation by many workers [*125, 126*].

For the synthesis reaction to occur on a catalyst, either one or both of the reactants must be chemisorbed. The view that the dissociative adsorption of nitrogen is the slowest reaction step in ammonia synthesis is supported by evidence from independent sources, so that it seems reasonable to deduce that those metals which chemisorb nitrogen weakly (strong chemisorption would inhibit rather than enhance reactivity) are most active for the synthesis reaction. Perhaps the most convincing evidence lies in the fact that only those metals that are capable of chemisorbing nitrogen as atoms are active in ammonia synthesis, yet most metals are capable of adsorbing hydrogen. The transition metals chemisorb nitrogen most readily and this tendency increases from right to left across the Periodic Table, as is shown by an increase in the heat of adsorption, a decrease in activation energy for adsorption and an increased tendency to form nitrides. Experimental evidence using the flash filament technique (Section 3.2.8.1) shows that there are two forms of chemisorbed nitrogen, one of which is more weakly held than the other [127–129]. It has been inferred that the weak chemisorption is most probably nitrogen adsorbed as atoms. Furthermore, the heat of adsorption of nitrogen on most metals is large, and some metals, notably iron, require an activation energy for nitrogen to be adsorbed in its atomic form [130]. It is probable that the bond order between the metal and atomic nitrogen is of the order of 3, so that it is not surprising that high heats of adsorption and low coverages have been observed. Because the metals of Group VIII after osmium possess a smaller number of vacant d-orbitals than either Fe, Ru or Os, then the inability of those metals after osmium in the series to either chemisorb nitrogen as atoms or effect ammonia synthesis is understandable.

Metals of Groups VIA, VIIA and the first three of Group VIII conform to those conditions outlined above in that they chemisorb nitrogen as atoms. In some cases an activation energy is required, but reasonably strong binding to the metal is reflected in high heats of adsorption, but not sufficiently high as to preclude further surface reaction with other species such as hydrogen. Thus, the fact that iron is the most widely used commercial catalyst is not surprising, since, apart from possessing the requisite chemical properties, it is also the cheapest. Under the conditions of synthesis nitrides of iron are not formed and it is concluded that nitrogen atoms absorbed by the bulk metal do not play an important part in the synthesis reaction [131].

Catalysts used industrially for ammonia synthesis are usually reduced oxides of iron promoted with metal oxides such as alumina, silica, or zirconia, all of which are amphoteric in nature and are reduced only with difficulty. The addition of small amounts of alkali, such as K_2O or CaO, has also been practised, since it was found possible to further increase the yield of ammonia in such a way. A catalyst to which both alumina and alkali has been added is known as a doubly promoted catalyst. Much speculation has arisen concerning the function of promoters. It was supposed, and later confirmed (by surface area measurements and chemisorption experiments), that the main function of alumina in a singly or doubly promoted catalyst is to preserve a large surface area and to prevent the iron crystallites from sintering. The role which the alkali promoter plays is less understood. It has been suggested that the acidic character of the singly promoted catalysts ($Fe–Al_2O_3$) is neutralized by the addition of small amounts of alkali. Evidence [132] that imine and amine radicals are formed at singly promoted catalysts when hydrogen is admitted to such a surface previously covered with nitrogen indicates that the surface has an acidic nature. This has not been observed on doubly promoted catalysts and these seem to effect more easily the decomposition of amine radicals. Other workers have suggested that the addition of alkali modifies the electronic properties of the singly promoted catalyst in some way, thereby leading to enhanced activity.

8.4.1. Kinetics of ammonia synthesis and decomposition

As soon as it was known (i) that the rate of ammonia formation was substantially the same as the rate of nitrogen adsorption and (ii) from work on the exchange of ammonia with deuterium over metals that neither the adsorption of hydrogen nor the desorption of ammonia was particularly slow, Temkin and Pyzhev [133] proposed a kinetic scheme embodying the essentials of the synthesis and decomposition reactions. The following assumptions were made by them in the deduction of their equation describing the rate of ammonia synthesis.

1. The dissociative adsorption of nitrogen at the catalyst surface is the slowest step in the kinetic scheme.

2. Nitrogen atoms are more prevalent at the surface than any other adsorbed species containing nitrogen.

3. Neither hydrogen nor ammonia influences the rate of adsorption of nitrogen.

4. The activation energies E_a and E_d for adsorption and desorption respectively decrease linearly with increase in coverage θ.

The rate of adsorption of nitrogen over the entire surface may be written in terms of an Elovich equation (see Section 2.4.2),

$$r_a = k_a p \exp\left(-g\theta_N\right) \tag{124}$$

where k_a and g are kinetic constants, p is that partial pressure of nitrogen which determines the rate of condensation of nitrogen atoms at the surface, and θ_N is the fraction of surface covered by nitrogen atoms. Similarly, the rate of desorption of nitrogen may be represented

$$r_d = k_d \exp\left(h\theta_N\right) \tag{125}$$

If the dissociative adsorption of nitrogen is the rate-determining process, the net rate r_s of ammonia synthesis, given by the difference between the rates of adsorption and desorption, is thus

$$r_s = r_a - r_d = k_a p \exp\left(-g\theta_N\right) - k_d \exp\left(h\theta_N\right) \tag{126}$$

Now if the amount of nitrogen at the surface is determined by the equilibrium

$$N_2(ads) + 3H_2 \rightleftharpoons 2NH_3 \tag{127}$$

then the adsorbed nitrogen is not in equilibrium with gaseous nitrogen, but rather with the nitrogen pressure $p_{N_2}^*$, which would be exerted if equilibrium were established between gaseous nitrogen and the instantaneous partial pressures of hydrogen and ammonia. Thus

$$p_{N_2}^* = \frac{1}{K} \frac{p_{NH_3}^2}{p_{H_2}} \tag{128}$$

where K is the thermodynamic equilibrium constant for equation (127). Hence $p_{N_2}^*$ is the appropriate term to substitute for p in the rate of adsorption and synthesis equations (124) and (126). The coverage θ in the synthesis equation is found by equating the rates of adsorption and desorption. The result is the Temkin isotherm (see Section 2.3.7)

$$\theta_N = \frac{1}{(g+h)} \ln \frac{k_a}{k_d} p_{N_2}^* = \frac{1}{a} \ln c_0 \, p_{N_2}^* \tag{129}$$

where a and c_0 are the constants which appear in Table 2.1, Chapter 2. Substituting for θ_N in equation (126), the rate of synthesis is

$$r_s = k_a \, p_{N_2} \left(\frac{c_0 p_{NH_3}^2}{K p_{H_2}^3} \right)^{-g/a} - k_d \left(\frac{c_0 p_{NH_3}^2}{K p_{H_2}^3} \right)^{h/a} \tag{130}$$

$$= k'_a p_{N_2} \left(\frac{p_{H_2}^3}{p_{NH_3}^2} \right)^i - k'_d \left(\frac{p_{NH_3}^2}{p_{H_2}^3} \right)^{1-i}$$

where $k'_a = k_a(K/c_0)^i$, $k'_d = k_d(K/c_0)^{i-1}$ and $i = g/a$. At low efficiencies the rate of formation of ammonia is therefore

$$\frac{dp_{NH_3}}{dt} = k'_a p_{N_2} \left(\frac{p_{H_2}^3}{p_{NH_3}^2} \right)^i \tag{131}$$

The results of rate measurements over a wide range of conditions during ammonia synthesis can be expressed in terms of equation (131) with values of i between 0·50 and 0·67, although chemisorption measurements indicate values as high as 0·75. The velocity constant seems therefore to depend on various experimental parameters such as contact time, conversion efficiency and also the ratio of hydrogen to nitrogen present. As far as the value of i is concerned, rate measurements during ammonia decomposition also imply sensitivity to the ratio of hydrogen to nitrogen. Temkin-Pyzhev kinetics are generally obeyed over iron catalysts and the exponents x and y in the equation for the rate of ammonia decomposition

$$r_d = k p_{NH_3}^x p_{H_2}^y \tag{132}$$

vary from 0·5 to 0·9 and from −0·7 to −1·5 respectively, depending on the physical form of the catalyst. Three distinct activation energies have been reported [134, 135], a very low value being found between two higher values as indicated in Fig. 7. In the region of low activation energy the exponents x and y are -0.90 and $+1.35$ respectively, thus indicating a kinetic equation in which the partial pressure functions are the reciprocal of those normally observed. It has been suggested that in the region of low activation energy the surface is converted to a true nitride. When alkali is used as a promoter the region of low activation energy is absent, and it is possible that the alkali ions reduce the acidity of the surface for nitrogen, thus rendering the catalyst one which only weakly adsorbs nitrogen. The fact that velocity constants for ammonia synthesis depend on the hydrogen-to-nitrogen ratio and also that nitrogen chemisorption is affected by the presence or otherwise of hydrogen shows that the Temkin-Pyzhev theory fails to account for the influence that hydrogen adsorption has on the rate of synthesis.

Scholten and Zweitering [136, 137], using a vacuum microbalance (see

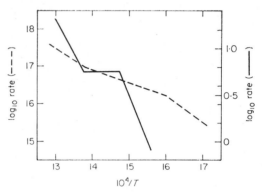

FIG. 7. Ammonia synthesis and decomposition. Temperature dependence of activation energy for decomposition reaction [136]. —— Unpromoted catalyst. - - - Promoted catalyst.

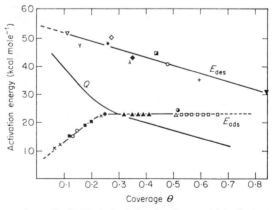

FIG. 8. Ammonia synthesis. Variation of activation energies of adsorption and desorption and heat of chemisorption, as a function of coverage when nitrogen is adsorbed [136].

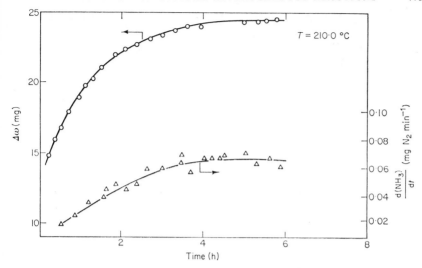

FIG. 9. Ammonia synthesis. Weight change $\Delta\omega$ and rate of ammonia formation [*136*].

Section 3.2.3.1), studied the chemisorption of nitrogen on iron and investigated its relation to the rate of ammonia synthesis. They found that the rate of chemisorption at constant pressure and at temperatures between 200 and 250°C is slow and depends strongly on surface coverage. Figure 8 shows that the activation energies for adsorption and desorption also depend on coverage. The temperature-independent factors were also calculated and it was apparent that a compensation effect was operating. It was therefore concluded that the rate of chemisorption cannot be strictly compared, on a quantitative basis, with the rate of synthesis. However, using the same microbalance, it was possible actually to measure the increase in weight of the catalyst and simultaneously measure the rate of ammonia synthesis [*138*]. Figure 9 shows that both weight and reaction rate become constant at the same time. The nitrogen coverage could also be compared with that value which would give a rate of nitrogen chemisorption equal to the observed rate of ammonia synthesis. It was found that when such a comparison was made good agreement was obtained, except for measurements at low temperatures where there may have been considerable chemisorption of hydrogen. This evidence points to nitrogen chemisorption as the rate-determining step in the ammonia synthesis.

Kummer and Emmett [*139*] showed that hydrogen–deuterium exchange (Section 3.3.7.2) with ammonia occurs rapidly far below synthesis temperatures on iron catalysts. Furthermore, they demonstrated that the exchange reaction

$$N_2^{28} + N_2^{30} = 2N_2^{29} \tag{133}$$

occurred at a rate comparable to the synthesis reaction, thus serving to justify the original assumption of Temkin and Pyzhev that N_2 chemisorption is probably the slowest step in the overall rate process. However, some considerable variation in the rates of exchange at different catalysts and the inhibitive effect which hydrogen produced caused some speculation whether nitrogen chemisorption was, in fact, the rate-determining step in the synthesis. Also, it was uncertain whether the exchange reaction followed the same reaction path as the catalytic synthesis.

In the preceding discussion no account was taken of the possible influence which H_2 may have on the adsorption of N_2. Tamaru [140], using a doubly promoted iron catalyst, has studied this problem and he found that the chemisorption of nitrogen is accelerated by hydrogen to an extent depending on the quantity of hydrogen simultaneously chemisorbed. By using a circulating gas flow method, Tamaru [141] (see Section 3.3.10) also succeeded in measuring the adsorption of both nitrogen and hydrogen whilst catalysis is actually proceeding. The ammonia produced during reaction was removed by a cold trap and the circulation rate of nitrogen and hydrogen was high: consequently the observed reaction rate corresponded closely to the initial reaction rate in the absence of ammonia. The results indicate that the adsorption of nitrogen on the catalyst is accelerated by hydrogen and that the rate of adsorption of nitrogen is about ten times the rate at which ammonia is produced. On the other hand, hydrogen adsorption is not markedly affected by the simultaneous adsorption of nitrogen, even when the quantity of nitrogen and hydrogen are comparable.

An interesting point which emerges from this work is that the observed rate of ammonia synthesis corresponds closely to the rate of adsorption of nitrogen in the absence of hydrogen measured by Scholten and Zweitering [136]. Since it was clearly demonstrated by Tamaru that hydrogen causes a marked acceleration in the rate of nitrogen adsorption, and accepting the well-established fact that hydrogen is adsorbed rapidly, it follows that nitrogen is more rapidly adsorbed by a surface containing preadsorbed hydrogen than on a bare surface. Experiments in which ammonia was removed from an equilibrated mixture of nitrogen and hydrogen prior to reacting with a catalyst surface covered with adsorbed nitrogen showed that adsorbed nitrogen does not react rapidly to form ammonia. If nitrogen chemisorption were slow compared with subsequent surface reaction, then by definition adsorbed nitrogen should form NH_3 rapidly. Consequently, it seems unlikely that nitrogen chemisorption is the only rate-determining process.

Taylor and co-workers [132] compared the rates of NH_3 and ND_3 syntheses by passing stoichiometric mixtures of N_2 and H_2, or D_2, over doubly promoted iron catalysts in a flow system. For a tubular reactor in which the gas flow is considered not to vary across the tube radius, the contact time τ, defined as the ratio of the reactor volume V to the volumetric flowrate u, is given by

$$\tau = \frac{V}{u} = \frac{V}{F/\rho} = \frac{PV}{RTF} \tag{134}$$

where F is the feed rate in units of mass per unit time and $\rho (=P/RT)$ is the absolute gas density. The contact time τ is therefore proportional to P/F. Expressing partial pressures in terms of the degree of conversion y to ammonia, equation (131) becomes, for a flow system,

$$\frac{dy}{d(1/F)} = \frac{a_1 P^{(1+i)}}{y^{2i}} \tag{135}$$

where a_1 is a constant.

At zero conversion $1/F$ must be infinity, so the solution to equation (135) is

$$y^{(2i+1)} = (2i + 1)a_1 P^{(1+i)} \frac{1}{F} \tag{136}$$

from which i can be determined by plotting $\log y$ versus $\log (1/F)$. A typical value is 0.8 at $300°C$. The intercept should be proportional to $P^{(1+i)}$, i.e. to $P^{1.8}$, but this is, in fact, not found to be true. Consequently, the assumptions of the theory must be further examined.

Taylor and co-workers [132] pointed out that the assumption of surface heterogeneity, implicit in the Temkin-Pyzhev treatment, is not essential to the theory of ammonia synthesis. On a homogeneous surface, the rate of dissociative adsorption of nitrogen (equal to the rate of ammonia synthesis) is

$$r_s = r_a = k_a p_{N_2}(1 - \theta_N)^2 \tag{137}$$

The fraction of surface covered by nitrogen atoms, determined by the equilibrium

$$N(ads) + 1.5H_2 \rightleftharpoons NH_3 \tag{138}$$

is, according to the Langmuir isotherm (see Section 2.3),

$$\theta_N = \frac{b p_N}{1 + b p_N} \tag{139}$$

where p_N is the partial pressure which would be exerted by nitrogen if it were in equilibrium with adsorbed nitrogen atoms. Using the equilibrium relation

$$K_N = \frac{p_{NH_3}}{p_N p_{H_2}^{1.5}} \tag{140}$$

one obtains

$$\theta_N = \frac{K \dfrac{p_{NH_3}}{p_{H_2}^{1.5}}}{1 + \dfrac{K p_{NH_3}}{p_{H_2}^{1.5}}} \tag{141}$$

where $K = b/K_N$. Hence equation (137) becomes

$$r_s = \frac{k_a p_{N_2}}{\left(1 + \dfrac{K p_{NH_3}}{p_{H_2}^{1.5}}\right)^2} \tag{142}$$

Expressing partial pressures in terms of the total pressure P and the conversion y to ammonia, the rate of synthesis at low conversions in a flow system now becomes

$$\frac{dy}{d\left(\dfrac{1}{F}\right)} = \frac{a_2 P}{\left(1 + \dfrac{a_3 y}{P^{0.5}}\right)^2} = \frac{k_1}{(1 + B_1 y)^2} \tag{143}$$

With the same boundary condition as before this, on integration, becomes

$$(1 + B_1 y)^3 - 1 = \frac{3 B_1 k_1}{F} \tag{144}$$

This rate equation fits the experimental data very well at a given pressure. The constant k_1 is also found to be directly proportional to the total pressure, but B_1, which should vary inversely with $P^{0.5}$, is found to be independent of pressure.

Another expression which has been substituted for θ_N is an adaption of one deduced by Brunauer and co-workers [142] from the basis of the Langmuir hypotheses. Brunauer and co-workers [142] considered the surface to be divided up into elements of size dS and obtained an expression for the coverage by equating the rates of adsorption and desorption

$$kp(1 - \theta_s) = k' \theta_s \exp\left(\frac{Q_s}{RT}\right) \cdot \tag{145}$$

where

$$Q_s = Q_0 - iS \tag{146}$$

thus assuming a linear decrease in heat of adsorption with coverage as in the Temkin isotherm (Section 2.3.7). The total fraction of surface covered can therefore be calculated by summing the contributions of each of the surface elements δs, and one writes

$$\theta = \int_0^1 \theta_s dS = \int_0^1 \frac{\dfrac{k}{k'} p \exp\left(\dfrac{-Q_s}{RT}\right) dS}{1 + \dfrac{k}{k'} p \exp - \dfrac{Q_s}{RT}} = \frac{RT}{i} \ln\left\{\frac{1 + a_0 p}{1 + a_0 p \exp - \dfrac{i}{RT}}\right\} \tag{147}$$

where a_0 is a constant $(= k/k')$. Applied to the dissociative adsorption of nitrogen this becomes

$$\theta_N = \frac{2RT}{i} \ln \left\{ \frac{1 + \dfrac{a_0 K p_{NH_3}}{p_{H_2}^{1\cdot5}}}{1 + \dfrac{a_0 K p_{NH_3} \exp\left(-\dfrac{i}{RT}\right)}{p_{H_2}^{1\cdot5}}} \right\} \tag{148}$$

When $a_0 K p_{NH_3}/p_{H_2}^{1\cdot5} \gg 1 \gg \{a_0 K p_{NH_3}/p_{H_2}^{1\cdot5} \exp(-i/2RT)\}$, this reduces to the Temkin isotherm. Contrary to the Temkin isotherm the above isotherm is applicable over the full range of coverage and should therefore be more useful. At low values of θ_N equation (148) may be written

$$\theta_N = \frac{2RT}{i} \ln \left(1 + \frac{a_0 K p_{NH_3}}{p_{H_2}^{1\cdot5}}\right) \tag{149}$$

and when substituted into the synthesis equation (137), becomes

$$r_s = r_a = \frac{k_a p_{N_2}}{\left(1 + \dfrac{a_0 K p_{NH_3}}{p_{H_2}^{1\cdot5}}\right)^{2i}} \tag{150}$$

If $i=1$ in the above equation, it is identical in form to equation (142). Hence, in this case, it is possible to obtain the same rate equation by assuming either a homogeneous or a heterogeneous surface.

The failure of the constant B_1 in equation (144) to conform to the experimental results is easily accommodated by a simple modification to the above theory. Taylor and co-workers [132] suggested that the principal nitrogen-containing species at the surface may be imine radicals. The equilibrium determining the amount of nitrogen at the surface would then be

$$NH(ads) + H_2 \rightleftharpoons 2NH_3 \tag{151}$$

The equation corresponding to (131) is therefore

$$r_s = k p_{N_2} \left(\frac{p_{H_2}}{p_{NH_3}}\right)^{2i} \tag{152}$$

or, corresponding to the rate equation (142) (using the more accurate form of the isotherm given by equation (148)),

$$r_s = \frac{k p_{N_2}}{\left(1 + \dfrac{a_0 K p_{NH_3}}{p_{H_2}}\right)^2} \tag{153}$$

When equations (152) and (153) are expressed in terms of the conversion y and the total pressure P, they become, for a flow system,

$$\frac{dy}{d\left(\dfrac{1}{F}\right)} = \frac{a_4 P}{y^{2i}} \tag{154}$$

and

$$\frac{dy}{d\left(\dfrac{1}{F}\right)} = \frac{a_5 P}{(1 + B_2 y)^2} = \frac{k_2}{(1 + B_2 y)^2} \tag{155}$$

respectively. Hence all the experimental facts are now satisfied. k_2, in equation (155), is directly proportional to P as before, but in addition B_2 is independent of total pressure, as found experimentally.

The most striking evidence in support of the hypothesis that imine radicals rather than nitrogen atoms are the most abundant surface species comes from a comparison of the results for synthesis using H_2 and D_2. The ratio of the parameters a_5 (a_H using hydrogen and a_D using deuterium), as determined from the kinetic equation (155), is 2·7 at both 250°C and 300°C. For the mechanism involving adsorbed nitrogen atoms this ratio, written in terms of partition functions, is

$$\frac{(a_H)_N}{(a_D)_N} = \left(\frac{f_{H_2}}{f_{D_2}}\right)^{1\cdot 5} \frac{f_{ND_3}}{f_{NH_3}} \tag{156}$$

but for the mechanism involving adsorbed imine radicals is

$$\frac{(a_H)_{NH}}{(a_D)_{NH}} = \frac{f_{H_2}}{f_{D_2}} \frac{f_{ND_3}}{f_{NH_3}} \frac{f_{NH(ads)}}{f_{ND(ads)}} \tag{157}$$

Whereas equation (156) is independent of the nature of the surface, since the partition functions for adsorbed nitrogen cancel, equation (157) depends on the partition function of the adsorbed species and hence depends on the nature of the surface. The partition functions for the gaseous species are readily calculated, but it is necessary to assume that the vibrational frequencies associated with adsorbed NH are identical with those found in organic imines. Thus

$$\frac{f_{ND(ads)}}{f_{NH(ads)}} = \frac{f_{vi\ ND}}{f_{vi\ NH}} \exp\left(\frac{\Delta E_0^0}{RT}\right) \tag{158}$$

where ΔE_0^0 is the difference in zero point energies between the two species and can be calculated from the difference in vibrational energies in the ground state. The ratio $(a_H)_{NH}/(a_D)_{NH}$ turns out to be 2·4 at 300°C and 2·7 at 250°C, in excellent agreement with the observed value. In contrast, the ratio $(a_H)_N/(a_D)_N$ is grossly in error and furthermore depends strongly on temperature, whereas experiments indicate the ratio should be independent of temperature.

Tamaru's work [141], previously discussed, demonstrating that nitrogen chemisorption is not the only rate-determining surface process, is thus clearly supported by the theoretical and experimental work of Taylor and co-workers, who showed unequivocally that a mechanism involving imine radicals best explains the results. All the experimental results of Taylor et al [132] relating to the kinetic behaviour, pressure dependence and isotope effect are consistent with this theory.

8.4.2. Stoichiometry of ammonia synthesis

Horiuti [143] and later Kwan [144] have emphasized the importance of stoichiometry in elucidating the slowest step in a reaction sequence. The rate of the slowest step may not always be identifiable with the overall reaction rate of the process. If the stoichiometric number n is defined as the number of times the slowest step occurs for each complete sequence of the reaction, then for the forward step in an equilibrium reaction

$$v' = nV' \tag{159}$$

where v' is the overall rate of the forward process and V' is the rate of the slowest step in the sequence of elementary reactions contributing to the overall forward reaction. Similarly for the reverse reaction

$$v'' = nV'' \tag{160}$$

To illustrate the significance of n, each step in the ammonia synthesis is written separately,

$$N_2 \rightarrow 2N(ads) \tag{161}$$

$$H_2 \rightarrow 2H(ads) \tag{162}$$

$$N(ads) + H(ads) \rightarrow NH(ads) \tag{163}$$

$$NH(ads) + H(ads) \rightarrow NH_2(ads) \tag{164}$$

$$NH_2(ads) + H(ads) \rightarrow NH_3(ads) \tag{165}$$

$$NH_3(ads) \rightarrow NH_3 \tag{166}$$

If nitrogen chemisorption is the slowest step in the above sequence, n will be unity, for this step only occurs once during the completion of the reaction sequence. If either of the steps (163), (164) or (165) is the slowest step, then n would be two, for each of these elementary processes would occur twice during the sequence, since two nitrogen atoms are formed in the first supposed rapid step. Similarly, if (166) is the slowest step n would have a value of 3. It is evident, then, that the determination of n would help to eliminate some of the ambiguity of deciding which is the rate-determining elementary step.

It will be shown in the following analysis that the stoichiometric number may be expressed in terms of the change in Gibbs free energy for the equilibrium reaction. The question therefore arises whether the classical equilibrium constant k (devised from the law of mass action) is applicable when the overall reaction consists of a number of independent reaction sequences. Kwan [144] has shown that the appropriate form of the law is

$$\prod_{i=1}^{S} \left(\frac{k_i'}{k_i''} \right)^{n_i} = K \tag{167}$$

where k'_i/k''_i is the ratio of velocity constants for the forward and reverse reaction of step i in a sequence of S steps, and n_i is the number of times the ith step occurs. Thus the classical law is really a special case of the above general equation.

Now from the definition of the stoichiometric number the steady rate of the overall reaction is

$$V = \frac{v'(s) - v''(s)}{n_s} \quad s = 1 \ldots S \tag{168}$$

Suppose that the reaction is practically completed as soon as, say, the rth step in the sequence has taken place. Then the forward rate V' and the reverse rate V'' of the overall reaction are

$$V' = \frac{v'(r)}{n_r} \text{ and } V'' = \frac{v''(r)}{n_r} \tag{169}$$

respectively. The forward and backward rates, $v'(s)$ and $v''(s)$ of any step s in the sequence can be expressed, from statistical thermodynamics, as

$$\ln \frac{v'(s)}{v''(s)} = \frac{-\Delta G(s)}{RT} \tag{170}$$

where $\Delta G(s)$ is the increment in Gibbs free energy for the sth step. For any step t occurring after the rth step $\Delta G(t)=0$, since the reaction is assumed complete after the rth step. On the other hand, from equations (169) and (170)

$$\ln \frac{V'}{V''} = \frac{-\Delta G(r)}{RT} \tag{171}$$

The change in Gibbs free energy ΔG for the overall reaction will be the total sum of the increments in free energy for each step. If the sth step occurs n_s times, then the total change in free energy is

$$\Delta G = \sum_{s=1}^{S} n_s \,\Delta G(s) \tag{172}$$

and, because there is no further change in free energy after the rth step,

$$\Delta G = n_r \Delta G(r) \tag{173}$$

From equation (171) therefore

$$\ln \frac{V'}{V''} = \frac{-\Delta G}{n_r RT} \tag{174}$$

By measuring the forward rate and the net rate near equilibrium, Horiuti [145] found n to be 2, thus conflicting with the view that nitrogen chemisorption is the rate-determining step during synthesis. However, Mars *et al.* [*146*] also determined the stoichiometric number by similar measurements near to equilibrium conditions and found n to be unity.

Another approach considered by Mars and co-workers [*146*] is to transform the kinetic equation for synthesis into a form applicable near equilibrium conditions. The rate of synthesis may be written

$$r_s = k' \, p_{N_2}^a \, p_{H_2}^{3a} - k'' \, p_{NH_3}^{2a} \tag{175}$$

where

$$\frac{k'}{k''} = \frac{(p_{NH_3}^{2a})_e}{(p_{N_2}^a)_e \, (p_{H_2}^{3a})_e} = K^a \tag{176}$$

and the subscript e refers to equilibrium conditions. Comparing the ratio with that in equation (167) it is obvious that a is the reciprocal of n. If we define Δ as the quantity $((p_{NH_3})_e - p_{NH_3})$, measuring the displacement from equilibrium, then, from the stoichiometry of the overall reaction $((p_{N_2})_e - p_{N_2}) = -\Delta/2$ and $((p_{H_2})_e - p_{H_2}) = -3\Delta/2$. Substituting into (175), expanding the resulting equation (neglecting terms of higher order than Δ) and making use of (176) we obtain

$$r_s = k'(p_{N_2}^a)_e \, (p_{H_2}^{3a})_e \, a\Delta \left\{ \frac{1}{2(p_{N_2})_e} + \frac{9}{2(p_{H_2})_e} + \frac{2}{2(p_{NH_3})_e} \right\} \tag{177}$$

Now the rate of the forward reaction at equilibrium is $k'(p_{N_2}^a)_e \, (p_{H_2}^{3a})_e$ and this is equal to twice the number, ν, of molecules reacting per unit time in either direction at equilibrium. Hence

$$\frac{r_s}{\Delta} = k_s = a\nu \left\{ \frac{1}{(p_{N_2})_e} + \frac{9}{(p_{H_2})_e} + \frac{4}{(p_{NH_3})_e} \right\} \tag{178}$$

Provided ν can be measured independently, a (and hence n) can be calculated from the experimental value of the rate constant for the synthesis reaction and the partial pressures at equilibrium. To determine ν an exchange reaction must be chosen which follows exclusively the same reaction path as the synthesis reaction. Two possibilities exist,

$$N_2^{30} + N^{14}H_3 \rightarrow N_2^{29} + N^{15}H_3 \tag{179}$$

and

$$N_2^{28} + N_2^{30} \rightarrow 2N_2^{29} \tag{133}$$

For the first exchange reaction,

$$k'_1 = \nu'_1 \left(\frac{2}{p_{NH_3}} + \frac{1}{p_{N_2}} \right) \tag{180}$$

while, for the second exchange reaction,

$$k'_2 = \nu'_2 \left(\frac{1}{p_{N_2}} \right) \tag{181}$$

If it is assumed that $\nu = \nu'_1 = \nu'_2$ then by comparing k' with k_s the value of n may be found.

The experimental requirement is therefore to measure the rate of the synthesis reaction and the rates of both exchange reactions simultaneously. Purified synthesis gas (a mixture of 3 parts hydrogen to 1 of nitrogen) is brought near to equilibrium by passing it through a reactor containing catalyst at a temperature greater than that at which the synthesis reaction is to be studied. A stoichiometric mixture of hydrogen and an enriched nitrogen isotope is then injected into the synthesis gas, now near to chemical equilibrium, but far from isotopic equilibrium. This mixture is then passed first through a reactor containing a small amount of catalyst (by means of which the rate of synthesis and the rate of the first exchange reaction is measured), and secondly through a reactor containing a much larger amount of catalyst so that the rate of the much slower second exchange reaction can be measured. The rate of chemical reaction is determined by chemical estimation of the amount of ammonia present in the outlet stream of the first reactor while the rates of both exchange reactions are determined by mass spectrometric analysis of the outlet stream from each of the two reactors. From such experiments, Mars and co-workers [146] found the stoichiometric number to be unity, thus confirming earlier views that nitrogen chemisorption is the rate-determining step during ammonia synthesis.

8.5. Catalytic Cracking

The need to husband the oil resources of the world has resulted in the development and use, on a very large scale, of catalysts capable of transmuting high molecular weight hydrocarbons into more useful hydrocarbons of a lower molecular weight which are employed as fuels for combustion engines, as starting materials for a wide variety of plastics and pharmaceuticals and as industrial solvents. The basic requirements of such a catalyst are manifold. First and foremost, the catalyst must possess a high cracking activity which may be described as the ability to convert a gas oil to a high octane number fuel. Secondly, the catalyst must not produce excessive carbonaceous deposits on the surface, resulting in loss of activity. Other important requirements include ease of regeneration, ability to withstand thermal shock, resistance to poisoning by nitrogen and sulphur compounds nearly always present in the feed, and considerable material strength. In view of these requirements, coupled with a low production cost, it is not surprising that only a few commercially successful cracking catalysts have been developed, the chief one amongst these being silica–alumina. It is necessary to consider the chemical properties and composition of cracking catalysts and also to establish the physical structure to understand something of the nature of catalytic cracking and to enable the further development of specific catalysts for particular needs.

8.5.1. Catalyst composition and chemical properties

It has been known for some time that the activity of a cracking catalyst is inextricably related to its chemically acidic nature. The close resemblance between acidic catalysts such as aluminium chloride, hydrogen fluoride, sulphuric acid, phosphoric acid and boron trifluoride, active in a variety of hydrocarbon reactions, led to the realization that a silica–alumina cracking catalyst could possibly be classified as an acidic catalyst. It will emerge in this discussion that the acidity attributed to cracking catalysts arises from a Lewis type of acid in which the compound is deficient in electrons and/or a Brönsted type of acid with dissociable protons.

The acidity of uncalcined clays and silica–alumina gels has been measured in accordance with their base-exchanging properties [147–150]. This approach has been criticized by Oblad et al. [151] on the grounds that these substances are calcined at temperatures of 800°C before use as cracking catalysts, and it has been established that calcination destroys the base-exchanging properties of such materials. Nevertheless, measurements using uncalcined silica–alumina are important because of the chemical reactions which occur in the gel state during preparation. Many difficulties of interpretation arise when the acidity of a cracking catalyst is determined in aqueous solution. For example, the base-exchanging capacity of uncalcined clays is dependent on pH, and a hysteresis effect has been noted when a measurement is made at a low pH after previously measuring at a high pH [150]. This phenomenon was attributed to a change at the catalyst surface which creates base-exchanging capacity during the actual measurement. Hence it is obvious that the acidity, as measured in aqueous solution, depends on the environment.

Oblad et al. [151] avoided some of the above difficulties by adsorbing, from the vapour phase, bases such as ammonia and quinoline at the surface of cracking catalysts. A distinct advantage is that adsorption can be studied at temperatures at which the catalyst will crack hydrocarbons. By adsorption and desorption experiments in a flow system using a nitrogen stream saturated with quinoline, it was demonstrated that quinoline was reversibly adsorbed at a temperature of 300°C on silica gel (inactive for cracking hydrocarbons), whereas, under the same conditions, a silica–alumina cracking catalyst tenaciously retained some quinoline after desorption. The retention was attributed to chemisorbed quinoline, and it was shown that this quinoline could be extracted unchanged by boiling with aqueous HCl, thus indicating that quinoline is chemisorbed without rupture of the molecule. The fraction of the total quinoline sorbed which could not be removed easily on desorption was found to be independent of the quinoline partial pressure, whereas the amount of reversibly sorbed quinoline increased with increase in the partial pressure of quinoline. These observations are therefore in accord with the view that some of the quinoline is chemisorbed. Similarly, the effect of temperature on the adsorption of quinoline clearly shows that a small quantity of the total quinoline is held tenaciously. Similar conclusions to these were drawn by Tamele [152], who adsorbed ammonia on cracking catalysts. Tamele [152] also titrated a silica–alumina catalyst, suspended in a non-aqueous medium such as benzene, with n-butylamine and compared its acid strength with a number of other acids (of known strength in aqueous media) in benzene, titrating with a series of bases. It was found that the silica–alumina catalyst was comparable in acid strength to very strong acids, but that the amount of acid (in terms of acid sites per unit surface area of catalyst) is very small. The hydrogen content of cracking catalysts after calcination is of some importance in attempting to correlate acidity with the chemical nature of the acid sites. It is quite clear that, after calcination at 800°C, no water is present at the catalyst surface, but hydrogen is present as hydroxyl groups attached to silicon and aluminium atoms in the lattice structure. Thus there is more than sufficient hydrogen present to account for the acidity in terms of a Brönsted type of acid.

Two experimental techniques have enabled the acid centres at the surface of cracking catalysts to be identified as the catalytically active sites. In the

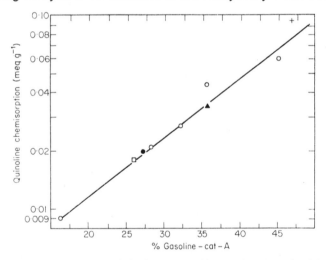

FIG. 10. Cracking reactions. Quinoline adsorption as a function of activity [151].

first place there is a direct correlation between measured acidity and catalytic activity. Figure 10 shows the acidity, as measured by quinoline chemisorption, as a function of the cracking activity of various catalysts [151]. The possible errors which may arise by measuring acidity in aqueous media is demonstrated by the fact that active silica–magnesia catalysts give a basic reaction in water, whereas quinoline chemisorption indicates considerable acidity. As shown in Fig. 11, Tamele [*152*] also found a direct relation between the activity of a number of silica–alumina catalysts and their acidity. The activities were measured by cracking, at 500°C, isopropyl benzene and determining the degree of conversion. The acidity was estimated by titration of the

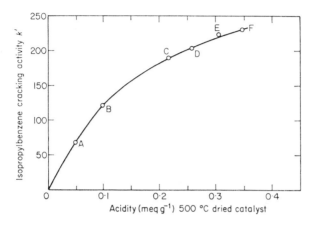

FIG. 11. Cracking reactions. Dependence of isopropylbenzene cracking activity with acidity of catalyst [*152*].

catalyst, in a non-aqueous medium, by anhydrous n-butylamine. Another important observation which led to the identity of the active sites was that some organic nitrogen compounds, basic in character, were poisons for the catalytic cracking of hydrocarbons. Thus quinoline and other bases are effective poisons for the catalytic cracking of cumene by a silica–alumina catalyst [*151*]. Figure 12 shows that an exponential relation obtains between the amount of poison adsorbed by the catalyst and its deactivation. This probably corresponds to a selective type of poisoning, as discussed in Chapters 3 and 4, when the poison is preferentially adsorbed on active centres which constitute only a small fraction of the surface. The interpretation of this effect is that the nitrogenous base is chemisorbed at acid centres, and since the disappearance of catalytic activity is concomitant with such chemisorption, then the acid centres are, in fact, responsible for catalytic activity. An objection which has been raised to this interpretation is that the order of effectiveness as poisons should correlate with the basicity of these compounds. Thus piperidine should be the most effective poison, since it is the strongest base, but other considerations, such as the stability of the poison at high temperatures and the fact that, for large molecules, van der Waal's attractive forces as well as chemical forces play an important role in the adsorption of the base, upset such a correlation of the effectiveness of the poison with basicity. The

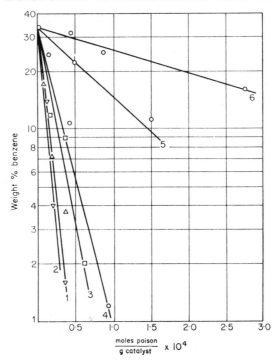

FIG. 12. Poisoning effect of organic nitrogen compounds in cracking reactions [*151*].

deactivation of some cracking catalysts can also be accomplished by the addition of inorganic basic ions. Bitepazh [*147*] found that an uncalcined silica–alumina catalyst, after ion exchange with sodium ions, was inactive as a cracking catalyst.

Various attempts have been made to explain the acidity of cracking catalysts in terms of their chemical and crystallographic structures. Noteworthy are the researches of C. L. Thomas [*148, 153*] and of Oblad, Milliken and Mills [*151, 154*]. The structures are composed of silicon, aluminium, oxygen and small amounts of hydrogen. Pauling [*155*] originally proposed that, in any crystal lattice containing both negative and positive ions, a net negative charge can be created by the isomorphous substitution of a positive ion whose valency is lower than that of the substituted positive ion. In many naturally occurring silica aluminates an aluminium ion has been substituted for a silicon ion in a silica lattice constituted from silica tetrahedra, so that there is a positive valence deficiency in the structure. This valence deficiency is satisfied by the presence, in the vicinity of the point in the lattice at which substitution has occurred, of a positive ion which thus tends to stabilize the structure. The evidence presented by Milliken *et al.* [*154*] suggests that most silica–alumina cracking catalysts are mixtures of silica and alumina particles with the silicon and aluminium ions sharing oxygen ions within the lattice. Thus the chemical properties of alumina in its various crystal forms will be manifested by the mixed oxide structure, while the crystal habits of silica will be only of secondary importance in determining the nature of the catalyst. The structures of basic alumina are boehmite, bayerite and hydrargillite and in these crystals aluminium has a coordination number of six, sharing valence bonds with six oxygen atoms. Since the ionic size of the aluminium ion is relatively small (*ca.* 0·5 Å) it is able to change from a four-coordinated state to a six-coordinated state, fitting between either four or six closely packed oxygen ions. The various crystal forms of alumina are simply arrangements of either or both of the coordination states of aluminium. When aluminium is present with a coordination number of four it behaves as an acid. The salts of such an acid are ring structures of six alumina tetrahedra enclosing a cation. Silica, on the other hand, always has a coordination number of four in quartz, tridymite and crystobalite the different crystalline forms being due to different arrangements of the silica tetrahedra.

The manner in which a cracking catalyst is prepared may be expected to have some bearing on its chemical structure. A synthetic catalyst can be prepared by precipitating sodium silicate with sodium aluminate, adding either sulphuric acid or sodium hydroxide to control the pH. After ageing and washing, sodium ions in the hydrogel are exchanged for ammonium ions and the gel then dried and finally calcined. The changes occurring during such a process have been closely studied by Oblad *et al.* [*151*], who concluded that the stable tetrahedral structure of silica enforced a corresponding coordination of the oxygen ions around an aluminium ion, the crystal habit being stabilized by cations such as sodium or ammonium ions. Since cristobalite has a similar structure to potassium aluminate (Fig. 13), well-defined cristobalite rings are formed at these points of particle contact where oxygen ions are shared between silica and alumina. This cristobalite structure is retained throughout the subsequent treatment of the gel. The stability of such a structure is, however, dependent on the relative micelle sizes, the amount of aluminium present, the pH of the gel and the basicity of the stabilizing cations. For example, in gels containing more than about 15% aluminium the ability of the ammonium ion to stabilize the cristobalite structure begins to fail below a pH of 6. Under these conditions the structure reverts to that of aluminium hydroxide (hydrargillite). Thus it seems that the reactions of alumina in the presence of silica are reversible and that the acidity of the calcined gel is dependent on the environment. It can be envisaged that, in the gas phase, molecules in close proximity to the surface of a silica–alumina catalyst may induce ionic rearrangement and therefore an acid centre. C. L. Thomas [*148*] and Tamele [*156*] favour an acid complex of this type ($HOAlO_3$), with a dissociable

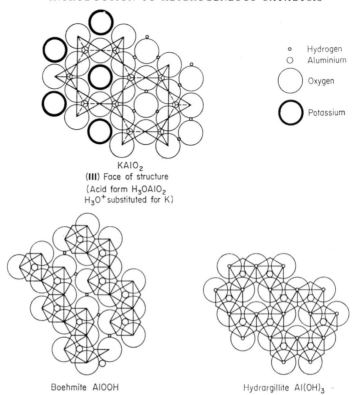

o   Hydrogen
O   Aluminium

Oxygen

Potassium

KAlO₂
(III) Face of structure
(Acid form H₃OAlO₂
H₃O⁺ substituted for K)

Boehmite AlOOH   Hydrargillite Al(OH)₃

FIG. 13. Structures of some cracking catalysts. Alumina cell structures.

proton, in the calcined catalyst. Despite this, it appears from the work of Oblad, that the nature of the materials obtained after high-temperature treatment of synthetic zeolites is similar to a mixture of γ-alumina and silica particles, with oxygen ions common to both alumina and silica at points of contact between particles. This latter structure corresponds to the anhydride of $HAlO_2$ and contains no dissociable protons and would therefore be thought of as a Lewis rather than a Brönsted acid. The active principle is thus the existence of aluminium ions in a strained lattice with an electron deficiency (a Lewis acid). Electron donors from the gas phase are therefore capable of forming complexes with the Lewis acid and hydrogen, which originates either from a part of the aluminate structure or from an external source. Such interaction requires the mobility of oxygen within the crystal lattice of the catalyst, so that a stabilized structure containing four-coordinated aluminium may be formed. This mobility has been demonstrated by oxygen-exchange experiments. The adsorbed electron donor now behaves as the stabilizing ion for the four-coordinated aluminium ion.

The picture of the active catalyst site which thus emerges is that a hydrocarbon in the gas phase donates an electron to a Lewis acid centre with the consequent formation of a carbonium ion at the catalyst surface, thus stabilizing the acid centre. The tendency of aluminium to revert to its six-coordinated state provides the driving force for desorption. However, for hydride ion transfer, which also takes place during cracking reactions, the simultaneous presence and coexistence of a Lewis acid and a Brönsted acid is required at a dual site. It has been pointed out that the transfer of protons can be envisaged without resort to a Brönsted acid formulation. To effect such transfer only minor structural changes in γ-alumina are necessary. The catalyst is largely a network of oxygen ions which can accommodate, and therefore transfer, protons. It is thus quite possible that both Brönsted and Lewis acid centres are present at the surface.

8.5.2. Mechanism of cracking reactions

The mechanism by which cracking reactions take place has been deduced largely by observation of the distribution of cracked products. The hypothesis that carbonium ions are involved in cracking reactions was arrived at by application of the above structural principles and the fact that some hydrocarbon reactions in solution bear a resemblance to gas reactions which are catalysed by silica–alumina. The reactions occurring in catalytic cracking are fundamentally different from thermal cracking reactions, since, in the former case, carbonium ions are probably formed whereas free radicals are involved in thermal cracking. The complexity of cracking reactions has somewhat obscured an understanding of the mechanism of the primary and secondary reactions, since skeletal isomerization and hydrogen transfer result in products differing from the primary products. It has been suggested that a small amount of olefin, arising from a thermal cracking reaction, is necessary to initiate formation of a carbonium ion: the fact that a catalytic reaction is occurring does not exclude the simultaneous occurrence of a small amount of thermal cracking.

Taylor *et al.* [157] believed that a cracking reaction could not, in fact, occur unless a carbon atom in the hydrocarbon approaches the surface within a radius corresponding to chemical interaction with the catalyst acid sites. This would necessitate the breaking of a carbon–hydrogen bond prior to the formation of a carbonium ion. The exchange reaction between methane and deuteromethanes on silica–alumina was therefore studied and it was found to occur at 345°C, a temperature considerably lower than that at which cracking reactions take place. This means that carbon–hydrogen bonds are quite easily broken and that a dehydrogenation reaction is probably the primary step in catalytic cracking. A catalyst–carbon linkage could therefore be formed and a carbonium ion could be established by this mechanism. The exchange reaction may be thought of as the simultaneous transfer of a deuteride ion and a proton between catalyst and substrate,

$$CD_4 + HA \longrightarrow \overset{+}{C}D_3 + A^- + HD \tag{182}$$

where HA represents an acid site at the catalyst surface. If this site were a Lewis type acid, the hydrogen could be transferred from its interstitial position in the crystal lattice. On the other hand, if the acid were a Brönsted type of acid, the dissociable proton would simply be exchanged. A similar mechanism was thus postulated [158] for the first step in the cracking of paraffins as for the exchange reaction (182). The carbonium ion is formed by the simultaneous loss of a hydride ion from the paraffin and a proton from the catalyst,

$$R_1-CH_2-\underset{\underset{H}{|}}{CH}-R_2 + HA \longrightarrow R_1-CH_2-\overset{+}{C}H-R_2 + A^- + H_2 \tag{183}$$

the carbonium ion being in close association with the acid centre (designated A^-). This step is consistent with the general conclusions just drawn, i.e. the hydrocarbon donates electrons to the acid site, thus forming a carbonium ion and consequently stabilizing the valency-deficient acid centre.

An alternative hypothesis, proposed by C. L. Thomas [148], for the formation of a carbonium ion is that a small amount of olefin may be formed in the gas phase, possibly by thermal cracking, or it may be present in the initial feed. The olefin is then supposed to accept a proton from the acid centre as in the previous case.

$$R_1-HC = CH-R_2 + HA \longrightarrow R_1-HC-\underset{\underset{H}{|}}{\overset{+}{C}H}-R_2 + A^- \tag{184}$$

Schmerling [159–161] also proposed that this reaction is the first step in the catalytic alkylation of paraffins by olefins. Comparing both mechanisms, since the same intermediate structure is formed it would be expected that paraffins would crack at the same rate as olefins. On the other hand, it is known that the catalytic cracking of olefins occurs at considerably lower temperatures than paraffins. The evidence is thus in favour of the latter reaction in which a carbonium ion is formed from an olefin initially present in the gas phase [162]. Propagation of the reaction could then occur by the carbonium ion reacting with the paraffin to be cracked (say, $R_3(CH_2)_3R_4$) to yield a new paraffin molecule and a carbonium ion of the original paraffin

$$R_1-CH_2-\overset{+}{C}H-R_2 + R_3-CH_2-CH_2-CH_2-R_4 \longrightarrow R_1-CH_2-CH_2-R_2$$
$$+ R_3-\overset{+}{C}H-CH_2-(CH_2)_2-R_4 \tag{185}$$

The next logical step would be the decomposition of the carbonium ion in accordance with the general rule of least rearrangement. Thus, scission at the

β-carbon atom is the most likely, since this involves a rearrangement of electrons only, resulting in an olefin and a new carbonium ion,

$$R_3\overset{+}{-}CH-(CH_2)_3-R_4 \longrightarrow R_3-CH = CH_2 + \overset{+}{C}H_2-CH_2-R_4 \quad (186)$$

This new carbonium ion is a primary carbonium ion and would be likely to rearrange in accordance with Markownikoff's rule to give the more stable secondary carbonium ion by a simple intramolecular proton shift,

$$\overset{+}{C}H_2-CH_2-R_4 \longrightarrow CH_3-\overset{+}{C}H-R_4 \quad (187)$$

Subsequent steps are repetitions of the above scheme, the secondary carbonium ion forming an olefin and a primary carbonium ion, the latter then rearranging to another secondary carbonium ion. This continues until the carbon chain is so short that scission at the β-carbon atom to give an olefin is a very slow process. The residual carbonium ion can, however, exchange a proton with another larger paraffin molecule to produce a new carbonium ion and so propagate the carbon chain further,

$$CH_3-\overset{+}{C}H-CH_3 + R_5-CH_2-R_6 \longrightarrow CH_3-CH_2-CH_3 + R_5-\overset{+}{C}H-R_6 \quad (188)$$

A scheme similar to that above was proposed by Greensfelder, Voge and Good [158] to explain the catalytic cracking of cetane and it accounted extremely well for the observed distribution of cracked products.

By observing the distribution of products obtained by cracking hexane [158] the relative reactivities of primary, secondary and tertiary carbon atoms was found to be in the ratio 1:2:20. Quaternary carbon atoms, however, retard the rate of cracking. The formation of large amounts of isoparaffins and iso-olefins during catalytic cracking is explained by an intramolecular rearrangement of a secondary carbonium ion,

$$CH_3-CH_2-\overset{+}{C}H-CH_2-CH_2-R \longrightarrow \overset{+}{C}H_2-CH(CH_3)-CH_2-CH_2-R$$

$$\downarrow$$

$$CH_3-C(CH_3)=CH_2 + \overset{+}{C}H_2R \longleftarrow CH_3-\overset{+}{C}(CH_3)-CH_2-CH_2-R$$

$$\downarrow RH$$

$$CH_3-CH(CH_3)-(CH_2)_2R + \overset{+}{R} \quad (189)$$

No isomers of the parent hydrocarbon are observed when paraffins are cracked, but isomerization of the carbonium ion of the original paraffin does occur and this leads to isomeric products with a smaller number of carbon atoms than the parent hydrocarbon. Some hydrogen is produced during the primary reaction in which the carbonium ion is formed, but most of it probably arises through the dehydrogenation of naphthenes (cycloalkanes) which themselves are formed by cyclization and condensation reactions. Extensive condensation reactions eventually result in carbonaceous deposits which are always found on the catalyst surface after cracking reactions. Ethane and methane, also formed during catalytic cracking, are not as easy to account for, though the possible formation of methyl and ethyl carbonium ions should not be excluded, in spite of their instability. C. L. Thomas [148] showed that aromatic hydrocarbons can be formed by a repetition of hydrogen transfer reactions. First a carbonium ion (derived from the parent hydrocarbon) reacts by transforming a hydride ion to an olefin to produce an olefinic carbonium ion

$$R—CH_2—CH_2—CH = CH—R + R—\overset{+}{C}H—R$$

$$\longrightarrow R—CH_2—CH_2—\overset{+}{C}=CH—R + R—CH_2—R + R—CH_2—R \qquad (190)$$

An intramolecular rearrangement within the olefinic carbonium ion follows, involving the migration of a proton, and this more stable carbonium ion then transfers a proton to another olefin, thus resulting in a diolefin and a new carbonium ion. The diolefin produced would readily dimerize and form aromatics,

$$R—CH_2—\overset{+}{C}H—CH=CH—R + R—CH=CH—R$$

$$\longrightarrow R—CH=CH—CH=CH—R + R—CH_2—\overset{+}{C}H—R \qquad (191)$$

The catalytic cracking of olefins is similar in principle to the cracking of paraffins, but takes place more readily. A distinguishing feature is that a large fraction of the products contains isomers of the parent hydrocarbon. This is in contrast to paraffin cracking, during which only the products of cracking isomerize. It is therefore evident that intramolecular rearrangements within olefinic carbonium ions occur involving transfer of hydrogen, the tendency being to produce a more stable ion. Hence reactions such as (191) are quite feasible.

Naphthenic hydrocarbons [163] will also crack in the presence of silica–alumina catalysts. However, even at 500°C, cyclohexane hardly decomposes and there is a high concentration of hydrogen in the product gas. This indicates that, at this temperature, dehydrogenation occurs at a faster rate than cracking. At a higher temperature, the rate of cracking exceeds the rate of dehydrogenation, as evidenced by the small amount of hydrogen present when isopropylcyclohexane and amylcyclohexane crack. Some work has been reported on the cracking of bicyclic naphthenes, especially decalin [163, 164]. The tendency of decalin to crack more readily than cyclohexane is probably due to the presence of two tertiary carbon atoms in decalin as compared to the relatively more stable secondary carbon atoms in cyclohexane [148]. The products of cracking are mainly C_4-hydrocarbons, olefins and also some geometrical isomers of decalin which are the precursors of propane, propylene and monocyclic naphthenes. No benzene is produced and only small amounts of alkylbenzenes are formed, but the presence of naphthalene in the products is of interest. It was suggested [164] that naphthalene might have been formed form tetralin rather than decalin. Cyclo-olefins crack more readily than naphthenes and the products formed are mainly isomers of the parent hydrocarbon, but with a different number of C atoms in the ring [164, 165]. Thus, cyclohexane produces methylcyclopentene in good yield over a silica–alumina–thoria catalyst at 300°C. Aromatic products are also produced, as might be expected from previous considerations.

When aromatic hydrocarbons are subjected to catalytic cracking [166, 167] the cardinal feature distinguishing these reactions is the selectivity with which alkyl groups are detached from substituted benzenes. The larger the alkyl group the easier it is to crack the reactant, yet the selectivity for cleavage remains high. For example, when a mixture of amylbenzenes was cracked, the product contained 34% benzene and 29% amylenes [166]. Further reactions of the resultant olefins involving cracking, polymerization and hydrogen transfer are possible, so that the final product contains a variety of hydrocarbons very different from the original feed. Another feature is the reversible character of the cracking reaction. At elevated pressures, but lower temperatures, propylene and benzene react over a silica–alumina–zirconia catalyst to produce mono- and di-isopropylbenzene in high yield [167]. One suggested mechanism [148] involves an alkylaromatic carbonium ion which then decomposes to yield an olefin.

(192)

Greensfelder, Voge and Good [158] proposed an alternative mechanism in which the benzene ring exchanged a carbonium ion for a proton from the catalyst

$$
\text{C}_6\text{H}_5\text{-CH}\!\!\begin{array}{c}\text{CH}_3\\\text{CH}_3\end{array} + \text{H}^+ \longrightarrow \text{C}_6\text{H}_6 + \overset{+}{\text{CH}}\!\!\begin{array}{c}\text{CH}_3\\\text{CH}_3\end{array}
$$

$$
\downarrow
$$

$$
\text{H}^+ + \text{CH}\!\!\begin{array}{c}\text{CH}_3\\\text{CH}_2\end{array} \tag{193}
$$

The ease with which the alkyl carbonium ion is formed thus determines the extent of dealkylation. Because of the difficulty with which a methyl carbonium ion is formed, toluene is hardly cracked at all. The reaction between xylene and benzene to give toluene does, however, provide evidence for the transitory existence of a methyl carbonium ion [168].

8.5.3. Physical structure of cracking catalysts

The techniques described in Chapter 4 clearly reveal some of the structural characteristics of cracking catalysts. A low-temperature adsorption–desorption isotherm over a wide range of relative pressures will, for example, give information concerning surface area, pore volume, average pore radius and the pore size distribution. Furthermore, from the nature of the hysteresis loop it is possible, in some cases, to deduce something of the geometry of the pore structure. If such studies are supported by X-ray and electron microscope measurements, there emerges a reasonably clear picture of the pore structure.

The naturally occurring virgin silica–magnesia catalysts are found to contain very small pores, a fuller study revealing an average pore radius of less than 20 Å, such small pores accounting for almost the entire surface area [169]. On the other hand, the synthetic silica–alumina catalysts contain pores which are larger, although still with a fairly small average pore radius of less than 25 Å [169]. The clay mineral catalysts have a wider pore size distribution than either the natural or synthetic silica–alumina catalysts and contain some fairly large-diameter pores [169, 170]. The surface areas of the synthetic catalysts are generally much greater than those of the clay catalysts.

The structure of the diatomaceous earth catalysts is of interest. They are the siliceous remains of microscopic marine organisms. The adsorption–desorption isotherm shows an unexpected hysteresis effect [169, 171]. The isotherm approaches the saturated vapour pressure asymptotically and is characteristic of either non-porous or alternatively large pore structures. However, a detailed fine analysis of the hysteresis loop shows that the desorption curve joins the adsorption curve fairly sharply in the vicinity of 0·5 relative pressure, so that the presence of a small pore structure is indicated. An electron microscope examination might be expected to supplement this information. However, a uniform pore structure, with pores of about 1000 Å, is revealed [169, 176]. The electron micrograph at high magnification does, however, indicate a large central pore structure containing a regular arrangement of smaller pore structures and these, in turn, contain even smaller pores [169]. Such small pores would account for the hysteresis effect observed.

A study [169] of vacuum sintering and steam sintering of cracking catalysts shows that, for those which are vacuum sintered, the pore structure remaining is essentially similar to the initial structure. Steam sintering affects silica–magnesia preparations less than silica–alumina catalysts, since, as previously discussed, there is a general reorganization of the silica–alumina structure with heat treatment and this is intimately connected with the water content. The effect of steam sintering on silica–alumina catalysts is to enlarge the pore structure, and hence steam at high temperature may be the principal cause in the deterioration of silica–alumina cracking catalysts during use. The non-porous Linde silica is remarkably stable to thermal effects and there is little loss of area when heated to 1000°C.

8.6. Catalysis of Electrode Reactions

Since electron transfer processes between electrode and electrolyte occur during the operation of a galvanic cell, it should be possible, in principle, to assist such transfer by the use of suitable catalysts. The catalyst will participate in the process of adsorption, electron transfer and surface reaction and the best catalyst will be one which enhances the rates of adsorption and surface reaction. However, the catalyst should not adsorb the substrate with such a high heat of adsorption as to inhibit the formation of the activated complex and subsequent disruption to products. Thus, the choice of a catalyst will, in general, pose similar problems to those encountered in the selection of catalysts for gas-phase reactions. With the advent of the fuel cell for producing power has come an intensive effort to solve some of the less-tractable problems in catalysis, such as the selection of a suitable catalyst that will function under conditions appropriate to the reversible electrode, viz. at chemical equilibrium and in the presence of electrolyte.

8.6.1. Catalyst activity

Parsons [173], in demonstrating the dependence of electrode reactions on the free energy of adsorption of the reactant, showed that the exchange current during electrolytic hydrogen evolution is high for those metals which are able to adsorb atomic hydrogen readily. The exchange current, which is the electrode current at the reversible electrode potential, is a convenient measure of catalytic activity. It may be evaluated experimentally by applying the Tafel [174] equation and finding the slope and the intercept to a plot of polarization against the logarithm of the current at the cathode. As the standard free energy of hydrogen adsorption at different metals approaches zero, so the exchange current reaches a maximum. Metals with high heats of adsorption show a decreasing exchange current as the negative value of the standard free energy of adsorption of hydrogen increases. The slow reaction of adsorbed hydrogen atoms and ions appears to control the rate of hydrogen evolution and the correlation between exchange current and free energy of adsorption lends support to this.

Transition metals are, in fact, the most frequently encountered fuel cell catalysts. The d-band structure of a metal is related to its work function and its adsorptive properties: the heat of adsorption decreases with increasing percentage d-character (see Section 6.3.1.1) and, consequently, with increasing work function. Hence it is not surprising that Conway and Bockris [175] found a correlation between the exchange current and the work function at the metal electrode. Figure 14 shows that the exchange current increases exponentially with work function. A similar relation between the exchange current and the d-band character of the metal is also displayed and is shown in Fig. 15. If the rate of the electrochemical reaction is controlled by a slow electron transfer process, the exchange current should increase exponentially with decrease in work function, for to discharge an ion requires a metal whose work function

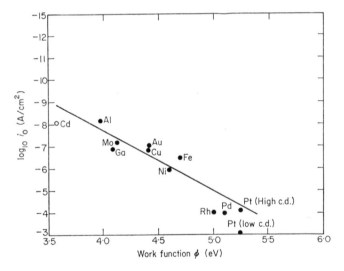

FIG. 14. Dependence of H_2 evolution exchange current on electronic work function for some electrode reactions [175].

is lower than the ionization potential of the species being discharged at the electrode. Hence, if the work function is decreased, the energy necessary to cause discharge, given by the difference between the ionization potential and the work function, is correspondingly increased and the activation energy decreased.

Of the group VIII metals, palladium and platinum are the most active catalysts for the anodic oxidation of hydrogen, carbon monoxide, hydrocarbons and alcohols. Each of these metals exhibits a low heat of adsorption and they are well-known catalysts for hydrogenation reactions. Their d-band character suggests that their catalytic properties for anodic oxidation derive

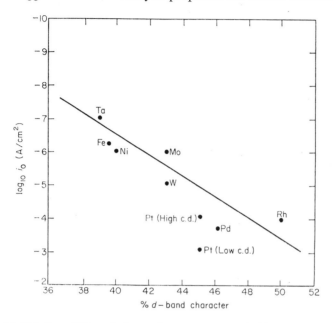

FIG. 15. Relation of H_2 evolution exchange current on % d-band character for some electrode reactions [175].

from covalent bonding between the fuel and metal. These catalysts are usually impregnated on porous carbon electrodes or, alternatively, on metal grids or sintered metals, so as to increase the surface area for catalysis.

8.6.2. Porous electrodes

The porous diffusion electrodes referred to usually alter the kinetics of the overall electrode process. Not only is there a decrease in concentration near the vicinity of a normal electrode, caused by slow mass transport and resulting in concentration polarization, but also the porous structure of the electrode impedes progress of the gas along the pore and electrochemical reaction is therefore affected. Problems of pore diffusion and reactivity have already been discussed in Section 4.5. Similar methods may be applied to diffusion electrodes, for the problem is essentially that of diffusion accompanied by a first-order chemical reaction [176]. Two important conclusions are derived from such an analysis. (1) A porous electrode polarizes twice as fast as a non-porous electrode yielding the same current. This is due to the fact that the slope to the Tafel equation is twice that for a non-porous electrode (cf. effect of mass transport on the activation energy for isothermal reaction in pores, Chapter 4). (2) High current densities are obtainable as a direct consequence of increased surface area available for electrochemical reaction.

In a porous electrode the stability of the gas–liquid boundary is important. This is determined by a three-phase equilibrium between gas, electrolyte and porous solid, and has been discussed by Eisenberg [177] and Justi [178]. In order to establish a stable equilibrium between gas, electrolyte and electrode it is desirable to have a gradient in pore diameters in a direction normal to the electrode surface. Bacon [179] employed a fine pore structure near the electrolyte boundary and a coarser pore structure near the gas boundary. Such an arrangement permits pressure fluctuation in the gas supply, for if the gas pressure increased suddenly the electrolyte would, in accordance with the Kelvin equation discussed in Section 4.3.1, be displaced from the coarse pores towards the smaller-diameter pores near the electrolytic solution. On the other hand, if the pressure were to decrease momentarily, the gas would recede from the fine pores towards the coarser pores, where the larger capillary diameters are capable of accommodating higher gas pressures.

8.6.3. Comparison of electrode systems

Fuels such as hydrogen, hydrocarbons and carbon monoxide are catalytically oxidized at an anode by similar metals and the pattern of activity is uncommonly similar. A comparison of hydrogen and carbonaceous fuels at electrodes will therefore be of importance in evaluating and establishing a hierarchy of catalyst reactivities. The oxygen electrode is also included in the comparison, a. reactivities at the cathode affect the overall performance of the cell.

8.6.3.1. Hydrogen Electrode

Anthony [180] showed that a platinum-plated carbon anode when immersed in 6N potassium hydroxide gave a current density of 24·3 mA cm^{-2} at a cell voltage of 0·4 V (the cathode being a 4% Pd on an oxygenated carbon electrode). Other transition metals and metal oxides were also studied and it was possible to compare their performances. A rhodium-plated carbon anode was almost as effective as platinum, although palladium gave a much smaller current density of 10·5 mA cm^{-2} at a cell votage of 0·4 V. Mixed oxides and spinels gave only small current densities.

Current densities much higher than reported by Anthony have been obtained by employing porous anodes. Justi and Winsel [181] obtained 750 mA cm^{-2} with polarizations of only 40 mV using an electrode which consisted of a homoporous nickel macropore skeleton containing a Raney nickel catalyst embedded in the pores.

Young and co-workers [182] correlated the H_2 open circuit potentials obtained for the platinum metal catalysts with their d-band structure. It was assumed that open circuit polarization, defined as the voltage loss from the reversible potential at open circuit, is a measure of the energy loss caused by irreversible chemisorption at the electrode. As Table 8.3 shows, a uniform

TABLE 8.3. Open-circuit hydrogen potentials[a] on group VIII and group Ib metals [182].

Group	VIII			Ib
	Metal E_0(V)	Metal E_0(V)	Metal E_0(V)	Metal E_0(V)
1	Fe 0·575	Co 0·745	Ni 0·830	Cu 0·385
2	Ru 0·870	Rh 0·850	Pd 0·836	Ag 0·380
3	Os 0·859	Ir 0·850	Pt 0·845	Au 0·305

[a]Versus standard hydrogen potential.

periodicity in the open circuit potential is evident, the second or third member of each row showing a maximum potential. A sharp decrease in open circuit potential is evident for the group Ib metals. Such behaviour is consistent with the manner in which the d-band structure alters through the series.

Potential measurements in alloy electrodes have further strengthened the belief that the d-band structure of the metal catalyst influences the performance of the electrode. For example, the potential of a platinum–hydrogen electrode was decreased considerably by alloying with gold. It is supposed that the gold contributes electrons to the d-band vacancies in platinum, thus rendering the catalyst relatively inactive. The maximum activity was obtained with alloys of platinum and iridium which have between 0·5 and 2 d-band vacancies per atom.

A similar approach to that for metals was made with metal oxide catalysts correlating open circuit potentials with the electronic configuration of the metal ion of the oxide. The potential obtained depends on the number of unpaired electrons associated with the free cation of the oxide, maximum activity being obtained for ferric oxide, which has five unpaired electrons per atom of iron. Now five unpaired electrons is known to be a very stable configuration, so that it is surprising that ferric oxide has the highest activity. A possible explanation is that the electron transfer process is not rate determining. Provided sufficient energy is available to cause electron transfer to take place at all, maximum activity would be observed.

8.6.3.2. Carbonaceous Fuels

A study of the results obtained for the chemisorption of hydrocarbons on metals provides an answer as to why hydrocarbon fuels follow the same pattern of activity as hydrogen. The bond which is broken when ethylene is adsorbed on platinum is a carbon–carbon π-bond and this is replaced by a carbon–metal bond. If the rate-determining process in the half-cell is associated with the formation and rupture of the carbon–metal bond, and as the bond energy is related to the d-character of the metal catalysts which are active for hydrogen oxidation, they should also be active for the oxidation of hydrocarbons. Saturated hydrocarbons are also adsorbed by metals, a carbon–metal bond being formed so that similar behaviour to that of hydrogen may be expected.

Higher currents are obtained from partially oxidized fuels than from hydrocarbons. For example, silver-catalysed porous nickel electrodes have yielded current densities of 200 mA cm^{-2} at 0·75 V when ethylene glycol is used as the fuel. Platinum, palladium, rhodium and iridium give rise to current densities of 10 mA cm^{-2} when methanol is oxidized. Formic acid and formaldehyde are also oxidized by the noble metals. The fact that higher current densities are obtained with partially oxidized fuels is due to the fact that such oxidations, although involving a free energy change less than that involved in the complete oxidation of a hydrocarbon, occur under reversible conditions whereas the oxidation of a fuel such as propane is highly irreversible.

Young and Rozelle [182] correlated open-circuit potentials with electronic configuration for catalysts known to be active in oxidizing ethylene, acetylene and carbon monoxide. Similar results are found for these fuels as for hydrogen. Metal oxides such as manganese dioxide and cuprous oxide, in spite of their catalytic activity in oxidizing CO, are not active fuel-cell catalysts.

8.6.3.3. Oxygen Electrode

Berl [183] made an exhaustive study of the reduction of oxygen at carbon cathodes and concluded that hydroxyl and perhydroxyl ions are initial products. In the presence of protons these will combine to form hydrogen peroxide, which, in turn, decomposes to oxygen and water. Assuming such a mechanism operates at an oxygen electrode, a catalyst should, in addition to adsorbing oxygen, be capable of decomposing hydrogen peroxide. The most active oxygen electrode catalysts will both adsorb oxygen and decompose hydrogen peroxide. A Raney silver-impregnated nickel matrix having contiguous pores of two different sizes has been reported by Justi [181] and is an excellent oxygen electrode. This electrode yielded 150 mA cm^{-2} at a polarization of 0·64 V from the theoretical value. Similar performances have been achieved using colloidal silver in a silver matrix. Silver gauze, although giving an open-circuit potential near to the reversible potential in alkaline solution, did not perform at all well under load and was very much inferior to that of platinum gauze, which, at $1 \cdot 0$ mA cm^{-2}, polarized only 200 mV from the theoretical value. The poor activity of silver gauze as compared with colloidal silver is ample evidence of the importance of catalyst state and surface area.

Nickel peroxidized after impregnation with lithium hydroxide [184] is also effective as a catalyst at 200°C. Other mixed metal oxides with the spinel structure have given good performances when impregnated on a porous carbon electrode in alkaline solution. A study of the catalytic activity of a number of mixed oxides and their corresponding spinels [185] showed that in every case the spinel had the superior catalytic activity. Defect structures (see Section 5.2.3) are introduced during the high-temperature preparation of spinels and this is probably the reason for their high activity.

8.7. Catalysis of the Oxidation of Elemental Carbon

An unusual example of heterogeneous catalysis is to be found in the oxidation of carbon when the latter occurs as graphite or diamond or as most of its numerous amorphous forms. A wide diversity of solids will, upon addition to elemental carbon, produce marked catalysis of the gasification—the term used to describe the conversion of carbon to gaseous CO and CO_2. The process is of technological interest for two main reasons. Firstly, graphite is a convenient material to use as a moderator in nuclear reactors. Secondly, its mechanical strength at high temperatures surpasses that of most metals and alloys. In order, therefore, for gaseous corrosion of graphite to be successfully combated, the mechanism of catalysis must first be elucidated.

The study of the influence of solid catalysts on the gasification of graphite by oxidizing gases such as O_2, CO_2, H_2O and the oxides of nitrogen has evoked extensive researches, and these have established that a large number of the elements (particularly the transition metals, Fe, Co and Ni, the alkali metals and certain other elements from Groups I and II of the Periodic Table) function as efficient catalysts. Figure 16, taken from the work of Sykes and Thomas [186], illustrates the extent of the catalysis obtained, and the importance of the mode of addition, as well as the concentration, of the catalyst.

It was once thought [187–189] that solids which functioned as catalysts in the oxidation of carbonaceous solids did so because they were able to form intermediates which, in turn, served to oxidize the carbon in their vicinity and, in so doing, returned to their original condition, only to be reconverted to the active intermediate by the oxidizing gas. Thus, a metal catalyst, for example, could exert its catalytic influence by being first converted to an oxide (or a series of oxides).

By about a decade ago considerable evidence had accumulated to suggest that the intermediate compound theory might be erroneous for a variety of

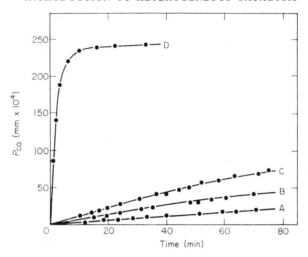

FIG. 16. Effect of iron on rate of formation of CO during graphite oxidation: A, Pure graphite; B, Fe:C = 1:4800; C, Fe:C = 1:2900; D, Fe:C = 1:580 [*186*].

reasons. One of the most compelling reasons adduced by Long and Sykes [*190–192*] in their detailed studies of the effect of specific catalysts on the reactions

$$C(s) + CO_2(g) \rightleftharpoons 2CO(g)$$
$$C(s) + H_2O(g) \rightleftharpoons CO(g) + H_2(g)$$
$$3C(s) + 2O_2(g) \rightleftharpoons 2CO(g) + CO_2(g)$$

was that the general course of the gasification processes did not appear to be modified by the catalytic impurities. There were also other factors which seemed inexplicable in terms of the intermediate compound theory. Thus, impurities added to purified carbons interacted with the oxidizing gases to a much greater extent than did the natural impurities originally present in the carbons [*192*]. In the light of these factors, Long and Sykes formulated [*192*] their electronic theory, the essence of which can be summarized as follows. The evolution of carbon monoxide, which is present incipiently (see Fig. 17(a)) at the surface of any carbonaceous solid on to which oxygen has been adsorbed, is facilitated if the carbon–carbon bonds which must be broken are weakened. This weakening should occur if the carbon matrix transfers an electron to, say, a transitional metal ion (Fig. 17(b)) or if a covalent bond is formed between the matrix and, for example, an alkali metal atom M (Fig. 17(c)). Long and Sykes estimated [*191*], using molecular orbital theory, that the changes in bond strength brought about by such electronic rearrangements are of a significant magnitude.

The electronic theory has done much to place the likely modes of action of various specific catalysts into rational perspective; it is indisputable that, under certain circumstances [*190, 193, 194*], it offers a logical interpretation of observed phenomena. However, in recent years, especially as a result of the painstaking studies of Duval and his co-workers [*195–197*], evidence has emerged which again favours the intermediate-compound theory. Moreover, the results of Hennig [*198*], Dawson [*199*] and others [*200–204*], who showed

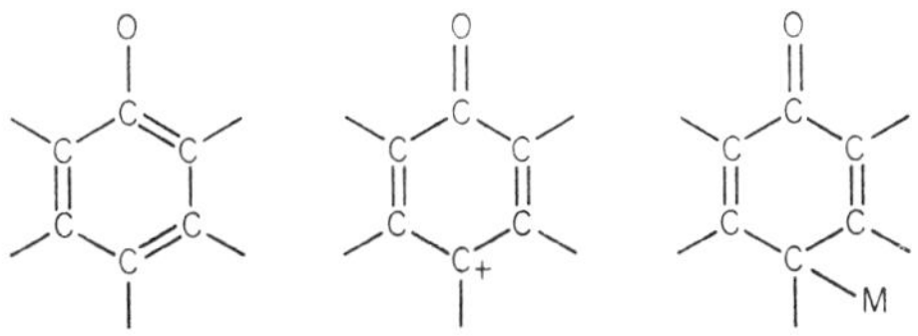

FIG. 17. Oxidation of carbon. Illustration of electronic theory of Long and Sykes.

that many metals (including certain noble ones, of colloidal dimensions) can act as extremely good oxidiation catalysts, have confronted the electronic theory, and, indeed, all other theories, with a new challenge.

In view of the foregoing remarks, it is of interest to consider a selection of some recent results.

So far as the role of iron and other transition metals is concerned, controversies have continued unabated for over a decade as to whether it is necessary to have this impurity present in the form of the free metal, as the carbide, or as the oxide. By employing microcinematography [*200, 201*], definite evidence has been obtained against the view that the oxide possesses the greatest catalytic efficiency, for it is certain that the catalytic pitting of the basal {000l} planes of graphite crystals is progressively diminished as the iron particles gradually become oxidized. It would, however, be premature to contend that the free metal or the carbide is the effective form of the catalyst, for it may be that a thin layer of defective (non-stoichiometric) oxide is the dominating agency, and that only when the non-stoichiometry is destroyed will there be no catalysis.

Closer studies of the nature of the lattice defects in graphite [*205, 206*] and the way in which radiation damage modifies the pattern of catalysis [*199*] are making it increasingly apparent that not only is it important to know the chemical state of the catalytic impurity, it is also imperative to know the nature of the crystalline imperfection with which the catalyst interacts. For example, in the presence of single vacancies in the {000l} planes (see Fig. 18), colloidal iron is a much more active catalyst than in the absence of such vacancies [*203*].

Some of the most significant results appertaining to catalysis of carbon–gas reactions have been obtained by Duval and Amariglio. They showed [*197*] that silver, manganese and barium are extremely effective catalysts, more so

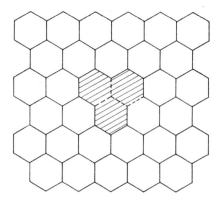

FIG. 18. Shaded area represents a single vacancy in a basal plane of graphite.

than the transition metals. (The oxidation rates at low temperature were increased by factors ranging from 10^3 to 10^4 in the presence of manganese and silver.) Similar results have been obtained by independent workers [201].

Although it cannot be stated unequivocally that one theory is distinctly more acceptable than the other, it must be concluded that, at present, the intermediate-compound theory is rather more favoured than it has been for about thirty years. According to Duval [207], there are two convincing pieces of information which militate against the acceptance of the electronic theory. Firstly, experimental evidence seems to suggest that the activation energy of oxidation is independent of the concentration of the catalytic impurity. Secondly, the metals that seem to be most active are capable of forming higher oxides which are reducible, by carbon, to lower oxides. On the other hand, Hennig [198, 208] points out that from the observed motion of catalyst particles as they channel across a graphite surface, bonds are formed (similar to those postulated by Long and Sykes) between the catalyst and a number of reactive carbon atoms at the metal–carbon interface.

Again, both Duval [207] and Hennig [198] have observed a dependence of catalytic activity upon the particle size of the metal catalyst, a feature which Duval has used to introduce the Kobozev theory of ensembles [209] (see p. 306), successfully applied in other fields, to carbon chemistry. Others [210, 211] have adduced evidence showing that each catalytic system must be considered individually, for it seems that, with a given catalytic impurity, the nature of the active form depends upon which oxidation reaction is occurring. Yet others [212, 213] have endeavoured to relate the catalytic activity of the first-row transition metal oxides to the corresponding crystal-field stabilization energies. Undoubtedly, much work still remains to be done before the numerous obscurities and enigmas associated with carbon–gas reactions are dispelled.

References

1. F. S. Stone, *Adv. Catalysis* **13**, 1 (1962).
2. G. H. Twigg, *Proc. R. Soc.* **A88**, 92 (1946).
3. L. Ya. Margolis, *Adv. Catalysis* **14**, 429 (1963).
4. J. K. Dixon and J. E. Longfield, "Catalysis", Vol. VII, p. 325. Reinhold, New York (1960).
5. J. Zawadzki, *Disc. Faraday Soc.* **8**, 140 (1950).
6. M. Bodenstein, *Z. Elektrochem.* **47**, 501 (1944).
7. O. A. Uyehara and K. M. Watson, *Ind. Engng Chem.* **35**, 541 (1943).
8. T. Baron, W. R. Manning and H. F. Johnstone, *Chem. Engng Progr.* **48**, 125 (1952).
9. M. I. Temkin and N. V. Kulkova, *Dokl. Akad. Nauk SSSR* **105**, 1021 (1955).
10. L. Ya. Margolis and S. Z. Roginskii, *Izv. Akad. Nauk SSSR* **9**, 107 (1957).
11. F. S. Stone, *in* "Chemistry of the Solid State", ed. by W. E. Garner, Ch. 15. Butterworths, London (1955).
12. W. E. Garner and F. J. Veal, *J. chem. Soc.*, 1487 (1935).
13. T. Ward, *J. chem. Soc.*, 1244 (1947).
14. W. E. Garner, F. S. Stone and P. F. Tiley, *Proc. R. Soc.* **A211**, 472 (1952).
15. R. M. Dell and F. S. Stone, *Trans. Faraday Soc.* **50**, 501 (1954).
16. R. Rudham and F. S. Stone, *in* "Chemisorption", ed. by W. E. Garner, p. 205. Butterworths, London (1957).
17. E. R. S. Winter, *Adv. Catalysis* **10**, 196 (1958).
18. E. R. S. Winter, *J. chem. Soc.*, 2726 (1955).
19. S. J. Teichner and J. A. Morrison, *Trans. Faraday Soc.* **51**, 961 (1955).
20. S. J. Teichner, R. P. Marcellini and P. Rue, *Adv. Catalysis* **9**, 458 (1957).
21. R. P. Marcellini, R. E. Ranc and S. J. Teichner, Proc. 2nd Int. Congr. Catalysis. Editions Technip, Paris (1961).
22. R. P. Eischens and W. A. Pliskin, *Adv. Catalysis* **9**, 662 (1957).
23. G. Blyholder, Proc. 3rd Int. Congr. Catalysis. North Holland Publishing Co., Amsterdam (1964).
24. J. Coue, P. C. Gravelle, R. E. Ranc, P. Rue and S. J. Teichner, Proc. 3rd Int. Congr. Catalysis. North Holland Publishing Co., Amsterdam (1964).
25. M. E. Dry and F. S. Stone, *Disc. Faraday Soc.* **28**, 192 (1959).
26. N. P. Keier, S. Z. Roginskii and I. S. Sazenova, *Izv. Akad. Nauk SSSR Otd. fiz. Nauk* **21**, 183 (1957).
27. L. Ya. Margolis, *Usp. Khim.* **28**, 615 (1959).
28. Th. Wolkenstein, *Adv. Catal.* **12**, 189 (1960).
29. G. D. Lyubarskii, *Dokl. Akad. Nauk. SSSR* **110**, 112 (1956).
30. L. Ya. Margolis, *Izv. Akad. Nauk SSSR Otd. khim. Nauk* 1173 (1958).
31. F. S. Stone, R. Rudham and R. L. Gale, *Z. Elektrochem.* **63**, 129 (1959).
32. G. H. Twigg, *Proc. R. Soc.* **A188**, 93 (1946).
33. G. H. Twigg, *Proc. R. Soc.* **A188**, 113 (1946).
34. A. Orzechowski and B. E. McCormack, *Can. J. Chem.* **32**, 388 (1954).
35. A. I. Kurilenko, *Zh. fiz. Khim.* **32**, 1043 (1958).
36. L. Ya. Margolis and S. Z. Roginskii, *Izv. Akad. Nauk SSSR Otd. khim. Nauk* 282 (1956).
37. E. T. McBee, H. B. Hass and P. A. Wiseman, *Ind. Engng Chem.* **37**, 432 (1945).
38. O. V. Isaev, L. Ya. Margolis and I. S. Sazenova, *Dokl. Akad. Nauk SSSR* **88**, 79 (1953).

39. V. M. Belousov, Ya. B. Gorokhovatskii, M. Ya. Rubanik and A. V. Gershingorn, *Kinet. Katal.* **3**, 221 (1962).
40. E. Kh. Enikeev, O. V. Isaev, and L. Ya. Margolis, *Kinet. Katal.* **1**, 431 (1960).
41. G. L. Simard, J. F. Steger, R. J. Arnott and L. A. Siegel, *Ind. Engng Chem.* **37**, 432 (1945).
42. V. A. Roiter, N. A. Stukanovskaya and N. S. Volkooskaya, *Ukr. khim. Zh.* **24**, 35 (1958).
43. L. Ya. Margolis and E. G. Plyshevskaya, *Izv. Akad. Nauk SSSR Otd. khim. Nauk* 697 (1953).
44. D. J. Berets and H. Clark, *Adv. Catalysis* **9**, 204 (1957).
45. P. Mars and D. W. van Krevelin, *Chem. Engng Sci.* **3**, Sp. Suppl. 41 (1954).
46. P. H. Calderbank, *Ind. Chem.* **28**, 291 (1952).
47. P. H. Pinchbeck, *Chem. Engng Sci.* **6**, 105 (1957).
48. A. F. d'Allesandro and A. Farkas, *J. Colloid Sci.* **11**, 653 (1956).
49. P. Mars, Kinetics of Oxidation Reactions on Vanadium Pentoxide Catalysts. Thesis, University of Delft (1958).
50. J. A. Campbell, *J. Soc. chem. Ind.* **48**, 93 (1929).
51. S. Bransom, L. Hanlon and B. Smythe, *Trans. Faraday Soc.* **52**, 672 (1956).
52. W. J. Thomas, *Trans. Faraday Soc.* **53**, 1124 (1957).
53. W. J. Thomas, *Trans. Faraday Soc.* **55**, 624 (1959).
54. P. W. Selwood, *J. Am. chem. Soc.* **79**, 3346 (1957).
55. P. W. Selwood, Proc. 2nd Int. Conf. Catalysis, Vol. 2, p. 1795. Editions Technip, Paris (1961).
56. R. Eischens and W. A. Pliskin, *Adv. Catalysis* **10**, 1 (1958).
57. L. H. Little, N. Sheppard and D. J. C. Yates, *Proc. R. Soc.* **A259**, 242 (1960).
58. H. L. Pickering and H. C. Eckstrom, *J. phys. Chem.* **63**, 512 (1959).
59. G. H. Twigg and E. K. Rideal, *Trans. Faraday Soc.* **36**, 533 (1940).
60. E. K. Rideal and G. H. Twigg, *Proc. R. Soc.* **A171**, 55 (1939).
61. C. Kemball, *J. chem. Soc.* 735 (1956).
62. G. C. A. Schuit and L. L. van Reijen, *Adv. Catalysis* **10**, 242 (1958).
63. G. I. Jenkins and E. K. Rideal, *J. chem. Soc.*, 2490 (1955).
64. G. I. Jenkins and E. K. Rideal, *J. chem. Soc.*, 2496 (1955).
65. E. K. Rideal, *J. chem. Soc.*, 309 (1922).
66. H. Zür Strassen, *Z. phys. Chem.* **A169**, 81 (1934).
67. O. Beeck, *Rev. mod. Phys.* **17**, 61 (1945).
68. O. Beeck, *Disc. Faraday Soc.* **8**, 118 (1950).
69. G. C. Bond, "Catalysis by Metals". Academic Press, London (1962).
70. G. H. Twigg, *Disc. Faraday Soc.* **8**, 152 (1950).
71. P. G. Ashmore, "Catalysis and Inhibition of Chemical Reactions". Butterworths, London (1963).
72. G. D. Halsey, Jr., *J. phys. Chem.* **67**, 2038 (1963).
73. G. D. Halsey, Jr., *J. chem. Phys.* **17**, 758 (1949).
74. O. Beeck, A. E. Smith and A. Wheeler, *Proc. R. Soc.* **A177**, 74 (1940).
75. C. D. Wagner, J. N. Wilson, J. W. Otvos and D. P. Stevenson, *J. chem. Phys.* **20**, 338, 1331 (1952).
76. J. Turkevich, D. O. Schissler and A. P. Irsa, *J. Phys. Colloid Chem.* **55**, 1078 (1951).
77. T. I. Taylor, *in* "Catalysis", ed. by P. H. Emmett, Vol. 5, p. 527 (1957).
78. G. C. Bond, *Trans. Faraday Soc.* **52**, 1235 (1956).

79. J. Horiuti, Proc. 2nd Int. Congr. Catalysis, Vol. 1, p. 1191. Editions Technip, Paris (1961).
80. T. Keii, *J. chem. Phys.* **22**, 144 (1954).
81. G. C. Bond and J. Newham, *Trans. Faraday Soc.* **56**, 1851 (1960).
82. C. Kemball and J. R. Anderson, *Proc. R. Soc.* **A226**, 472 (1954).
83. G. C. Bond and J. Turkevich, *Trans. Faraday Soc.* **50**, 1335 (1954).
84. H. C. Rowlinson, R. L. Burwell, Jr., and R. H. Tuxworth, *J. phys. Chem.* **63**, 1016 (1954).
85. G. C. Bond, D. A. Dowden and N. Mackenzie, *Trans. Faraday Soc.* **54**, 1537 (1958).
86. G. C. Bond and P. B. Wells, Proc. 2nd Int. Congr. Catalysis, Vol. 1, p. 1135. Editions Technip, Paris (1961).
87. J. E. Douglas and B. S. Rabinowitch, *J. Am. chem. Soc.* **74**, 2486 (1952).
88. G. C. Bond, *J. chem. Soc.* 4288 (1958).
89. Y. F. Yu, J. J. Chessick and A. C. Zettlemoyer, *J. phys. Chem.* **63**, 1626 (1959).
90. E. B. Maxted and M. Josephs, *J. chem. Soc.* 2635 (1956).
91. H. A. Smith and H. T. Meriwether, *J. Am. chem. Soc.* **71**, 413 (1949).
92. P. W. Selwood, *J. Am. chem. Soc.* **79**, 4637 (1957).
93. R. Suhrmann, *Adv. Catalysis* **9**, 88 (1957).
94. M. Y. Kagan and S. D. Friedman, *Dokl. Akad. Nauk SSSR* **68**, 697 (1949).
95. B. M. W. Trapnell, *Adv. Catalysis* **3**, 1 (1951).
96. A. Farkas and L. Farkas, *Trans. Faraday Soc.* **33**, 827 (1937).
97. A. Amano and G. Parravano, *Adv. Catalysis* **9**, 716 (1957).
98. R. K. Greenhalgh and M. Polanyi, *Trans. Faraday Soc.* **35**, 520 (1939).
99. J. Horiuti, G. Ogden and M. Polanyi, *Trans. Faraday Soc.* **30**, 663 (1934).
100. F. Hartog, J. H. Tebben and P. Zwietering, Proc. 2nd Int. Congr. Catalysis, Vol. 1, p. 1229. Editions Technip, Paris (1961).
101. J. R. Anderson and C. Kemball, *Adv. Catalysis* **9**, 51 (1957).
102. S. W. Weller, L. J. E. Hofer and R. B. Anderson, *J. Am. chem. Soc.* **70**, 799 (1948).
103. L. J. E. Hofer and W. C. Peebles, *J. Am. chem. Soc.* **69**, 893 (1947).
104. J. T. McCartney, L. J. E. Hofer, B. Seligman, J. A. Lecky, W. C. Peebles and R. B. Anderson, *J. phys. Chem.* **57**, 730 (1953).
105. C. J. Davisson and L. H. Germer, *Phys. Rev.* **43**, 292 (1933).
106. G. J. H. Dorgelo and W. M. H. Sachtler, *Naturwissenschaften* **20**, 576 (1959).
107. M. V. C. Sastri and T. S. Viswanathan, *J. Am. chem. Soc.* **77**, 3967 (1955).
108. M. McD. Baker and E. K. Rideal, *Trans. Faraday Soc.* **51**, 1597 (1955).
109. J. Bagg and F. C. Tompkins, *Trans. Faraday Soc.* **51**, 1071 (1955).
110. D. F. Klemperer and F. S. Stone, *Proc. R. Soc.* **A243**, 375 (1957).
111. W. K. Hall, R. J. Kokes and P. H. Emmett, *J. Am. chem. Soc.* **82**, 1027 (1960).
112. R. B. Anderson, *in* "Catalysis", ed. by P. H. Emmett, Vol. 4, p. 257 (1956).
113. H. H. Storch, N. Golumbic and R. B. Anderson, "Fischer-Tropsch and Related Synthesis". Wiley, New York (1956).
114. R. B. Anderson, R. A. Friedel and H. H. Storch, *J. chem. Phys.* **19**, 313 (1951).
115. E. F. G. Herington, *Chemy Ind.* 346 (1946).
116. S. W. Weller and R. A. Friedel, *J. chem. Phys.* **17**, 801 (1949).
117. S. W. Weller and R. A. Friedel, *J. chem. Phys.* **18**, 157 (1950).
118. R. A. Friedel and R. B. Anderson, *J. chem. Phys.* **72**, 1212 (1950).
119. H. B. Bruner, *Ind. Engng Chem.* **41**, 2511 (1949).
120. F. Haber, *Z. angew. Chem.* **27**, 473 (1914).

121. W. Nernst, W. Jost and G. Jellinek, *Z. Elektrochem.* **13**, 521 (1907).
122. W. Nernst, W. Jost and G. Jellinek, *Z. Elektrochem.* **14**, 373 (1908).
123. A. T. Larsen and R. L. Dodge, *J. Am. chem. Soc.* **45**, 2918 (1924).
124. A. T. Larsen and R. L. Dodge, *J.A.C.S.* **46**, 367 (1924).
125. K. G. Denbigh, *Trans. Faraday Soc.* **40**, 352 (1944).
126. F. Horn, *Chem. Engng Sci.*, Symposium on Chemical Reaction Engineering, **14**, 77 (1962).
127. G. Ehrlich, *J. chem. Phys.* **23**, 1543 (1955).
128. G. Ehrlich, *J. chem. Phys.* **24**, 482 (1956).
129. G. Ehrlich, *J. phys. Chem.* **60**, 1388 (1956).
130. D. O. Hayward and B. M. W. Trapnell, *In* "Chemisorption", ed. by W. E. Garner, 2nd Ed. Butterworths, London (1964).
131. S. R. Logan, R. L. Moss and C. Kemball, *Trans. Faraday Soc.* **56**, 144 (1960).
132. A. Ozaki, H. S. Taylor and M. Boudart, *Proc. R. Soc.* **A258**, 47 (1960).
133. M. I. Temkin and V. Pyzhev, *Acta physicochim. U.S.S.R.* **12**, 327 (1940).
134. K. S. Love and P. H. Emmett, *J. Am. chem. Soc.* **63**, 3297 (1941).
135. K. S. Love and S. Brunauer, *J. Am. chem. Soc.* **64**, 745 (1942).
136. J. J. F. Scholten and P. Zweitering, *Trans. Faraday Soc.* **53**, 1363 (1957).
137. J. J. F. Scholten, Thesis, University of Delft (1959).
138. J. J. F. Scholten, J. A. Konvalinka and P. Zweitering, *Trans. Faraday Soc.* **56**, 262, (1960).
139. J. T. Kummer and P. H. Emmett, *J. chem. Phys.* **19**, 289 (1951).
140. K. Tamaru, *Trans. Faraday Soc.* **59**, 979 (1963).
141. K. Tamaru, Proc. 2nd Int. Congr. Catalysis, Vol. 1, p. 325. Editions Technip, Paris (1961).
142. S. Brunauer, K. S. Love and R. G. Keenan, *J. Am. chem. Soc.* **56**, 35 (1934).
143. J. Horiuti, *Adv. Catalysis* **9**, 339 (1957).
144. T. Kwan, *J. Catalysis* **1**, 199 (1962).
145. S. Enomoto and J. Horiuti, *Proc. Japan Acad.* **28**, 493 (1952).
146. C. Bokhoven, M. J. Gorgels and P. Mars, *Trans. Faraday Soc.* **55**, 315 (1959).
147. Yu. A. Bitepazh, *J. gen. Chem U.S.S.R.* **17**, 199 (1947).
148. C. L. Thomas, *Ind. Engng Chem.* **41**, 2564 (1949).
149. H. Weil-Malherbe and J. J. Weiss, *J. chem. Soc.* 2164 (1948).
150. C. E. Marshall, "Colloid Chemistry of the Silicate Minerals". Academic Press, New York (1949).
151. A. G. Oblad, T. H. Milliken, Jr., and G. A. Mills, *Adv. Catalysis* **3**, 199 (1951).
152. M. W. Tamele, *Disc. Faraday Soc.* **8**, 270 (1950).
153. C. L. Thomas, *J. Am. chem. Soc.* **66**, 1586 (1944).
154. T. H. Milliken, Jr., A. G. Oblad and G. A. Mills, Gordon Research Conf. (1949).
155. L. Pauling, "The Nature of the Chemical Bond". Cornell University Press, Ithaca (1948).
156. M. W. Tamele and L. R. Reyland, Am. Chem. Soc. Meeting, Abstr., p. 79–O (March 1949).
157. G. Parravano, E. F. Hammel and H. S. Taylor, *J. chem. Soc.* 2269 (1948).
158. B. S. Greensfelder, H. H. Voge and G. M. Good, *Ind. Engng Chem.* **41**, 2573 (1949).
159. L. Schmerling, *J. Am. chem. Soc.* **66**, 1422 (1944).
160. L. Schmerling, *J. Am. chem. Soc.* **67**, 1778 (1945).

161. L. Schmerling, *J. Am. chem. Soc.* **68**, 275 (1946).
162. V. Haensel, *Adv. Catalysis* **3**, 179 (1951).
163. B. S. Greensfelder, H. H. Voge and G. M. Good, *Ind. Engng Chem.* **37**, 1038 (1945).
164. H. S. Bloch and C. L. Thomas, *J. Am. chem. Soc.* **66**, 1589 (1944).
165. B. S. Greensfelder and H. H. Voge, *Ind. Engng Chem.* **37**, 983 (1945).
166. C. L. Thomas, J. Hoekstra and J. T. Pinkston, *J. Am. chem. Soc.* **66**, 1694 (1944).
167. B. S. Greensfelder, H. H. Voge and G. M. Good, *Ind. Engng Chem.* **37**, 1168 (1945).
168. R. D. Hansford, C. G. Myers and A. N. Sachanen, *Ind. Engng Chem.* **37**, 671 (1945).
169. H. E. Ries, Jr., *Adv. Catalysis* **4**, 87 (1952).
170. T. D. Oulton, *J. Phys. Colloid Chem.* **52**, 1296 (1948).
171. H. E. Ries, M. F. L. Johnson and J. S. Melik, *J. Phys. Colloid Chem.* **53**, 638 (1949).
172. H. L. Ritter and L. C. Drake, *Ind. Engng Chem. Anal. Ed.* **17**, 782 (1945).
173. R. Parsons, *Trans. Faraday Soc.* **54**, 1053 (1958).
174. J. Tafel, *Z. Phys. Chem.* **50**, 641 (1905).
175. B. E. Conway and J. O'M. Bockris, *J. chem. Phys.* **26**, 532 (1957).
176. C. E. Heath and W. J. Sweeney, *in* "Fuel Cells", ed. W. Mitchell, Jr., Ch. 3. Academic Press, New York (1963).
177. M. Eisenberg and H. Silverman, *Electrochim. Acta* **6**, 93 (1962).
178. E. Justi, M. Pilkuhn, W. Scheibe and A. Winsell, "Hochbelastbare Wasserstoff-Diffusion-Elektroden". Akad. Wiss. Mainz (1960).
179. F. T. Bacon, *Beama J.* **61**, 6 (1954).
180. P. P. Anthony, Ph.D. thesis, Ohio State University (1957).
181. E. Justi and A. Winsell, *Naturwissenschaften* **47**, 289 (1960).
182. G. J. Young and R. B. Rozelle, *in* "Fuel Cells", ed. by G. J. Young, p. 23. Reinhold, New York (1960).
183. W. G. Berl, *Trans. electrochem. Soc.* **83**, 253 (1943).
184. F. T. Bacon, *in* "Fuel Cells", ed. by G. J. Young, p. 51. Reinhold, New York (1960).
185. R. A. Humphrey, Ph.D. thesis, Ohio State University (1955).
186. K. W. Sykes and J. M. Thomas, *J. Chim. phys.* **58**, 70 (1961).
187. J. D. Lambert, *Trans. Faraday Soc.* **34**, 1080 (1938).
188. C. Kroger, *Z. angew. Chem.* **52**, 129 (1939).
189. A. W. Fleer and A. H. White, *Ind. Engng Chem.* **28**, 1301 (1936).
190. F. J. Long and K. W. Sykes, *Proc. R. Soc.* **A193**, 377 (1948).
191. F. J. Long and K. W. Sykes, *J. Chim. phys.* **47**, 361 (1950).
192. F. J. Long and K. W. Sykes, *Proc. R. Soc.* **A215**, 100, 111 (1952).
193. J. D. Blackwood, *Rev. Pure appl. Chem.* **4**, 25 (1954).
194. P. L. Walker, F. Rusinko and L. G. Austin, *Adv. Catalysis* **11**, 133 (1959).
195. X. Duval, *J. Chim. phys.* **58**, 3 (1961).
196. C. Heuchamps, Thesis University of Nancy (1960).
197. H. Amariglio, Thesis University of Nancy (1962).
198. G. R. Hennig, *J. inorg. nucl. Chem.* **24**, 1129 (1962).
199. I. M. Dawson and E. A. C. Follett, *Proc. R. Soc.* **A274**, 386 (1963).
200. J. M. Thomas, *in* "Chemistry and Physics of Carbon", ed. by P. L. Walker, Ch. 3. Dekker Press, New York (1965).

201. J. M. Thomas and P. L. Walker, "Proc. Symp. Carbon", Paper XIII–13, Tokyo (1964).
202. H. Marsh, T. E. O'Hair and R. Reed, *Trans. Faraday Soc.* **61,** 285 (1965).
203. E. A. C. Follett, *Carbon* **1,** 329 (1964).
204. A. E. B. Presland and J. A. Hedley, Proc. 5th Int. Congr. Electron Microscopy, Vol. 1, p. 110. Academic Press, New York (1962).
205. S. Amelinckx and P. Delavignette, *J. nucl. Mater.* **5,** 17 (1962).
206. E. E. G. Hughes, J. M. Thomas, H. Marsh and R. Reed, *Carbon* **1,** 339 (1964).
207. X. Duval, Remarks made at the Gordon Research Conference on Coal Science, New Hampton, U.S.A. (July 1962).
208. G. R. Hennig, "Proceedings of Fourth Conference on Carbon", p. 145. Pergamon Press, New York (1960).
209. N. I. Kobozev, *Acta physicochim. U.R.S.S.* **21,** 294 (1946).
210. H. Harker, "Proceedings of Fourth Conference on Carbon", p. 125. Pergamon Press, New York (1960).
211. J. R. Gallagher and H. Harker, *Carbon* **2,** 163 (1964).
212. J. F. Rakszawski and W. E. Parker, *Carbon* **2,** 53 (1964).
213. E. A. Heintz, W. E. Parker and J. F. Rakszawski, *Carbon* (in press).

CHAPTER 9

Design of Catalytic Reactors

9.1. Statement of the Problem

Many important industrial processes may be classified as heterogeneous gas–solid catalysed reactions. The presence of the solid catalyst accelerates the conversion of reactants into products, but unless a considered choice is made concerning reaction conditions, catalyst geometry and reactor dimensions, the potential value of the process as an economic venture will be reduced. Problems such as the transfer of heat to or from a reactor will substantially influence its ultimate design. Similarly, the rate of transport of gaseous molecules, not only to and from the catalyst surface but also within the complex pore structure of the solid itself, will affect both the dimensions of the reactor unit required and the selection of catalyst particle size. The kinetics of the heterogeneous reaction will obviously determine the mass of catalyst, and hence the reactor volume, which is necessary to achieve a specified conversion within a given time in the case of a batch process, or within a given reactor length in the case of a flow process.

It is the object of reactor design theory to assess the dimensions of reactor required and the amount of catalyst to use for an efficient conversion of a specified input of reactant into a desired product. This may be accomplished provided that certain operating conditions, such as the initial temperature, pressure and reactant concentrations, are chosen and a decision made concerning the type of reactor to be used. For example, a batch or flow reactor may be used in which the conversion may be effected isothermally or adiabatically. Such operating variables constitute the design conditions. Different sets of design conditions will produce different estimates for the size of reactor. The optimum design will be that which is most economical in a pecuniary sense. Specialized mathematical procedures, such as the theory of dynamic programming introduced by Bellman [1], can be employed to optimize the design. In practice, the final choice of operating conditions is often made on the basis of merely a few design calculations. Such calculations depend directly upon (a) the available kinetic data, (b) mass-transfer processes and (c) heat-transfer processes.

9.2. Kinetics

9.2.1. The rate of chemical reaction

For a homogeneous reaction occurring between ideal gases at constant temperature and volume, the rate of chemical reaction may be defined as

$$r_V = -\frac{1}{V}\frac{dn_A}{dt} \tag{1}$$

where V is the constant volume of the closed reaction system, and n_A is the number of moles of a reactant A present at time t. The rate of reaction is some function f of the thermodynamic activities of the reactants. Thus, for an ideal gas,

$$r_V = kf\{a_A\} = kf\left\{\frac{n_A}{n_T}P\right\} = kf\left\{\frac{RT}{V}n_A\right\} \tag{2}$$

where n_T is the total number of moles of the reacting mixture, P the total pressure and T the absolute temperature. Hence, taking a first-order gas reaction as an example, one writes from equations (1) and (2)

$$-\frac{dn_A}{dt} = (kRT)n_A = k_c n_A \tag{3}$$

where k_c is the velocity constant expressed in concentration units.† Equation

† The units of the velocity constant depend on (i) the units in which the rate of reaction is measured and (ii) the units in which concentration or activity are measured. Unless otherwise stated, the units in which the first-order rate constant is expressed are T^{-1} and the subscript after k is dropped forthwith.

(3) is readily integrated and yields the usual expression for the time the reactants take to reach a certain specified conversion. For a heterogeneous reaction, the rate is proportional to the surface area S of the catalyst, so the equation

$$rs = -\frac{1}{S}\frac{dn_A}{dt} \tag{4}$$

is the analogue of equation (1). It is obvious that equation (3) may also be substituted into equation (4) for heterogeneous reactions, but the units of the velocity constant will differ. More frequently the rate is defined per unit mass of catalyst rather than per unit surface area. This is because, for a given catalyst preparation, the surface area is usually directly proportional to the mass of catalyst. But such effects as porosity, degree of sintering and particle size will cause a variation in surface area per unit mass for the same catalyst if it has been prepared and pretreated in different ways. To estimate the extent of the effective surface active in a reaction requires refined techniques in the measurement of chemisorption and it is impracticable to undertake such measurements whilst preparing design data. In the absence of a quick and easily adaptable method, rates based on unit mass of catalyst may be used. Alternatively, rates of reaction based on unit volume of reactor are employed and the volume requirement for reaction calculated. The result is then multiplied by catalyst bulk density to give the mass of catalyst required.

9.2.1.1. Tubular Reactors

The equations for either homogeneous or heterogeneous reactions taking place in a flow system are entirely different from those occurring at constant volume in a batch reactor. In a flow system one can no longer represent any change in the reaction parameter as a function of time. Rather it is expressed as a function of position in the reactor. Consider an infinitesimally small section, thickness δz, of a tubular reactor of uniform cross-sectional area A_c. Assume that the gas flows through the element with a constant volumetric velocity u (corresponding to a linear velocity $u_l = u/A_c$). When the catalyst particle diameter is small in comparison with the tube radius a flat velocity profile obtains in the tube (henceforth referred to as plug-type flow). For this type of flow the value of u_l is independent of the tube radius. The effect of velocity profile and longitudinal mixing on the mean residence time within continuous flow reactors has been discussed fully by Danckwerts [2]. However, the principles upon which the design of such reactors is based is most easily explained, without extraneous complications, by reference to plug-type flow without longitudinal diffusion.

If c_z represents the concentration of reactant at a point z along the axis of flow, then the differential equation representing the conservation of mass within the element is written:

$$\frac{\partial c}{\partial t} A_c \, \delta z = u \frac{\partial c}{\partial z} \delta z - r_v A_c \, \delta z \tag{5}$$

The first term on the right-hand side of the equation represents the net mass flow through the element due to a concentration gradient, while the second

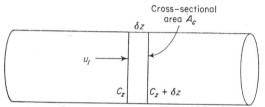

FIG. 1. Conservation of mass in a flow reactor.

term is the mass of reactant removed as a result of reaction in the volume element $A_c \delta z$. Under steady-state conditions these two terms will exactly balance and the term on the left-hand side, representing the accumulation of mass within the element, will be zero. Hence equation (5) becomes:

$$\frac{u}{A_c} \frac{dc}{dz} = u_l \frac{dc}{dz} = r_v \tag{6}$$

or, in terms of the fractional conversion x,

$$F \frac{dx}{dz} = r_v A_c \tag{7}$$

where F is the feed rate to the reactor inlet and is expressed in units of moles per unit time. For a heterogeneous catalytic reaction, if δx is the fractional conversion of feed in an element containing a mass of catalyst δm, equation (7) may be modified, if desired, to

$$F \frac{dx}{dm} = r_m \tag{8}$$

in which the rate is now defined per unit mass of catalyst.

If the reaction involves a change in volume, then equations (6), (7) or (8) cannot be integrated without taking account of such a variation. Suppose that a reactant A is converted to products and that there is a volume change on reaction: this is a direct result of there being a difference in number of moles between products and reactants in the stoichiometric equation defining the overall reaction. Let δ represent the net difference between the number of moles of product and reactant per mole of A in the stoichiometric equation, then

$$\delta = \frac{p + q + \ldots - a - b - \ldots}{a} \tag{9}$$

for a reaction in which a moles of A and b moles of B produce p moles of P and q moles of Q.

If the volume of the reacting system varies linearly with conversion then, we may write

$$n_T = n_T{}^0 (1 - \delta x_A) \tag{10}$$

where $n_T{}^0$ is the initial total number of moles of the reacting system and x_A is the fractional conversion of A in the reactor. Writing down a material balance for component A across the reactor, the number of moles of A may be expressed as a function of the fractional conversion

$$n_A = n_A{}^0 - x_A n_A{}^0 = n_A{}^0 (1 - x_A) \tag{11}$$

where $n^0{}_A$ is the initial number of moles of reactant A at the inlet to the reactor. From equations (10) and (11) the mole fraction of A may be expressed in terms of the fractional conversion and the coefficient δ (which is a measure of the volume change on reaction at constant pressure), so that

$$\frac{n_A}{n_T} = \frac{n_A{}^0(1 - x_A)}{n_T{}^0(1 - \delta x_A)} \tag{12}$$

If pure A is fed to the reactor, then $n_A{}^0 = n_T{}^0$. If the reaction rate is expressed in terms of the moles of reactant converted per unit time per unit mass of catalyst then, for a first-order reaction,

$$r_m = kp_A = \frac{kn_A P}{n_T} \tag{13}$$

where P is the total pressure in the reactor.

Substituting equations (12) and (13) into the integrated form of (8)

$$m = \frac{F}{kP} \int_0^{x_A} \frac{(1 - \delta x_A)}{(1 - x_A)} dx_A = \frac{F}{kP} \left[(1 - \delta) \ln \frac{1}{1 - x_A} - \delta x_A \right] \tag{14}$$

Hence the mass of catalyst required for a given fractional conversion may be calculated from the above equation. One has only to divide this quantity by the bulk density ρ_b of the catalyst to obtain the volume requirement for the reactor. It should be noted that if the stoichiometry is such that δ is zero (no change in volume during reaction), then equation (14) reduces to

$$m = \frac{F}{kP} \ln \frac{1}{1 - x_A} \tag{15}$$

the form of which is similar to that obtained for a first-order reaction in a static system. In general, whatever the form of the rate expression, it may be substituted in either of the fundamental rate equations (8) or (7) and subsequently integrated.

9.2.1.2. The Plug-flow Model and Its Limitations

At this point, it is as well to examine critically those assumptions which are adopted in deriving kinetic equations based on the plug-flow model and to indicate briefly what limitations there are in applying such a model to hetero-geneous catalysed reactions occurring in packed tubular reactors. In Section 9.2.1.1 it was shown how design equations are obtained when the gas flows through the tube with a velocity profile which is uniform across the tube diameter. Such a system can be characterized by a particular residence time (calculated by dividing the volume of the reactor by the volumetric gas velocity) and we are now concerned with those factors which cause the gas to depart from a flat velocity profile, thus producing a situation in which some gas resides in the tube for longer periods than the residence time (as calculated by the plug-flow model), and some gas for shorter periods. Departure from the plug-flow model means that the simple equations derived in the last section cannot be used and must be modified according to the nature of the velocity profile.

Deviations from plug flow may be caused by transverse temperature gradients. If an exothermic reaction occurs in a tubular reactor from which heat is removed by means of some external cooling arrangement, then because the gas at the centre of the tube will be at a higher temperature than at the wall there will be a transverse temperature gradient and the temperature profile will assume a parabolic shape. A flat velocity profile will therefore not obtain in the tube. If the reactor is operated adiabatically, although there is no removal of heat in a radial direction, then because the gas near the wall has a lower velocity than the gas at the centre of the tube, (and has therefore reacted more because of the longer time spent there), for an exothermic reaction the temperature in the centre of the bed will be lower than that at the wall and an inverted parabolic temperature profile will develop. For exothermic reactions carried out under non-adiabatic conditions in which there is heat removal at the tube wall, the effect due to the transverse temperature gradient and the effect due to the velocity profile become superimposed, with the result that the temperature profile assumes a trough near the centre of the tube with a slight maximum near the tube wall. When radial temperature gradients do exist, because the rate of reaction varies exponentially with the reciprocal of absolute temperature there is likely to be a considerable variation in reaction rate across the tube diameter (a factor which can be of the order of 4000 or more for most simple reactions). However, approximate methods are available for dealing with design data when there are transverse tempera-ture gradients and these are dealt with in Section 9.3.2. Catalyst particles with high thermal conductivity and low porosity tend to reduce these adverse effects. Only when it is certain that there is a good approximation to isothermal conditions can temperature gradients in the radial and longitudinal directions be ignored and the plug-flow model applied with reasonable confidence.

Transverse and longitudinal diffusion and convection also cause deviations from the plug-flow model. While longitudinal diffusion tends to encourage gas to spend less time in the reactor than it would do had plug-flow conditions been maintained (because there is a concentration gradient from reactor inlet to exit and this enhances flow in the direction of bulk flow), transverse diffusion tends to reduce any variation in concentration across the diameter and therefore brings the reactor performance nearer to the plug-flow model. In packed tubular reactors, provided that the tube length is more than about 100 particle diameters, longitudinal and radial diffusion effects may be safely neglected. Section 9.3.1 shows how longitudinal and radial dispersion effects may be incorporated in some design equations.

A further effect which usually invalidates the assumption of plug flow is the existence of a transverse velocity gradient due to the viscous drag effect at the tube wall. In empty tubes a parabolic velocity profile obtains for laminar-type flow, but in packed tubes, since the gas velocity is reduced to zero at the surface of particles, the effect of the packing is to smooth out the velocity profile across the tube. Thus the distribution of velocities across the tube diameter is such that there is a fair approximation to plug flow provided the diameter of the tube is at least 30 particle diameters.

9.2.2. The overall rate of reaction

We recall that the observed net rate in a catalytic reaction depends on the relative rates of five quite distinct processes. The correct rate expression to substitute in equations (6), (7) or (8) will therefore be determined by the overall rate. The mass transport of gases to and from the surface and also into the interior pore structure of the solid must be taken into account. To allow for the influence of diffusion in pores, the rate expression is simply multiplied by a fraction termed the effectiveness factor (see p. 476). The way the rate equation is modified to account for gas-phase diffusion is discussed in Section 9.2.3. Only the three processes of adsorption, surface reaction and desorption are considered in this section.

On the basis of the rate equations developed in Chapter 2 suitable expressions may be obtained for substitution into the rate equations (7) or (8). As a result, for reactions occurring in a flow system, a relationship obtains between the mass of catalyst to use in a given volume of reactor and the conversion of the reactant into product. Once again it is important to stress the difference from the case of a static system. In a flow system, for instance, the rate of adsorption of reactant does not steadily decrease to zero after a sufficient time has elapsed for all the vacant active sites on the surface to be occupied.

TABLE 9.1.

Reaction		Mechanism	Basic rate equation for mechanism	Driving force	Adsorption term
1. $A \rightleftharpoons P$	(i)	Adsorption of A controls rate	$r = k\left(p_A c_V - \dfrac{c_A}{K}\right)$	$p_A - \dfrac{p_P}{K}$	$1 + \dfrac{K_A}{K}p_A + K_P p_P$
	(ii)	Surface reaction controls rate, single site mechanism	$r = kc_A - k'c_P$	$p_A - \dfrac{p_P}{K}$	$1 + K_A p_A + K_P p_P$
	(iii)	Surface reaction controls rate, adsorbed A reacts with adjacent vacant site	$r = kc_A c_V - k'c_P$	$p_A - \dfrac{p_P}{K}$	$(1 + K_A p_A + K_P p_P)^2$
	(iv)	Desorption of P controls rate	$r = k\left(\dfrac{c_P}{K_P} - p_P c_V\right)$	$p_A - \dfrac{p_A}{K}$	$1 + \dfrac{K_A}{K}p_A + K_P p_P$
	(v)	A dissociates when adsorbed and adsorption controls rate	$r = k\left(p_A c^2{}_V - \dfrac{c_A}{K_A}\right)^2$	$p_A - \dfrac{p_P}{K}$	$\left(1 + \left(\dfrac{K_A}{K}p_A\right)^{1/2} + K_P p_P\right)^2$
	(vi)	A dissociates when adsorbed and surface reaction controls rate	$r = kc_A{}^2 - k'c_P c_V$	$p_A - \dfrac{p_P}{K}$	$\left(1 + (K_A p_A)^{1/2} + K_P p_P\right)^2$
2. $A + B \rightleftharpoons P$ Langmuir-Hinshelwood mechanism (adsorbed A reacts with adsorbed B)	(i)	Adsorption of A controls rate	$r = k\left(p_A c_V - \dfrac{c_A}{K_A}\right)$	$p_A - \dfrac{p_P}{Kp_B}$	$1 + \dfrac{K_A p_P}{Kp_B} + K_B p_B + K_P p_P$
	(ii)	Surface reaction controls rate	$r = kc_A c_B - k'c_P c_V$	$p_A p_B - \dfrac{p_P}{K}$	$(1 + K_A p_A + K_B p_B + K_P p_P)^2$

		$r=$	driving force	adsorption term
(iii)	Desorption of P controls rate	$r = k\left(\frac{c_P}{K_P} - p_P c_V\right)$	$p_A p_B - \frac{p_P}{K}$	$1 + K_A p_A + K_B p_B + KK_P p_A p_B$
(iv)	A dissociates when adsorbed only half of which react and adsorption of A controls rate. A adsorbed on dual site	$r = k\left(p_A c_V^2 - \frac{c_A}{K'_A}\right)$	$p_A - \left(\frac{p_P}{K p_B}\right)^2$	$\left(1 + \frac{K_A^{1/2}}{K}\cdot\frac{p_P}{p_B} + K_B p_B + K_P p_P\right)^2$
(v)	B dissociates when adsorbed only half of which react and adsorption of A controls rate. A adsorbed on dual site	$r = k\left(p_A c_V - \frac{c_A}{K_A}\right)$	$p_A - \frac{p_P}{K\sqrt{p_B}}$	$1 + \frac{K_A}{K}\frac{p_P}{\sqrt{p_B}} + (K_B p_B)^{1/2} + K_P p_P$
(vi)	A dissociates when adsorbed and adsorption of A controls rate	$r = k\left(p_A c_V^2 - \frac{c_A}{K_A}\right)^2$	$p_A - \frac{p_P}{K p_B}$	$\left(1 + \left(\frac{K_A p_P}{K p_B}\right)^{1/2} + K_B p_B + K_P p_P\right)^2$
(vii)	A dissociates when adsorbed and surface reaction controls rate	$r = kc_A^2 c_B - k' c_P c_V$	$p_A p_B - \frac{p_P}{K}$	$(1 + (K_A p_A)^{1/2} + K_B p_B + K_P p_P)^3$

Note that:

1. The expression for the rate of reaction in terms of partial pressures is proportional to the driving force divided by the adsorption term.

2. To derive the corresponding kinetic expressions for a bimolecular–unimolecular reversible reaction proceeding via an Eley-Rideal mechanism (adsorbed A reacts with gaseous or physically adsorbed B), the term $K_B p_B$ should be omitted from the adsorption term. When the surface reaction controls the rate the adsorption term is not squared and the term $K_B p_B$ is omitted.

3. To derive the kinetic expression for irreversible reactions simply put K large, thus omitting the second term in the driving force.

4. If two products are formed (P and Q) the equations are modified by (a) multiplying the second term of the driving force by p_Q, and (b) adding $K_Q p_Q$ within the bracket of the adsorption term.

5. For reversible reactions bimolecular in both directions mechanisms 2(vi) and 2(vii) are improbable, since they require that the reverse reactions occur by the products P and Q forming an activated complex with three active vacant sites.

On a given part of the surface (a given position along the length of the reactor) the rate of adsorption will be constant for steady-state conditions. It will be sufficient, in the treatment which follows, to develop equations for the rates of adsorption, surface reaction and desorption for one type of reaction and express these equations in a form suitable for design calculations. Equations for other reaction mechanisms may be deduced analogously (see Table 9.1).

Consider, for heuristic purposes, an equilibrium reaction

$$A + B \underset{k'}{\overset{k}{\rightleftharpoons}} P \tag{16}$$

which occurs at the surface of a catalyst. The net rate of reaction is the difference between the rates of the forward and reverse reactions. If it is assumed that, in the case under consideration, the forward rate is determined by the simultaneous presence of adsorbed A and B molecules, i.e. a Langmuir-Hinshelwood mechanism applies, then this means that the rate is proportional to the number, N_{AB}, of pairs of adjacent sites occupied by A and B and this is given by the product of the number N_A of adsorbed A molecules and the fraction of adjacent sites occupied by B, $\theta_B/(1-\theta_A-\theta_B)$.

If we let c_A represent the number of molecules of A adsorbed per unit area of catalyst, then, as in Chapter 2, this is equivalent to a surface concentration

$$c_A = \frac{N_A}{S} \tag{17}$$

where N_A is the number of molecules of type A absorbed on a surface of area S. Thus the concentration c_{AB} of pairs of sites occupied by A and B is

$$c_{AB} = c_A \frac{\theta_B}{1 - \theta_A - \theta_B} \tag{18}$$

where θ is the fraction of available surface covered. If the fraction of the surface occupied by A and B is comparatively small, then the right-hand side of the equation reduces to $c_A\theta_B$. Just as the surface concentration of A or any other adsorbed atoms or molecules can be expressed by an equation such as (17), so the total concentration of active sites and the total concentration of vacant sites may be written

$$c_S = \frac{N_S}{S} \tag{19}$$

and

$$c_V = \left(N_S - \frac{\sum_j N_j}{S}\right) \tag{20}$$

respectively, where $\sum_j N_j$ is the total number of adsorbed species j of any kind occupying an area S of catalyst. For low coverage the rate of the forward reaction is therefore

$$kN_A \frac{\theta_B}{1 - \theta_A - \theta_B} = kS\, c_A \frac{\theta_B}{1 - \theta_A - \theta_B} \approx kS \frac{c_A c_B}{c_S} \tag{21}$$

The reverse rate of the surface reaction is proportional to the number of pairs of sites formed by adsorbed product molecules and adjacent vacant sites and is therefore

$$k'N_P(1 - \sum_j \theta_j) = \frac{k'S\, c_P c_V}{c_S} \tag{22}$$

Hence the net rate of the surface reaction will be

$$r_S = k_S \frac{c_A c_B}{c_S} - k'_S \frac{c_P c_V}{c_S} \tag{23}$$

where k_S and k'_S is now written in place of kS and $k'S$.

In the case of an Eley-Rideal mechanism the rate of the surface reaction is

$$r_S = k_S c_A a_B - k'_S \frac{c_P c_V}{c_S} \tag{24}$$

where a_B is the activity of the B molecules which may either be physically adsorbed or remain in the gas phase.

Writing the rate of adsorption of a reactant A in the form

$$r_\alpha = k_A p_A (1 - \sum_j \theta_j) \tag{25}$$

and expressing the free surface in terms of the surface concentration of vacant sites, this becomes

$$r_\alpha = k_A p_A c_V \tag{26}$$

Similarly, the rate of desorption of A from the surface will be

$$r_\delta = k'_A c_A \tag{27}$$

If, as in many catalytic reactions, one or more of the reactants is reversibly adsorbed at the surface, then there will be an equilibrium between adsorbed and desorbed molecules. The net rate of adsorption of A will therefore be

$$r_a = r_\alpha - r_\delta = k_A p_A c_V - k'_A c_A \tag{28}$$

Identical expressions are applicable to other reactants reversibly adsorbed. It is also obvious how an adsorption equation in terms of surface concentrations would be rewritten, say, for the dissociative adsorption of A_2. Hougen and Watson [3] have discussed the various cases in detail. The third column of Table 9.1 gives examples of various basic rate equations for some selected mechanisms.

In an exactly analogous way the rate of desorption of product P from the surface may be written

$$r_d = k'_P c_P - k_P p_P c_V \tag{29}$$

The overall rate of the surface reaction can obviously be expressed in terms of the partial pressures of reactants and products by elimination of the terms c_A, c_B and c_P through equations such as (27) and (28) and substitution into equation (23). However, the final equation obtained is unmanageable, containing several constants, which, for practical reasons, cannot be determined independently. It should also be noted that the partial pressures in equations such as (28) and (29) are the values corresponding to a position immediately adjacent to the surface. If there is no resistance to diffusion, then the partial pressures will be those for the bulk gas phase. If there is resistance to diffusion, then one has to calculate the correct value of partial pressures in the bulk of the gas, as described in Section 9.2.3.

The overall reaction rate is usually determined by the slowest reaction occurring in the whole sequence of rate processes constituting the reaction mechanism. Consider the bimolecular reaction (16) occurring by an Eley-Rideal mechanism and further consider three cases in which either one of the rate processes, adsorption, surface reaction, or desorption is slow compared to the other two and hence determines the overall rate.

9.2.2.1. *Adsorption of Reactant A the Slowest Rate Process*

Under such circumstances the rates of the other processes, adsorption (and desorption if reversible) of B, the surface reaction and the desorption (and adsorption if reversible) of product P will be rapid compared with the slow adsorption of A. It is therefore not a drastic assumption to write the surface concentrations of B and P in terms of the equilibria resulting from such comparatively rapid adsorption and desorption. Hence

$$k_B p_B c_V = k'_B c_B \tag{30}$$

from which

$$(c_B)_e = K_B p_B c_V \tag{31}$$

where K_B is the ratio k_B/k'_B and the subscript e now refers to an equilibrium surface concentration. Similarly

$$(c_P)_e = k_P p_P c_V \tag{32}$$

Equating the forward and reverse rates for the surface reaction gives

$$c_A = \frac{c_P c_V}{K_S c_B} \tag{33}$$

Substituting these surface concentrations into equation (28) for the net rate of adsorption which, for this case, is the rate-determining process,

$$r_a = k_A c_V \left\{ p_A - \frac{K_P}{K_A K_B} \frac{1}{K_S} \frac{p_P}{p_B} \right\} \tag{34}$$

Making use of equations (31), (32) and (33), it follows that the equilibrium constant K for the homogeneous reaction can be inserted in place of the equilibrium constants for adsorption and surface reaction. Equation (34) then becomes

$$r_a = k_A c_V \left\{ p_A - \frac{1}{K} \frac{p_P}{p_B} \right\} \tag{35}$$

Now the total number of adsorption sites can be written in terms of the number of vacant and occupied sites. Thus

$$N_S = N_V + N_A + N_B + N_P \tag{36}$$

and so

$$c_S = c_V + c_A + c_B + c_P \tag{37}$$

From the expressions (31), (32) and (33) for the surface concentrations of B, P and A it follows that the term c_V is

$$c_V = \frac{c_S}{1 + K_B p_B + K_P p_P + \dfrac{K_A}{K} \dfrac{p_P}{p_B}} \tag{38}$$

The rate equation is therefore

$$r_a = \frac{k_A c_S \left(p_A - \dfrac{1}{K} \dfrac{p_P}{p_B} \right)}{1 + K_B p_B + K_P p_P + \dfrac{K_A}{K} \dfrac{p_P}{p_B}} \tag{39}$$

New experimental techniques have enabled adsorption and desorption to be measured in a flow apparatus at high temperatures and as reaction is actually proceeding [4]. Hence separate evaluations of K_A, K_B and K_P are possible. The partial pressures of A, B and P can be expressed in terms of the fractional conversion of reactant entering the reactor. For example, if the reactants are fed to the reactor in equimolar proportions and at a total pressure P then the partial pressures in terms of x are

$$p_A = \left\{ \frac{1-x}{2-x} \right\} P = p_B, \text{ and } p_P = \left\{ \frac{x}{2-x} \right\} P \tag{40}$$

It is quite obvious that a final rate equation such as (39), in terms of the conversion, can therefore be substituted into the rate equation (8) and integrated either analytically or graphically.

9.2.2.2. Surface Reaction Slow Compared to Adsorption and Desorption

If the net rate of adsorption of reactant and the net rate of desorption of product are assumed to be fast in comparison with the interaction of the adsorbed species at the surface, then, proceeding as before, the equilibrium surface concentration for A, B and P can be substituted into the rate equation (24) for the surface reaction. After eliminating c_V, the final result is

$$r_S = \frac{k_S c_S K_A K_B \left\{ p_A p_B - \frac{1}{K} p_P \right\}}{(1 + K_A p_A + K_B p_B + K_P p_P)^2} \tag{41}$$

9.2.2.3. Desorption the Slowest Rate Process

In this case all the rate processes except that for desorption of the product are considered to be at equilibrium. The rate equation corresponding to such a mechanism is

$$r_d = \frac{k_d c_S K \left(p_A p_B - \frac{p_P}{K} \right)}{1 + K_A p_A + K_B p_B + K_P p_P + K_P K p_A p_B} \tag{42}$$

9.2.2.4. General Rate Equations

There are, of course, numerous mechanisms by which catalysis can occur. For example, a Langmuir-Hinshelwood or an Eley-Rideal mechanism (see Chapters 1 and 2) may operate and any one of the rate processes, adsorption, surface reaction or desorption may be rate controlling. A mechanism involving dissociation of the adsorbate is quite common in catalysis and either unimolecular or bimolecular surface reactions can occur. The rate equation corresponding to any one particular mechanism may be deduced by similar methods to the three cases considered in the previous section. In each case the rate expression involves a kinetic term multiplying the ratio of (a) a term in the numerator which is a measure of the driving force encouraging the reaction to proceed, and (b) a term in the denominator which is a measure of the retardation due to the adsorption of reactants and/or products. Table 9.1 shows, for some selected mechanisms, the form of the overall rate expression obtained in relation to the fundamental rate equation assumed for the mechanism under consideration.

In principle it is then necessary only to substitute the appropriate rate equation into an equation of continuity such as (7) or (8); an integration (usually numerical) will then provide an answer representing the mass of catalyst necessary for a specified conversion.

9.2.2.5. Initial Rates as a Function of Pressure

Although the application of rate data to the design of catalytic reactors has been emphasized, it is as well to point out that equations such as (39), (41) and (42) may give a useful lead to the reaction mechanism. For example, if the reactants are present initially in equimolar proportions and we are interested in initial reaction rates r^0, then equation (39) reduces to the form

$$r_a{}^0 = \frac{k_A c_S p_A{}^0}{1 + K_B p_A{}^0} = \frac{a_1 p_A{}^0}{1 + b_1 p_A{}^0} \tag{43}$$

where a_1 and b_1 are constants. The total pressure

$$P = p_A{}^0 + p_B{}^0 = 2p_A{}^0 \tag{44}$$

may conveniently be regarded as the reaction parameter if P happens to be a quantity which can be easily measured. Similarly equation (41) reduces to

$$r_S{}^0 = \frac{k_S c_S K_A K_B (p_A{}^0)^2}{\{1 + (K_A + K_B)p_A{}^0\}^2} = \frac{a_2 (p_A{}^0)^2}{(1 + b_2 p_A{}^0)^2} \tag{45}$$

Equation (43) reduces to

$$r_a{}^0 = \frac{k_a c_S K (p_A{}^0)^2}{\{1 + (K_A + K_B)p_A{}^0 + K K_P (p_A{}^0)^2\}} \tag{46}$$

and if K is large compared to K_A, K_B and K_P

$$r_a{}^0 = \frac{k_a c_S}{K_P} = a_3 \tag{47}$$

Figure 2 shows the dependence of initial rate on total pressure for the three cases considered. Other mechanisms will give different functions, though they may be difficult to distinguish graphically. Nevertheless, the study of initial rates as a function of total pressure often eliminates certain mechanisms unequivocally. Thaller and Thodos [5] have reported an example in which the reaction rate is determined by one of two mechanisms depending on the

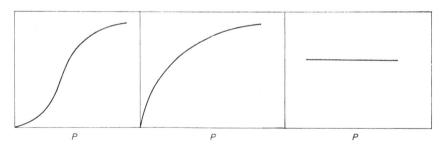

FIG. 2. Initial rates as a function of total pressure. Left, equations (45) and (46), centre, equation (43); right, equation (47).

temperature. Over a certain range of temperature the rate could not be described by using either one of the kinetic equations independently. This situation was carefully considered by Bischoff and Froment [6], who wrote down the general equations for each step in a model process. The model adopted was a unimolecular catalytic decomposition,

$$A \rightleftharpoons P + Q \tag{48}$$

By considerations similar to those outlined above they obtained a nonlinear rate expression in terms of the partial pressures of A, P and Q. Following Yang and Hougen [7], the equation was reduced to the form of an initial rate equation. The relations they obtained were similar in form to equation (46) (desorption the slowest rate process), and to equation (45) (surface reaction slow). However, it was clearly shown that if more than one step in the overall process is considered slow in comparison with other steps then the initial rate curves are easily confused with those obtained when only one step is rate determining.

From the previous remarks concerning the various rate equations applicable to different mechanisms, it is obvious that a marathon research programme would have to be undertaken to obtain sufficient precise data before the reaction mechanism could be deduced. This is not only because of the number of parameters that may be arbitrarily chosen for each rate equation, but also because a least squares method of determining the constants in the equation may yield more than one function which fits the data. Firstly it should be recognized that there is often insufficient time available for an extensive research programme prior to designing a commercial reactor. Secondly, even if sufficient time is available, one need not necessarily be concerned with reaction mechanism, for the prime object in reactor design is to obtain an empirical rate equation which describes adequately the effect of the important variables. When designing reactors, therefore, it is better to select the simplest form of rate equation which closely fits the experimental results over the desired range of operation.

9.2.3. The effect of mass transfer through the gas phase to the surface

For a reactant to be adsorbed or for a product to be desorbed, the species in question must be transported through the gas phase to or from the catalyst surface. The rate at which this occurs depends on conditions of temperature, pressure and gas velocity relative to the surface. When conditions are not turbulent, the rate of mass transfer may be relatively slow and would then actually impede the progress of the reaction. In industrial reactors, this situation is to be avoided, since, at atmospheric pressure and above, the slowest process is often molecular diffusion. When the reaction occurs in a flow system, the gas velocity is usually large enough for mass transport to occur by turbulent diffusion. Under such circumstances the overall reaction rate is usually independent of the rate of mass transport. If N is the rate of mass transfer per unit area and k_D is a mass-transfer coefficient, then N may be written in terms of the driving force effecting mass transfer. The driving force will be the partial pressure difference existing between the gas phase and the gas–solid interface. Thus

$$N = k_D(p - p_i) \tag{49}$$

where the partial pressures in the gas phase and at the interface are p and p_i respectively.

From equation (49), k_D may be regarded as the rate at which mass is transferred per unit external surface area of catalyst per unit partial pressure difference between the bulk gas and the interface. This mass-transfer coefficient may, in some circumstances, be measured [8] or, alternatively, as will emerge later, correlated in terms of the physical parameters of the system [9].

If, for a given component, $(p-p_i) \approx p$, then partial pressures in the gas phase can be used for substitution into the rate equation. At the other extreme, diffusion to and from the surface may be very slow compared with adsorption, surface reaction and desorption. There are obviously many cases lying somewhere between these régimes.

Consider a unimolecular reaction

$$A \rightleftharpoons P \tag{50}$$

in which the diffusion of reactant and product are slow compared with the other rate processes. As before, the surface reaction is assumed to be at equilibrium so that

$$\frac{c_P}{c_A} = K_S \tag{51}$$

Since the surface concentrations c_A and c_P of A and P respectively, are directly proportional to the respective partial pressures p_{Ai} and p_{Pi} at the interface, then

$$\frac{p_{Pi}}{p_{Ai}} = K_S \tag{52}$$

Under steady-state conditions in a flow system, the rate of reaction at any point along the length of the reactor will be balanced by the rates of mass transfer of reactants and products. Hence the overall rate, measured in units of moles per unit mass per unit time, is given by:

$$r = (k_D)_A \, S_x(p_A - p_{Ai}) = - (k_D)_P \, S_x(p_P - p_{Pi}) \tag{53}$$

where S_x is the area available for mass transfer, in this case, the external surface area of the catalyst. Eliminating p_{Ai} and p_{Pi},

$$r = \frac{1}{\dfrac{1}{(k_D)_A \, S_x} + \dfrac{1}{K_S(k_D)_P \, S_x}} \left(p_A - \frac{p_P}{K_S} \right) \tag{54}$$

If K_D is considered as a diffusion coefficient defined by

$$\frac{1}{K_D} = \frac{1}{(k_D)_A} + \frac{1}{K_S(k_D)_P} \tag{55}$$

the overall rate is

$$r = K_D S_x \left(p_A - \frac{p_P}{K_S} \right) \tag{56}$$

The corresponding expression for a bimolecular process is obtained in a similar way, but the form of the equation is much more complicated.

In the steady state the diffusion rate balances the rate of chemical reaction, so that equation (53) becomes

$$k_D S_x(p_A - p_{Ai}) = k p_{Bi} \tag{57}$$

where k is the first-order chemical rate constant expressed in units appropriate to those in which the overall reaction rate is measured. Eliminating p_{Bi}, the net rate is

$$r = \frac{1}{\frac{1}{k_D S_x} + \frac{1}{k}} p_B \tag{58}$$

The condition that mass transfer through the gas phase may be neglected in comparison with chemical reaction is therefore

$$k \ll k_D S_x \tag{59}$$

To obtain numerical estimates of the relative importance of mass transfer through the gas phase, one or more of the following three methods may be used.

9.2.3.1. *The Resistance Method for Estimating the Effect of slow Mass Transfer through the Gas Phase*

Suppose the chemical reaction rate r is represented by a first-order kinetic equation. Under steady-state conditions this rate will be balanced by the rate of mass transfer,

$$k_D S_x(p - p_i) = r = k p_i \tag{60}$$

By eliminating p_i, the net reaction rate is

$$r = \frac{1}{\frac{1}{k_D S_x} + \frac{1}{k}} p = k_0 p \tag{61}$$

where k_0 is termed the overall rate coefficient. Thus, both transport through the gas phase and chemical reaction may be regarded as offering a resistance to diffusion.

When $k_D S_x \to \infty$ the reaction rate is controlled by the chemical reaction, and when $k_0 \to \infty$ the rate is limited by diffusion. For a bimolecular reaction with second-order kinetics the rate of mass transfer is (see equation (49))

$$N S_x = k_D S_x(p - p_i) \tag{62}$$

and the chemical reaction rate

$$r_c = k p_i^2 \tag{63}$$

Eliminating p_i then gives an expression for the net rate

$$r = k_D S_x \left\{ 1 + \frac{k_D S_x}{2kp} + \sqrt{\left[\left(1 + \frac{k_D S_x}{2kp} \right)^2 - 1 \right]} \right\} p \qquad (64)$$

This equation may be written in a more convenient form

$$r = k_D{}^* S_x p \qquad (65)$$

but it must be remembered that, in this case, $k^*{}_D$ is itself dependent on the partial pressure of reacting gas.

For substitution into a rate equation the constants must be known. Generally, the chemical rate constant will have been determined under conditions where mass transfer is unimportant, while the mass-transfer coefficient is either measured [8, 9] or estimated from empirical correlations discussed later. The reverse problem of evaluating the rate coefficients from experimental data is much more difficult, except, of course, when either mass transport to the surface or chemical reaction is unequivocally the rate-determining process. For design purposes, therefore, the equations are easily solved. Even if a quite complicated mechanism is invoked, involving consecutive nth order reactions, the problem can be solved. The numerical problem is a difficult and tedious one, but modern high-speed digital computers facilitate a rapid solution. In general, for any reaction j, the rate of mass transfer is balanced by the chemical reaction rate,

$$k_D S_x (p - p_i) = k p_i{}^n \qquad (66)$$

and therefore

$$r_j = k p_i{}^n = k \left(p - \frac{r_j}{k_D S_x} \right)^n \qquad (67)$$

Values of $r_j = r_j(k, k_D S_x, p)$ can be found by numerical methods using the above equation, and the method may be extended to any number of consecutive reactions.

9.2.3.2. *The Correlation Method for Estimating the Effect of Mass Transfer through the Gas Phase*

If both the overall reaction rate r and the value of the mass-transfer coefficient k_D is known, the driving force can be calculated from an equation such as (60). The numerical value of $p - p_i$, which is the pressure difference between the bulk gas phase and the interface, will then determine whether or not mass transfer through the gas phase is a rate process which should be considered in comparison with rates of adsorption, surface reaction and desorption. If $(p - p_i)/p \ll 1$ then resistance due to mass transfer is negligible. On the other hand, if $p - p_i$ is a large fraction of p, then mass transfer is important and values of the partial pressure at the interface should be used when calculating overall reaction rates.

If the mass-transfer coefficient can be calculated, then it is clear that one may reliably estimate the effect of diffusion. The mass-transfer coefficient depends entirely on the physical properties of the system. The way in which the solid is packed, the nature of the diffusing component, and other properties of the gas such as viscosity and density, will determine its value. If turbulent gas conditions prevail within the reactor, then mass transport will be facilitated, but stagnant conditions which exist in batch-type reactors will impede mass transfer. Several investigators [10, 11, 12, 13] have obtained empirical correlations of k_D for different gases in fixed bed flow systems. Other correlations [14, 15] have been found for fluidized beds.

Colburn [16] and Chilton [17] combined dimensionless groups of the variables affecting the prevailing physical conditions in flow systems and defined a mass-transfer factor,

$$j_D = \left(\frac{k_D M}{G} p_f\right) \left(\frac{\eta}{\rho \bar{D}_i}\right)^{2/3} \tag{68}$$

M is the mean molecular weight, $G(=\rho u_i)$ the mass flowrate per unit cross-sectional area, η the gas viscosity, $\bar{D}_i$ an average diffusivity for component i in the presence of all other components in the system, and ρ the absolute gas density. The term p_f is often referred to as the drift factor which, for equimolal diffusion, is equal to the total pressure P. The value of p_f is obtained by computing the logarithmic mean of $(P+\delta_j p_j)$, where p_j is the partial pressure of component j and δ_j is the net difference in the number of moles of products and reactants per mole of j present. The full significance of p_f for complex gas mixtures in which reaction is occurring has been reviewed by Hougen and Watson [18]. The second dimensionless group on the right of the equation is the Schmidt group, designated (Sc), which depends on the ratio of viscous and diffusive forces. Hougen and Wilke [19] obtained a correlation between the mass-transfer factor and the dimensionless Reynolds group describing conditions of flow in a packed bed. They showed that, for particles of a given shape,

$$j_D = f\left(\frac{d\rho u_i}{\eta}\right) = f(Re) \tag{69}$$

where the group (Re) is defined in terms of the mass flow $G(=\rho u_i)$ of gas per unit area, the diameter d of the solid particles and the gas viscosity η, following the classical experiments of Reynolds. The precise function to use in equation (69) depends on the flow conditions [20]. For Reynolds numbers greater than 350 in beds of spherical particles,

$$j_D = 1 \cdot 06 (Re)^{-0 \cdot 41} \tag{70}$$

and for Reynolds numbers less than 350

$$j_D = 1 \cdot 82 (Re)^{-0 \cdot 51} \tag{71}$$

It is quite evident that if the mass-transfer coefficient is defined as in equation (49), then from equations (68) and (69),

$$(p - p_i) = \frac{r}{k_D} = r \frac{M p_f}{G} \frac{(Sc)^{2/3}}{j_D} \tag{72}$$

Provided that an estimate can be obtained for the two dimensionless groups, the drift factor p_f and the reaction rate, a numerical value for $(p - p_i)$ is obtained. If experimental rate data is available in the form of an observed rate for a given conversion and partial pressure, then the value of r for a given p will give the value of p_i. In this way the importance of mass transfer through the gas phase can be assessed.

9.2.3.3. *The Reactor Unit Method for Estimating the Effect of Mass Transfer through the Gas Phase*

Hurt [10] developed a method in which the size of a reactor required to achieve a specified conversion may be calculated and the effects of diffusion and surface reaction are separable. The separate contributions to the total reactor size may then be compared and their relative importance assessed. The method has its limitations, and is only applicable to first-order and psuedo-first-order reactions. Its chief merit is that it provides both a method which is easy to use and a concept of physical magnitudes involved.

Consider an element, thickness δz, cross-sectional area A_c, of a reactor packed with a catalyst having a bulk density ρ_b. The mass of catalyst in this elementary volume is

$$m = \rho_b A_c \delta z \tag{73}$$

On the basis of equation (49) the number of moles of a gaseous component transferred within the element is

$$NS_x m = k_D(p - p_i) S_x m = k_D S_x (p - p_i) \rho_b A_c \delta z \tag{74}$$

where S_x is the external surface area per unit mass of the catalyst, not including the area of pores, and ρ_b is the bulk density of the catalyst. From the ideal gas laws a change δn in number of moles may be written in terms of a corresponding partial pressure change and the total number of moles n_T. Thus

$$\delta n = \delta \left(n_T \frac{p}{P} \right) \tag{75}$$

Adapting equation (75) to the case of a flowing gas, then if F is the total molar flowrate of the gas it follows that the change in molar flowrate of a given component is written

$$\delta \left(F \frac{p}{P} \right) = \delta \left(G_m A_c \frac{p}{P} \right) \tag{76}$$

where G_m is the total molar flowrate per unit cross-sectional area A_c. This change in molar flowrate through the cross-sectional area A_c is balanced by mass transfer. Hence,

$$\delta \left\{ G_m A_c \frac{p}{P} \right\} = k_D S_x (p - p_i) \, \rho_b A_c \delta z \tag{77}$$

If a change in molar volume occurs on reaction, G_m must first be expressed in terms of the partial pressure of reactant before equation (77) can be integrated. If δ is the change in number of moles on reaction per mole of a particular reactant then

$$G_m = G_m{}^0 \frac{\left(1 + \dfrac{\delta p^0}{P} \right)}{\left(1 + \dfrac{\delta p}{P} \right)} \tag{78}$$

where the superscript zero refers to conditions at the reactor inlet.

For a reaction in which there is no change in molar volume, G_m is constant, so equation (77) when integrated gives

$$\int_{p_1}^{p_2} \frac{dp}{(p - p_i)} = \frac{\rho_b k_D S_x M P}{\rho u_t} z \tag{79}$$

where ρ is the gas density. The limits of integration represent the partial pressures in the bulk gas phase at two points in the reactor separated by a distance z along the reactor length. The expression inside the integrand in this equation represents the ratio of the change in partial pressure to the driving force tending to effect such a change. Chilton and Colburn [17], in an analysis of mass-transfer processes in distillation and absorption columns, defined this quantity as the number of transfer units, N_T. It is a measure of the resistance to mass transfer. The group $\rho_b k_d S_x M P / \rho u_t$ has the dimensions of reciprocal length and is defined as the height of a transfer unit, H_T. Equation (79) can therefore be represented,

$$z = H_T N_T \tag{80}$$

Provided that one is dealing with a first-order reaction involving a single reactant and product, a similar approach is possible for dealing with the resistance to chemical reaction. It may therefore be represented by an equation comparable to (80). If either one of the rate processes adsorption, surface reaction or desorption is the rate-determining equation describing the effective chemical reaction

$$A \rightleftharpoons P \tag{81}$$

first order in each direction, then, adopting the procedure outlined in Section 9.2.2, the rate can very often be represented in terms of a linear driving force.

For example, if the surface reaction is the important rate process, the equation analogous to (41) for the above case is

$$r_S = \frac{k_S c_S K_A \left(p_A - \frac{1}{K} p_P \right)}{1 + K_A p_A + K_P p_P} \tag{82}$$

and if both K_P and K_A are small, this reduces to

$$r_S = k_S c_S K_A \left(p_A - \frac{1}{K} p_P \right) = \bar{k}_S \left(p_A - \frac{1}{K} p_P \right) \tag{83}$$

where $\bar{k}_S$ represents the value of $k_S c_S K_A$. Introducing the symbol $p_A{}^*$, which is defined as the partial pressure of A in equilibrium with a pressure p_P of product P,

$$p_A{}^* = \frac{p_P}{K} \tag{84}$$

then the rate may be expressed in the form of a linear driving force

$$r_S = \bar{k}_S (p_A - p_A{}^*) \tag{85}$$

One difficulty now arises. The partial pressure of A to use in these equations should be the partial pressure p_{Ai} at the interface, since mass transfer, by definition, is, in the case under discussion, considered a slow process. But $p_A{}^*$ is related to the partial pressure of the product in the gas phase. If the reaction is an irreversible one, $p_A{}^*$ is small in comparison with p_A and p_{Ai}, and in this case either value may be used.

Equating the change in molar flowrate to the extent of reaction within the element δz, then for a reaction in which there is no change in molar volume,

$$G_m A_c \frac{\delta p}{P} = r_S \rho_b A_c \delta z = \bar{k}_S (p_i - p^*) \rho_b A_c \delta z \tag{86}$$

Over a distance z in the reactor,

$$\int_{p_1}^{p_2} \frac{dp}{p_i - p^*} = \frac{\bar{k}_S \rho_b P}{G_m} z \tag{87}$$

and, by analogy with equation (80),

$$z = H_C N_C \tag{88}$$

N_C is termed the number of catalytic units; it represents the integral of the ratio of the change in partial pressure to the chemical driving force effecting such a change. H_C has the dimensions of length and is termed the height of a catalytic unit.

If the overall reaction is regarded as the total change in moles of a constituent and is the sum of the contributions from changes produced by mass transfer and chemical reaction, then the height of H_R of a reactor unit is defined by

$$H_R = z/N_R \tag{89}$$

where N_R is the number of reactor units in the length z being considered. Rearranging equations (79) and (87), writing them in differential form and adding gives

$$H_T + H_C = \frac{dz}{dp}\left\{(p - p_i) + (p_i - p^*)\right\} = \frac{dz}{dp}(p - p^*) \tag{90}$$

Comparing the differential form of equations (79) and (87) with equation (90), the number of reactor units N_R is given by

$$N_R = \int_{p_1}^{p_2} \frac{dp}{p - p^*} \tag{91}$$

and so

$$H_R = H_T + H_C \tag{92}$$

This latter equation embodies the very essence of the idea that effects due to mass transfer and chemical reaction are separable. Provided always that the driving force for an irreversible chemical reaction may be represented by a linear function, then equation (92) may be employed to estimate the value of H_C from separate assessments of H_R and H_T. Hence, the relative importance of resistance to reaction by mass transfer through the gas phase will have been estimated. It is again worth emphasizing that the method is only limited to first-order or pseudo-first-order irreversible reactions.

Values of H_T may be calculated from independent correlations of the important operating variables. Since

$$H_T = \frac{G_m}{k_D S_x \rho_b P} \cdot \tag{93}$$

then from correlations of the mass-transfer factor given by equations (69) and (70)

$$\frac{H_T}{\left(\frac{\eta}{\rho D}\right)^{2/3}} = \frac{1}{\rho_b S_x j_D} \tag{94}$$

Hurt [10] obtained empirical correlations of H_T and the Reynolds group for catalyst pellets of various dimensions. The most reliable mass-transfer data were obtained for the adiabatic humidification of air, passing partially dried air through a packed bed of silica gel wetted with water. By measuring the rate of evaporation of solid naphthalene particles into air or hydrogen, the data were extended to cover the various sizes and shapes of catalyst particles normally encountered. In this way, values of H_T were found for a wide variety of operating conditions and particle size representing the physical conditions obtaining inside a reactor. H_C is evaluated by applying the results of kinetic experiments to equation (87). In the absence of mass transfer effects

$$H_C = \frac{z}{N_C} = \frac{z}{\int_{p_1}^{p_2} \dfrac{dp}{p - p^*}} \tag{95}$$

and, since $p^* \ll p$ for an irreversible reaction,

$$H_C = \frac{z}{\ln \dfrac{p_2}{p_1}} \tag{96}$$

with p_1 and p_2 corresponding to the inlet and exit partial pressure of the reactant in a reactor of length z. H_R is now found by addition and its relative importance to H_T is now known. H_R is obviously a specific quantity for a given chemical reaction and varies considerably with temperature since it involves an overall rate constant k_R defined from

$$H_R = H_T + H_C = \frac{G_m}{k_D S_x \rho_b P} + \frac{G_m}{\bar{k}_s \rho_b P} = \frac{1}{k_R} \frac{G_m}{\rho_b P} \tag{97}$$

As yet there are no reliable empirical correlations of H_R with varying reaction conditions.

9.2.3.4. Experimental Methods for Estimating the Effect of Mass Transfer through the Gas Phase

Two simple experimental methods are available for determining the conditions under which the resistance to mass transfer of the gas film adjacent to the surface has an influence on the measured reaction rate. In the first method, a number of flow experiments are analysed in which the flowrate of a given stoichiometric ratio of reactants is varied, but the ratio of catalyst mass to flowrate is constant. From the rate equation (8),

$$\frac{m}{F} = \int_0^x \frac{dx}{r_m} = f(x) \tag{98}$$

it is evident that the integral (which is a function, f (x), of the conversion x) should remain constant for various values of F, provided that m/F remains constant. If diffusion through the gas phase is important the rate of reaction, r_m, will be affected, and the conversion will no longer remain constant for various rates of flow. Below a certain flowrate the conversion will start to decrease and it is at this point that the rate of transport through the gas phase becomes significant and compares with the chemical reaction rate.

In the second method, conversions are compared at given ratios of m/F for various values of F. A decrease in conversion is evidence of a reaction rate which is influenced by diffusion of reactants or products to or from the catalyst surface. Figure 3 shows a situation in which diffusion affects the reaction rate at low velocities, but at velocities higher than that corresponding to A, gas diffusion is unimportant.

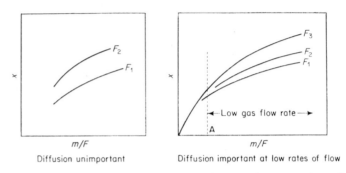

FIG. 3. The effect of mass transfer through the gas phase on the measured reaction rate.

9.2.4. The effect of mass transfer in pores

The general problem of reaction rates being affected by the diffusion of reactants and products in and out of pores and by the transfer of heat when non-isothermal conditions obtain, was discussed in Section 4.5. For the design of catalytic reactors it is important to know to what extent mass and heat transfer in pores is responsible for the observed reaction rate. Such information ameliorates the choice of the catalyst pellet size such that, under acceptable reaction conditions, only the exterior surface of the catalyst is accessible to reactant molecules. If the internal surface of the catalyst is unavailable, there is, in some cases, a serious reduction of surface area. This may be quite unacceptable for reactions carried out on a large scale, but, of course, may be used to advantage in kinetic experiments where one may wish to eliminate the effects of mass and heat transfer. If, therefore, independent kinetic data are used as a basis for design, it is essential to know the extent of in-pore mass and heat-transfer effects before any attempt can be made to substitute any functions describing the reaction rate into a rate equation.

For design purposes, it is useful to define an effectiveness factor η which is the ratio of the reaction rate r_p impeded by mass and heat-transfer effects to the unimpeded chemical reaction rate r. It is then only necessary to multiply the chemical rate by the value of η to obtain the reaction rate under conditions where mass and heat-transfer effects are significant. First we consider the case in which the temperature in the pellet interior is equal to the temperature at the periphery of the pellet, i.e. no heat-transfer effects. For a spherical catalyst particle it was shown in Chapter 4 that the ratio of the two reaction rates, which in this case also determines the fraction of surface available for reaction, is given by

$$f_s = \frac{3}{\phi_s} \left(\frac{1}{\tanh \phi_s} - \frac{1}{\phi_s} \right) \tag{99}$$

For the practical catalyst pellet the same ratio was given simply by $(\tanh h)/h$. In both cases the modulii ϕ_s and h are written in terms of the pore volume v_p and external surface area S_x. Thus

$$\phi_s = R \left(\frac{k_v}{D_e} \right)^{1/2} = \frac{3v_p}{S_x} \left(\frac{k_v}{D_e} \right)^{1/2} \tag{100}$$

where k_v is the rate constant per unit volume, R is the radius of the spherical catalyst pellet and D_e is the effective diffusivity of the pellet. For the practical pellet of undefined geometry, $3h = \phi_s$ provided the model adopted is one in which the experimental surface area and pore volume is equated to the surface area and pore volume of a pellet containing pores of average length L and pore radius R. It is therefore possible to define a single effectiveness factor η given by the same ratio which defines f. Hence one may write

$$\eta = \frac{r_p}{r} = \frac{1}{h} \tanh h \tag{101}$$

in which h is the modulus defined according to the model adopted by Wheeler for the practical catalyst pellet. Equations (99) and (100) have important practical implications. The effectiveness of the internal surface of the porous pellet approaches unity asymptotically as the pellet radius or chemical rate constant are made small, or as the diffusion coefficient is made large. Conversely, the effectiveness becomes small for large particles, large rate constants, or small values for the effective diffusivities.

As discussed in Section 4.5.5, substantial temperature gradients can occur within a porous catalyst pellet, since the rate at which heat is dissipated may not be sufficiently fast to allow isothermal conditions. In this case mass and heat transfer within a spherical pellet occur simultaneously and the effectiveness factor is now defined as the ratio of the actual rate to that which would occur if the whole of the interior of the pellet were exposed to reactant at the same concentration and temperature as that existing at the periphery. Under steady-state conditions, the rate of diffusion of reactant across a spherical boundary distance a from the periphery (at which point the concentration of reactant is c) is equal to the rate of reaction within the bounding surface. The heat released or consumed by reaction is also transferred across the same boundary, so that the rate of heat transfer is $D_e(\partial c/\partial a)\Delta H$. This can be equated to the product of a heat-transfer coefficient κ_e and a temperature gradient

$\partial T/\partial a$. Integrating from the periphery to a point a within the particle interior, there results

$$\Delta T = T - T_0 = -\frac{\Delta H \, D_e}{\kappa_e}(c_0 - c) \qquad (102)$$

If, as in Section 4.5.5, the differential equations for mass and heat transfer within the spherical particle are written, then equation (102) may be used to eliminate one of the variables. Numerical solutions have been obtained for these equations and are illustrated in Fig. 4. In Chapter 4, solutions were obtained in terms of the Thiele modulus, but here it is more useful to define a new dimensionless modulus Φ. The solutions are thus given in the form of a family of curves of the effectiveness factor η versus the dimensionless modulus defined by

$$\Phi = -\frac{R^2}{D_e}\frac{r}{c_0} = \phi_s^2 \, \eta \qquad (103)$$

where r is the measured reaction rate per unit catalyst volume and ϕ_s is the Thiele modulus for a spherical particle. The parameters for which solutions are given are the two independent parameters $\beta = -c_0 \, \Delta H \, D_e/\kappa_e T$ and $\gamma = E/RT_0$. As seen from equation (103) the parameter β represents the maximum temperature difference that could exist in the particle relative to the particle surface $(T-T_0)_{max}/T_0$. For an isothermal reaction therefore, $\beta=0$, while for an exothermic reaction it is positive and for an endothermic reaction is negative. For values of $\beta>0$, the effectiveness factor can exceed unity since the increase in rate caused by the positive temperature gradient may not be entirely compensated by the negative concentration gradient. The rate within the pellet may thus be much greater than it would be if the concentration and temperature inside the pellet were the same as that at the exterior.

To evaluate the effectiveness factor the effective diffusivity must be known. It is usually calculated in terms of the diffusive flux per unit total cross-section of the porous solid by application of the kinetic theory of gases. Allowance for the volume fraction of the voids and the tortuosity of the pores must be made. By comparing values for the diffusion coefficient, calculated from kinetic theory, with experimental diffusivity data for a particular catalyst support tortuosity factors may be estimated and hence the effective diffusivity determined. Such methods have been reviewed by Satterfield and Sherwood [21]. Provided the reaction rate r has been measured under the same conditions as those for which the value of η is derived, the dimensionless parameter Φ may be calculated from equation (103). This parameter, in contrast to the Thiele modulus containing an intrinsic rate constant (often unknown), contains only observable quantities. η may therefore be evaluated from plots such as those shown in Fig. 4. It should be noted from the shapes of the curves that, for highly exothermic reactions, there are three conditions under which the rate of heat release equals the rate of heat removal. This is discussed at greater length in Section 9.5. However, for porous catalyst pellets, the region of multiple solutions corresponds to combinations of large values of β and γ seldom encountered in practice.

For the isothermal case, to evaluate η experimentally it is sufficient to use

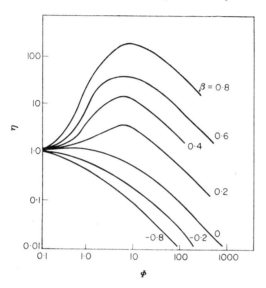

FIG. 4. Effectiveness factor η as a function of the dimensionless modulus Φ: first-order reaction in a spherical particle. $\gamma = E/RT = 20.$ $\beta = c_0 \Delta H D_e / \kappa_e T_0.$

rate data known to be dependent on pore diffusion, thus giving a numerical value for r_p, and also the rate r corresponding to a condition where there is no resistance to diffusion in pores. One experimental method of estimating η would therefore be to measure the observed rate as a function of particle size and hence the modulus h. A plot of $\eta(=(\tanh h)/h)$ as a function of h will therefore give values of η corresponding to particular values of r_p. When the specific rate is no longer a function of the diameter of the catalyst pellet (as at $\eta \approx 1$), its value is assumed not to be influenced by diffusion effects. Any other value of r on the curve is one which is dependent on in-pore diffusion. The ratio of these two rates, r_p and r, is thus the effectiveness factor for that particular particle size. For small particles (large h) the experimental rate is inversely proportional to particle size.

For design purposes, it is apparent that the rate of chemical reaction, independent of in-pore diffusion effects, when multiplied by the effectiveness factor appropriate to the catalyst, yields a rate which can then be used for subsequent calculations. As in the case of diffusion through the gas phase, there is a useful criterion to decide whether or not mass- and heat-transfer effects are important. For the isothermal case, referring to equation (101), when η is unity the total surface is available for reaction. For tanh h to be approximately equal to h, h must be small. A criterion for negligible resistance to in-pore diffusion is therefore

$$h = \frac{R}{3} \left(\frac{k_v}{D_e}\right)^{1/2} \ll 1 \tag{104}$$

or, alternatively, that η is confined to the range 0·3 to 6 [22]. For the general non-isothermal case, Weisz and Hicks [23] give the criterion

$$\exp\left(\frac{\gamma\beta}{1+\beta}\right) < 1 \tag{105}$$

9.3. Mass and Heat Transfer in Solid Catalyst Beds

9.3.1. Mass transfer in packed beds

When a gas flows through a packed bed of solid material, as it does in a catalytic reactor, in addition to chemical reaction mass is transferred by molecular diffusion and bulk flow. It is possible to write down differential equations describing transport conditions within the bed by considering the conservation of mass within an elementary volume element of the containing vessel.

Suppose some solid material is packed in a cylindrical tube of radius R and length L. Consider the mass entering and leaving an elementary volume as depicted in Fig. 5. This elementary volume is a cylindrical shell of thickness

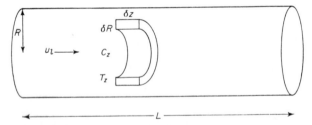

FIG. 5. Conservation of mass and heat within an elementary cylindrical volume.

δR and length δz and is concentric with the cylinder axis about which there is general symmetry. We assume that equimolar counterdiffusion is occurring and write down, in turn, the mass which is entering the volume longitudinally and radially in a time δt.

Mass entering by longitudinal bulk flow

$$u_l 2\pi R \delta R c_z \tag{106}$$

Mass entering by radial diffusion:

$$- D_e 2\pi R \delta z \left(\frac{\partial c}{\partial R}\right)_R \tag{107}$$

Mass entering by longitudinal diffusion:

$$- D_e 2\pi R \delta R \left(\frac{\partial c}{\partial z}\right)_z \tag{108}$$

D_e is an effective diffusion coefficient, which, in general, will be different for the radial and longitudinal directions, but is, herein, assumed the same. Next, the mass leaving this elementary volume in time δt is accounted for. Mass leaving by longitudinal bulk flow:

$$u_l 2\pi R \delta R c_{z+\delta z} = u_l 2\pi R \delta R \left\{ c_z + \left(\frac{\partial c}{\partial z}\right)_z \delta z \right\} \tag{109}$$

Mass leaving by radial diffusion:

$$- D_e 2\pi (R+\delta R)\delta z \left(\frac{\partial c}{\partial R}\right)_{R+\delta R} = - D_e 2\pi (R+\delta R)\delta z \left\{ \left(\frac{\partial c}{\partial R}\right)_R + \left(\frac{\partial^2 c}{\partial R^2}\right)_R \delta R \right\} \tag{110}$$

Mass leaving by longitudinal diffusion:

$$- D_e 2\pi R \delta R \left(\frac{\partial c}{\partial z}\right)_{z+\delta z} = - D_e 2\pi R \delta R \left\{ \left(\frac{\partial c}{\partial z}\right)_z + \left(\frac{\partial^2 c}{\partial z^2}\right)_z \delta z \right\} \tag{111}$$

Consider, for the moment, that no chemical reaction occurs within the bed, and that mass is transferred by equimolar counterdiffusion only. The mass accumulating within the element in time δt is $(\partial c/\partial t) 2\pi R \delta R \delta z$ and is equal to the algebraic sum of the quantities (106) to (111) inclusive entering and leaving the element. Neglecting second-order differences, the conservation equation is therefore

$$\frac{\partial c}{\partial t} = D_e \left\{ \frac{\partial^2 c}{\partial R^2} + \frac{1}{R}\frac{\partial c}{\partial R} + \frac{\partial^2 c}{\partial z^2} \right\} - u_l \frac{\partial c}{\partial z} \tag{112}$$

The diffusivity D_e is really an effective diffusivity, for it embraces both molecular diffusion and eddy diffusion due to turbulent conditions. It may be defined by analogy with Boussinesq's concept of eddy viscosity [24]. The overall rate of diffusion per unit area is given by

$$N = - (D + \epsilon) \frac{\partial c}{\partial z} \tag{113}$$

where D is the molecular diffusivity and ϵ the eddy diffusivity. Measurements and correlations of the effective diffusivity have been made [25, 26] for catalysts as a function of the particle diameter d_p and the tube diameter R. The results were expressed in terms of a dimensionless group (Pe), the Peclet number, which is given by $u_l d_p / D_e$. For Reynolds numbers between 30 and 700 it was found that (Pe) is a particular function f of the square of the ratio of particle diameter to tube diameter,

$$(Pe) = \frac{u_l d_p}{D_e} = f\left\{\left(\frac{d_p}{2R}\right)^2\right\} \tag{114}$$

Thus, for a given ratio of particle diameter to tube diameter, the function f gives the Peclet number and hence the effective diffusivity.

If, in addition to transport of matter by longitudinal and radial diffusion, chemical reaction occurs at a rate $r(=kc^n)$ per unit reactor volume, the conservation equation becomes

$$\frac{\partial c}{\partial t} = D_e\left(\frac{\partial^2 c}{\partial R^2} + \frac{1}{R}\frac{\partial c}{\partial R} + \frac{\partial^2 c}{\partial z^2}\right) - u_l\frac{\partial c}{\partial z} - kc^n \tag{115}$$

For a reaction in which there is a change in molar volume, the linear velocity will vary along the reaction path. In this case equation (115) should be written

$$\frac{\partial c}{\partial t} = \frac{D_e}{u_l}\left(\frac{\partial^2 (u_l c)}{\partial R^2} + \frac{1}{R}\frac{\partial (u_l c)}{\partial R} + \frac{\partial^2 (u_l c)}{\partial z^2}\right) - \frac{\partial (u_l c)}{\partial z} - kc^n \tag{116}$$

The steady-state solution to this equation has been found for certain conditions [27]. In particular, Vignes and Trambouze [28] solved, by a numerical method, the equation for a second-order reaction under isothermal conditions. Laminar-flow conditions were assumed and for all points in the reaction volume the concentration of both reactants equated. A similar approach for isothermal catalytic reactions is possible. Baron, Manning and Johnstone [29] studied the isothermal catalytic oxidation of SO_2 by coating the wall of a reactor tube with a vanadium pentoxide catalyst. In this case they neglected longitudinal diffusion. Since no chemical reaction occurred within the volume of the tube, but only at the tube wall, for plug-type flow the appropriate steady-state equation is

$$\frac{\partial c}{\partial z} = \frac{D_e}{u_l}\left(\frac{\partial^2 c}{\partial R^2} + \frac{1}{R}\frac{\partial c}{\partial R}\right) \tag{117}$$

with the boundary conditions

$$c = c_0 \text{ at } z = 0 \text{ for all values of } R \tag{118}$$

$$\frac{\partial c}{\partial R} = 0 \text{ at } R = 0 \text{ for all values of } z \tag{119}$$

$$\frac{\partial c}{\partial R} = -\frac{kc}{D_e} \text{ at } R = \frac{d}{2} \text{ for all values } 0 < z < L \tag{120}$$

where d and L are the tube diameter and length respectively. The boundary condition represented by equation (120) arises because, at the tube wall, chemical reaction is exactly balanced by diffusion. The solution may be accomplished analytically by using the Laplace transform operator method. The result is an equation giving c in terms of an infinite series containing the two independent variables z and R.

In catalytic reactors conditions are very often non-isothermal, so the steady-state solution of equation (117)

$$u_l \frac{\partial c}{\partial z} = D_e \left\{ \frac{\partial^2 c}{\partial R^2} + \frac{1}{R} \frac{\partial c}{\partial R} \right\} - r \tag{121}$$

is required in conjunction with an equation for heat conservation. The solution of the simultaneous equations is accomplished by standard numerical methods and gives the concentration of reactant as a function of the two independent variables corresponding to radial position and length. If the reactor is a narrow tube, the change in concentration from axis to tube wall at any given position will be small, especially if the bulk flowrate is high. Under these circumstances equation (121) reduces to the rate equation (6). Equation (121) is therefore to be regarded as an equation for use whenever diffusive mass transfer in a direction perpendicular to flow is concomitant with a steady-state chemical reaction. The complete solution of an equation such as (115) in conjunction with an analogous equation for heat transfer will give an indication of stability in packed beds. This question is discussed in Section 9.5.

9.3.2. Heat transfer in packed beds

Because the rate of a chemical reaction is an exponential function of temperature it is obvious that the conversion of reactant is very much influenced by heat transferred to or from the wall of the reactor and the heat released during reaction. Furthermore, it is not always possible, or even desirable, to operate chemical reactors under isothermal conditions. The highly exothermic nature of some reactions may enhance either the reaction rate, or the equilibrium yield, or both. For such reactions it is better for the heat of reaction to be absorbed by the reactants, as would be the case in a reactor operated adiabatically. Some reactions are so exothermic that the temperature increase over the first portion of the reactor length may be such that the catalyst activity is impaired. A compromise between adiabatic and isothermal operation is made in these circumstances.

By analogy with mass-transfer effects in packed beds, a general equation for heat transfer may be deduced. Referring to Fig. 5, an equation for the conservation of heat within an elementary cylindrical volume is deduced as follows:

Heat entering element by longitudinal flow:

$$\rho u_l c_p 2\pi R \delta R (T - T_0)_z \tag{122}$$

Heat entering by radial conduction:

$$- \kappa_e 2\pi R \delta z \left(\frac{\partial T}{\partial R}\right)_R \qquad (123)$$

Heat entering by longitudinal conduction:

$$- \kappa_e 2\pi R \delta R \left(\frac{\partial T}{\partial z}\right)_z \qquad (124)$$

Heat out of element by longitudinal flow:

$$\rho u_l 2\pi R \delta R (T - T_0)_{z+\delta z} \qquad (125)$$

Heat out by radial conduction:

$$- \kappa_e 2\pi (R + \delta R) \,\delta z \left(\frac{\partial T}{\partial R}\right)_{R+\delta R} \qquad (126)$$

Heat out by longitudinal conduction:

$$- \kappa_e 2\pi R \delta R \left(\frac{\partial T}{\partial z}\right)_{z+\delta z} \qquad (127)$$

Heat out by chemical reaction:

$$- 2\pi R \delta R \delta z r \Delta H \qquad (128)$$

ρ is the gas density, c_p the specific heat at constant pressure and κ_e the effective thermal conductivity of the bed assumed the same in both longitudinal and radial directions. T_0 is any arbitrary temperature to which the heat content of the system may be referred. The heat accumulating within the element is calculated by defining a mean specific heat $(c_p)_m$ and a mean density ρ_m. The heat capacity of the system is then given by

$$(c_p)_m \rho_m = c_p \rho \psi_b + (c_p)_s \rho_s (1 - \psi), \qquad (129)$$

where ψ_b is the void space per unit reactor volume. The subscript s refers to the solid. The resulting equation for the conservation of heat is then

$$(c_p)_m \rho_m \frac{\partial T}{\partial t} = \kappa_e \left(\frac{\partial^2 T}{\partial R^2} + \frac{1}{R}\frac{\partial T}{\partial R} + \frac{\partial^2 T}{\partial z^2}\right) - \rho u_l c_p \frac{\partial T}{\partial z} - r\Delta H \qquad (130)$$

Because there is more than one mechanism by which heat may be transferred within the packed bed, κ_e is a property of the system which depends on variables such as temperature, gas flowrate, particle diameter and porosity, and the thermal conductivity of the gas and solid. In fact, κ_e will, in general, depend on radial position, and an additional term $(\partial \kappa_e / \partial R)(\partial T / \partial R)$ must then be added to the right-hand side of equation (130).

For most design problems the steady-state solution of equation (130) is required,

$$\rho u_l c_p \frac{\partial T}{\partial z} = \kappa_e \left(\frac{\partial^2 T}{\partial R^2} + \frac{1}{R}\frac{\partial T}{\partial R}\right) - r\Delta H \qquad (131)$$

This latter equation represents an approximation, since the temperature of the gas and the solid are assumed identical at every point in the reactor volume. If gas and solid temperatures differ, the procedure is to write an energy balance equation for the gas and for the solid and then solve the two resulting simultaneous partial differential equations numerically. Considerable simplification results if the gas is sufficiently well mixed for the gas temperature T_g to be constant over any given cross-section. Then only the conservation equation for the solid need be considered, but this will involve an additional term due to a quantity of heat $k_g 2\pi R \delta R \delta z (T - T_g)$ transferred from the gas.

Wilhelm and co-workers [30] have discussed the solution for a case in which longitudinal heat transfer is neglected and the heat liberated by chemical reaction is considered to be a linear function of temperature. The conservation equation which Wilhelm obtained is

$$\kappa_e \left(\frac{d^2 T}{dR^2} + \frac{1}{R} \frac{dT}{dR} \right) - k_g(T - T_g) + (\alpha + \beta T) = 0 \qquad (132)$$

the last bracketed term representing the heat liberated by reaction. The kinetic coefficients α and β may be found from separate experiments designed specifically to examine the reaction rate. The effective thermal conductivity κ_e, and the heat-transfer coefficient k_g (based on the reactor volume), can be found from experiments in which the reactor tube is packed with inert material having similar thermal properties to the catalyst. Equation (132) may be reduced to a Bessel equation of order zero and may thus be solved analytically for a particular set of boundary conditions.

When it is safe to ignore the variation in temperature across the tube radius, a fairly simple approach to the problem of heat transfer is possible. An equation for the conservation of energy may be written by considering an elementary section, thickness δz, of the tube of cross-sectional area A_c. The surface area of the wall per unit length will be $2A_c/R = A_w$ (say). Therefore the amount of heat exchanged with surroundings is:

$$k_w A_w (T_m - T_w) \delta z \qquad (133)$$

where T_m is the mean temperature over the cross-section, T_w is the wall temperature and k_w a heat-transfer coefficient.

The heat produced by reaction in the element is:

$$- \Delta H r A_c \delta z \qquad (134)$$

where r is the reaction rate and ΔH the enthalpy of reaction. The energy balance is therefore

$$\sum_i m_i c_p \delta T = k_w A_w (T_m - T_w) \delta z - \Delta H r A_c \delta z \qquad (135)$$

where the summation on the left of the equation represents the net increase in enthalpy of all the components of the reaction mixture. Finally, a rate equation such as

$$F \, \delta x = r A_c \delta z \tag{136}$$

must be written. These simultaneous equations can be solved by standard numerical procedures. If the reactor is adiabatic, then no heat is exchanged with the surroundings and the energy conservation equation is

$$-\Delta H r A_c \delta z = F \Delta H \delta x = \sum_i m_i c_p \delta T \tag{137}$$

In this case the solution is simpler, since a relation between temperature and conversion is first found, and then equation (137) subsequently integrated.

9.4. Design Calculations

In the previous sections equations were derived to provide a numerical estimate of those factors which contribute to the design of a gas–solid catalytic reactor. From such an analysis emerges the relative importance of chemical reactivity, mass transfer by diffusion, and also heat transfer. The object was to combine suitable rate equations with equations for the conservation of mass and heat. The simultaneous solution of sets of such equations will enable a calculation of concentration and temperature profiles within a packed reactor, and hence an estimate of reactor size to achieve a specified conversion. In this section an outline is given of an approach to the solution of flow-reactor design problems.

9.4.1. Isothermal conditions

For reactions carried out under isothermal conditions the design problem is comparatively simple. An example is the catalytic hydrogenation of ethylene studied by Wynkrop and Wilhelm [31]. Their results were obtained using a copper oxide–magnesia catalyst packed in a tubular reactor maintained at various constant temperatures between 35 and 80°C. The reactants were mixed before flowing through the catalyst bed at 1 atm pressure. The rate expression they obtained corresponds to a reaction which is first order with respect to the partial pressure of hydrogen,

$$r = k p_{H_2} \tag{138}$$

Substituting this into a material balance equation such as (8) and integrating

$$\frac{1}{F} \int_0^m \mathrm{d}m = \int_0^x \frac{\mathrm{d}x}{k p_{H_2}} \tag{139}$$

where m is the required mass of catalyst to achieve a fractional conversion x. From the stoichiometric equation

$$C_2H_4 + H_2 \rightleftharpoons C_2H_6 \qquad (140)$$

p_{H_2} may be written in terms of the conversion x. Equation (139) then gives

$$m = \frac{F}{kp}(2\ln x - x) \qquad (141)$$

By choosing the required conversion level the mass of catalyst necessary to achieve this conversion may be directly calculated. The units of k must, of course, be those appropriate to equation (141). For reactions other than first order, the velocity constant should therefore be transposed. If the original kinetic data were not obtained in a flow system the rate of reaction may be expressed as moles converted per unit time per unit volume: in this case it is essential when using equation (141) to refer the rate to unit mass of catalyst before any numerical calculations are made.

9.4.2. Adiabatic conditions

In an adiabatic reactor the heat evolved by reaction is entirely absorbed by the reactant gases. For an exothermic reaction, therefore, the temperature will increase along the reaction path. In the case of an equilibrium reaction, as will be pointed out later, there is some advantage in operating two or more adiabatic reactors in series. Both the single adiabatic reactor and two adiabatic reactors in series are discussed here.

9.4.2.1. Single Adiabatic Reactor

The catalytic oxidation of SO_2 is an exothermic reaction which has been widely studied and applied to the design of industrial reactors for the sulphuric-acid industry. The conservation of heat requires that the heat of reaction is entirely absorbed by the reacting gases so that

$$\int_{T_0}^{T} \sum_i m_i c_{pi} dT = -F\Delta H \int_0^x dx \qquad (142)$$

In conjunction with this equation a rate equation must be considered. It does not matter much in which form the rate data are available. The kinetic studies of Uyehara and Watson [32] and of Hurt [33] showed that the rate of reaction at a platinum surface is given by

$$r = \frac{k\left(p_{SO_2}\,p_{O_2}^{1/2} - \dfrac{p_{SO_3}}{K}\right)}{\left(1 + \left(\dfrac{K_{O_2}}{p_{O_2}}\right)^{1/2} + K_{SO_3}\,p_{SO_3}\right)^2} \qquad (143)$$

where K is the equilibrium constant of the homogeneous reaction and K_{O_2} and K_{O_2} are the adsorption equilibrium constants for O_2 and SO_3, respectively. This expression implies a mechanism in which the surface interaction of adsorbed SO_2 with adsorbed oxygen atoms is rate controlling. Experimental tests have shown the importance of diffusion, so that the partial pressures in this equation are those at the interface, i.e. the thin film of gas immediately adjacent to the surface. These values of p_i may be computed from the gas partial pressures as previously illustrated in Section 9.2.3. The next step is to express the gas partial pressures in terms of the conversion x and obtain a rate equation in terms of x. The rate equation which results, which will not be a simple function of x, is then substituted into the conservation equation

$$\frac{1}{F}\int_0^m dm = \int_0^x \frac{dx}{r} = \int_0^x \frac{dx}{f(x)} \tag{144}$$

and must be solved simultaneously with the heat conservation equation (142). Because the reaction rate is not a simple function of x, the simultaneous equations are best solved by a numerical method. If equations (142) and (144) are written in finite difference form they become

$$\Sigma m_i c_{pi} \Delta T = - (F \Delta H) \Delta x \tag{145}$$

and

$$\frac{\Delta m}{F} = \frac{\Delta x}{r} \tag{146}$$

Standard numerical methods may be used to solve such simultaneous equations.

In this way the complete course of conversion and temperature may be found along the reactor length. Figure 6 shows typical conversion and

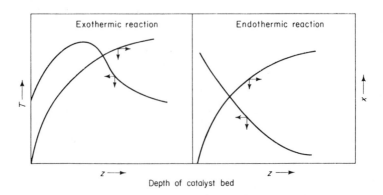

FIG. 6. Temperature T and conversion x profiles in an adiabatic-flow reactor.

temperature profiles to be expected in an adiabatic-type flow reactor when an exothermic reaction, such as the catalytic oxidation of SO_2, or an endo-thermic reaction, such as the dehydrogenation of ethyl benzene, occurs.

9.4.2.2. Two Adiabatic Reactors in Series

It is well known that the catalytic oxidation of SO_2 has rate maxima at certain optimum temperatures depending on the initial gas composition and the

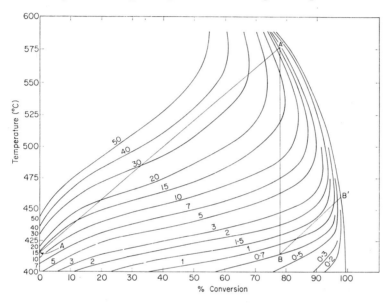

FIG. 7. Rate of catalytic oxidation of SO_2 as a function of temperature and conversion. Figures on each curve represent the reaction rate $r \times 10^{-1}$ g moles sec^{-1} g^{-1}.

physical properties of the catalyst. Because the reaction rate constant and the position of adsorption equilibrium depend on temperature, an optimum temperature can be chosen for operation such that the overall reaction rate is a maximum. Figure 7 shows rate data as a function of temperature for various conversions. These experimental values were obtained by Eklund [34] using a vanadium pentoxide catalyst and an inlet gas composition of 7% SO_2, 11% O_2 and 82% N_2. Kubota et al. [35] have obtained more recent data. If high conversions are to be reached at a substantial rate, it is evident that the temperature must be kept in the region of 400–500°C. It is impossible to achieve a high conversion in a single adiabatic reactor because the temperature increases to such an extent along the length of the reactor that the position of equilibrium is unfavourable. One solution is to use a number of reactors in series with intercooling between each stage.

Consider, for example, two adiabatic reactors in series, the gas inlet to each reactor being controlled at a temperature of 415°C. The molar composition of the gas entering the reactor is 7% SO_2, 11% O_2 and 82% N_2 and the flowrate 170 g moles sec^{-1}. We proceed to calculate the weight of catalyst for each reactor, so that 78% of the SO_2 is converted in the first reactor and 20% in the second reactor. The first step is to write down an equation for the conservation of heat in the first reactor. Over the relevant temperature range the mean specific heats for SO_2, SO_3, O_2 and N_2 are 12·2, 18·1, 7·9 and 7·3 cal (g mole)$^{-1}$ deg C^{-1} respectively. It is convenient to regard the gas mixture as an ideal gas, assigning to it an average molecular weight. The difference in enthalpy between the products P and reactants R will therefore be

$$H_P - H_R = (T - 415)\left\{ [(170 \times 0·07 - \alpha) \times 12·2] + \left[\left(170 \times 0·11 - \frac{\alpha}{2} \right) \times 7·9 \right] + \right.$$
$$\left. [170 \times 0·821 \times 0·73] + [\alpha \times 18·1] \right\} \qquad (147)$$

where α g moles SO_2 and $\frac{\alpha}{2}$ g moles of O_2 have been converted to α g moles of SO_3. Now the percentage conversion x may be written

$$x = \frac{\alpha}{170 \times 0·07} \times 100 \qquad (148)$$

so that

$$H_P - H_R = (T - 415)(1312 + 1·9x) \qquad (149)$$

The heat conservation equation for adiabatic conditions is

$$H_P - H_R = \Delta H^0_{415}\,\alpha = 23{,}290\,\alpha = 2795x \qquad (150)$$

where ΔH^0_{415} is the standard enthalpy of reaction at 415°C. Hence,

$$T = 415 + \frac{2795x}{1312 + 0·23x} \approx 415 + 2·11x \qquad (151)$$

Equation (151) provides a linear relation between temperature and conversion in the first reactor and is drawn on the conversion chart in Fig. 7 as the line AA'. The rate equation for the first reactor is

$$\frac{m_1}{F} = \frac{m_1}{170 \times 0·07} = \int_0^{0·78} \frac{dx}{r} \qquad (152)$$

where m_1 is the mass of catalyst necessary to achieve a 78% conversion. To solve this equation using the rate data in Fig. 7, it is necessary to plot $1/r$ as a function of x. The solution is accomplished by reading off from the chart the rate corresponding to the intersection of the straight line with various conversion ordinates. The results of such a calculation are plotted in Fig. 8. The area under the curve between the origin and the ordinate $x=0.78$ gives the value of the integral as 3.36×10^5 g sec (g mole)$^{-1}$. From equation (152), m_1 is therefore 4000 g.

A similar calculation is undertaken for the second reactor. A useful approximation is to use the same heat conservation equation for the second

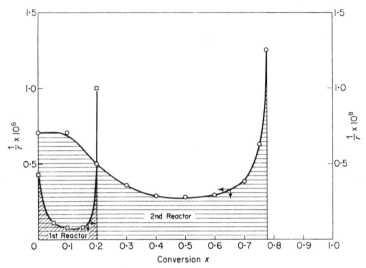

FIG. 8. Areas under curves give values of integrals in equations (152) and (153).

reactor as for the first reactor. This is only possible because the reactants are highly diluted with N_2. Equation (151) is therefore plotted on the conversion chart, the line BB' starting at a conversion of 78%. This will be the inlet condition for the second adiabatic reactor. The same procedure is followed as before. $1/r$ is plotted as a function of x by noting the rate corresponding to the intersection of the straight line BB[1] with selected ordinates. The result is incorporated in Fig. 8. For this second reactor the rate equation is

$$\frac{m_2}{F} = \frac{m_2}{170 \times 0.07(1 - 0.78)} = \int_0^{0.20} \frac{dx}{r} \tag{153}$$

The area under the curve $1/r$ as a function of x between $x=0$ and $x=0.20$ (corresponding to $x=0.78$ and $x=0.98$ on the original abscissa) is 49×10^5 g sec (g mole)$^{-1}$. From equation (153), therefore, $m_2=12,000$ g.

9.4.3. Non-adiabatic conditions

When a catalytic reaction occurs under non-adiabatic conditions in a tubular reactor, because there is transfer of heat with the surroundings via the tube wall, radial temperature gradients will manifestly influence the course of the reaction in the packed bed. Any satisfactory design method should aim to predict both radial and longitudinal temperature gradients, the concentration profile and the mean conversion of reactants into products.

Heat-transfer experiments have shown that the radial-temperature profile in packed tubes is parabolic. Furthermore, most of the resistance to heat transfer is near the tube wall. For high Reynolds numbers it is useful· to assume that all the resistance to heat transfer is in a thin layer adjacent to the tube wall. With this approximation, it is only necessary to specify a heat-transfer coefficient k_w based upon the mean temperature of the reaction mixture. Under these conditions the design proceeds by analogy with the discussion on heat transfer in packed beds in Section 9.3.2. The two simultaneous equations to solve are the material energy balance equation (136) and the energy balance equation (137). An approximate calculation of this kind gives a lower conversion for the same depth of catalyst bed when compared with a more rigorous calculation in which it is recognized that there will be temperature gradients across the entire tube radius. Once it is recognized that heat will be transferred radially from the tube centre to the tube wall, the equation describing longitudinal and radial heat transfer will be equation (131) developed in Section 9.3.2. Similarly, the concentration profile will be described by equation (121), Section 9.3.1. The simultaneous solution of these, together with an appropriate rate equation will yield both temperature and concentration profiles in the reactor. Hence the mass of catalyst required to achieve a specified mean conversion can be calculated.

One of the most convenient numerical methods of solving such problems is to write the differential equations in finite difference form. If the bed is divided up into l longitudinal increments of size Δz and n radial increments of size ΔR, then

$$z = l\Delta z \text{ and } R = n\Delta R \tag{154}$$

With the above notation the differential equation (131) for heat transfer becomes

$$T_{n,l+1} = T_{n,l} + \frac{\Delta z}{(\Delta R)^2 \rho u_l c_p}\left(\frac{T_{n+1,l} - T_{n,l}}{n} + T_{n+1,l} - T_{n,l} + T_{n-1,l}\right) - r\Delta H \Delta z \tag{155}$$

The differential equation (121) describing the concentration profile becomes, in terms of the conversion x,

$$x_{n,l+1} = x_{n,l} + \frac{\Delta z}{(\Delta R)^2}\frac{D_e}{u_l}\left(\frac{(x_{n+1,l} - x_{n,l})}{n} + x_{n+1,l} - 2x_{n,l} + x_{n-1,l}\right) + \frac{r}{c_0(u_l)_0}\Delta z \tag{156}$$

where c_0 is the inlet reactant concentration and $(u_l)_0$ the linear velocity at the tube entrance.

Provided that sufficient rate data are available equations (155) and (156) may be solved numerically by standard methods.

The results of such a calculation are shown in Figs. 9 and 10. It is clear that the conversion is higher at the centre of the tube than at the tube wall. Furthermore, the temperature reaches a maximum and then decreases at points further along the reactor. The decrease in temperature is due to the radial transfer of heat to the wall and surroundings becoming larger than the heat evolved by chemical reaction. The resulting data are probably best

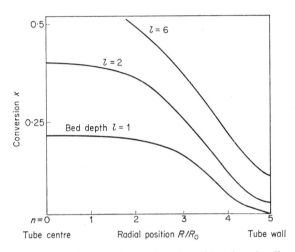

FIG. 9. Conversion profiles as a function of length and radius.

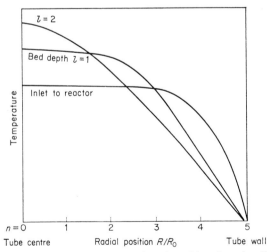

FIG. 10. Temperature profiles as a function of length and radius.

represented by taking the mean temperature and conversion over a given cross-section of tube. The mean temperature T_m over the radius at a given position along the reactor is given by

$$T_m = \frac{\int_0^{R_0} T(\rho u_l c_p) 2\pi R \, dR}{\rho u_l c_{pm} \pi R_0^2} = \frac{2 \int_0^{R_0} T c_p R \, dR}{c_{pm} R_0^2} = \frac{2}{c_{pm}} \int_0^1 T c_p n \, dn \quad (157)$$

where R_0 is the tube radius and $n = R/R_0$. Similarly, the mean conversion at the point l is

$$x_m = \frac{2 \int_0^{R_0} x R \, dr}{R_0^2} = 2 \int_0^1 x n \, dn \quad (158)$$

If the quantities $T c_p n$ and xn (which can be evaluated from the temperature and concentration profiles) are plotted against n, the area under the curves will give T_m and x_m for each position l along the tube. The mass of catalyst to use for a specified conversion is then easily computed by reading off the bed depth from a plot of x_m as a function of the reactor length z. If the bulk density of the catalyst is ρ_b then the mass of catalyst required is

$$m = \pi R^2_0 z \rho_b \quad (159)$$

where z will correspond to the bed depth ($l \Delta z$) which produces the required conversion.

9.5. Thermal Stability of Packed Bed Reactors

9.5.1. Stable thermal conditions

In continuous, exothermic, processes a thermal balance exists between the heat evolved by chemical reaction and the heat lost to the surroundings. If the system considered is a continuous gas–solid catalytic reaction, heat is produced at the catalyst surface and is lost from the surface of the catalyst particles to the flowing gas stream and thence, if the reactor is not an adiabatic one, to the tube wall and the surroundings. The rate at which heat is evolved by chemical reaction will depend on the kinetics of chemical reaction. For the purposes of the following discussion, an Arrhenius form of rate equation will be assumed. If the inlet gas temperature is not too high, the rate of chemical reaction will be rate-determining, but, as the temperature increases, ultimately diffusion of gas to the surface will be rate-controlling. Hence, at the lower temperatures, the rate of heat release from the solid increases exponentially with temperature, whereas at higher temperatures, the rate of heat release depends on the square root of the temperature. In the steady state the rate of chemical reaction balances the rate of diffusion. From equation (61), Section 9.2.3, we therefore conclude that the heat evolution curve, represented as a function of temperature, is described by the relation

$$\frac{dq_1}{dt} = r\Delta H = \frac{p\Delta H}{\dfrac{1}{k} + \dfrac{1}{k_D S_x}} \tag{160}$$

where ΔH is the exothermic heat of reaction, k the reaction rate constant and k_D the mass-transfer coefficient as defined by equation (49), Section 9.2.3. When the rate of chemical reaction is rate-controlling

$$\frac{dq_1}{dt} = kp\Delta H \tag{161}$$

whereas in the régime of diffusion control

$$\frac{dq_1}{dt} = k_D S_x p\Delta H \tag{162}$$

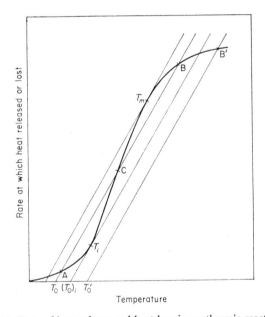

FIG. 11. Rate of heat release and heat loss in exothermic reactions.

The chemical rate constant k is an exponential function of reciprocal temperature (the classical Arrhenius law) whereas the mass transfer coefficient k_D is, in these circumstances, a function of $\sqrt{T}$. The rate of heat release as a function of temperature is therefore a sigmoid-shaped curve as depicted in Fig. 11. Now the rate of heat loss from the surface will, to a first approximation, be proportional to the temperature difference between the solid and the gas if convective heat transfer is the most significant mechanism. Hence the heat loss curve in Fig. 11 is a straight line the slope of which depends on the numerical value of the heat-transfer coefficient k_g. The intercept on the T-axis depends on the initial inlet gas temperature T_0. The equation depicting the rate of heat loss is, therefore,

$$\frac{dq_2}{dt} = k_g(T - T_0) \tag{163}$$

Let us examine the consequence of simultaneous heat generation and heat loss. The principles involved have been scrupulously analysed by Frank-Kamenetsky [36] in his treatise on heat exchange in chemical kinetics and Semenov in his discussions [37] of the self-ignition temperature in combustion processes. If the inlet gas temperature is at T_0, the points A and B in Fig. 11 represent stable steady states of low and high conversion respectively. A momentary rise or fall in temperature will cause the heat loss to become greater or smaller than heat generation and the points A or B, respectively, will be regained. On the other hand, if the initial gas temperature is at T'_0 the temperature rises until the stable steady state B' is attained, where the rate is controlled by diffusion. Now refer again to the gas inlet temperature T_0. The heat release curve crosses the heat loss curve at point C. This point represents an unstable steady state. Even though heat lost is equal to heat evolved at this point, any small variation in temperature will result in either a downward or upward displacement in temperature to a stable steady state, depending in which direction the temperature is displaced. Hence the region between A and B is not attainable in practice and is an unstable condition. If the flowing gas is initially hot and the solid bed of catalyst is cold, the temperature of the solid will rise to the point A. However, if the catalyst bed is initially hot with respect to a colder incoming gas, the temperature will continue to increase to the upper stable steady state B. The point T_i at which the heat loss line is tangential to the heat-release curve represents a critical condition. T_i has been referred to [38] as the minimum ignition temperature of the surface and the point (T_0^0) the minimum gas temperature for ignition of the cold surface. Here, the term ignition implies a self-accelerating rise in temperature and is analogous to the self-ignition temperature in combustion processes. Finally, the tangent to the curve at T_m reveals that it is the lowest temperature at which there is a rapid reaction rate under the prevailing conditions of flow and inlet gas temperature.

9.5.2. Predicting the autothermal region

Van Heerden [39] has considered the conditions under which an exothermic process has more than one stationary state and compared his conclusions with four typical catalytic heterogeneous processes used industrially. The conditions and parameters used in his calculations were typified by the ammonia synthesis, SO_2 and naphthalene oxidations, and the conversion of CO to CH_4 with steam. Following the analogy of combustion processes, van Heerden predicted the conditions which dictate the success of an autothermal process, i.e. one in which the heat of reaction sustains the conversion of reactants into products.

Consider an adiabatic flow reactor of length L. The heat generated by reaction is completely absorbed by the flowing gas, so that at a point z along the reactor

$$\Delta HA \exp\left(\frac{-E}{RT}\right) c = \rho u_l c_p \frac{dT}{dz} = \frac{\rho c_p}{\tau} \frac{dT}{d\xi} \tag{164}$$

provided plug-flow conditions are maintained. $\tau \; (=LA_c/u=L/u_l)$ is the residence time of gas in the reactor and $\xi (= z/L)$ is the fraction of the total reactor length L at which point the concentration of reactant is c. If the gas-inlet temperature is T_0 and the inlet concentration c_0 the conservation of energy demands that

$$\rho u_l c_p(T - T_0) = u_l \Delta H(c_0 - c) \tag{165}$$

For graphical representation of the heat-generation term it is convenient to use dimensionless quantities. We therefore define a dimensionless parameter

$$\phi = \frac{RT}{E} \tag{166}$$

Further simplification results if an adiabatic temperature rise T_a is defined. This is the increase in temperature of the gas stream if the heat of reaction is completely absorbed by the unused reactant. Hence

$$\Delta T_a = \frac{c_0 \Delta H}{\rho c_p} \tag{167}$$

and in dimensionless form

$$T/\Delta T_a = \chi \; \text{(say)} \tag{168}$$

It is also convenient to define a temperature T_e at which the fraction of reactant converted is $1/e$ for a given residence time τ, whence

$$A\tau = \exp\left(\frac{E}{RT_e}\right) = \exp\left(\frac{1}{\phi_e}\right) \tag{169}$$

where A represents the temperature-independent kinetic factor. Substituting these dimensionless parameters into equations (164) and (165) we obtain

$$\frac{d\chi}{d\xi} = (1 + \chi_0 - \chi) \exp\left\{\frac{1}{\phi_e} - \frac{1}{\phi}\right\} = (1 + \chi_0 - \chi) \exp\left\{\frac{1}{\phi_e} - \frac{1}{\phi_a \chi}\right\} \tag{170}$$

where

$$\phi_a = \frac{R\Delta T_a}{E} \tag{171}$$

Now the quantity $d\chi/d\xi$ is really the rate of heat generation $_g$ per maximum quantity of available heat and is denoted Q_g. Since the maximum amount of heat available is $u_l c_0 \Delta H$ then

$$\frac{d\chi}{d\xi} = \frac{1}{\Delta T_a}\frac{dT}{d\xi} = \frac{\rho u_l c_p \dfrac{dT}{d\xi}}{u_l c_0 \Delta H} = Q_g \qquad (172)$$

Equation (170) is thus equivalent to a dimensionless form of heat-generation curve for the adiabatic reactor and, as shown in Fig. 12, has an asymmetrical bell shape.

Dimensionless parameters may also be introduced into an equation for the rate of heat loss per maximum amount of available heat—denoted Q_l. The rate of heat dissipation depends on the mass flowrate (ρu_l) and the temperature difference $(T-T_0)$. Hence

$$Q_l = \frac{\rho u_l c_p (T - T_0)}{u_l c_0 \Delta H} = \frac{T - T_0}{\Delta T_a} = \chi - \chi_0 \qquad (173)$$

This linear equation is depicted in Fig. 12 by the straight line of unit slope.

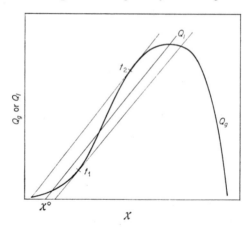

FIG. 12. Heat-generation and heat-loss curves in dimensionless form.

The problem of predicting a region in which a reaction is self-sustaining thus reduces itself to finding the range over which the three parameters ϕ_e, ϕ_a and ϕ_0 will result in the Q_l line intersecting the Q_g curve at more than one point. This region will obviously lie within the area enclosed by the two straight tangents t_1 and t_2. Hence the condition is that the line Q_l must lie within the two tangents to the curve Q_g. This will only occur provided that χ_0 (and hence ϕ_0), ϕ_a and ϕ_e have been correctly chosen. Since these parameters will be fixed by both the operating conditions and the reaction kinetics, whether or not a particular reaction can be operated in a self-sustaining region depends entirely on whether there is a sufficient amount of heat exchanged along the reaction path. In an adiabatic fluidized-bed reactor there is almost a complete exchange of heat at all points along the reaction path. In a flow reactor, however, often there is insufficient exchange of heat. If heat is exchanged between outlet and inlet gas streams, then the required autothermal

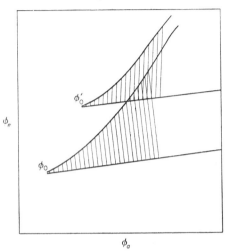

FIG. 13. Prediction of autothermal regions from a plot of ϕ_e, the dimensionless temperature (at which fraction converted is l^{-1}), and ϕ_a, the dimensionless adiabatic temperature rise.

condition can be realized. Let us examine how this prediction can be made. Figure 13 shows the relation between ϕ_e and ϕ_a for various values of the parameter ϕ_0. The relation is calculated by finding the two tangents to Q_g for a particular value of ϕ_0. For each value of ϕ_0 two relationships will be obtained corresponding to each of the two tangents. The upper curves bounding the shaded areas represent the relation between ϕ_e and ϕ_a for the condition imposed by the tangent t_1, and the lower line represents the relation obtained for the condition imposed by t_2. There can be more than one stationary solution to the equations only within a shaded area and hence this determines the autothermal region. As outlined at the beginning of this section, there are two stable stationary solutions corresponding to a high and low degree of conversion. The state corresponding to a high degree of conversion can be realized only provided that $T > T_i$.

Van Heerden expressed activation energies, values of T_e derived from observed kinetics and values of ΔT_a calculated from thermochemical data in terms of ϕ_e and ϕ_a for four selected catalytic reactions and considered their usual operating conditions (ϕ_0). Only the point corresponding to naphthalene oxidation fell within an autothermal region. The three points locating the other processes do not coincide with an autothermal region. This implies that with the catalysts available the reactions could not be operated auto-thermally without depending on heat exchange between the exit and inlet gas streams. Hence, for the NH_3, SO_2 and CO oxidations, the installation of a heat exchanger is necessary if self-supporting conditions are to be realized without resort to the introduction of a little air to supply the required sensible heat. A similar problem exists for the manufacture of CS_2, from sulphurs and carbon or some hydrocarbons, although Sykes et al. [40] and Thomas [41] proposed the introduction of air or oxygen to render the system thermally self-supporting.

The net result of using a heat exchanger to exchange heat between inlet and outlet gases is to magnify the heat effect within the reactor and displace the heat-loss curve to the right. To obtain an intersection between the curves Q_g and Q_l at high conversions it may be necessary, in some cases, to add heat to the system deliberately. The examples quoted come into this category. Figure 14 shows the effect of adding heat according to two different schemes. Heat may be added at a constant rate such that Q_l is displaced to the right and an intersection with the curve Q_g obtained at a fairly high conversion. If the constant rate of heat addition is Q, then a heat balance gives

$$Q_l = \frac{\rho u_l c_p (T - T_0)}{u_l c_0 \Delta H} - \frac{Q}{u_l c_0 \Delta H} = \chi - \chi_0 - Q' \qquad (174)$$

and so the line still has a slope of unity, but is displaced to higher values of χ by an amount Q'. Alternatively, the rate at which heat is added to the system may be chosen such that it is proportional to the difference in temperature between outlet and inlet streams, in which case

$$Q_l = \frac{\rho u_l c_p (T_{out} - T_0)}{u_l c_0 \Delta H} - \frac{k_g (T_{out} - T_0)}{u_l c_0 \Delta H} = (1 - k'_g \Delta T_a)(\chi_{out} - \chi_0) \qquad (175)$$

This results in the Q_l line haveing a slope smaller than unity by an amount $k'_g \Delta T_a$, where $k'_g (= k_g / u_l c_0 \Delta H)$ is an overall dimensionless heat-transfer co-efficient. Either method of heat addition may result in the Q_l and Q_g curves intersecting at a favourable high conversion.

The general problem of stability in both adiabatic and non-adiabatic fixed-bed reactors has been investigated by Amundson et al. [42, 43]. By selecting a simple model and writing the conservation equations for mass and heat with respect to both individual particles and the flowing fluid, sets of partial differential equations similar to equations (116) and (130) were obtained and solved by special techniques. It was concluded that there are some conditions within a packed bed when temperature and concentration profiles are extremely sensitive, not only to small changes in inlet conditions but also to changes in particle temperature.

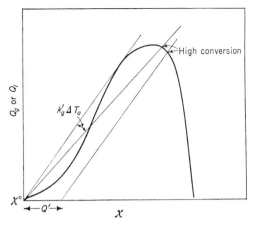

FIG. 14. Addition of heat to render a process autothermal.

9.6. Fluidized Bed Reactors

The use of fluidized bed reactors for gas–solid reactions has certain advantages over batch and tubular type reactors. Apart from the mechanical advantage gained by the ease with which solids may be conveyed, the high wall-to-bed heat-transfer coefficient enables heat to be abstracted or absorbed by the reactor with little difficulty. Furthermore, because of the movement of solid particles, the whole of the gas in the reactor is substantially at the same temperature and, with the solids, forms a continuous homogeneous phase. Another advantage is that the external surface area offered to the gas is greater than that for a fixed bed and so reactions limited by diffusion in pores will yield higher conversions in a fluidized bed. It is outside the scope of this book to enter into a discussion of the mechanics of fluidization and reference is made to standard works and researches of various authors [44–46]. It will be sufficient to say that when a gas is passed upward through a bed of solid there is a pressure drop across the bed which increases steadily with the gas flow. A point is eventually reached when the upward drag on the solid particles by the gas is equal to the weight of the particles. If the gas flow is increased further, the upward drag increases and lifts the particles, thereby increasing the voidage of the bed. The fixed bed then continues to expand until it attains the loosest packing arrangement. Any further increase in gas velocity causes particles to separate from one another and to be freely suspended. The whole bed is now in the fluidized state. Any increase in gas flow is no longer matched by a corresponding increase in pressure drop, since the velocity of gas flow through the interstices between the particles is decreased as a result of bed expansion. Increasing the gas flow beyond the point of incipient fluidization results in an increase in the voidage of the bed. A point is eventually reached when gas bubbles form within the bed. The fluidized bed then appears to be rather like a boiling liquid. The gas bubbles which form move upwards through the solid particles, which are now in a state of continuous motion. For solid-catalysed gas reactions a knowledge of the residence time distribution of the gas in the bed is of primary importance in predicting performance.

Various models have been proposed to account for the behaviour of fluidized beds. In each model two phases are distinguished: the bubble or lean phase, which is identified with the rising gas bubbles containing little or no solid particles, and the emulsion or dense phase, which has an approximately constant voidage and in which the solid particles are thoroughly mixed. Gas is able to cross flow from emulsion phase to bubble phase. May [47] considered the consequences of a chemical reaction occurring in a fluidized bed reactor by writing two simultaneous equations for the conservation of mass within the bubble phase and the dense phase. Consider a differential element of height δz in a reactor of cross-sectional area A_c and total height L. May [47] tacitly assumed that the mass transferred by cross flow was not a function of bed height and so an average concentration gradient $(c_d - c_b)/L$ is written over the whole length of the bed. For the bubble phase, the conservation equation is simply

$$u_b \frac{\partial c_b}{\partial z} = u_c \frac{c_d - c_b}{L} \qquad (176)$$

where u_b is the volumetric bubble velocity, and u_c the total volumetric cross-flow rate. The subscripts d and b are used to denote the concentration c in the dense phase and bubble phase respectively. This equation merely states that the change in concentration in the bubble phase is a result of the transfer of mass between dense and bubble phase.

In order to write the conservation equation for the dense phase, consider the material entering and leaving an element of volume $A_c \delta z$:
mass entering by bulk flow:

$$u_d(c_d)_z \qquad (177)$$

Provided that the particle motion is regarded as a diffusive flux of material, then we can write,

mass entering by eddy dispersion:

$$DA_t \left(\frac{\partial c_d}{\partial z} \right) \qquad (178)$$

where u_d is the volumetric gas velocity in the dense phase, D is a diffusion coefficient based on the dispersion of the dense phase by the eddy current motion within the bed and A_t is the area available for mass transfer by eddy dispersion. We now account for the total mass leaving the volume element. Additional mass has been transferred within the dense phase in the volume element due to cross flow and chemical reaction.

Mass leaving by bulk flow:

$$u_d(c_d)_{z+\delta z} = u_d \left(c_d + \frac{\partial c_d}{\partial z} \delta z \right) \tag{179}$$

Mass leaving by eddy dispersion:

$$DA_t \left(\frac{\partial c_d}{\partial z} \right)_{z+\delta z} = DA_t \left(\frac{\partial c_d}{\partial z} + \frac{\partial^2 c_d}{\partial z^2} \delta z \right) \tag{180}$$

Mass transferred by cross flow:

$$u_c \frac{c_d - c_b}{L} \delta z \tag{181}$$

Mass of reactant consumed by chemical reaction:

$$k c_d \rho_b A_c \delta z \tag{182}$$

The term $(\rho_b A_c \delta z)$ is the mass of catalyst present in the elementary volume considered. In the steady state the equation for the conservation of mass within the volume element must therefore be

$$DA_t \frac{\partial^2 c_d}{\partial z^2} - u_d \frac{\partial c_d}{\partial z} - u_c \frac{c_d - c_b}{L} - k c_d \rho_b A_c = 0 \tag{183}$$

So far the significance of A_t, the area across which mass is transferred by eddy motion, has been ignored. May and others have implicitly assumed that it is the area offered for mass transfer by the dense phase within the volume element. Unless the total cross-sectional area of the bed is completely occupied by the dense phase, then A_t has not the same meaning as A_c used in the reaction-rate term. Provided that the bubble diameters are not too large, it is justifiable to assume that they are equal and the equation reduces to that given by May.

Defining c_L as the total concentration of material (gas and solids) leaving the bed,

$$(u_d + u_b)c_L = u_d(c_d)_L + u_b(c_b)_L \tag{184}$$

where $(c_d)_L$ and $(c_b)_L$ are, respectively, the concentrations of the dense and bubble phases leaving the bed. The boundary conditions for this problem are

$$c_b = c_0 \text{ at } z = 0 \tag{185}$$

$$\frac{\partial c_d}{\partial z} = -\frac{u_d(c_0 - c_d)}{DA_t} \text{ at } z = 0 \tag{186}$$

and

$$\frac{\partial c_d}{\partial z} = 0 \text{ at } z = L \tag{187}$$

The second boundary condition arises because, at the entrance to the bed, where there is no chemical reaction, mass transferred by dispersion is exactly balanced by mass transferred due to bulk flow. May solved these equations for particular conditions and concluded that the fraction of reactant converted is closer to that expected for a bed in which there is relatively no mixing of solids than one in which there is complete backmixing (see Fig. 15);

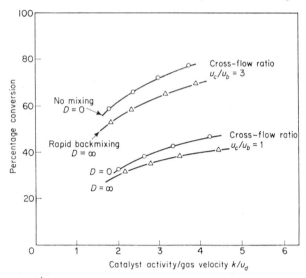

FIG. 15. Chemical reaction in a fluidized bed: effect of solids backmixing.

this is despite the efficient mixing observed in large-diameter beds. The corollary is that gas–solid contacting in the reaction zone is much less efficient than particle mixing. A consequence of this is that, for a given conversion, a greater mass of catalyst has to be used if solids backmixing (large D) is appreciable. It also emerged that for two parallel reactions

$$A \xrightarrow{k_1} P \tag{188}$$

$$B \xrightarrow{k_2} Q \tag{189}$$

occurring simultaneously, the extent of gas–solid contact affects the selectivity. When equations (176), (183) and (184) are solved simultaneously for two parallel first-order reactions, with $k_1 > k_2$, it is seen from Fig. 16 that, under conditions of good contact, the reaction with the greatest velocity constant gives the largest specific conversion. However, if contact is poor, the reaction selectivity decreases. The contact efficiency may be defined arbitrarily as the ratio of the weight of catalyst to achieve a specified conversion (under conditions of plug flow) to the weight required for the same conversion under the actual prevailing flow conditions. Thus, although the same total conversion may be realized, more catalyst is required when the contact efficiency is poor. There is proportionately more of the less reactive material B converted at the expense of the more reactive A.

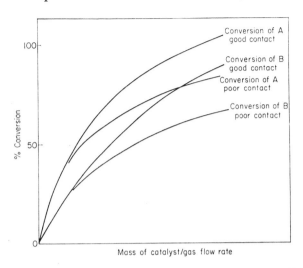

Fig. 16. Chemical reaction in a fluidized bed: effect of contact efficiency.

The difficulty with the above approach, in which flow and diffusive parameters are assigned to a model, is that the assumptions do not conform strictly to the pattern of behaviour within the bed. It is doubtful whether either solid dispersion or gas mixing can be looked upon as a diffusive flux. Furthermore, the residence time of material in the bed is not very well described by the particular combination of terms in equation (183). It has been shown that motion of solid particles around gas bubbles is similar to the flow of an ideal fluid around solid spheres. Rowe [48] examined the nature of both solid particle and gas flow in fluidized beds and applied his conclusion to the design of a fluidized bed in which a slow first-order reaction occurred. This more realistic approach aims at calculating, from a knowledge of the mechanics, a mean residence time for the particles at which surface the reaction occurs. Only slow reactions are limited by aerodynamic conditions and Rowe chose a solid–gas reaction whose rate of reaction would be affected by the flow pattern in the bed. He concluded that unless the particle size of the solid material is correctly chosen most of the reactant gas passes through the bed in the form of bubbles and has insufficient time to react. The most effective way of increasing the contact efficiency of gas with solid is to increase the particle size. It is desirable that as much of the gas pass through the emulsion phase as possible, for this results in increased contact of gas with solid. By doubling the particle size the time required to convert the reactants completely into products is halved. Provided that the gas flow is still sufficiently fast to cause bubble formation, heat transfer to the walls will remain high enough to retain the advantages of fluid bed operation.

9.7. Optimum Design of Catalytic Reactors

9.7.1. Continuous variation of a parameter along the reaction path

It was shown in Section 9.4.2 that an optimum temperature may be chosen to operate the SO_2 converter such that the overall reaction rate is a maximum. Now, this will only correspond to a maximum rate for a particular conversion. The reactor could be a large one containing a large amount of catalyst giving high conversions or may contain a small amount of catalyst with a correspondingly smaller conversion. For a given reaction mixture there will be an optimum temperature giving a maximum rate for each level of conversion. For reversible reactions taking place under non-isothermal conditions, therefore, it should be possible to select a temperature profile such that the rate is a maximum at each and every level of conversion. In an exothermic reaction the effect of an increase in temperature is not only to reduce the conversion but also to increase the reaction rate. Hence there will also be an optimum temperature for the operation of an isothermal reactor.

The effects are immediately apparent. By the selection of a suitable temperature profile, an increase in the output of reaction product per unit time per unit volume may be achieved. Also, where several side reactions occur, an optimum sequence of temperatures can be chosen such that this results in an improvement in the fraction of reactant converted to the desired product. The optimum temperature profile chosen, which effects an increase in the output of the reactor, is not necessarily compatible with a sequence of temperatures aimed at producing the maximum yield. One must therefore carefully decide whether the output or the yield is most likely to produce a significant economic advantage.

Denbigh [49] has proposed that the maximum amount of reaction occurs in a given time when the parameter affecting reaction rate is adjusted so that the rate is a maximum at each and every stage of the reaction. To achieve this, the parameter in question must be varied continuously along the reaction path. Suppose that the rate of reaction r is some function $F(c_0, x, \theta)$ of the inlet reactant concentrations c_0, the conversion x, and any parameter θ affecting the reaction rate. It is required to determine that relation between x and θ which will make either the time parameter t for batch reactions or the length parameter z for flow reactions, a minimum for a given degree of conversion. For a flow reactor the length L of the vessel containing the catalyst may be written

$$L = \int_0^x F(c_0, x, \theta) \, dx \qquad (190)$$

The calculus of variations shows that the necessary condition for L to be a minimum is for a variation in the integral to be zero. Under these conditions the Eulerian characteristic equation resulting from this condition is simply

$$\left(\frac{\partial F}{\partial \theta}\right)_x = 0 \qquad (191)$$

and the necessary condition for the total length L of the reactor to be a minimum is

$$\left(\frac{\partial r}{\partial \theta}\right)_x = 0 \qquad (192)$$

at every value of the length parameter. For a batch reactor the condition for the total time of reaction to be a minimum is for equation (192) to be satisfied at each instant during reaction.

9.7.2. Optimum temperature profile for a reversible reaction

From the remarks in Section 9.7.1 it follows that the output of the reactor (the amount of product obtained per unit time per unit volume or mass of catalyst) may be limited by both the velocity of reaction and the position of equilibrium. An increase in temperature may have a favourable effect on the velocity of reaction, but may impede the conversion of reactant due to the position of chemical equilibrium. Many industrial catalytic exothermic reactions fall into this category. For example, if the only consideration during ammonia synthesis or SO_2 oxidation is the fraction of reactants converted, then Le Chatelier's principle demands that the process be operated at low temperatures. Concomitant with a low temperature is a slow rate of reaction, so a large amount of catalyst would be required to achieve a high conversion.

Denbigh [49, 50], Bilous and Amundson [51, 52], Horn [53] and Aris [54] have contributed much to problems concerning optimum processes. Consider the reversible exothermic equilibrium which Denbigh, and also Bilous and Amundson, discussed

$$A + B \underset{k'_v}{\overset{k_v}{\rightleftharpoons}} C + D \qquad (193)$$

Since the equilibrium constant $K(=k_v/k'_v)$ decreases with increasing temperature, a high conversion at the reactor outlet demands a low temperature. On the other hand, at and near to the reactor inlet, where the conversion is in any case small, equilibrium considerations may be neglected and a high temperature may be allowed to promote an increase in the reaction rate. It follows that a decreasing temperature gradient along the length of the reactor will produce a high conversion in a short length. Consider the rate equation (6), Section 9.2.1, for a flow reactor. If we express the concentration in moles per unit mass and the rate of reaction in moles per unit volume per unit time, then to make the equation dimensionally consistent the concentration must be multiplied by the absolute gas density. Hence equation (6) becomes

$$\rho u_l \frac{dc}{dz} = r = kf(\rho c) \qquad (194)$$

which, for the reversible reaction (193), second order in both directions, is

$$\rho u_l \frac{dc}{dz} = k_v(\rho c)^2 - k'_v \left\{\rho(c_0 - c)\right\}^2 = \left(\frac{MP}{RT}\right)^2 \left\{k_v c^2 - k'_v(c_0 - c)^2\right\} \qquad (195)$$

The product (ρu_l) on the left-hand side of the equation is the mass velocity of gas per unit area and is independent of temperature, since any change in ρ is compensated by an opposite change in the superficial gas velocity u_l. On the right-hand side of the equation ρ has been expressed in terms of the temperature T, the total pressure P and the average molecular weight M of the reactants. Rearranging this equation and integrating at constant T and P gives

$$\left(\frac{MP}{R}\right)^2 \frac{2c_0}{\rho u_l} z = \frac{T^2 \sqrt{K}}{k_v} \ln \frac{c(\sqrt{K} - 1) + c_0}{c(\sqrt{K} + 1) - c_0} \qquad (196)$$

Remembering that both K and k_v are exponential functions of temperature, the left-hand side of equation (196), λ, may be plotted as a function of temperature for given inlet and outlet conditions corresponding to a reactor

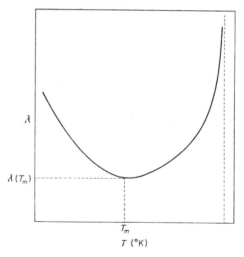

FIG. 17. Optimum isothermal temperature for a reversible exothermic reaction.

of a particular length z and hence containing an amount of catalyst $(\rho_b A_c)$. The function $\lambda(T)$ has a minimum as shown in Fig. 17. The point at which the minimum occurs corresponds to the temperature at which an isothermal reactor should be operated if it is to be designed to contain the minimum amount of catalyst.

For most exothermic catalytic reactions, isothermal operation is not the most satisfactory way of ensuring the use of the least amount of catalyst. It should be appreciated that although a reactor may be designed for a minimum volume for specific operating conditions, this minimum may not be the smallest value that the volume may take for it may be possible to choose superior operating conditions. Rewriting equation (195) in the integral form and rearranging,

$$\lambda = \left(\frac{MP}{R}\right)^2 \frac{2c_0}{\rho u_l} z = -2c_0 \int_{c_0}^{c} \frac{T^2 dc}{k_v c^2 - k'_v(c_0 - c)^2} \qquad (197)$$

Since λ depends on how T varies with z, the value of the integral on the right-hand side of equation (197) will depend on the temperature profile along the reactor length. The problem now is to determine that function $z(T)$ which will make the value of this integral the least possible. This is a simple problem in the calculus of variations. The characteristic Eulerian equation which determines the necessary condition is

$$\frac{\partial}{\partial T} \left\{ \frac{k_v c^2 - k'_v(c_0 - c)^2}{T^2} \right\} = 0 \qquad (198)$$

from which is obtained

$$\frac{c^2}{(c_0 - c)^2} = \frac{E - 2RT}{E' - 2RT} \frac{k_v}{k_v'} = \frac{E - 2RT}{E' - 2RT} \exp \frac{\Delta H}{RT} \qquad (199)$$

where E and E' are the activation energies in the forward and reverse direction, and ΔH is the exothermic heat of reaction. A step-by-step procedure is necessary for the calculation since this equation determines the temperature for a given concentration which, in turn, depends on position according to equation (195). The temperature profile may then be found by writing equation (195) in the form of finite differences and solving it simultaneously with equation (199). The resulting relation between z and T determines the optimum temperature gradient along the reactor. It is found that the length of a reactor operating with the optimum temperature gradient is substantially smaller than an isothermal reactor for the same fraction of reactants converted. The form which equation (199) takes is similar to that deduced by Denbigh [49], who calculated, for the synthesis of ammonia, the temperature corresponding to the maximum velocity in a static system.

9.7.3. Optimum temperature profile for consecutive reactions

Very often a waste product is produced during catalytic reactions. For example, during the catalytic chlorination of hydrocarbons the first reaction product may be the desired material. If the monochloro-substituted derivative is the required product to be marketed, then the process should be arranged to maximize the yield of this component. Consider the consecutive reactions

$$A \xrightarrow{k_{v1}} B \xrightarrow{k_{v2}} C \tag{200}$$

where the amount of B is to be maximized and the product C is waste. Suppose the activation energy E_2 for the formation of C from B is greater than that (E_1) for the formation of B from A. Intuitively, it seems that at the point in the reactor where the concentration of B is high the temperature must be low in order to impede any further reaction of B. On the other hand, at the inlet to the reactor the temperature should be high in order to facilitate the conversion of A to B: here the amount of B that has been formed will be small and so little C is produced. Hence there should be a decreasing temperature gradient along the reactor length.

The formal mathematics required to solve such a problem are more complex than for the case of the reversible reaction considered in Section 9.7.2. This is because the yield of B depends on an infinite set of variables T_i, the instantaneous temperature at any point z_i along the reactor length. B must therefore be maximized with respect to temperature for all values of z_i. This constitutes a relatively complex problem in the calculus of variations and will not be considered here. It is sufficient to say that a decreasing temperature gradient is necessary to optimize the process. Figure 18 shows the essential form of Bilous and Amundson's results in a convenient form. For any given process time θ there is a decreasing optimum temperature gradient exemplified by the difference between the initial optimum temperature describing the inlet conditions and the final optimum temperature describing the exit conditions. The calculations show that although both the initial and final optimum temperatures are lowered by an increase in θ, the decrease in temperature between inlet and exit is hardly affected. In Fig. 19 the optimum yield is also compared with the temperature profile. Such curves enable one to estimate, for a given weight of catalyst, the temperature drop necessary to give the optimum yield. The actual numerical values will depend on the reaction orders, the kinetic constants k_{v1} and k_{v2}, and the activation energies E_1 and E_2. The general conclusions are: (1) the optimum yield decreases with an increase in E_1/E_2 and K_{v2}/k_{v1}; (2) higher yields are obtainable when the second reaction is of second order; (3) the optimum temperature gradient is always negative for $E_1 < E_2$ and for $E_1 > E_2$, although the authors pointed out that the method they employed may not necessarily give a physical solution for the case $E_1 < E_2$; (4) the temperature drop along the reactor increases with the ratio E_1/E_2. The temperature gradient is much steeper if the second reaction is of second order.

The optimum isothermal temperature is much easier to find. In this case we are interested in calculating the temperature at which an isothermal reactor would operate if the yield of B is to be maximized. For a consecutive reaction,

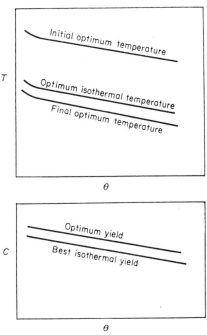

FIG. 18. Optimum temperature and yield profiles for consecutive reactions. θ is the process time.

such as (200), in which we will suppose that there is no change in molar volume, the concentration of B reaches a maximum value c_{Bmax} at a point z_m along the reactor. At constant temperature the concentration profile is found by solving simultaneously the equation (cf. equation (6))

$$\frac{u_l dc_A}{dz} = - k_{1v}c_A \tag{201}$$

and

$$\frac{u_l dc_B}{dz} = k_{1v}c_A - k_{v2}c_B \tag{202}$$

with the boundary conditions $c_A = c_A{}^0$ and $c_B = c_B{}^0$ at $z = 0$, where c_A and c_B refer to the concentrations of A and B, respectively, at a point z. The maximum value of c_B is found to be

$$c_{Bmax} = \frac{k_{v1}c_A{}^0}{k_{v2}} \left(\frac{k_{v1}{}^2 c_A{}^0}{k_{v2}[k_{v1}c_A{}^0 + (k_{v1} - k_{v2})c_B{}^0]} \right)^{\frac{k_{v1}}{k_{v1}-k_{v2}}} \tag{203}$$

To find the temperature at which c_{Bmax} is a maximum for reactors of fixed size (containing known amounts of catalyst), since k_{v1} and k_{v2} are both functions of the absolute temperature T, we require

$$\frac{dc_{Bmax}}{dT} = \left(\frac{\partial c_{Bmax}}{\partial k_{v1}} + \frac{\partial c_{Bmax}}{\partial k_{v2}} \frac{dk_{v2}}{dk_{v1}} \right) \frac{dk_{v1}}{dT} = 0 \tag{204}$$

If k_{v1} and k_{v2} are of the Arrhenius form, an expression for the optimum temperature as a function of reactor size can be obtained. The expression is very cumbersome and the result is best illustrated by comparing it with the optimum temperature profile in Fig. 18. The optimum isothermal temperature is fairly close to the exit temperature when a temperature gradient is employed. Furthermore, when the reactor is operated under optimum isothermal conditions, the yield is not much less than the optimum yield obtained employing the optimum temperature.

9.7.4. Optimum catalyst concentrations in bifunctional catalyst systems

Some classes of concurrent and consecutive reactions are catalysed by the presence of two or more catalysts. Such reactions include dehydrogenation and isomerization sequences which are of importance in the petroleum industry. The question of mass transfer and chemical reaction in bifunctional catalysts was fully discussed in Chapter 7 and it was pointed out that for a reaction

$$A \overset{X}{\to} B \overset{Y}{\to} C \tag{205}$$

in which A is transformed into B by catalyst X and B into C by catalyst Y, there must exist an optimum ratio of X to Y to give the maximum output of C. The problem was treated in fairly general terms for a number of different reaction sequences by Gunn and Thomas [55], who showed that one could choose the ratio X/Y to maximize the output of desired product, the ratio depending upon the relative magnitude of the various kinetic constants and diffusion coefficients. Furthermore, it emerged that if catalysts X and Y could be incorporated together in single catalyst pellets, then catalyst Y was brought to regions where B was formed, thus reducing considerably any resistance which B may suffer in being transported from X to Y and therefore enhancing the output of desired product even further.

Since, at the inlet to the reactor, no B is formed, it is evident that one could dispense with catalyst Y at the reactor entrance. *Mutatis mutandis*, once A has been consumed there is no point in having any X present. Therefore there must exist an optimum catalyst composition gradient along the reactor. This intuitive reasoning was confirmed by calculations [55] in which the reactor volume requirements for a single reactor was compared with the reactor volume requirements for two reactors in series, the first containing only catalyst X and the second containing only catalyst Y. It was found that a single reactor containing a mixture of X and Y along its length was far superior to two reactors in series, one containing X and the other Y.

9.7.5. Practical determination of optimum temperature sequence

Having outlined some of the general principles in Section 9.7.3, it is pointed out that the calculation of an optimum temperature profile can be extremely difficult and tedious if an attempt is made at an analytical solution. For any catalytic reaction, except those described by the simplest kinetics, it is far more realistic to plot a maximum rate curve as a function of temperature and concentration from practical rate data. It is relatively simple to estimate the optimum temperature gradient along the reactor length from such a maximum rate curve. Consider the SO_2 oxidation using a vanadia catalyst. Figure 19 shows the same rate curves, obtained by Eklund, as were considered in Section 9.4.2. The maximum rate curve is easily plotted by drawing through the inflection points of each of the constant rate curves the curve DD'. Along this curve the rate of reaction is a maximum for any given conversion. If it were desired to operate an SO_2 reactor along this reaction path, then the inlet gas would have to be at a temperature of about 700°C, clearly an impractical proposition for a catalyst which readily sinters. Alternatively, if the inlet gas is at an acceptable temperature of 590°C an isothermal path represented by EE' could be followed until the maximum rate curve is reached, after which a falling temperature gradient from E' to D' would be employed. Although the mass of catalyst required for 98% conversion would be less than for a two-stage adiabatic reactor, as discussed in Section 9.4.2, the elaborate and expensive arrangement for heat transfer required to follow the maximum rate curve only serves to emphasize those advantages offered by an adiabatic reactor. By plotting $1/r_{max}$ as a function of conversion, a simple graphical integration shows that the mass of catalyst required to achieve 78% conversion (following the course EE'F) is only 1260 kg as compared to 4000 kg for the same conversion in a single adiabatic reactor (following the course AA').

For the adiabatic reactor considered in Section 9.4.2, at 78% conversion the rate of reaction had decreased considerably and to such an extent that equilibrium conditions (shown by curve CC' in Fig. 19) had almost been achieved. At this point the exit gases were cooled before they entered the second adiabatic reactor. By this means a high rate of conversion was re-attained. This suggests that an approximate optimum temperature profile could be realized by using several adiabatic reactors in series. One would operate the first adiabatic reactor until the temperature is such that the maximum rate curve is reached. Once the maximum rate curve is intersected, it is desirable to cool the reactants to such a level that the same rate of reaction is maintained at the entrance to the second adiabatic reactor. In a sense, this is an arbitrary operation, for the more adiabatic reactors there are in series then the closer is the approach to the maximum rate curve. The path (A-1-2-3-4-5-6-7-8-9) shows such a scheme for five reactors. However, this would require considerable expenditure on intercooling equipment. In

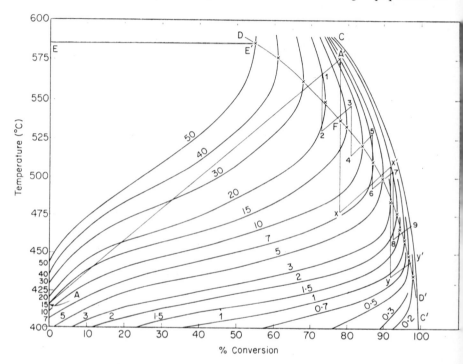

FIG. 19. Procedure for finding a suitable sequence of adiabatic catalytic reactors to approximate to the optimum temperature profile.

practice, three reactors in series would not be inappropriate. In this case the reaction path followed would be AA', XX', YY'. It is therefore concluded that, for many industrial heterogeneous catalytic reactions, equally good results may be achieved by operating the reactor at the minimum isothermal temperature— a much more economic operation.

The high cost and extreme difficulty in using the necessary heat transfer and control equipment to maintain the correct optimum temperature gradient may be an enormous factor in deterring the installation of appropriate heat exchangers in reactors. Added to this is the serious problem of being able to remove sufficient heat from the reactor to attain the desired temperature gradient. The requirements for heat transfer may be so great as to oppose the advantages of a falling temperature gradient. On the other hand, provided that the reaction is not very exothermic, operation of an isothermal reactor will achieve a comparable result but without the attendant difficulty of removing excessive amounts of heat from the reactor outlet. But it should always be borne in mind that to operate an isothermal reactor also requires expensive heat-transfer equipment. Hence, the practical difficulties and expense of operation should be carefully considered before concluding that an optimum temperature profile or an optimum isothermal temperature will gain a pecuniary advantage. More often it is better to use a sequence of adiabatic reactors.

References

1. R. Bellman, "Dynamic Programming". Princeton University Press, Princeton, New Jersey (1957).
2. P. V. Danckwerts, *Chem. Engng Sci.* **2**, 1 (1953).
3. O. A. Hougen and K. M. Watson, "Chemical Process Principles", Vol. 3, pp. 908 *et seq.* Wiley, New York (1955).
4. K. Tamaru, *Trans. Faraday Soc.* **55**, 824 (1959).
5. L. H. Thaller and G. Thodos, *A.I.Ch.E.Jl.* **6**, 369 (1960).
6. K. B. Bischoff and G. F. Froment, I. and E.C. Fundamentals **1**, 195 (1962).
7. K. H. Yang and O. A. Hougen, *Chem. Engng Progr.* **46**, 146 (1950).
8. J. de Acetis and G. Thodos, *Ind. Engng Chem.* **52**, 1003 (1960).
9. R. D. Bradshaw and C. O. Bennett, *A.I.Ch.E.Jl.* **7**, 48 (1961).
10. D. M. Hurt, *Ind. Engng Chem.* **35**, 522 (1943).
11. O. A. Hougen and C. R. Wilke, *Trans. Am. Inst. chem. Engrs* **41**, 445 (1945).
12. B. W. Gamsen, *Chem. Engng Progr.* **47**, 19 (1951).
13. S. Ergun, *Chem. Engng Progr.* **48**, 227 (1952).
14. W. Resnick and R. R. White, *Chem. Engng Progr.* **45**, 337 (1949).
15. K. N. Kettenring and J. M. Smith, *Chem. Engng Progr.* **46**, 139 (1950).
16. A. P. Colburn, *Trans. Am. Inst. chem. Engrs* **29**, 174 (1933).
17. T. H. Chilton and A. P. Colburn, *Ind. Engng Chem.* **26**, 1183 (1953).
18. O. A. Hougen and K. M. Watson, "Chemical Process Principles", Vol. 3, pp. 979 *et seq.* Wiley, New York (1955).
19. A. Wheeler, *Adv. Catalysis* **3**, 250 (1950).
20. E. W. Thiele, *Ind. Engng Chem.* **31**, 916 (1939).
21. C. N. Satterfield and T. K. Sherwood, "The Role of Diffusion in Catalysis". Addison-Wesley, Reading, Massachusetts (1963).
22. P. B. Weisz, *Z. phys. Chem.* NF**11**, 1 (1957).
23. P. B. Weisz and J. S. Hicks, *Chem. Engng Sci.* **17**, 265 (1962).
24. J. Boussinesq, Mém près par div savants a l'Acad Sci. Paris **23**, 46 (1877).
25. R. A. Bernard and R. H. Wilhelm, *Chem. Engng Progr.* **46**, 233 (1950).
26. R. W. Fuhien and J. M. Smith, *A.I.Ch.E.Jl.* **1**, 28 (1955).
27. T. Baron, *Chem. Engng Progr.* **48**, 118 (1952).
28. J. P. Vignes and P. J. Trambouze, *Chem. Engng Sci.* **17**, 73 (1962).
29. T. Baron, W. R. Manning and H. F. Johnstone, *Chem. Engng Progr.* **48**, 125 (1952).

30. R. H. Wilhelm, W. C. Johnston and F. S. Acton, *Ind. Engng Chem.* **35**, 562 (1943).
31. R. Wynkrop and R. H. Wilhelm, *Chem. Engng Progr.* **46**, 300 (1950).
32. O. A. Uyehara and K. M. Watson, *Ind. Engng Chem.* **35**, 541 (1943).
33. D. M. Hurt, *Ind. Engng Chem.* **35**, 527 (1943).
34. R. B. Eklund, "The Rate of Oxidation of Sulphur Dioxide with a Commercial Vanadium Catalyst". Almguist and Wiksell, Stockholm (1956).
35. H. Kubota, T. Akchata and T. K. Shindo, *Can. J. chem. Engng* **39**, 64 (1961).
36. D. A. Frank-Kamanetsky, "Diffusion and Heat Transfer in Chemical Kinetics". Akademii Nauk SSSR, Moscow (1948).
37. N. N. Semenov, "Some Problems of Chemical Kinetics and Reactivity", Vol. II, Part IV. Pergamon Press, London (1958).
38. K. J. Cannon and K. G. Denbigh, *Chem. Engng Sci.* **6**, 155 (1957).
39. C. Van Heerden, *Chem. Engng Sci.* **8**, 133 (1958).
40. A. Owen, K. W. Sykes and D. J. D. Thomas, *Trans. Faraday Soc.* **49**, 1207 (1953).
41. W. J. Thomas, *Ind. Chem.* 589 (Dec. 1959).
42. Liu Shean-Lin and N. R. Amundson, *A.I.Ch.E.Jl.* **2**, 117 (1956).
43. Liu Shean-Lin, R. Aris and N. R. Amundson, I. and E.C. Fundamentals **2**, 12 (1963).
44. D. F. Othmer, "Fluidisation". Reinhold, New York (1956).
45. M. Leva, "Fluidisation". McGraw-Hill, New York (1959).
46. J. F. Davidson and D. Harrison, "Fluidised Particles". Cambridge University Press (1963).
47. W. G. May, *Chem. Engng Progr.* **55**, 49 (1959).
48. P. N. Rowe, *Chem. Engng Progr.* **60**, 75 (1964).
49. K. G. Denbigh, *Trans. Faraday Soc.* **40**, 352 (1944).
50. K. G. Denbigh, *Chem. Engng Sci.* **8**, 125 (1958).
51. O. Bilous and N. R. Amundson, *Chem. Engng Sci.* **5**, 81 (1956).
52. O. Bilous and N. R. Amundson, *Chem. Engng Sci.* **5**, 115 (1956).
53. F. Horn, *Chem. Engng Sci.* (Symp. Chemical Reaction Engineering) **14**, 77 (1962).
54. R. Aris, "The Optimal Design of Chemical Reactors". Academic Press, New York (1961).
55. D. J. Gunn and W. J. Thomas, *Chem. Engng Sci.* **20**, 89 (1965).

Author Index

Numbers in parentheses are the reference numbers and are given to assist locating in the text references for which the authors' names are not given. Numbers in italics are the pages on which the references are listed

A

Abaledo, C. R., 132 (330), *173*
Acton, F. S., 485 (30), *515*
Adachi, G., 266 (67, 70), 276 (67, 70), 269 (67, 70), *295*
Adam, N. K., 87 (142)
Aerts, E., 259 (42), *295*
Affrossman, S., 146 (405, 406), 147 (405), *175*
Aigrain, P., 257 (35), *295*
Akchata, T., 489 (35), *516*
Alexander, L. E., 161 (490), *177*
Allen, F. G., 70 (18), 74 (18, 62), 75 (77), *166, 167*
Allen, J. A., 70 (39), *166*
Allpress, J. G., 159 (486), *177*, 308 (22), *317*
Alpert, D., 77 (81), *167*
Amano, A., 400 (97), *447*
Amariglio, H., 133 (337), *173*, 443 (197), *449*
Amberg, C. H., 15 (4), *63*, 102 (210), 162 (519), *170, 178*
Amelinckx, S., 245 (15), 247 (15, 19), 265 (15), 267 (15), *294*, 443 (205), *450*
Amenomiya, Y., 107 (223, 224, 225, 226), 108 (223), 109 (223), 110 (223), 111 (224, 226), *171*
Amundson, N. R., 226 (47), *240*, 348 (17, 18), *363, 364*, 500 (42, 43), 507 (51, 52), *516*
Anderson, J. R., 72 (44), 85 (120), 150 (432), 156 (44), 157 (44), 158 (44), 159 (44), 160 (44), *166, 168, 176*, 395 (82), 400 (101), *447*
Anderson, J. S., 71 (40), 137 (355), *166, 174*
Anderson, R. B., 98 (183), *170*, 402 (102, 104), 403 (112), 404 (112, 113, 114), 405 (114, 118), 408 (112), *447*
Andrew, E. R., 127 (292, 294), 128 (294, 296), 129 (292, 294), 129 (303, 304), *172, 173*
Anthony, P. P., 438 (180), *449*
Aris, R., 500 (43), 507 (54), *516*
Arlman, E. J., 129 (307, 308), *173*, 279 (124, 125), 284 (125), *297*, 307 (20), *317*
Armstrong, R. A., 70 (6), *65*
Arnell, J. C., 184 (8, 9), *239*
Arnott, R. J., 381 (41), *446*
Ashley, K. D., 188 (13), *239*
Ashmead, D. R., 149 (427), *176*
Ashmore, P. G., 53 (114), *65*, 389 (71), *446*
Aston, J. G., 47 (103), *65*
Austin, L. G., 443 (194), *449*
Aylmore, D. W., 79 (91), 96 (168), *167, 169*
Azim, S., 156 (475), *177*, 266 (79), 268 (79), *296*

B

Bacon, F. T., 438 (179), 441 (184), *449*
Bagg, J., 72 (46), 156 (46), 158 (46), *166*, 266 (82), *296*, 403 (109), *447*

Subject Index

A

Acceptor bond, 276
  impurity (and levels), 256
Acetylene hydrogenation, 396
  chemisorption of reactants during, 396
  kinetics of, 397
  selectivity and, 397
Acidity and catalytic activity, 425
Active surface area, 98
  using radio isotopes, 100
Activation energy
  influence of diffusion on, 234
  true and apparent, 301
Adsorption hysteresis, *see also* Hysteresis, 197–204
Adsorption
  energetics, 17
  isotherms, 32–48
  nature of, 14
  physical adsorption and chemisorption, 15
  rates of, 49
  specificity in, 17
  study of, 61
  time of, 68, 75
Ammonia decomposition, 408
Ammonia synthesis, 408–421
  adsorption of reactants during, 164, 408, 414
  catalysts for, 409
  comparison of rate with rate of ND_3 formation, 418
  compensation effect in, 413
  extent of nitrogen coverage during, 165, 413
  imine radical formation during, 417

  kinetics of, 411
  nitride formation and, 409
  rate determining step in, 410
  stoichiometry of, 419
Area of adsorbed molecules, 85
Arnell's equation for permeability, 185
Atactic, isotactic and syndiotactic polymers, 280
Autothermal region of operating chemical reactors, 496

B

Balandin's multiplet theory of catalysis, 9, 305
B E T theory of adsorption, 45
  and surface area determinations, 82
  current status of, 48
Bifunctional catalysis, 338
Bulk flow in pores, 214
Brönsted acid sites on catalyst surfaces, 423
Brunauer-Emmett-Teller theory, *see* B E T theory of adsorption
Burgers circuit, 243
Burgers vector of a dislocation, 243
  and strain energy, 244

C

Calorimetric measurements of heats of adsorption, 102
  of CO adsorptions, 368
  of CO oxidation, 113, 368–375
Capillaries, shapes of, 197–204
Carbon, catalysis of the oxidation, 441–444